ADVANCES IN SWINE IN BIOMEDICAL RESEARCH

Volume 2

ADVANCES IN SWINE IN BIOMEDICAL RESEARCH

Volume 2

Edited by

Mike E. Tumbleson

University of Illinois
Urbana, Illinois

and

Lawrence B. Schook

University of Minnesota
St. Paul, Minnesota

Springer Science+Business Media, LLC

Library of Congress Cataloging in Publication Data

Advances in swine in biomedical research / edited by Mike E. Tumbleson and Lawrence B. Schook.
  p.   cm.
  "Proceedings of an International Symposium on Swine in Biomedical Research, held October 22–25, 1995, at the University of Maryland, University College Park, Maryland"—T.p. verso.
  Includes bibliographical references and index.
  ISBN 978-1-4613-7699-6   ISBN 978-1-4615-5885-9 (eBook)
  DOI 10.1007/978-1-4615-5885-9
  1. Diseases–Animal models—Congresses. 2. Swine as laboratory animals—Congresses. I. Tumbleson, M. E. (Mike E.) II. Schook, Lawrence B. III. International Symposium on Swine in Biomedical Research (1995: University of Maryland)
  [DNLM: 1. Disease Models, Animal—congresses. 2. Swine—physiology—congresses. 3. Genetic Engineering—congresses. QY 60.S8 A244 1996]
  RB125.A36   1996
  619'.9—DC20
  DNLM/DLC   96-38650
  for Library of Congress   CIP

Proceedings of an International Symposium on Swine in Biomedical Research, held October 22–25, 1995, at the University of Maryland, University College Park, Maryland

ISBN 978-1-4613-7699-6

© 1996 Springer Science+Business Media New York
Originally published by Plenum Press, New York in 1996
Softcover reprint of the hardcover 1st edition 1996

We wish to give special thanks to

Jenny Malakowsky
and
Treena Schmidt

for their assistance in
preparing this publication

PREFACE

Similarities in structure and function between pigs and human beings include size, feeding patterns, digestive physiology, dietary habits, kidney structure and function, pulmonary vascular bed structure, coronary artery distribution, propensity to obesity, respiratory rates, tidal volumes and social behaviors. Since the pig is an omnivore, it provides an adaptable model to evaluate chronic and acute exposures to xenobiotics such as alcohol, caffeine, tobacco, food additives and environmental pollutants. Swine have been used successfully as models to evaluate alcoholism, diabetes, absorption, digestion, total parenteral nutrition, organ transplantation, atherosclerosis, exercise, hypertension, hemorrhagic hypotension, melanoma, gingivitis, obstructive and reflux nephropathy, osteochondrosis, dermal healing and septic shock.

A severe and worsening shortage of organs and tissues for transplantation in patients with severe organ failure has encouraged the consideration of interspecies or xenotransplantation. In developing programs toward this end, the pig generally is viewed as the preferred donor because of its size, physiology and availability. The pig harbors relatively few diseases which could be transmitted inadvertently to human patients. The ability to genetically modify swine to ameliorate the consequences of the human immune response offers a further significant advantage.

Another important consideration for an animal model is that basic biologic background information be available for investigators to design future prospective studies. Animal models should accurately and precisely emulate human conditions; survive sufficient time to be functional; provide for multiple and serial samples; and have a gestation period adequate for intervention. When considering human fetal research, one must appraise the ethical implications; therefore, it is paramount that an animal model be developed.

Limitations on the development of animal biotechnologic advances may restrict opportunities for improving human health. With respect to biologic interactions, interrelationships between human health and animal biotechnology are symbiotic as well as synergistic. The relationship of bioscience and biotechnology to developing countries is not so much one of aggressive impact and doubtful benefits, but of indispensability.

Mike Tumbleson, University of Illinois
Larry Schook, University of Minnesota

ACKNOWLEDGMENTS

The coeditors wish to extend appreciation to members of the organizing and scientific committees. They provided assistance in developing the sound technical program, as well as aiding in peer review of the submitted papers. Participating individuals were:

- Craig Beattie, USDA/ARS/MARC
- Richard Binns, The Babraham Institute
- David Brown, University of Minnesota
- Harold Gonyou, Prairie Swine Centre
- Eric Johnson, University of Minnesota
- Charles Louis, University of Minnesota
- Arthur Matas, University of Minnesota
- James Mickelson, University of Minnesota
- Thomas Molitor, University of Minnesota
- Nancy Monteiro-Riviere, North Carolina State University
- Michael Murtaugh, University of Minnesota
- Jack Odle, North Carolina State University
- Reinhard Pabst, Hannover Medical School
- Linda Panepinto, Panepinto & Associates
- Victor Perman, University of Minnesota
- Jeffrey Platt, Duke University
- Peter Reeds, USDA/ARS/CNRC
- Mark Rutherford, University of Minnesota
- Gail Scherba, University of Illinois
- Steve Squinto, Alexion Pharmaceutical
- Bert Stromberg, University of Minnesota
- James Terris, Uniformed Services University
- Matthew Wheeler, University of Illinois
- Federico Zuckermann, University of Illinois

CONTENTS

Volume 2

Methods and Techniques

Nutrition

CONTENTS

Volume 1

Overview

General

Transgenics

Immunology and Infectious Diseases

SWINE RESEARCH BREEDS, METHODS, AND BIOMEDICAL MODELS

Michael P. Murtaugh, PhD,[1] Nancy A. Monteiro-Riviere, PhD,[2] and Linda Panepinto, BS[3]

[1] Veterinary PathoBiology
University of Minnesota
St. Paul, Minnesota 55108-6010
[2] College of Veterinary Medicine
Cutaneous Pharmacology and Toxicology Center
North Carolina State University
Raleigh, North Carolina 27606
[3] Panepinto & Associates
Box 192 Redstone Canyon
Masonville, Colorado 80541

In this overview, we depict the use of swine as biomedical models. The growing acceptance of swine as a standard laboratory animal is due largely to the development, characterization and widespread availability of specialized pigs. A historical review of miniature swine breed development worldwide is provided by Panepinto. Miniature swine have been established from naturally occurring populations found in various areas of the world and from genetic selection of two or more existing breeds. Naturally occurring miniature breeds include Yucatan pigs from the Yucatan peninsula of Mexico and Vietnamese pigs from southeast Asia.

Methods for handling and housing miniature swine are described but research is necessary to establish standardized procedures. Restraint produces profound short term stress for 5 to 7 min as evidenced by behavior, catecholamines and other indicators. Handling procedures which provoke noted differences in swine vocalization did not result in differences in levels of stress hormones during short term restraint (Friedman). However, pigs which show calm behavior are likely to exhibit low catecholamine levels (Ikeda).

The demand for standardized research swine will lead to further characterization of existing breeds and perhaps to the development of additional breeds. For example, the Göttingen pig was developed over 30 years ago but recently was rederived under controlled microbiological conditions and now is maintained in a full barrier facility (Bollen). Minisib pigs were developed following a 10 generation series of crosses involving 4 distinct breeds from Europe and Asia. The resulting animals are small and robust but have a polyploid karyotype ranging from 36 to 38 diploid chromosomes (Tikhonov). Information on breed similarities and differences will be useful.

Advances in Swine in Biomedical Research, edited by Tumbleson and Schook
Plenum Press, New York, 1996

Efforts are underway to improve pig uniformity and availability by increasing reproductive performance. Sinclair pigs have an average litter size of 7 piglets of which about 6 survive to weaning (Bouchard); other breeds are similar. Superovulation yields of 25 to 30 ova in miniature pigs are similar to domestic swine stock (Hampshire) and may provide the means for increasing uniformity. Similarly, methods for *in vitro* fertilization and maturation are under development to reduce the incidence of polyspermic penetration of oocytes and increase survival (Funahashi).

The behavior of pigs facilitates their use in research. Centuries of domestication have resulted in animals which are genetically and experimentally amenable to control. Pigs are social, establish hierarchies and rarely exhibit aggression against familiar animals. Swine readily learn operant tasks, including tolerance of bleeding. Their behavior is stereotypic, which has facilitated research on causes and development of such behaviors to improve pig management and to model human behavior (Gonyou).

Pigs are an excellent model to study cutaneous pharmacology and toxicology of human skin. Porcine skin is morphologically and functionally similar to human skin and has become the accepted model for cutaneous absorption and toxicity studies. Models have been developed to predict skin penetration and irritation potential of vesicants (Reifenrath); mechanisms of penetration were deduced by Monteiro-Riviere for vesicant uptake and transdermal iontophoretic drug delivery. In particular, the critical role of the cutaneous vasculature in topical uptake was established. Development of an isolated perfused porcine skin flap allows for *in vitro* cutaneous pharmacology, toxicology and phototoxicology studies in a viable system including a functional microcirculation (Monteiro-Riviere). Transplacental drug delivery has been modeled in pregnant miniature sows for applications in fetal antiarrhythmic drug therapy (Weist).

Cardiovascular similarities between humans and swine have contributed to the use of swine in the development of improved diagnostic and surgical procedures for ventricular septal defect (Swindle) and myocardial infarction (Borrego). Interestingly, Swindle established a colony of Yucatan miniature pigs with a hereditary ventricular septal defect which closely resembles the defect in humans and may be useful in elucidating its genetic and developmental basis. Collection of blood and lymph in conscious pigs for a period of weeks without disruption of lymph flow was accomplished using an external thoracic duct venous shunt (Mendenhall).

Swine can serve as environmental monitors for the determination of lead bioavailability. Comparison of lead absorption from lead salts and contaminated soil is not confounded by coprophagy, high rates of biliary excretion, or developmentally dependent changes in active transport mechanisms as observed in rodents (Casteel). Extracorporeal detoxification was demonstrated using lyophilized swine liver microsome preparations (Ryabinin). Since preparations are stable for a period of years they may have clinical applications in liver failure, toxicant poisoning and burn injury. Also, swine models were described for hypergonadotropism associated infertility in men using Meishan boars selected for high FSH concentration (Lunstra) and for osteopenia using immobilized pigs (Bouchard).

THE PIG AS A MODEL FOR CUTANEOUS PHARMACOLOGY AND TOXICOLOGY RESEARCH

Nancy A. Monteiro-Riviere, PhD[1*] and Jim Riviere, DVM, PhD[2]

Cutaneous Pharmacology and Toxicology Center
[1] Department of Companion Animal and Special Species Medicine
[2] Department of Anatomy, Physiological Science and Radiology
College of Veterinary Medicine
North Carolina State University
Raleigh, North Carolina 27606

1. ABSTRACT

The pig has been a well accepted model for cutaneous absorption and toxicity studies because the integument is morphologically and functionally similar to human skin. We will review how the pig is utilized in this field, with particular reference to studies conducted in our laboratory. There has been considerable debate as to whether the topical effect of an antiinflammatory drug is due to direct penetration or is secondary to systemic absorption and recirculation. We addressed this question by studying the mechanism of topical penetration of piroxicam, a nonsteroidal antiinflammatory drug, in the pig. The pivotal role of the cutaneous vasculature on absorption will be further discussed and contrasted to *in vitro* systems. Pigs have been accepted as a model for studying iontophoretic drug delivery in humans. Transdermal delivery permits administration of therapeutic compounds that would not normally penetrate the skin. The pig is a principal animal model used in this field. The pathway a drug follows during delivery has been controversial and is of toxicological significance. We have defined the precise anatomical pathway which mercuric chloride traverses the stratum corneum during iontophoresis. In addition, new research methodologies have been applied in cutaneous toxicology, partly because of the driving force to replace *in vivo* animal models, with more humane *in vitro* approaches. The isolated perfused porcine skin flap (IPPSF), developed in our laboratory, is a perfused skin model which allows for *in vitro* cutaneous pharmacology and toxicology studies to be conducted in a viable system that

* Reprint requests to: Dr. Nancy Monteiro-Riviere, Cutaneous Pharmacology and Toxicology Center, North Carolina State University College of Veterinary Medicine, 4700 Hillsborough Street, Raleigh, North Carolina 27606, (919) 829-4426.

has a morphological and functional microcirculation. A porcine skin model developed for use in phototoxicology will be characterized. Finally, studies involving use of this preparation to assess the toxicity of chemical vesicants will be discussed. These examples are illustrative of the dominant role of the pig as an animal model for human skin biology.

2. INTRODUCTION

The fields of cutaneous pharmacology and toxicology research have expanded dramatically due to recent advances in transdermal and topical drug delivery. It is important to have a relevant animal skin model that can predict human drug responses. This is especially appropriate in preclinical drug studies designed to develop a dosage formulation for human clinical trials and in the general field of risk assessment. In addition, an accurate animal model is indispensable in toxicology experiments studying potent chemicals where human trials are unethical and surrogate animal data is the only alternative.

Skin is a complex, integrated, dynamic organ that has many functions that go far beyond the barrier between the body's well regulated "milieu interieur" and the external environment. In mammals, insulation and temperature regulation are considered by some to be its primary function. The presence of fur or hair and an extensive blood supply which facilitates this heat regulation role also interact with the skin's barrier properties. Also, skin can be viewed as an endocrine organ and as a critical component of the immune system. Species differences in all of these functions may interact with its barrier properties and thus affect the selection of a proper animal model for predicting human absorption. When barrier, pelage, vascular, endocrine and immunological properties are considered *en masse*, pig skin is similar to human skin. The reader can appreciate its widespread acceptance in these other fields by simply consulting the other chapters in this text. Since the optimal animal model should be representative of all of these functions, it is little wonder that the pig has gained such widespread acceptance in skin biology. Our intent is to review and summarize the similarities of pig skin to that of human skin, to examine the putative mechanism of topical penetration of piroxicam, to determine the precise pathway of iontophoretic transdermal delivery, as well as the creation of the isolated perfused porcine skin and how it is used in studying the pathogenesis of sulfur mustard vesication and phototoxicology.

3. ADVANTAGES OF PIG SKIN

The porcine integument is morphologically,[1-4] histochemically,[5-7] biochemically and biophysically similar to human skin and has been utilized as a model for drug toxicity and percutaneous absorption studies. Pig skin resembles human skin in having a sparse hair coat, a relatively thick epidermis, similar epidermal turnover kinetics, lipid composition, and carbohydrate biochemistry, lipid biophysical properties and the arrangement of dermal collagen and elastic fibers.[2,8-12] Reported differences in the pig include a unique interfollicular muscle that spans the triad of hair follicle,[13] the presence of apocrine sweat glands only on the body surface[2,4] and a thicker stratum corneum.[1,14] From the perspective of percutaneous absorption studies, the most important factors that make pig skin predictive to that of humans include the sparse hair coat, epidermal thickness and properties of the stratum corneum lipids. Numerous investigators have reviewed the comparability of chemical absorption in animals and humans and the pig consistently ranks with nonhuman primates as the best animal model for percutaneous absorption.[14,15] Additionally, humans and pigs are of comparable body mass in contrast to smaller rodents. Therefore, the ratio of the topical dosing site area to the total body mass for a local application is similar in pigs and humans.

In rodents, this ratio represents a significant portion of skin surface area which in itself should result in significant systemic absorption. Because of these similarities and its widespread acceptance, we will focus in this paper on the use of the pig to study selected problems in chemical absorption and toxicity.

4. PIROXICAM ABSORPTION

Drugs are applied topically to the skin to exert either a local effect or for systemic therapy as in transdermal drug delivery systems. For most compounds, there is little debate as to whether the observed effect is due to direct penetration (most topical dermatologics) or secondary to achieving effective systemic concentrations (transdermal delivery systems). However, there is controversy surrounding the mechanism of action of certain topically applied antiinflammatory drugs. Topical delivery of nonsteroidal antiinflammatory drugs (NSAIDs) directly to the site of localized pain or inflammation would be most beneficial since it could prevent a number of side effects (such as nausea or gastrointestinal ulceration) associated with their oral administration. There is considerable debate as to whether any observed effect of NSAIDs (we will use piroxicam as a model drug) is due to direct penetration or occurs as a result of systemic absorption and subsequent recirculation.

Previous studies with piroxicam, a NSAID, in male rats is suggestive that topical administration resulted in a high concentration of the drug in the underlying musculature. These high concentrations could not be attributed entirely to entry from the systemic circulation.[16] The cutaneous microvasculature does not function as an infinite sink that removes all topically applied drugs to the systemic circulation.[17] This has been suggested by other investigators.[18-20] Further, as will be discussed later, coiontophoresis of vasoactive compounds has been shown to modulate the systemic absorption of lidocaine.[21] Using this information, we hypothesized that the cutaneous blood vessels play a pivotal role in determining local vs systemic absorption of topically applied drugs. Depending on the anatomy, topically applied drugs may either penetrate into deeper tissues or be absorbed completely into the systemic circulation. To test this hypothesis, the Yorkshire pig was used because there are regions of the skin which are supplied by either direct cutaneous (vessels do not perfuse deeper tissues before reaching systemic circulation) or by musculocutaneous (vessels traverse through the underlying musculature) arteries. Thus in a single animal, one could study the effect of different patterns of the subcutaneous vasculature on topical piroxicam penetration.

Piroxicam (4-hydroxy-2-methyl-N-2-pyridinyl-2H-1,2-benzothiazine-3-carbox-amide-1,1-dioxide) is a reasonably potent NSAID. Dermal penetration through pig skin was investigated *in vivo* and *in vitro* (diffusion cells), the latter consisting of avascular skin and capable of only assessing barrier properties. This study will be presented in detail below, although the original manuscript[22] should be consulted for complete experimental details.

4.1. Materials and Methods

Eight weanling female Yorkshire pigs (*Sus scrofa*) weighing 40 to 60 lbs, clipped of hair at the application sites 24 hr prior to use, were immobilized with xylazine and ketamine and placed in a dorsal recumbent position. As illustrated in Fig. 1, 4 separate areas were defined with sites A and C being perfused by cranial musculocutaneous arteries; whereas, B and D are perfused by caudal direct cutaneous arteries. A single 10 cm^2 area was dosed with 100 mg of ^{3}H-piroxicam formulated in a Carbopol® based gel with a final concentration of the drug (purity of greater than 98%) being 0.5% (w/w). Independent studies in rats, dogs and rhesus monkeys demonstrated this tritium label to be stable and useful in identifying

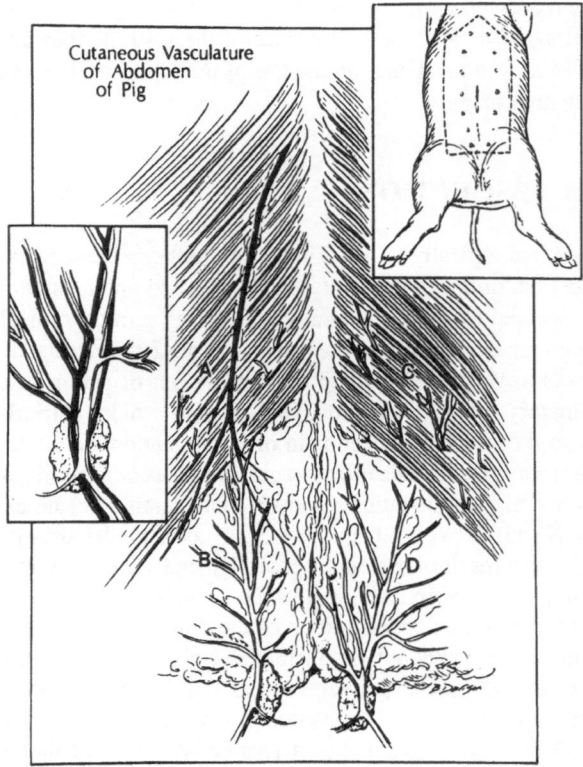

Figure 1. Diagram illustrating the cutaneous vasculature of the ventral surface of the pig. The cranial sites supplied by the musculocutaneous arteries are represented as A and C, while the caudal sites supplied by the direct cutaneous arteries are represented by B and D. From Monteiro-Riviere et al.[22]

specific metabolites of piroxicam.[23] A rigid nonocclusive shield was placed over all sites in order to prevent any cross contamination between sites. Dosing was randomized so each site was dosed twice over the course of the study.

After dosing, pigs were allowed to recover from the anesthesia and housed separately in stainless steel metabolism chambers for the 12 hr exposure. This time point was selected based on preliminary IPPSF studies which showed piroxicam flux plateauing at 8 hr. Thus, tissue concentrations at 12 hr should reflect a postabsorptive equilibrium. Following exposure, pigs were again sedated, protective dressings were removed and the dosing site swabbed to remove surface nonabsorbed drugs. The area was blotted dry and tape stripped 12 times. Two core biopsies were collected from each site (dosed sites last to prevent contamination). The first core, to be used for serial depth penetration sections, was embedded in OCT compound, quenched in an isopentane well cooled by liquid nitrogen and stored at -80C. The remaining core was divided through the subcutaneous layer and placed in separate scintillation vials containing 15 ml of Soluene® 350 for determination of total penetrated radioactivity. Upon completion of sampling, the pig was euthanized with an IV injection of Beuthanasia-D®. The OCT embedded tissue core (n = 4 per pig) was mounted in a cryostat and sectioned at 60 mm. All swabs, tape strips, core sections and solubilized tissue were aliquoted into cups. Samples were oxidized and the resulting tritium content assayed. Data was smoothed using a Lagrange interpolation of 1 degree and plotted as absolute DPM (mean

Table 1. *In vivo* absorption of piroxicam in cranial and caudal sites (mean ± SE)

	Cumulative mass (mg)[a]	Mean penetration depth (μm)	% Applied compound[b]
Cranial	8.36 ± 2.00	1743 ± 354	0.00430 ± 0.00140
Caudal	2.69 ± 0.51	1215 ± 104	0.00018 ± 0.00006

[a]Estimated from the AUC of the penetration profile at the depth of 9480 μm
[b]Discrete compound values estimated at 9480 μm

± SEM, n = 4 pigs per group) vs tissue depth. Samples for light microscopy (LM) were fixed in 10% neutral buffered formalin and processed routinely.

For the *in vitro* diffusion cell studies, freshly excised dermatomed skin (400 μm) samples taken from cranial and caudal sites from additional pigs were mounted in standard Franz diffusion cells with an exposed surface area of 3 cm^2. To initiate each experiment, the gel was applied as a thin film (10 mg/cm^2) to the stratum corneum surface. For the indicated surface area, the total dose of piroxicam applied was 150 μg. The receiver compartment contained 5 ml of phosphate buffered saline (pH 7.4) which was removed and replaced at each sampling time point (2, 4, 6, 8, 12, 24 and 48 hr). Solutions were analyzed with conventional liquid scintillation techniques, counting triplicate aliquots from each sample for 10 min on a liquid scintillation analyzer. Statistical analysis was performed using the Student's t Test.

4.2. Results

The passive *in vivo* penetration of piroxicam was variable, with a higher penetration in the cranial sites compared to the caudal sites. This was illustrated by a higher cumulative compound mass and mean penetration depth (MPD) in the cranial sites (Table 1). Fig. 2

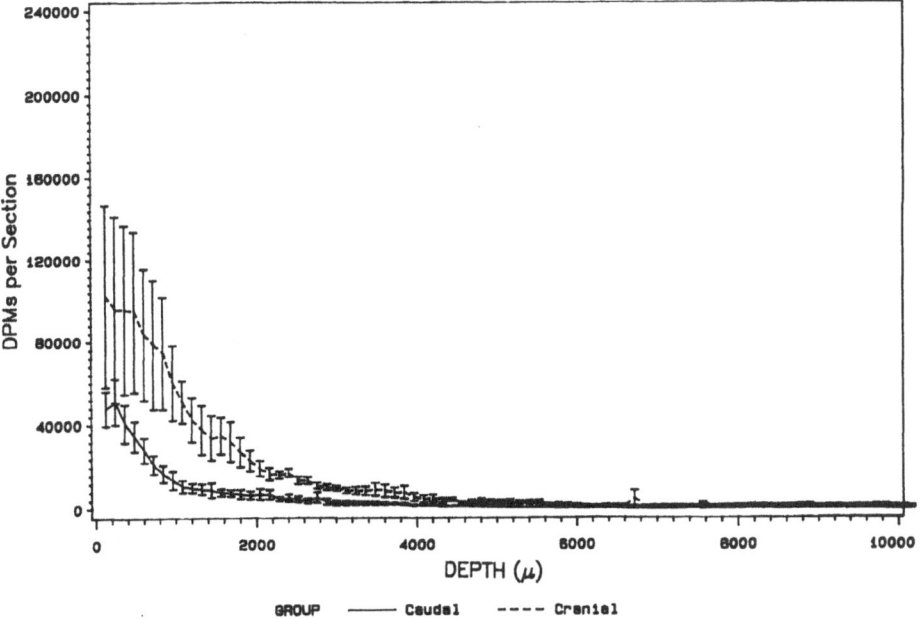

Figure 2. Absolute mean penetration profiles of ^{3}H-piroxicam vs depth from cranial and caudal dosing sites (n = 8). From Monteiro-Riviere *et al.*[22]

Table 2. Depth Correlation with cutaneous structure

Approximate depth (μm)	Microanatomy
0 to 15	stratum corneum[a]
16 to 45	epidermis
46 to 1749	dermis
1750 to 4850	subcutaneous fat
4851 to 10000	muscle

[a]Stratum corneum may be slightly thinner due to tape stripping

compares the absolute mean penetration profiles of piroxicam for cranial and caudal sites. Nondosed sites were assayed for ^{3}H-piroxicam using frozen serial sections and solubilized tissue cores. Radioactivity only slightly above normal background levels was noted at discrete depths, indicating minimal compound crossover from the dosed site. A difference was found between area penetration profiles in the absolute and normalized applied activity at most depths from 900 to 9480 μm. Using LM, these depths correlated to microscopic structures found in the deeper dermis, e.g., hair follicles, apocrine sweat glands, the subcutaneous layer, e.g., adipose tissue, blood vessels and nerves, and the skeletal muscle layers (Table 2). In the normalized absorbed activity profiles, however, differences between cranial and caudal areas were found at more discrete depths. There were differences among cranial and caudal sites in mean cumulative mass, percent of applied compound and percent of absorbed compound. In contrast, *in vitro* permeability studies indicated there was no difference in permeability of skin harvested from either cranial or caudal areas to piroxicam (Table 3).

4.3. Discussion

Topical penetration of piroxicam is not dependent upon absorption into systemic circulation, since only background concentrations of piroxicam were detected at nondosed, remote sites. Secondly, optimal penetration to deeper structures is dependent upon anatomical configuration of cutaneous vasculature. Penetrating musculocutaneous vessels were needed to provide a conduit for the penetration of drug to the subdermal levels. The lack of differences in extent and rates of penetration in the *in vitro* diffusion study using skin from these same sites rules out skin permeability differences as being the cause of enhanced tissue delivery beneath the cranial site. Therefore, a constant fraction of piroxicam is absorbed across the stratum corneum/epidermal barrier. In the site perfused by direct cutaneous vasculature, most of this absorbed drug enters the systemic blood and is eliminated from the animal. In contrast, the site perfused by musculocutaneous arteries retained a larger fraction of the absorbed drug within deeper tissues perfused by these vessels resulting in a smaller fraction of the absorbed dose being available to the systemic circulation.

Table 3. *In vitro* absorption of piroxicam using diffusion cells of the cranial and caudal sites (mean ± SE)[a]

	Flux (0 to 12 hr) ($\mu g/cm^2/hr$)	Cumulative amount transported (μg/24 hr)
Cranial	0.0062 ± 0.0009	0.410 ± 0.050
Caudal	0.0068 ± 0.0003	0.440 ± 0.080

[a]n = 7

These findings have a significant impact on the utilization of simple *in vitro* systems and physiochemical models based on stratum corneum penetration to predict topical drug penetration. These approaches are useful to determine the rate and extent of topical drug penetration across the stratum corneum and epidermis into the dermal tissue. However, the fraction of this "dermally deposited" drug subsequently absorbed into the systemic circulation is dependent upon both the physiology and anatomy of the dermal microcirculation. The penetrating musculocutaneous vasculature functions as a "convective" force driving the absorbed drug to deeper tissue beds. Drugs in these vessels are not absorbed into the systemic circulation, but rather distributed into deeper tissues on the way to the systemic circulation. Differences exist in the magnitude of cutaneous blood flow at different sites within species and at the same sites between species.[24] These differences reflect a vascular phenomenon that may impact on local drug delivery.

Previously, we have shown the function of cutaneous vasculature is important in predicting *in vivo* delivery of drugs administered topically by iontophoresis.[21,25] In these cases, cutaneous depot formation was favored (and transdermal flux decreased) by coionto-phoresing with a vasoconstrictor drug. In contrast, use of a vasodilator minimized the cutaneous depot (increased transdermal flux). Further modeling[17,26,27] suggests the primary role of the cutaneous microcirculation is to change the volume of dermal tissue being perfused by modulating blood perfusion through nonexchanging shunts vs capillary beds which actually perfuse the tissue. These events could allow perfusion of deeper tissue beds which would promote movement of topically applied drugs to these tissues. A knowledge of cutaneous microcirculation function and structure is essential to predict accurately topical drug delivery.

Structural differences exist between the skin of domestic and laboratory animals. Also, thickness of epidermis and dermis between species and within the same species in various regions of the body are different.[24] Therefore, it is important to study skin permeability using skin from a species that closely resembles that of man. Pig skin resembles human skin in both structure and function. Pig and human skin are similar in having a sparse hair coat, a relatively thick epidermis, similar epidermal turnover kinetics, lipid composition, enzyme histochemistry and carbohydrate biochemistry.[4,28,29] In conclusion, topical application of piroxicam can result in local delivery to deeper tissues without first being absorbed into the systemic circulation. Further, it would be expected that local application would be more efficacious at skin sites perfused by penetrating musculocutaneous vessels.

5. IONTOPHORESIS

Electrically assisted transdermal drug delivery, e.g., iontophoresis, electroosmosis and electroporation, has achieved a great deal of research attention and development focus as a potential strategy for transdermal drug delivery. This type of delivery involves transfer of ions or charged drugs through skin in the presence of an electric current. Transdermal delivery of this kind permits administration of therapeutic compounds that would not normally penetrate the skin. We have addressed two issues in iontophoresis which have been studied using porcine skin. The first is an extension of the above piroxicam study and involves determining the role of the cutaneous vasculature in the iontophoretic delivery of vasoactive drugs.[25,30,31] The second involves mapping the anatomical pathway of iontophoretic delivery.[32]

Classic theory suggests that charged drugs may be transported across a barrier such as the skin by applying an opposing electrical field. Thus, to deliver a positively charged drug (cation) such as lidocaine, arbutamine or LHRH, the drug is placed in the positively charged electrode (anode) and current applied. The electrical circuit is completed by the current being carried by all cations in the dosing electrode. Thus, if 2 cation drugs are placed in the same electrode, both drugs will carry the current and thus be delivered across the skin. This principle has been demonstrated in numerous *in vitro* studies using all types of animal and human skin. For example, if lidocaine is placed in the anode along with the cationic drug tolazoline in *in vitro* pig skin, less lidocaine will be delivered then if lidocaine was dosed alone. However, if the same study is done in *in vivo*, the opposite is seen and more drug is delivered in the combination. Why?

The explanation is identical to the difference noted in the *in vitro* and *in vivo* piroxicam studies. In *in vitro* studies, both lidocaine and tolazoline are "seen" by the electrical field as being equivalent charged ions and both carry the current across the skin. Thus as predicted by the Nernst equation, when 2 ions are present, the flux of either will be less than if present alone. When dosed *in vivo*, the same phenomenon occurs at the level of the stratum corneum barrier, but when tolazoline (vasodilator) penetrates the dermal vasculature, vasodilation occurs and greater systemic absorption occurs! As in piroxicam, transepidermal flux may be the same, but the underlying vasculature has a significant effect on the subsequent disposition of the compound.

This hypothesis was validated in additional studies which assayed the amount of iontophoresed lidocaine which penetrated the skin when codosed with either tolazoline or norepinephrine (vasoconstrictor). Core biopsy techniques were identical to that described above for piroxicam.[21] In these cases, compared to lidocaine alone, tolazoline resulted in less dermal lidocaine penetration (all is absorbed systemically) and norepinephrine caused more dermal penetration (less systemic absorption). This reflects these vasoactive compounds ability to modulate dermal depots of drug. These different lidocaine penetration profiles are depicted in Fig. 3. If one only used the *in vitro* diffusion cell data, very misleading results may be obtained. Therefore, utility of the pig as an *in vivo* animal model for studying the mechanism of iontophoretic drug delivery is clearly indicated.

The anatomical pathway through which a compound traverses the stratum corneum has not been defined precisely (Fig. 4). Studies using *in vitro* human abdominal skin suggest the pathway a compound travels is by the intercellular route.[33] Additional *in vitro* studies in hairless mouse skin, assessing ion and current fluxes using a vibrating probe, have implicated the appendageal pathways as the predominant route of electrically assisted drug delivery.[34] Also, cathodic iontophoretic transport of a fluorescein dye in dermatomed human skin showed transport can occur through a "pore" such as a hair follicle, a sweat duct or an imperfection in the skin.[35] Previous studies in our laboratory using *in vivo* and *in vitro* porcine skin demonstrated a pattern of morphological changes induced by lidocaine hydrochloride iontophoresis which suggest that epidermal alterations occur at focal areas and only occasionally are associated with hair follicles.[36] Assuming these changes are related to the presence of transported drug, these findings suggest interfollicular transport may be important and should not be discounted. A knowledge of the actual pathway a drug follows during delivery is of obvious toxicological significance as well as offers support for competing mathematical models of pore transport. Bronaugh *et al.*[37] have shown the hair density in human and pig skin is 11 hair follicles per cm^2, while that in rat is 289, mouse is 658 and hairless mouse is 75 hair follicles per cm^2. Use of hairy skin with a high density of follicles, e.g., rodents, may not be appropriate since the results would be biased toward follicular transport due to the sheer density of hair follicles. Even the use of hairless animals could be misleading because rudimentary follicles are present though the shafts are lacking.

The pathway of ion transport was investigated by iontophoresing mercuric chloride as a test compound. The purpose of this investigation was to: a) search for evidence of an iontophoretic pathway, b) determine if this pathway follows an intracellular or intercellular route and c) determine the extent or depth of the pathway.

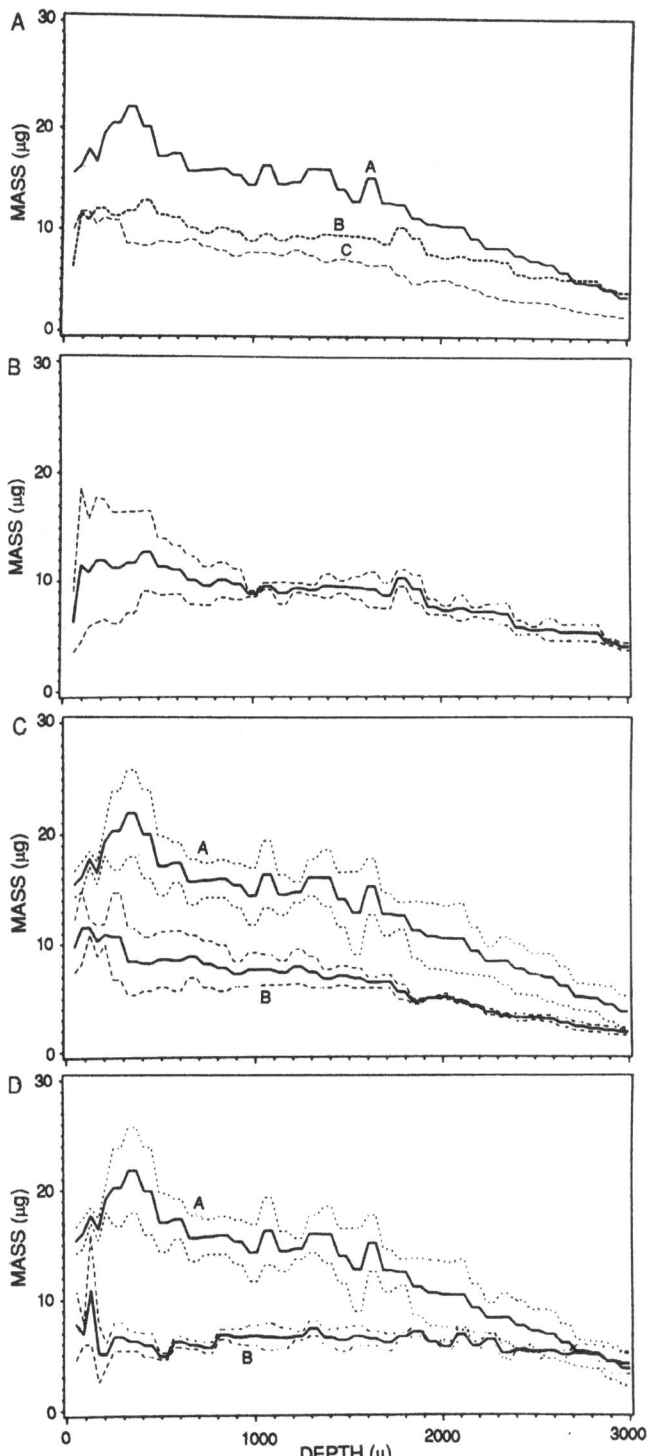

Figure 3. Depth of skin penetration of lidocaine delivered by iontophoresis. A. Mean (± SEM) depth profiles for lidocaine/norepinephrine (A), lidocaine alone (B) and lidocaine/tolazoline (C). B. Mean lidocaine depth profile. C. Mean depth profile following coiontophoresed of norepinephrine (A) or tolazoline (B). D. Mean depth profile for lidocaine/norepinephrine immediately following (A) or 4 hr after (B) termination of iontophoresis. From Riviere et al.[21]

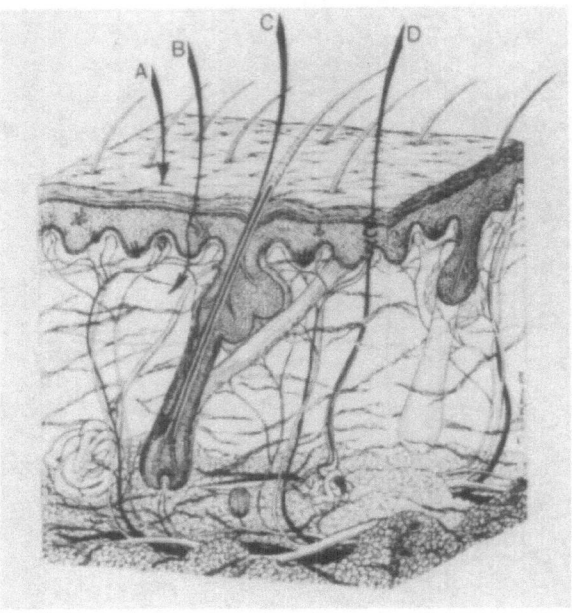

Figure 4. Schematic of skin illustrating its anatomical complexity and the 4 possible paths for chemical penetration. (A) intercellular, (B) intracellular, (C) through hair follicles and (D) through the openings of sweat glands. From Riviere and Monteiro-Riviere.[49]

5.1. Materials and Methods

Female weanling Yorkshire pigs (n = 4) weighing 15 to 25 kg were sedated with ketamine and xylazine and the hair trimmed from the caudolateral epigastric region of each pig. Iontophoretic electrodes, made of Porex and prepared as described previously,[36] had either a 4.5 or 3.0 cm^2 delivery area. The active electrode (anode) was saturated with a 7.4% w/v aqueous solution of mercuric chloride. The indifferent electrode (cathode) was saturated with a 10% aqueous sodium chloride solution and placed caudal to the active electrode. A voltmeter was used to ensure a constant current density of 200 mAmp/cm^2 for 1 hr. Passive transport was assessed with a no current driving the electrode. Following iontophoresis, pigs were euthanatized with Beuthanasia-D. The skin under each active electrode was dissected to the subcutaneous layer, rinsed 30 sec in phosphate buffered saline and blotted dry. The tissue sample was divided and exposed to the vapor of a 25% ammonium sulfide solution for varying times (0.5, 1.0, 1.5, 2.0, 5, 10, 15, 30, 45, 60 or 90 min; 60 min for passive) to form the immobile mercuric sulfide precipitant. Additional nondosed pig skin samples were developed in ammonium sulfide for 1 hr to serve as controls. All tissue samples were fixed in 2% paraformaldehyde and 2.5% glutaraldehyde in 0.2 M cacodylate buffer (pH 7.4) (light and transmission electron microscopy (TEM)) or 4% formaldehyde and 1% glutaraldehyde in 0.2 M phosphate buffer (pH 7.3) for dispersive-X ray microanalysis overnight at 4C. Samples were processed, sectioned and examined. The TEM tissue samples processed for energy dispersive X-ray microanalysis followed the standard TEM protocol with the omission of the osmium tetroxide postfixation. Energy dispersive X-ray microanalysis was carried out on an ETEC autoscan scanning electron microscope (20 KV, 1.0 nanoamp, 4000X

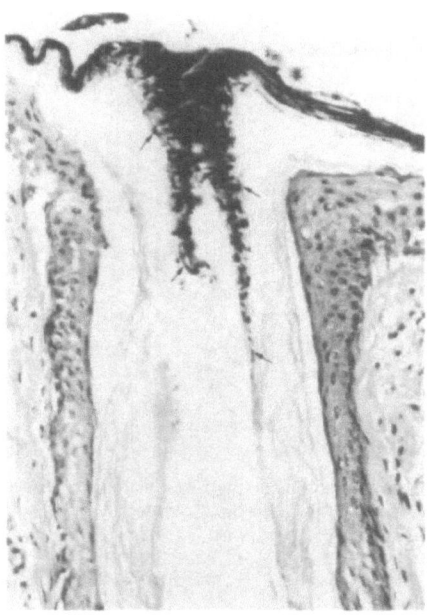

Figure 5. Light micrograph depicting the black mercuric sulfide precipitant (arrows) extending into the upper portion of a hair follicle. H&E. X425. From Monteiro-Riviere *et al.*[32]

magnification) interfaced to a Link System AN 10000 energy dispersive X-ray microanalysis system.

5.2. Results

Density of mercuric sulfide precipitate was directly proportional to the ammonium sulfide vapor development time. Light microscopic analyses visualized a significant amount of mercuric sulfide precipitant between the stratum corneum cell layers with penetration visible in up to 75% of the stratum corneum with the longest exposure times. In some cases, precipitate could be seen penetrating the orifice of hair follicles (Fig. 5). Passive delivery of mercuric chloride for 1 hr followed by a 1 hr development by ammonium sulfide showed the precipitate to penetrate uniformly through only the upper 25% of the stratum corneum. Additional nondosed pig skin control samples developed in ammonium sulfide for 1 hr showed no background precipitate.

TEM revealed the exact anatomical pathway of mercuric chloride transdermal iontophoresis. At short development times, the most superficial layers of the stratum corneum contained a fine granular deposit intracellularly, while in lower layers, precipitate was found as far down as 15 cell layers and localized within intercellular spaces. A few granules of precipitate were present within cells of the stratum granulosum, stratum spinosum, stratum basale and in the outer mitochondrial membrane of mitochondria in viable basal cells. In some instances, granules were present in superficial capillaries of papillary dermis and in fibroblast processes. At longer developments, deeper penetration could be visualized (21 layers) that was located between the individual stratum corneum cells (Fig. 6). In focal

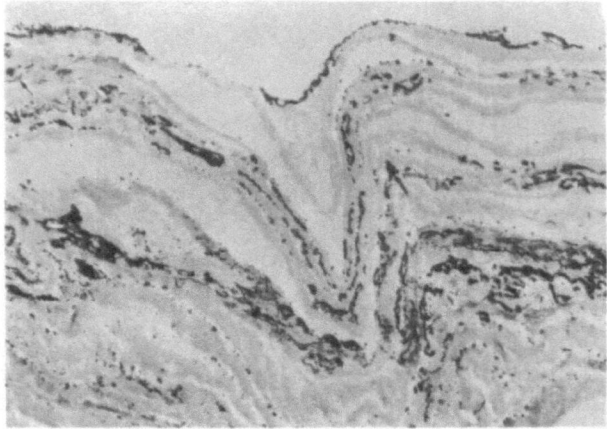

Figure 6. Transmission electron micrograph depicting black mercuric sulfide precipitant (large arrows) in intercellular space of stratum corneum cell layers. Note the cresent shaped deposits (small arrow) in stratum corneum layers. X7,000. From Monteiro-Riviere et al.[32]

regions, precipitant aggregated intercellularly with fine granules funneling intracellularly. However, the predominant route was intercellular. Passive delivery of mercuric chloride revealed precipitate penetrated to a depth of 12 corneum cell layers but no precipitate was present in viable epidermis or dermis. As with active transport, precipitate was observed primarily to be intracellular. Energy dispersive X-ray microanalysis showed intercellular precipitate was mercury and the spectrum from an adjacent precipitant free area demonstrated only background noise and no mercury peaks.

5.3. Discussion

Iontophoresis of mercuric chloride occurs via an intercellular pathway in *in vivo* porcine skin, a finding in agreement with *in vitro* human skin studies.[3] This technique of mercury precipitation, to localize the pathway of transdermal delivery, does not measure active flux of the absorbed ion; rather it only locates precipitated mercury which has remained after 60 min of iontophoretic administration. However, it is axiomatic the precipitate reflects mercury which has traversed this pathway. Localization of this precipitate in deep epidermal and dermal regions by TEM after iontophoresis, but not passive delivery, confirms the potential for transdermal delivery by iontophoresis. Results of passive mercury delivery are identical to those reported in human skin[38] and nude mouse.[39]

The pathway of mercuric chloride initially was intracellular through the first few layers of stratum corneum but, as the path progressed through the stratum corneum, it became primarily intercellular. The most superficial stratum corneum cells, called stratum disjunctum, are able to desquamate easily due to enzymatic degradation of the desmosome junctional complex. This may explain the intracellular location of precipitant at these superficial sites. This intracellular localization of mercuric chloride in the upper few layers of stratum corneum was demonstrated by passive delivery in human skin *in vitro*, in human plantar callus and with iontophoresis through human abdominal skin.[33,38]

Other investigators have shown the stratum corneum consists of 2 functionally different components. The upper few layers of loosely attached cells are considered to be functionally desquamated, while deeper stratum corneum cells are resistant to different forms of mechanical insult.[40,41]

The intercellular pathway for delivery of mercuric chloride by iontophoresis may provide an anatomical basis for interpreting potential size restrictions on defining the putative "pores" in mathematical models that assume an intercellular pathway. The maximal diameter would be intercellular space width and not that of a hair follicle shaft or sweat duct. Concerning follicular transport, if the hair follicle depicted in Fig. 5 were viewed from the surface, it would appear as a black spot and could be interpreted as an "unique" pathway for iontophoretic delivery. However, even with follicular transport, the final pathway is intercellular between hair follicle epidermal cells. Additionally, the actual surface area available for transport in a hair follicle reflects the area of the invagination and not just the opening on the surface. In all probability, intercellular space diameter would be an exaggerated upper limit since ion passage would occur between lipid sheaths which comprise the intercellular space.[42-44] As in passive delivery of lipid soluble drugs, the intercellular pathway suggests significant tortuosity which increases path length nonlinearly from that estimated on stratum corneum and epidermal size and thickness alone. These studies in the pig support the hypothesis in human that the predominant pathway for transdermal drug delivery in both passive and active systems is the intercellular route.[33,38] This finding has implications in interpreting mathematical models of iontophoretic transport which postulate pores as routes of entry.

6. ISOLATED PERFUSED PORCINE SKIN FLAP (IPPSF)

As discussed earlier and illustrated in the above studies, the pig is utilized widely as an animal model for human skin. However, conducting simple *in vitro* diffusion cell experiments may be misleading since the vasculature is not present. In contrast, using the intact pig is problematic since one can only detect absorbed compound by assaying blood or the entire animal for absorbed drug. Distribution through the body results in dilution and lowers those concentrations which could be detected. The systemic metabolism and elimination of the absorbed compound also confounds interpretation. Finally, *in vivo* studies are less humane, and for many toxic compounds, not possible. To overcome both of these limitations, we developed an isolated perfused tissue preparation which maintains the experimental control of an *in vitro* system but has the anatomical and physiological complexity inherent to *in vivo* skin. This model has been reviewed elsewhere[28,45,46] including the previous meeting of the present group.[47]

The IPPSF is a single pedicle, axial pattern tubed skin flap obtained from the ventral abdomen of female weanling swine. Two flaps, each lateral to the ventral midline, may be obtained from each pig in a single surgical procedure. A 2 step surgical procedure, depicted in Fig. 7 is used to create the flap (Stage I) and then harvest from the pig (Stage II). Pigs weighing 20 to 30 kg are premedicated with atropine sulfate and xylazine hydrochloride, induced with ketamine hydrochloride and inhalational anesthesia is maintained with halothane. Each pig is prepped for aseptic surgery in the caudal abdominal region. A 4 x 12 cm region of skin, previously demonstrated to be perfused primarily by the caudal superficial epigastric artery and its associated paired venae commitantes, is incised, dissected, tubed and the apposing edges sutured. Two da later, the second surgical procedure is performed to cannulate the artery and remove the flap from the pig. Two da was optimal from the perspective of lack of flap "leakiness," normal histological appearance and vascularization and animal housing economics.[48] The IPPSF is transferred to a perfusion chamber (Fig. 8) and the resulting wound flushed and allowed to heal. Then pigs can be returned to their previous disposition.

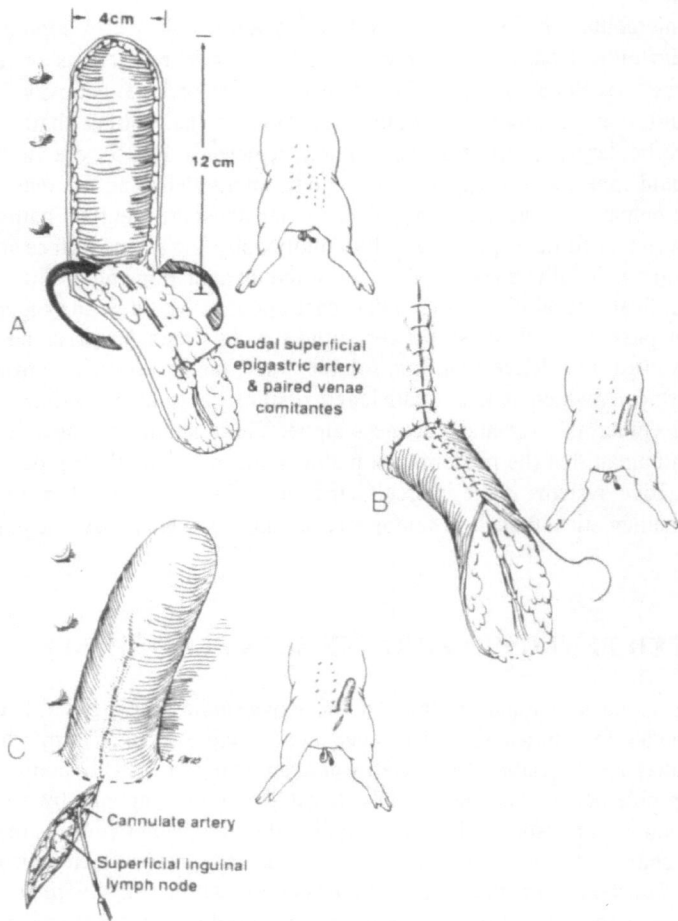

Figure 7. Surgical procedure for creating 2 IPPSFs. (A) First a single pedicle axial pattern skin flap is raised on both sides of the pig and (B) tubed completing the (Stage I) procedure. Two da later, (C) the superficial epigastric artery is cannulated (Stage II) and the flaps are transferred to the perfusion chamber. From Riviere et al.[46]

The perfusion chamber is a custom designed, temperature and humidity regulated chamber made specifically for this purpose. Perfusion pressure, flow, pH and temperature are monitored and recorded for future analyses. Temperature and relative humidity may be controlled independently according to the needs of specific experimental protocols. Normal experimental conditions are 37C with a 60 to 80% relative humidity. The perfusion media is a modified Krebs-Ringer bicarbonate buffer (pH 7.4, 350 mOsm/kg) containing albumin (45 g/l), and supplied with glucose (80 to 120 mg/ml) as the energy substrate. Albumin is included in the media primarily to provide the oncotic pressure needed to perfuse capillaries. However, as occurs *in vivo*, it also provides binding sites for relatively lipophilic drugs. Antiinfectives (penicillin G, amikacin) are included in the media to prevent overgrowth or bacteria from normal flora present on the skin surface. Also, heparin is included to prevent coagulation of the skin's vasculature from residual formed blood elements. Media is gassed

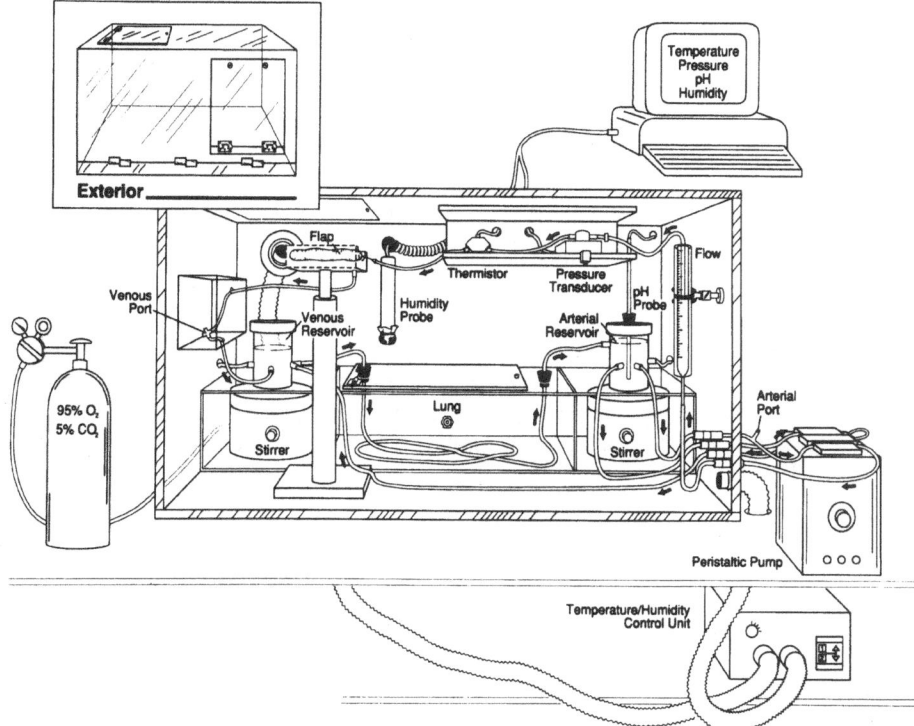

Figure 8. Schematic diagram of the IPPSF perfusion chamber. From Riviere *et al.*[93]

with 95% O_2/5% CO_2. Normal perfusate flow is maintained at 1 ml/min/flap (3 to 7 ml/min/100 g), with a mean arterial pressure of 30 to 70 mm Hg. Both recirculating and nonrecirculating (single pass) configurations are possible. Viability is assessed routinely by monitoring glucose utilization, vascular resistance (perfusate pressure ÷ flow) and histology.

Numerous studies have been conducted to assess percutaneous absorption of drugs and pesticides in the IPPSF and then compared to human studies. In all studies reported,[49-51] correlations between the IPPSF and human *in vivo* total absorption was high (R ≅ 0.9). Iontophoretic delivery of LHRH and arbutamine best illustrate the ability of the IPPSF to predict *in vivo* human plasma concentration time profiles.[52-54] In these cases, using both IPPSF and IV human pharmacokinetic studies, the mean and variance of human plasma profiles may be predicted accurately. The IPPSF also has been used to study the effects of vascular modulation on transdermal iontophoretic lidocaine delivery[17,25,27] and has supported the findings discussed above for *in vivo* studies. These percutaneous absorption studies recently were reviewed.[55] Finally, we "mapped" out the IPPSF's response to infusion with autonomic drugs.[56] Cutaneous metabolism of 1,25-dihydroxyvitamin D has been shown to reflect human data.[57] A human tumor bearing IPPSF was developed to assess regional delivery of anticancer drugs.[58] Additional studies with acids and bases,[59] paraquat[60] bis(2-chloroethyl sulfide),[61] chlorovinylarsine dichloride[62] and irritants such as sodium lauryl sulfate[63] have validated the IPPSF as a model for the assessment of direct chemical toxicity in skin.

6.1. Studies in Phototoxicology

Sunburn, the excessive cutaneous exposure to UVB radiation, is the most frequent phototoxic reaction.[64] Recently, with the identified increase in UVB exposure in certain geographical regions apparently secondary to ozone depletion, this topic is becoming more important. Phototoxicity after exposure to ultraviolet light mid wave UVB (290 to 320 nm) and long wave UVA (320 to 400 nm) has been studied with *in vitro* models, which include *Candida albicans*,[65] photohemolysis of red blood cells,[66] tissue culture[67] and isolated normal human fibroblasts.[68] Such ethical models lack the complexity of living skin systems and often are unreliable. *In vivo* models utilizing guinea pigs,[69] rabbits[70] and mice,[71] have been used to evaluate sunburn cells (SBC) in epidermis following UV radiation. However, most animal model skin is not anatomically comparable to human skin, sometimes resulting in false positive or false negative interpretations. UVB induced erythema [72,73] and SBC expression[73] in miniature pigs are comparable to humans. Because of the need for a complex *in vitro* model system, we characterized morphological and biochemical responses of the IPPSF to UVB radiation.[74]

6.1.1. Materials and Methods. The UVB source utilized in this study consisted of two 8 watt tubes emitting wavelengths of 280 to 320 nm, with the spectrum peak at 312 nm. Light intensity was measured using a UVB photodetector and found to be stable during the dosing period. *In vivo* and IPPSF UVB exposures were conducted. *In vivo* studies were conducted in sedated pigs as described above and the IPPSF's were created, harvested and perfused according to standard protocol. Two IPPSF protocols were used, flaps were either harvested and then exposed in the chamber to UVB or exposed while they were on the pig (*in situ*), harvested 16 hr after exposure and perfused for 8 hr. Treatments consisted of UVB doses of 1260 (n = 5), 630 (n = 7), 315 (n = 4) or 0 mJ/cm^2 (n = 4). UVB effects were monitored using vascular resistance, glucose utilization, and LM and TEM endpoints. SBC were quantitated on histological sections as an index of cellular damage. Epidermal cell proliferation, expressed by the growth fraction (GF) within the stratum basale, was quantitated by immunohistochemical staining of the proliferating cell nuclear antigen (PCNA). Sections were stained with monoclonal mouse PCNA.[75] GF was determined by dividing all cells in the active cycle (G$_1$, S, G$_2$ and M phases) by the total number of cells. Means for both endpoints were generated and the differences among UVB doses determined. IPPSF venous perfusate was analyzed for prostaglandin (PGE$_2$) using a radioimmunoassay.

6.1.2. Results. *In vivo* response to UVB was reported previously and consisted of erythema and edema and a dose related increase in SBC expression and a decrease in GF. Although the dosed sites of 5 pigs exhibited erythema, edema was present in only 3 pigs. Erythema was present following the dose and became more intense from 8 to 24 hr. Nonexposed control (0 mJ/cm^2) sites showed no change. LM of dosed sites revealed slight intercellular and slight to moderate intracellular epidermal edema, microvesicles and dermal

Table 4. Mean SBC ± SE per 278 µm of epidermis assessed at 8 and 24 hr *in vivo*

Dose(mJ/cm^2)	n	8 hr	SBC ratio	n	24 hr	SBC ratio
1260	4	8.47 ± 1.68	18.4	4	10.22 ± 0.96^b	26.9
630	4	6.52 ± 1.51	14.2	4	10.02 ± 1.99^b	26.4
315	3	4.38 ± 1.12	9.5	3	9.34 ± 0.57^b	24.6
0	3	0.46 ± 0.18	1.0	4	0.38 ± 0.18^a	1.0

^{a} and ^{b} differ significantly (p<0.05)

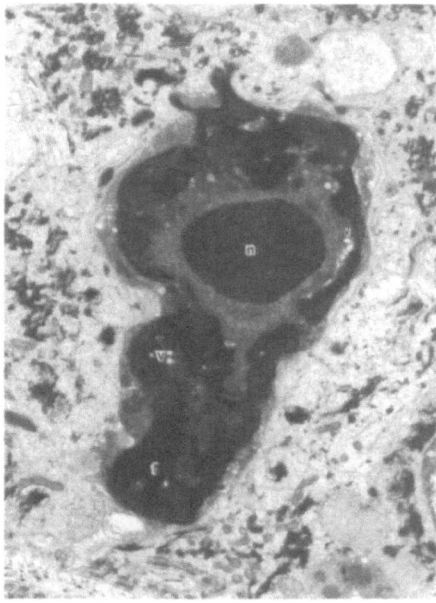

Figure 9. Transmission electron micrograph of a sunburn cells found in the stratum basale layer of pig skin following irradation with 1260 mJ/cm^2 of UVB. Note the pyknotic nucleus (n), cytoplasmic vacuoles (v) and condensed filaments (f). X12,300.

inflammatory cells. SBC, which appeared 8 hr postdose within epidermis, consisted of an eosinophilic cytoplasm with pyknotic nuclei. Mean SBC numbers, quantitated at 8 and 24 hr, followed a dose response (Table 4). Unlike 8 hr mean SBC counts, an increase was noted at 24 hr between dosed and control SBC counts. Ultrastructurally, SBC possessed a pyknotic nucleus, cytoplasmic vacuoles and condensed filament masses mixed with remnants of other cytoplasmic organelles (Fig. 9) 8 hr following dose. The PCNA antibody was localized to

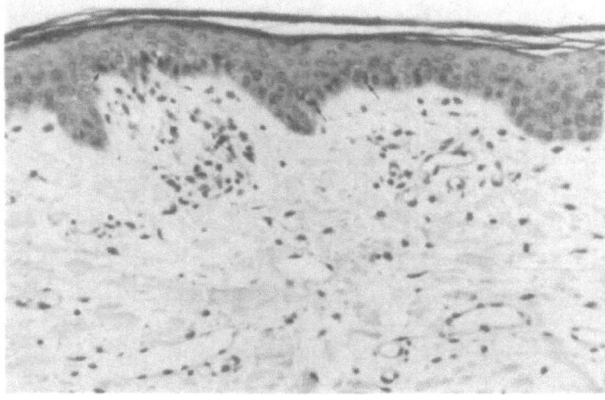

Figure 10. Light micrograph showing sunburn cells (arrows) in the IPPSF exposed to 1260 mJ/cm^2 of ultraviolet B radiation. X400. From Monteiro-Riviere et al.[74]

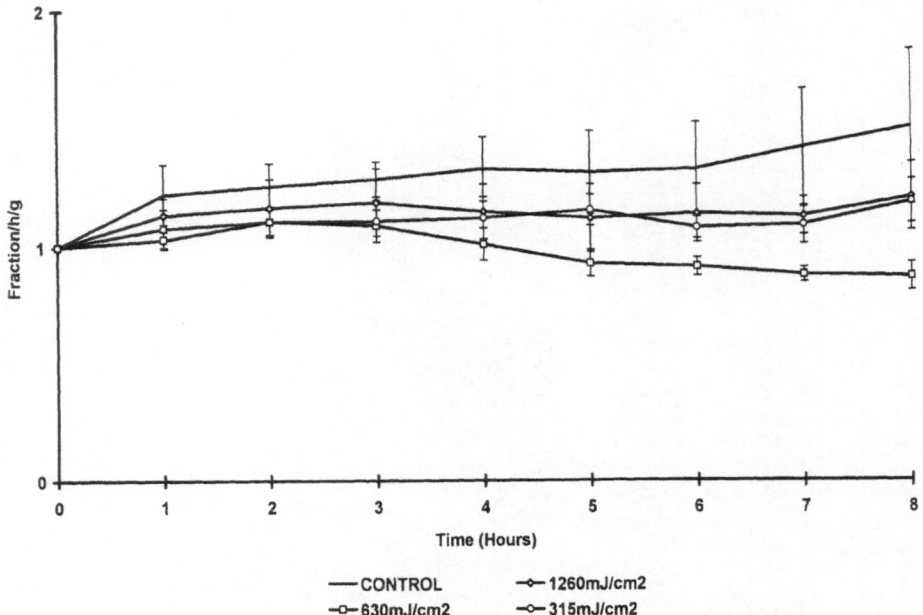

Figure 11. Normalized glucose utilization in IPPSFs exposed to ultraviolet radiation.

basal cells. After 8 hr, no changes in GF were observed, although mean GF decreased at 24 hr at all UV doses. A decrease in mean GF was found between 24 hr dosed and the nonexposed control sites.

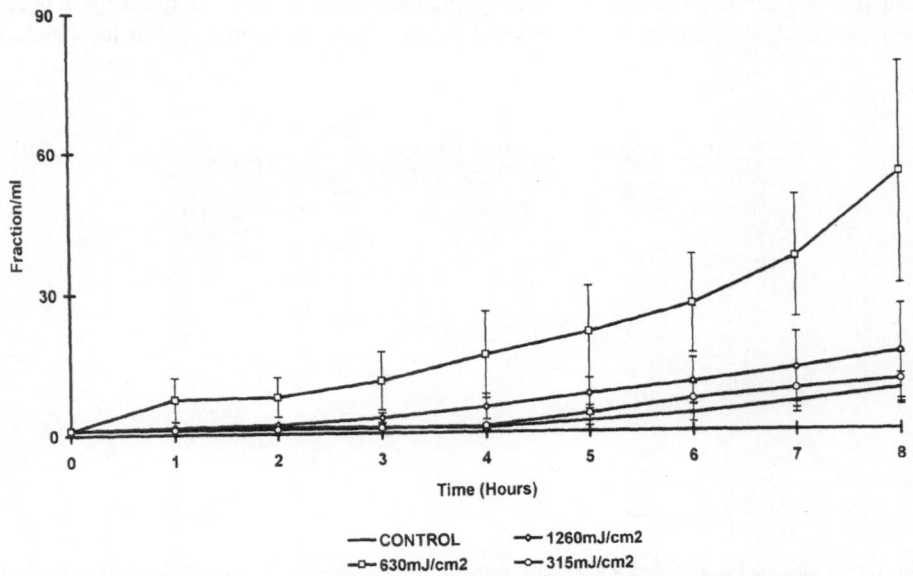

Figure 12. Normalized PGE$_2$ profiles in IPPSFs exposed to ultraviolet radiation.

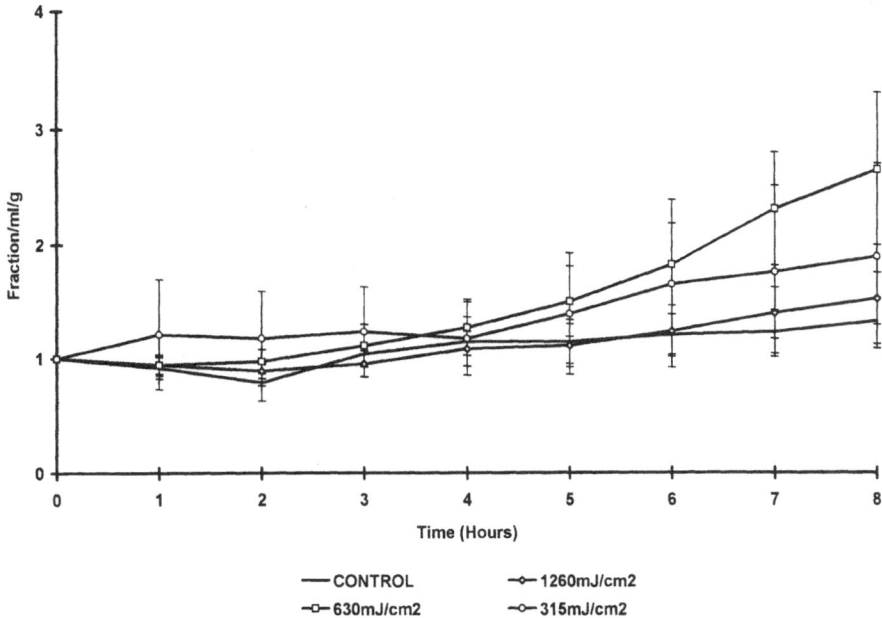

Figure 13. Normalized vascular resistance in IPPSFs exposed to ultraviolet radiation.

All dosed flaps exhibited slight intercellular and intracellular epidermal edema. The most notable feature in epidermis was the presence of SBC which appeared histologically (Fig. 10) and ultrastructurally similar to those found in *in vivo* exposure to UVB. Mean SBC numbers in the flap followed a dose response. When these values were compared to controls by use of a SBC ratio ($SBC_{Dose} \div SBC_{Control}$), this dose response was evident and similar to *in vivo* results. Mean GF of *in vitro* dosed IPPSFs likewise paralleled *in vivo* results.

Data for *in vitro* GU, PGE_2 and VR were normalized to the value of the initial (0 hr) data point to compare treatments. GU of all dosed flaps was lower than controls. GU of 1260 and 315 mJ/cm^2 dosed flaps remained steady while that of 630 mJ/cm^2 flaps decreased (Fig. 11). Venous PGE_2 concentration of all doses increased over 8 hr (Fig. 12). The 1260 mJ/cm^2 flaps were marginally higher than the 315 mJ/cm^2 and control flaps, while the 630 mJ/cm^2 dosed flaps steadily increased from 2 to 8 hr. VR profiles remained flat until 4 hr, when 630 and 315 mJ/cm^2 doses started to increase (Fig. 13). Terminal VR in 630 mJ/cm^2 dosed flaps was higher than 315, 1260 or control flaps. *In situ* exposed IPPSFs were comparable, except for an enhanced PGE_2 response and an increased weight gain during perfusion. LM revealed slight intercellular and slight to moderate intracellular epidermal edema, microvesicles, dermal infiltrate and SBC. Gross blisters were noted in the *in situ* flaps following perfusion as follows: 4/4 (1260), 0/4 (630), 3/5 (315) and 0/4 (0 mJ/cm^2). SBC were histologically and ultrastructurally similar to those found *in vivo* and *in vitro*. Mean SBC numbers and SBC ratios followed a dose response. On average, numbers of SBC in *in situ* flaps were less than those found in *in vivo* skin exposed to corresponding doses of UVB. As in the pig, mean GF decreased by a factor of 6 with increased dose. A difference in GF was noted between 1260 and 630 mJ/cm^2 flaps and control flaps.

6.1.3. Discussion. Based on the present study, this *in vitro* skin model appears to be ideal for studying UVB phototoxicity. Epidermal cellular damage was manifested by the presence of SBC scattered within the stratum basale. Although some SBC were present in controls, fewer were found in undosed IPPSF than in normal pig skin, indicating that SBC were not caused by ischemic changes or technical problems associated with the IPPSF. SBC *in vitro* and *in situ* appeared morphologically similar to those detected in *in vivo* exposures but were present in lower numbers. Mean SBC numbers were directly proportional to dose and increased in numbers at 24 compared to 8 hr. When these numbers were corrected for control values, *in vivo* and *in vitro* responses were identical. SBC found in pig epidermis were described as histologically[76] and ultrastructurally[77] similar to those in UVB exposed human skin. We confirmed these results in the *in vitro* IPPSFs.[74]

Epidermal cell proliferation assessed by PCNA staining and expressed as the GF were comparable at all control doses and increased marginally after 8 hr in low dose IPPSFs but were unchanged from control for other doses. However, at 24 hr in both *in vivo* and *in situ* exposures, GF decreased with increased dose, indicating a dose dependent cell death. This agrees with Hashimoto *et al.*[78] who found an initial decrease 48 hr after UVB followed by a terminal increase in cell proliferation in pig epidermis. Erythema, the classic response of skin to UV irradiation, was elicited in pig skin due to the perpendicular arrangement of vascular loops to the surface of the skin.[8] This was confirmed by anthracene UVA induced erythema in white pigs but not in mice.[79] Other investigators[72,73] have shown that UVB induced erythema in miniature pigs is comparable to that in humans. No erythema was caused by the *in vitro* irradiation of the IPPSF since the media lacked red blood cells. However, extravasated and capillary bound red blood cells were present in dermis and directly proportional to dose. This may have been due to a dose dependent release of PGE_2 that subsequently caused vasodilation and increased permeability, thereby releasing red blood cells from underperfused capillary beds.

Decreased GU in all groups of *in vitro* dosed IPPSF probably was due to cellular damage as manifested by the presence of SBC and decreased GF. PGE_2 concentration profiles of 1260, 315 and control *in vitro* dosed IPPSF showed little difference, while 630 mJ/cm^2 flaps contained higher concentrations. This dose also showed a relative decrease in GF. VR of 630 mJ/cm^2 *in vitro* dosed flaps was higher than other doses. The profile increased at 4 hr and was higher than 315, 1260 or control flaps. High concentrations of PGE_2 have been reported to increase vascular permeability, which may increase flap edema and VR.

The IPPSF qualitatively and quantitatively responds in a manner consistent with current human and animal model data. Because of its anatomical complexity, all aspects of the pathogenesis of UVB phototoxicity may be investigated humanely in a single preparation. The IPPSF may lend itself to the study of PUVA phototoxicity. This suggests the IPPSF may be a valuable *in vitro* model to study UVB induced phototoxicity or photosensitization, especially when interactions of prophylactic or therapeutic drugs are involved since the IPPSF has been shown to be predictive of *in vivo* drug absorption and cutaneous disposition.

6.2. Mustard Induced Vesication

The final area where the pig has been utilized as a model for skin research is the field of chemical induced vesication. In this case, experimental human studies are not possible and thus animal surrogates must be used. The mechanism of chemical induced cutaneous vesication has eluded investigators since World War I. The use of a potent cutaneous vesicant, sulfur mustard (bis(2-chloroethyl)sulfide, HD), in the Iran-Iraq war stirred a renewed interest. The biochemical basis and mechanism of HD-induced cutaneous vesication still remains unknown. HD is a lipophilic compound which penetrates the skin to cause erythema and blistering after a 4 to 24 hr latency period. In the presence of water, HD hydrolyzes to form hydrochloric acid and thiodiglycol. Humans exposed to the topical vesicant HD, after a delay of a few hr, usually develop erythema and fluid filled blisters which require a prolonged period to heal.[80-82]

Historically, vesicles or vesicle like lesions have been produced with HD in many diverse animal models, including bird skin, frog skin, canine mammary gland skin, rabbit ear skin and thermally burned reepithelialized guinea pig skin.[80] Microvesicles have been elicited with HD in the rabbit and guinea pig,[83] hairless guinea pig[84-86] and the human skin grafted athymic nude mouse model.[87-89] However, microvesicles have been described in pigs treated topically with neat butyl mustard[90] as well as with HD and lewisite (L) but no gross lesions were noticed.[91] The *in vitro* full thickness human skin organ culture model exposed to HD developed microscopic epidermal dermal separation.[92] However, no macroscopic blisters typical of human exposure have been reported in any of these models.[81-82] Part of the problem resides in the model systems used to study these events.

Since gross blisters never form with *in vitro* human and animal models, microvesicles are usually the accepted toxicologic end point. Part of the discrepancy seen in studies conducted in different species is a result of the complex pathogenesis of chemical vesication. The initiating biochemical lesion (DNA alkylation, glutathione depletion, inhibition of glycolysis, alkylation of basement membrane molecules) may be studied *in vitro*, but the formation of gross blisters probably requires other physiological factors not present in simple *in vitro* skin models. Although *in vivo* models may produce relevant lesions, mechanistic studies are difficult to conduct and the humane aspects of animal exposure preclude their widespread use.

Numerous studies conducted with vesicants demonstrated the suitability of the IPPSF as a useful model in this area. Studies with the HD monofunctional analogue 2-chloroethyl methyl sulfide (CEMS)[93] and L[62] demonstrated the IPPSF produced gross fluid filled blisters following vesicant exposure. TEM revealed separation between the lamina lucida and lamina densa in the epidermal dermal junction (EDJ), with intracellular vacuolization and mitochondrial swelling in the stratum basale and stratum spinosum cells. These changes were similar to those described after human exposure to sulfur mustard.[80-82]

Also, the IPPSF has been utilized to model the kinetics of the HD absorptive phase as it relates to pathogenesis of agent induced vesication.[94] These studies confirmed that HD rapidly penetrates skin and produces a vascular response which further modifies HD absorption. Also, they demonstrated this absorptive phase is characterized by a large degree of interflap variability which explains the resulting large variance observed "downstream" in vesication studies. Cutaneous metabolism of HD has been studied in the IPPSF.[95] The purpose of the work to be reviewed here[96,97] was to investigate pathogenesis of HD induced vesication by characterizing biochemical, physiological and morphological responses in the IPPSF and to determine the optimal HD dose to be used in vesicating studies. In addition, molecular epitopes of pig epidermal dermal junction were mapped and their role as putative targets for direct HD alkylation assesed. This work resulted in a shift in focus toward HD direct alkylation of the epidermal basement membrane as a significant factor in the pathogenesis of HD vesication.

6.2.1. Materials and Methods. The IPPSF was prepared, harvested and sampled as discussed above. Primary indicators of HD induced toxicity included changes in glucose utilization, vascular resistance, and LM and TEM. Each flap was perfused 1 hr prior to HD dosing to assess biochemical and morphological viability. In these studies, cumulative glucose utilization (CGU) was used to assess the effects of HD on IPPSF viability. All flaps were dosed with 200 μl of absolute ethanol (control) or HD within a 5.0 cm^2 Stomahesive dosing template using a positive displacement pipette. In this dose response study, flaps were dosed with absolute ethanol (n = 4) or HD ((0.2 mg/ml (n = 4), 0.5 mg/ml (n = 3), 1.25 mg/ml (n = 3), 2.5 mg/ml (n = 4), 5.0 mg/ml (n = 4) or 10.0 mg/ml (n = 5)) and perfused for 8 hr. Following flap perfusion, tissue samples were taken from the dosed area for LM and TEM and processed as described above. Tissue for immunochemistry was harvested and oriented

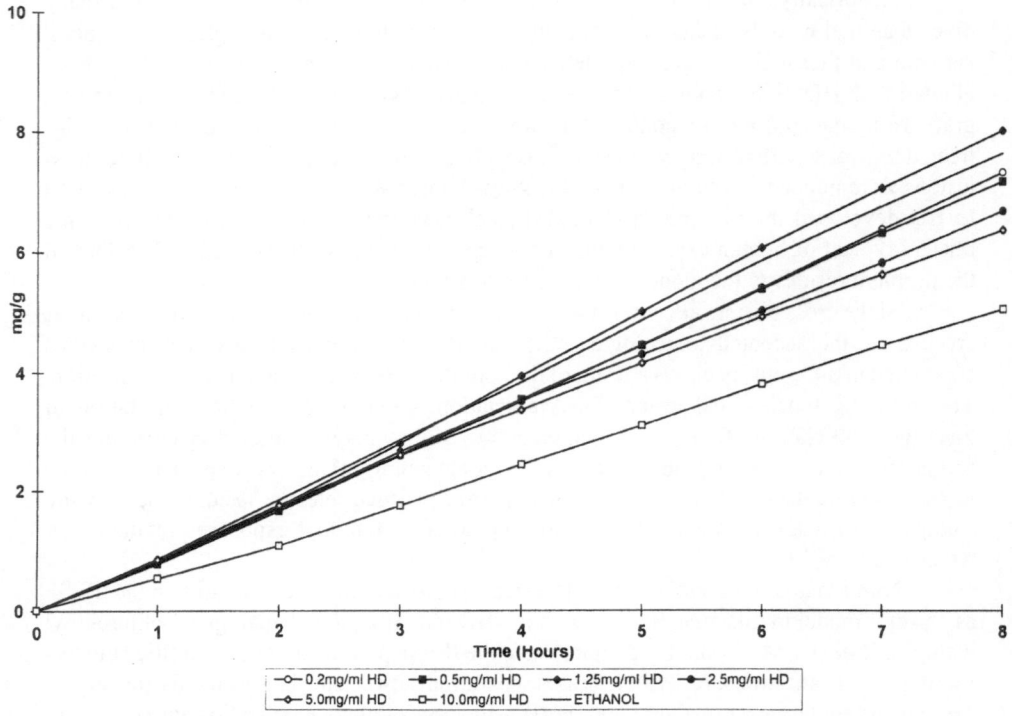

Figure 14. Cumulative glucose utilization in IPPSFs treated with HD and ethanol.

in an aluminum foil boat, quenched and embedded in OCT in an isopentane well immersed in liquid nitrogen.

Indirect immunohistochemistry (IH) (or immunofluorescence (IF)) and immunoelectron microscopy (IEM) were performed using rabbit antimouse Engelbreth-Holm-Swarm (EHS) tumor laminin, rabbit antimouse EHS tumor type IV collagen, rabbit antihuman fibronectin, monoclonal mouse antihuman GB3, human bullous pemphigoid antibody (BP, IgG titer >1:2560), and human epidermolysis bullosa acquisita antibody (EBA, IgG titer 1:320) on all ethanol and HD treated flaps. In addition, normal skin dissected from pig ventral abdomen acted as controls for flaps. The mouse antihuman monoclonal antibodies, L3d and 19-DEJ-1, did not cross react to EDJ epitopes in pig skin. IH was carried out using routine biotin/streptavidin methodology. IF was performed in a manner similar to IH. Cryosections were incubated in primary (GB3 or normal mouse serum control) and secondary (FITC labeled goat antimouse IgG) antibodies for 30 min each, mounted in glycerin and viewed on a Zeiss IM 35 inverted microscope equipped with epifluorescence. IEM was performed using a modification of the method described by Yaoita *et al.*[98]

6.2.2. Results. Mean CGU of all treatments increased in a linear manner (Fig. 14). Terminal CGU values were greatest in the 1.25 mg/ml dose, followed by the ethanol control, 0.2, 0.5, 2.5 and 5.0 mg/ml dosed flaps. The 10.0 mg/ml flaps demonstrated the most dramatic decrease in CGU when compared to controls. Comparison of CGU slopes showed differences between the 10.0 mg/ml dose and 1.25, 0.2 and 0.5 mg/ml doses.

One hr following dose, all treatments, except ethanol controls showed an increase in mean VR (Fig. 15). In controls, VR remained steady from 1 through 8 hr. Differences were found between treatment means of the 1.25 mg/ml dose and 0.5, 5.0 and 10.0 mg/ml doses. Also, differences were present between the 0.2 mg/ml dose and the 0.5 and 10.0 mg/ml doses

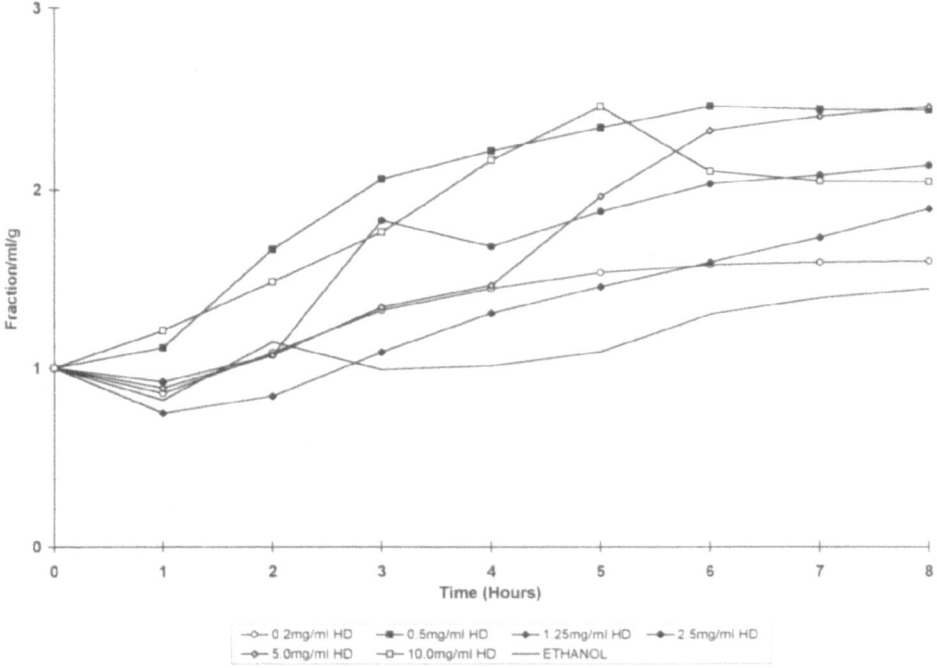

Figure 15. Normalized vascular resistance in IPPSFs treated with HD and ethanol.

and between controls and all dosed treatments. The time wise comparison of VR show some differences between ethanol and dosed treatments. From 6 to 8 hr postdose, the 0.5, 2.5, 5.0 and 10.0 mg/ml dosed treatments were different from ethanol controls.

Macroscopic (gross blisters) and microscopic observations of HD treated flaps showed a dose related effect in the IPPSF. Only those morphological alterations seen with

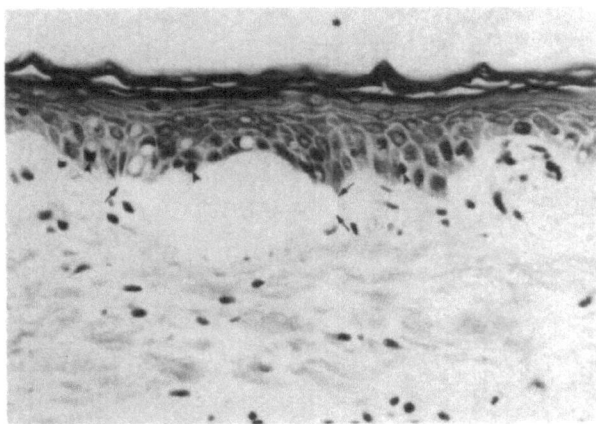

Figure 16. Light micrograph of a 10.0 mg/ml of HD induced blister (arrows) in the IPPSF. Note dark basal cells (arrow heads). H&E. X550.

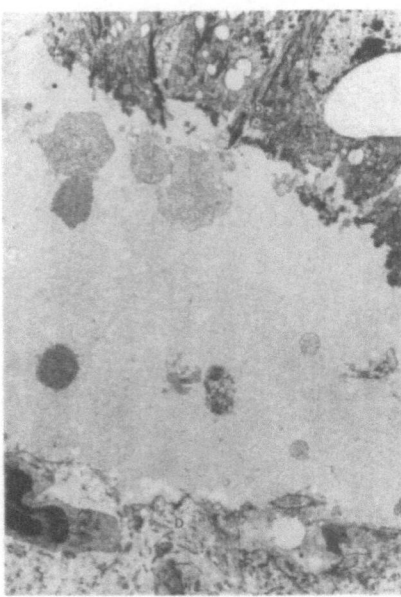

Figure 17. Transmission electron micrograph of a 10.0 mg/ml of HD induced blister in the IPPSF showing separation of the epidermal dermal junction. Note the separation occurs below the stratum basale cell (sb) and above the dermis (D). X9,800. From Monteiro-Riviere and Inman.[97]

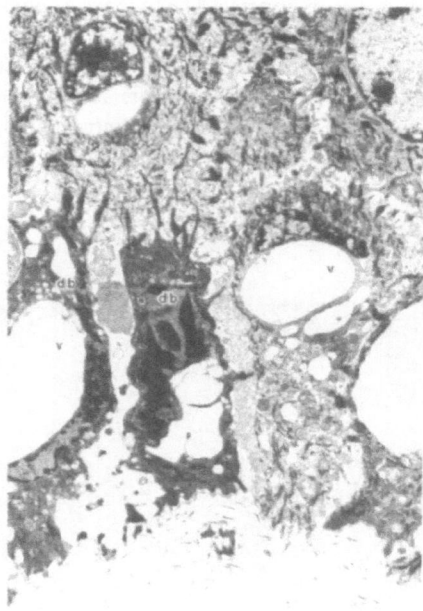

Figure 18. Transmission electron micrograph showing dark basal cells (db) in the stratum basale of an IPPSF dosed with 10.0 mg/ml of HD. Note the large cytoplasmic vacuoles (v). X7,200. From Monteiro-Riviere and Inman.[97]

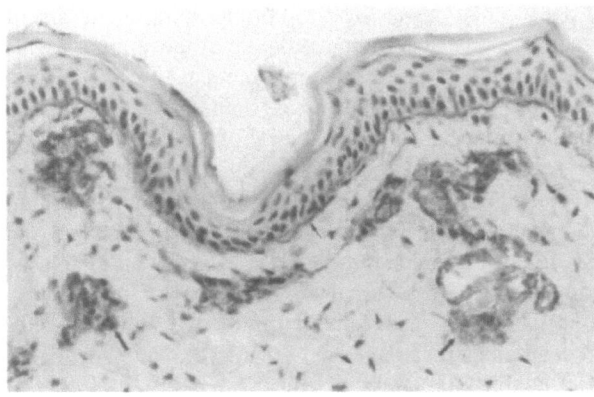

Figure 19. Immunohistochemistry of type IV collagen in normal pig skin. Note continuous staining of the epidermal dermal junction (small arrows) and capillary basement membrane (large arrows) H&E X450. From Monteiro-Riviere and Inman.[96]

the 10.0 mg/ml HD dose will be presented. The original manuscripts [96,97,99] should be consulted for details on all treatment groups.

Morphological effects in the highest dose of HD were more severe than in previous doses. All but one of the 10.0 mg/ml HD treated IPPSFs showed macroscopic blisters while all had microvesicles. LM evaluation of these flaps showed intracellular edema (hydropic degeneration or liquefaction degeneration), slight intercellular edema, focal epidermal necrosis, pyknotic basal cells and severe epidermal dermal separation (>20 cells affected) (Fig. 16). TEM showed that separation occurred at the epidermal dermal junction of the basement membrane (Fig. 17). Pyknosis was more severe than in any other of the HD treated IPPSFs, but was restricted to stratum basale and stratum spinosum cell layers (Fig. 18). In addition, there were dyskeratotic and karyolytic basal cells, large cytoplasmic vacuoles and dilated rough endoplasmic reticulum.

IH results were as follows. The laminin antibody bound to normal pig skin (NPS) formed a continuous linear label along the epidermal dermal junction (EDJ) and along the capillary basement membrane. In blistered areas of HD treated flaps, laminin stained the dermal side of the separation (blister floor), with occasional staining of the basal pole of the stratum basale cells (blister roof). In NPS, type IV collagen exhibited EDJ and capillary basement membrane staining similar to laminin (Fig. 19). Type IV collagen had a higher EDJ binding affinity in the ethanol flaps. In blistered areas of HD treated IPPSFs, staining was limited exclusively to dermis (Fig. 20).

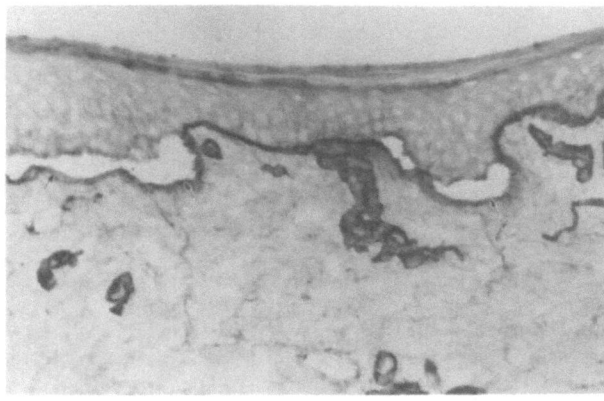

Figure 20. Immunohistochemistry of type IV collagen in the IPSF treated with 5.0 mg/ml of HD. Note staining along the floor of the blister (arrows). X450. From Monteiro-Riviere and Inman.[96]

Fibronectin stained a faint, fairly continuous band along the EDJ in NPS, ethanol flaps and nonblistered HD treated flaps. Staining was localized primarily to the dermis of HD blistered flaps, with stained fragments attached to the basal pole of the stratum basale cells. GB3 monoclonal antibody bound to the EDJ stained a continuous band in NPS, ethanol and nonblistered areas of HD treated flaps. In HD blistered areas, staining was localized to the dermis. BP antibody stained a faint, broken label along the EDJ of NPS. In contrast, all ethanol flaps and HD nonblistered areas exhibited a more intense and continuous staining of the EDJ. Antibody staining of HD blistered areas was limited to the basal pole of stratum basale cells. EBA antibody stained a continuous band along the EDJ in NPS, ethanol and HD nonblistered areas of the IPPSF. In HD blistered areas, antibody was bound to dermis. In the case of L3d and 19-DEJ-1, no staining of NPS was noted by IH.

The IEM data confirmed the IH results. With laminin, antibody binding in NPS exhibited a well defined, homogeneous pattern along the EDJ and around the dermal vasculature. Ethanol treatments and nonblistered areas of HD treated IPPSFs produced similar results. With HD induced blisters, staining occurred predominantly on the dermal interface with remnant staining on the basal pole of the stratum basale cells. Type IV collagen binding in NPS formed a discrete, fairly continuous band along the EDJ and capillary basement membrane. In contrast, staining in ethanol and HD nonblistered areas was more diffuse. Antibody localization in HD blistered areas was limited to the dermal side of the split (Fig. 21).

Although cross reactivity of fibronectin was low, slight antibody staining was found along the EDJ, the upper papillary dermis and the capillary basement membrane of NPS and HD treated nonblistered areas of the IPPSF. In HD IPPSFs, fibronectin bound predominantly to dermis with discrete areas of epidermal staining. GB3 antibody bound to the EDJ of NPS to form a discrete discontinuous pattern. In ethanol and nonblistered HD treated IPPSFs, staining was localized beneath the hemidesmosomes. In HD blistered areas, staining was localized primarily to the dermis, with occasional staining of the basal pole of the stratum basale cells. In NPS, BP stained a discrete discontinuous pattern along the EDJ. A more continuous staining pattern was observed in ethanol controls and nonblistered areas of HD treated flaps. In HD induced blistered flaps, BPA was localized to discrete areas within the hemidesmosomes. EBA stained a diffuse band along the EDJ of NPS, ethanol and nonblis-

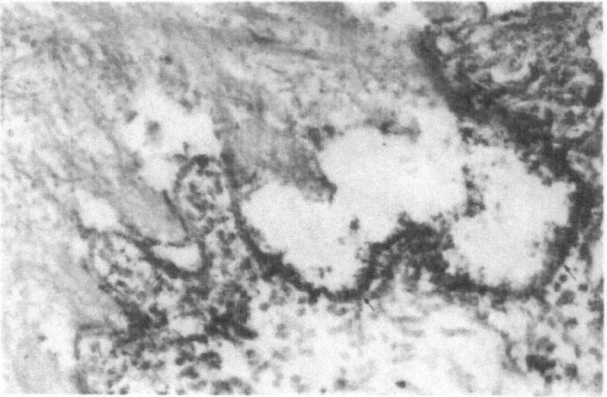

Figure 21. Immunoelectron microscopy of type IV collagen in the IPPSF treated with 10.0 mg/ml of HD. Note strong staining along the dermal side of the separation (floor of the blister) (arrows). X34,700. From Monteiro-Riviere and Inman.[96]

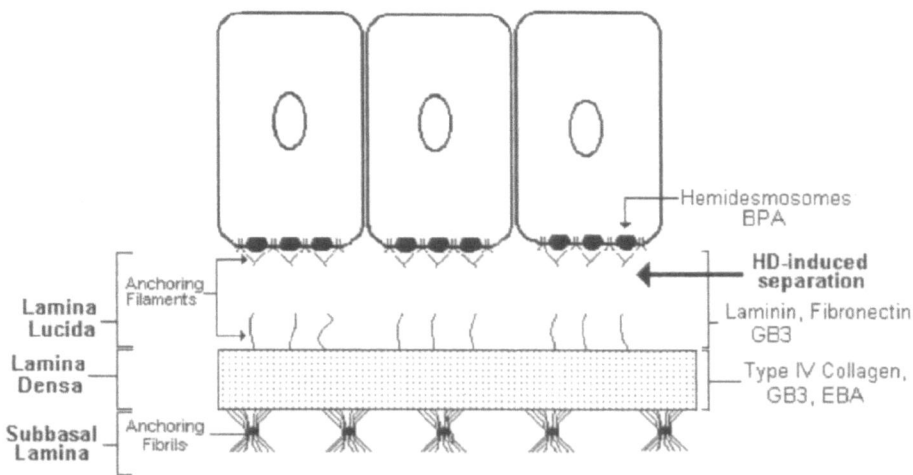

Figure 22. Schematic diagram illustrating the localization of HD-induced separation within the upper lamina lucida of the IPPSF in relation to six EDJ epitopes. Reprinted from Monteiro-Riviere and Inman.[96]

tered HD treated IPPSFs. In blistered HD IPPSFs, antibody was bound exclusively to dermis. No cross reactivity occurred in NPS with mouse antihuman monoclonal antibodies L3d and 19-DEJ-1. In summary, HD induced separation of the EDJ was determined by both IH and IEM to be localized to upper lamina lucida (Fig. 22).

6.2.3. Discussion of HD. A major concern in cutaneous vesicant research has been the inability to identify an *in vivo* or *in vitro* model that possesses a cutaneous tissue structurally and functionally similar to human skin and able to produce characteristic gross skin lesions typical of human exposure to HD. The IPPSF appears to be a useful model to study this phenomenon. Biochemical changes seen in the IPPSF agree with studies reported by other investigators. Most associate a decrease in GU and an inhibition of glycolysis as hallmarks of HD induced vesication.[92,100-102] To support this, treatment of IPPSFs with L and neat CEMS resulted in a decrease in CGU.[62,93] HD even decreased GU in the absence of gross blister formation, suggesting the magnitude of this effect may correlate with the occurrence of vesication. However, these findings link decreased glucose metabolism with vesication.

Others believe that changes in vascular permeability directly resulting from HD action at the microvasculature are involved in pathogenesis of HD induced blisters.[83,103-106] This hypothesis was supported in our studies, since changes in VR were an early event associated with HD treatment. Increased VR suggested that HD induced vasodilation and increased capillary permeability, which resulted in edema and perfusion resistance. This was confirmed in the HD kinetic studies discussed above[94] by time dependent changes in inulin and albumin physiological spaces, direct evidence of changes in vascular permeability. It is possible that a combination of decreased GU and increased capillary permeability results in vesication. L induced vesication is associated with less severe changes in both GU and VR, suggesting a different pathogenesis.[62] This different profile also supports the specificity of observed HD induced changes.

Gross and microscopic observations did show a correlation of dose concentration to blister and microvesicle formation. Blisters were similar to those found in HD exposed human skin.[81,82] Morphological observations included epidermal dermal separation, intra-cellular and intercellular epidermal edema and pyknotic nuclei. Microvesicles were caused by EDJ separation at the level of the lamina lucida. Focal distribution of lesions observed in our study may reflect the cell cycle specificity of epidermal basal cells to HD induced DNA toxicity and/or be due to uneven agent absorption in skin. Also, this is supported by the heterogeneous pattern of dark basal cell formation. This delay in blister formation is characteristic of HD vesication.

The EDJ of porcine skin has been described to be ultrastructurally similar to that of human skin.[3,28] EDJ consists of the basal cell plasma membrane (which includes the hemidesmosomes), lamina lucida, lamina densa (basal lamina) and subbasal lamina. Anti-bodies chosen for this study have been localized within human skin EDJ and were found to cross react with epitopes within porcine EDJ. The authors are unaware of any extensive mapping studies within the EDJ of normal pig skin. Stanley et al.[107] did report that laminin, type IV collagen and BP bound to EDJ of normal skin of weanling Yorkshire pigs. Rigel et al.[108] found that porcine skin exhibited a high degree of cross reactivity to murine laminin and to human type IV collagen and BPA. Mapping of additional antibodies within the EDJ allows further characterization of normal porcine skin and the IPPSF. In addition, specificity of these antibodies within pig skin EDJ provide the precise localization of the HD induced blister cleavage plane.

In the present study, indirect immunohistochemical staining has shown that antibod-ies to laminin, type IV collagen, fibronectin, GB3, BP and EBA cross react with EDJ epitopes in normal pig skin and in the in vitro IPPSF model. Laminin, type IV collagen and BP usually stained EDJ in the IPPSF more intensely than in NPS. IEM, in general agreement with IH, has shown the staining patterns in NPS are similar, though more discrete than in the IPPSF. However, GB3 antibody stained EDJ in nonblistered flaps more discretely than in NPS. The more diffuse staining pattern typically found along the EDJ of most IPPSFs may be explained by a slight diffusion of the EDJ epitopes during perfusion. Fibronectin staining of NPS capillary basement membrane was noted by IEM but not by IH, due to low cross reactivity of this antibody.

HD (and its analog, 2-chloroethyl methyl sulfide) and L caused the formation of gross blisters and microscopic EDJ separation in IPPSF similar to that described in vesicant exposure of human skin.[61,62,93,109] IH of HD blistered and nonblistered areas from the IPPSF dosing site revealed a slight to intense antibody staining. No difference in antibody staining affinity was found between IPPSF dosed with 0.2, 5.0 and 10.0 mg/ml HD and ethanol dosed IPPSFs. Therefore, HD does not appear to affect the integrity of the EDJ epitopes. Epitope mapping of HD blistered skin by IH localized the cleavage plane to be within the laminin and fibronectin, above type IV collagen, GB3 and EBA, and below BP antigenic sites. IEM precisely defined the plane of epidermal dermal cleavage. The laminin antibody bound primarily to dermis with focal areas of attachment to the basal pole of stratum basale cells. Woodley et al.[110] described the same pattern using IF in split human skin preparations. This indicated the cleavage plane in blistered HD treated IPPSFs was localized to the upper portion of lamina lucida. Type IV collagen, GB3 and EBA bound to the lamina densa and exclusively stained the dermal surface. Hemidesmosome associated BPA stained only epidermis. This is identical to the cleavage plane found in both experimental (suction, sodium chloride, heat shock, EDTA, various proteases) and diseased (junctional epidermolysis bullosa, autoimmune bullous diseases) states.

Characterization of EDJ in normal pig skin and in the IPPSF was determined by mapping of antibodies to laminin, type IV collagen, fibronectin, GB3, BP and EBA. These 6 antibodies did cross react with EDJ epitopes in NPS while L3d and 19-DEJ-1 did not. Antibody staining in the IPPSF was similar though generally more diffuse than in NPS. No differences in antibody affinity were found between ethanol flaps and nonblistered areas of

HD dosed flaps. The EDJ cleavage plane of the HD blistered IPPSFs was within the upper lamina lucida, the weakest component of the EDJ. This data, coupled with historical morphological studies, proves the pig skin EDJ has similarities to human skin and enhances the potential of the IPPSF as a model to study HD induced vesication and blistering diseases.

The observed events in pathogenesis of HD induced vesication has implications on the validity of a number of hypotheses presently in the literature. The original manuscript[96] should be consulted for this discussion. Pathogenesis of HD induced vesication is a multifaceted process which is associated with DNA damage and general cytotoxicity. The IPPSF appears to be a relevant model since the morphological appearance, biochemical, physiological and immunohistochemical effects are similar to that reported in humans. Importantly, gross blisters also form in this *in vitro* model. The major advantage of the IPPSF, in addition to the similarity to human lesions, is the experimental factors designed to address specific points in proposed mechanisms of HD vesication can be addressed in a biologically relevant preparation.

7. OVERALL CONCLUSION

All of the studies presented above further confirm the pig is an essential model in cutaneous research. Its previous acceptance as a model for chemical absorption across skin is supported. By extension, the IPPSF emerges as a useful hybrid tool, located in relative biological complexity between classic *in vitro* and *in vivo* systems. This system provides the major advantage of a viable, full thickness skin preparation with an intact vasculature, a relatively large surface area for dosing and control over experimental parameters and sample collection. Importantly, both absorption and toxicity studies may be conducted in the same preparation. This strength is illustrated nicely with both the iontophoresis and HD vesication studies, two applications which also shed light on the underlying mechanisms involved in these phenomenon. With the widespread increase of the pig as a model in transgenic studies, it is anticipated that its complete acceptance as a surrogate human skin model will be forthcoming. Although this should be viewed as a generally positive event, one must be cautious as there are differences (epidermal/dermal BM epitopes, skin appendages, biotransformation isoenzymes) which must be taken into account before direct extrapolation is possible. However, based upon many of the fields reviewed above, these may be relatively unimportant and masked by the overwhelming similarities already documented.

8. REFERENCES

1. Meyer, W., Schwarz, R., and Neurand, K., 1978, The skin of domestic mammals as a model for the human skin with special reference to the domestic pig, *Curr. Prob. Dermatol.* 7:39-52.
2. Montagna, W., and Yun, J.S., 1964, The skin of the domestic pig, *J. Invest. Dermatol.* 43:11-21.
3. Monteiro-Riviere, N.A., 1986, Ultrastructural evaluation of the porcine integument, in: *Swine in Biomedical Research,* Volume 1 (M.E. Tumbleson, ed.), Plenum Press, New York, pp. 641-655.
4. Monteiro-Riviere, N.A., and Stromberg, M.W., 1985, Ultrastructure of the integument of the domestic pig (*Sus scrofa*) from one through fourteen weeks of age, *Zbl. Vet. Med. C. Anat. Histolo. Embryol.* 14:97-115.
5. Meyer, W., Gorgen, S., and Schlesinger, C., 1986, Structural and histochemical aspects of epidermis development of fetal porcine skin, *Am. J. Anat.* 176:207-219.
6. Rigal, C., Pieraggi, M.T., Vincent, C., Prost, C., Bouissou, H., Snachez-Yuz, E., and Hernandez-Morro, B., 1991, Healing of full thickness cutaneous wound in the pig, *J. Invest. Dermatol.* 19:529-536.
7. Woolina, U., Berger, U., and Mahrle, G., 1991, Immunohistochemistry of porcine skin, *Acta. Histochem.* 90:87-91.
8. Forbes P., 1969, Vascular supply of the skin and hair in swine, *Adv. Biol. Skin* 9:419-432.

9. Meyer, W., Neurand, K., and Radke, B., 1981, Elastic fibre arrangement in the skin of the pig, *Arch. Dermatol. Res.* 270:391-401.

10. Meyer, W., Neurand, K., and Radke,B., 1982, Collagen fibre arrangement in the skin of the pig, *J. Anat.* 134:139-148.

11. Montagna, W., 1967, Comparative anatomy and physiology of the skin, *Arch. Dermatol.* 96:357-363.

12. Weinstein, G., 1966, Comparison turnover time of keratinous protein fractions in swine and human epidermis, in: *Swine in Biomedical Research*, (L.K. Bustad, and R.O. McClellan, eds.), Frayn, Seattle pp. 287-297.

13. Stromberg, M., Hwang, Y., and Monteiro-Riviere N.A., 1981, Interfollicular smooth muscle in the skin of the domestic pig (*Sus scrofa*), *Anat. Rec.* 201:445-462.

14. Bronaugh, R., Steward, R., and Congdon E., 1982, Methods for *in vitro* percutaneous absorption studies. II. Animals models for human skin, *Toxicol. Appl. Pharmacol.* 62:481-488.

15. Wester, R.C., and Maibach, H.I., 1977, Percutaneous absorption in man and animals, in: *A Perspective* (V.A. Drill, and P. Lazar, eds.), Academic Press, New York, pp. 111-126.

16. McNeill, S.C., Potts, R.O., and Francoeur, M.L., 1992, Local enhanced topical delivery of drugs: Does it truly exist? *J.Pharm. Res.* 9:1422-1427.

17. Riviere, J.E., and Williams, P.L., 1992, Pharmacokinetic implications of changing blood flow in skin, *J.Pharm. Sci.* 81:601-602.

18. Wada, Y., Etoh, Y., Ohira, A., Kimata, H., Koide, T., Ishihama, H., and Mizushima, Y., 1982, Percutaneous absorption and anti-inflammatory activity of indomethacin in ointment, *J. Pharm. Pharmacol.* 34:467-468.

19. Torrent, J., Izquierdo, I., Barbanoj, M.J., Moreno, J., Lauroba, J., and Jane, F., 1988, Anti-inflammatory activity of piroxicam after oral and topical administration on an ultraviolet-induced erythema model in man, *Curr. Ther. Res.* 44:340-347.

20. Guy, R.H., and Maibach, H.I., 1983, Drug delivery to local subcutaneous structures following topical administration, *J. Pharm. Sci.* 72:1375-1380.

21. Riviere, J.E., Monteiro-Riviere, N.A., and Inman, A.O., 1992, Determination of lidocaine concentrations in skin after transdermal iontophoresis: Effects of vasoactive drugs, *Pharm. Sci.* 9:211-214.

22. Monteiro-Riviere, N.A., Inman, A.O., Riviere, J.E., McNeill, S.C., and Francoeur, M.L., 1993, Topical penetration of piroxicam is dependent on the distribution of the local cutaneous vasculature, *Pharm. Res.* 10:1326-1331.

23. Hobbs, D.C., and Twomey, T.M., 1992, Metabolism of piroxicam by laboratory animals, *Drug Metabol. Dispos.* 9:211-214.

24. Monteiro-Riviere, N.A., Bristol, D.G., Manning, T.O., and Riviere, J.E., 1990, Interspecies and interregional analysis of the comparative histological thickness and laser Doppler blood flow measurements at five cutaneous sites in nine species, *J. Invest. Dermatol.* 95:582-586.

25. Riviere, J.E., Sage, B., and Williams, P.L., 1992, Effects of vasoactive drugs on transdermal lidocaine iontophoresis, *J. Pharm. Sci.* 80:615-620.

26. Williams, P.L. and Riviere, J.E., 1995, A biophysically-based dermatopharmacokinetic compartment model for quantifying percutaneous penetration and absorption of topically applied agents. I. Theory, *J. Pharm. Sci.* 84:599-608.

27. Williams, P.L, and Riviere, JE., 1993, A model describing transdermal iontophoresis delivery of lidocaine incorporating consideration of cutaneous microvasculature state, *J. Pharm. Sci.* 82:1080-1084.

28. Monteiro-Riviere, N.A., 1990, Specialized technique: Isolated perfused porcine skin flap, in: *Methods for Skin Absorption* (B.W. Kemppainen, and W.G. Reifenrath, eds.), CRC Press, Boca Raton, Florida, pp. 175-189.

29. Monteiro-Riviere, N.A., 1991, Comparative anatomy, physiology, and biochemistry of mammalian skin, in: *Dermal and Ocular Toxicology: Fundamentals and Methods* (D.W. Hobson, ed.), CRC Press, Boca Raton, Florida, pp. 3-71.

30. Riviere, J.E., Sage, B., and Monteiro-Riviere, N.A., 1989, Transdermal lidocaine iontophoresis in isolated perfused porcine skin, *J. Toxicol. Cutan. & Ocular Toxicol.* 8:493-504.

31. Sage, B.H., and Riviere, J.E., 1992, Model systems in iontophoresis-transport efficacy, *Adv. Drug Del.Rev.* 9:265-287.

32. Monteiro-Riviere, N.A., Inman, A.O., and Riviere, J.E., 1994, Identification of the pathway of transdermal iontophoretic drug delivery: Light and ultrastructural studies using mercuric chloride in pigs, *Pharm. Res.* 11:251-256.

33. Bodde, H..E., de Haan, F.H.N., Kornet, L., Craane-van Hinsberg, W.H.M., and Salomons, M.A., 1991, Transdermal iontophoresis of mercuric chloride *in vitro*: Electron microscopic visualization of pathways, *Proc. Int. Symp. Control. Rel. Bioact. Mater.* 18:301-302.

34. Cullander, C. and Guy, R.H., 1991, Sites of iontophoretic current flow into the skin: Identification and characterization with the vibrating probe electrode, *J. Invest. Dermatol.* 97:55-64.

35. Burnette, R.R., and Ongpipattanakul, B., 1988, Characterization of the pore transport properties and tissue alteration of excised human skin during iontophoresis, *J. Pharm. Sci.* 77:132-137.

36. Monteiro-Riviere, N.A., 1990, Altered epidermal morphology secondary to lidocaine iontophoresis: *In vivo* and *in vitro* studies in porcine skin, *Fundam. Appl. Toxicol.* 15:174-185.

37. Bronaugh, R.L., Stewart, R.F., and Congdon, E.R., 1982, Methods for in vitro percutaneous absorption studies II. Animal models for human skin, *Toxicol. Appl. Pharmacol.* 62:481-488.

38. Bodde, H.E., van den Brink, I., Koerten, H.K., and de Haan, F.H.N., 1991, Visualization of *in vitro* percutaneous penetration of mercuric chloride: Transport through intercellular space versus cellular uptake through desmosomes, *Control Rel.* 15:227-236.

39. Sharata, H.H., and Burnette, R.R., 1988, Effect of dipolar aprotic permeability enhancers on the basal stratum corneum, *J. Pharm. Sci.* 77:27-32.

40. Lundstrom, A., and Egelrud, T., 1988, Cell shedding from human plantar skin *in vitro*: Evidence of its dependence on endogenous proteolysis, *J. Invest. Dermatol.* 91:340-343.

41. Lundstrom, A., and Egelrud, T., 1990, Evidence that cell shedding from plantar stratum corneum *in vitro* involves endogenous proteolysis of the desmosomal protein desmoglein I, *J. Invest. Dermatol.* 94:216-220.

42. Swartzendruber, D.C., Wertz, P.W., Kitko D.J., Madison K.C., and Downing D.T., 1989, Molecular models of the intercellular lipid lamellae in mammalian stratum corneum, *J. Invest. Dermatol.* 92:251-257.

43. Madison, K.C., Swartzendruber, D.C., Wertz P.W., and Downing D.T., 1987, Presence of intact intercellular lipid lamellae in the stratum corneum, *J. Invest. Dermatol.* 88:714-718.

44. Menon, G.K., Feingold, K.R., and Elias, P.M., 1992, Lamellar body secretory response to barrier disruption, *J. Invest. Dermatol.* 98:279-289.

45. Bowman, K.F., Monteiro-Riviere, N.A., and Riviere, J.E., 1991, Development of surgical techniques for preparation of *in vitro* isolated perfused porcine skin flaps for percutaneous absorption studies, *Am. J. Vet. Res.* 52:75-82.

46. Riviere, J.E., Bowman, K.F., Monteiro-Riviere, N.A., Carver M.P., and Dix, L.P., 1986, The isolated perfused porcine skin flap (IPPSF). I. A novel *in vitro* model for percutaneous absorption and cutaneous toxicology studies, *Fundam. Appl. Toxicol.* 7:444-453.

47. Riviere, J.E., Bowman, K.F., and Monteiro-Riviere, N.A., 1986, The isolated perfused porcine skin flap: A novel animal model for cutaneous toxicologic research, in: *Swine in Biomedical Research,* Volume 1 (M.E. Tumbleson, ed.), Plenum Press, New York, pp. 657-666.

48. Monteiro-Riviere, N.A., Bowman, K.F., Scheidt, V.J., and Riviere, J.E., 1987, The isolated perfused porcine skin flap (IPPSF). II. Ultrastructural and histological characterization of epidermal viability, *In Vitro Toxicol.* 1:241-254.

49. Riviere, J.E., and Monteiro-Riviere, N.A., 1991, The isolated perfused porcine skin flap as an *in vitro* model for percutaneous absorption and cutaneous toxicology, *Crit. Rev. Toxicol.* 21:329-344.

50. Williams, P.L., Carver, M.P., and Riviere, J.E., 1990, A physiologically relevant pharmacokinetic model of xenobiotic percutaneous absorption utilizing the isolated perfused porcine skin flap (IPPSF), *J. Pharm. Sci.* 79:305-311.

51. Chang, S.K., Williams, P.L., Dauterman, W.C., and Riviere, J.E., 1994, Percutaneous absorption, dermatopharmacokinetics and related biotransformation studies of carbaryl lindane, malathion, and parathion in isolated perfused porcine skin, *Toxicol.* 91:269-280.

52. Heit, M., Williams, P.L., Jayes, F.L., Chang S.K., and Riviere J.E., 1993, Transdermal iontophoretic peptide delivery. *In vitro* and *in vivo* studies with luteinizing hormone releasing hormone (LHRH), *J. Pharm. Sci.* 82:240-243.

53. Riviere, J.E., William, P.L., Hillman, R., and Mishky, L., 1992, Quantitative prediction of transdermal iontophoretic delivery of arbutamine in humans using the *in vitro* isolated perfused porcine skin flap (IPPSF), *J. Pharm. Sci.* 81:504-507.

54. Williams, P.L., and Riviere, J.E., 1994, A "full-space" method for predicting *in vivo* plasma drug profiles reflecting both cutaneous and systemic variability. *J. Pharm. Sci.* 83:603-604.

55. Riviere, J.E., Monteiro-Riviere N.A., and Williams P., 1995, Isolated perfused porcine skin flap as an *in vitro* model for predicting transdermal pharmacokinetics, *Eur. J. Biopharm.* 41:152-162.

56. Rogers, R.A., and Riviere, J.E., 1994, Pharmacologic modulation of the cutaneous vasculature in the isolated perfused porcine skin flap, *J. Pharm. Sci.* 83:1682-1689.

57. Bikle, D.D., Halloran, B.P., and Riviere, J.E., 1994, Production of 1,25 dihydroxyvitamin D3 by perfused pig skin, *J. Invest. Dermatol.* 102:796-798.

58. Vaden, S.L., Page, R.L., Peters, B.P., Cline, J.M., and Riviere, J.E., 1993, Development and charac-
    terization of an isolated and perfused tumor and skin preparation for evaluation of drug disposition,
    Cancer. Res. 53:101-105.
59. Srikrishna, V., and Monteiro-Riviere, N.A., 1991, The effects of sodium hydroxide and hydrochloric acid
    on isolated perfused skin, *In Vitro Toxicol.* 4:207-215.
60. Srikrishna, V., Riviere, J.E., and Monteiro-Riviere, N.A., 1992, Cutaneous toxicity and absorption of
    paraquat in porcine skin, *Toxicol. Appl. Pharmacol.* 115:89-97.
61. Monteiro-Riviere N.A., King J.R., and Riviere J.E., 1991, Mustard induced vesication in isolated perfused
    skin: Biochemical, physiological, and morphological studies., in: *Proc. Ninth Med. Defense Biosci. Rev.*,
    Aberdeen Proving Ground, MD, USMRICD, pp. 59-162.
62. King, J.R., Riviere, J.E., and Monteiro-Riviere, N.A., 1992, Characterization of lewisite toxicity in
    isolated perfused skin, *Toxicol. Appl. Pharmacol.* 116:189-201.
63. Spoo, J.W., Rogers, R.A., and Monteiro-Riviere, N.A., 1993, Effects of formaldehyde, DMSO, benzoyl
    peroxide, and sodium lauryl sulfate on isolated perfused porcine skin, *In Vitro Toxicol.* 5:251-260.
64. Kornhauser, A., Wamer, W., and Giles, A., Jr., 1983, Light-induced dermal toxicity: Effects on the cellular
    and molecular level, in: *Dermatotoxicology* (F. Marzulli, and H. Maibach, eds.) Hemisphere Press,
    Washington, DC, pp. 323-355.
65. Daniels, F., Jr., 1965, A simple microbiological method for demonstrating phototoxic compounds, *J.
    Invest. Dermatol.* 44:259-263.
66. Fleisher, A.S., Harber, L.C., Cook, J.S., Baer, R.L., 1966, Mechanism of *in vitro* photohemolysis in
    erythropeoietic protoporphyria (EPP), *J. Invest. Dermatol.* 46:505-509.
67. Freeman, R.G., Murtishaw, W., and Knox, J.M., 1970, Tissue culture techniques in the study of cell
    photobiology and phototoxicity, *J. Invest. Dermatol.* 54:164-169.
68. Lasarow, R.M., Isseroff, R.R., and Gomez, E.C., 1992, Quantitative *in vitro* assessment of phototoxicity
    by a fibroblast-neutral red assay, *J. Invest. Dermatol.* 98:725-729.
69. Danno, K., and Horio, T., 1982, Formation of UV-induced apoptosis relates to the cell cycle, *Br. J.
    Dermatol.* 107:423-428.
70. Danno, K., and Horio, T., 1980, Histochemical staining of sunburn cells for sulphhydryl and disulphide
    groups: A time course study, *Br. J. Dermatol.* 102:535-539.
71. Woodcock, A., and Magnus, I.A., 1976, The sunburn cell in mouse skin: Preliminary quantitative studies
    on its production, *Br. J. Dermatol.* 95:459-468.
72. Forbes, P.D., Urbach, F., and Davies, R.E., 1977, Phototoxicity testing of fragrance raw materials, *Fd.
    Cosmet. Toxicol.* 15:55-60.
73. Sambuco, C.P., 1985, Miniature swine as an animal model in photodermatology: Factors influencing
    sunburn cell formation, *Photoderm.* 2:144-150.
74. Monteiro-Riviere, N.A., Inman, A.O., and Riviere, J.E., 1994, Development and characterization of a
    novel skin model for cutaneous phototoxicology, *Photodermatol. Photoimmunol. Photomed.* 10:235-243.
75. Foley, J.F., Dietrich, D.R., Swenberg, J.A., and Maronpot, R.R, 1991, Detection and evaluation of
    proliferating cell nuclear antigen (PCNA) in rat tissue by an improved immunohistochemical procedure,
    J. Histotech. 14:237-241.
76. Johnson, B.E., Mandell, G., and Daniels, F., Jr., 1972, Melanin and cellular reactions to ultraviolet
    radiation, *Nature New. Biol.* 235:147-149.
77. Olson, R.L., and Everett, M.A., 1975, Epidermal apoptosis: Cell deletion by phagocytosis, *J. Cutan. Path.*
    2:53-57.
78. Hashimoto, Y., Ohkuma, N., and Iizuki, H., 1991, Reduced superoxide dismutase activity in UVB-induced
    hyperproliferative pig epidermis, *Arch. Dermatol. Res.* 283:317-320.
79. Argenbright, L.W., and Forbes, P.D., 1982, Erythema and skin blood content. *Br. J. Dermatol.* 106:569-
    574.
80. Renshaw B., 1946, Mechanisms in production of cutaneous injuries by sulfur and nitrogen mustards, in:
    Chemical Warfare Agents and Related Chemical Problems, Technical Summary Report, Volume 1, Part
    III-VI, pp. 479-487.
81. Requena, L., Requena, C., Sanchez, M., Jaqueti, G., Aguilar, A., Sanchez-Yus, E., and Hernandez-Moro,
    B., 1988, Chemical warfare. Cutaneous lesions from mustard gas, *J. Am. Acad. Dermatol.* 19:529-536.
82. Willems, J.L., 1989, Clinical management of mustard gas casualties, *Ann. Med. Mil.* 3:1-61.
83. Vogt, R.F., Jr., Dannenberg, A.M., Jr., Schofield, B.H., Hynes, N.A., and Papirmeister, B., 1984,
    Pathogenesis of skin lesions caused by sulfur mustard, *Fundam. Appl. Toxicol.* 4:S71-S83.
84. Marlow, D.D., Mershon, M.M., Mitcheltree, L.W., Petrali, J.P., and Jaax, G.P., 1990, Sulfur mustard-in-
    duced skin injury in hairless guinea pigs, *J. Toxicol. Cut. Ocular Toxicol.* 9:179-192.

85. Petrali, J.P., Oglesby, S.B., and Mills, K.R., 1990, Ultrastructure correlates of sulfur mustard toxicity, *J. Toxicol. Cut. Ocular Toxicol.* 9:193-214.

86. Mershon, M.M., Mitcheltree, L.W., Petrali, J.P., Braue, E.H., and Wade, J.V., 1990, Hairless guinea pig bioassay model for vesicant vapor exposures, *Fundam. Appl. Toxicol.* 15:622-630.

87. Papirmeister, B., Gross, C.L., Petrali, J.P., and Hixson, C.J., 1984, Pathology produced by sulfur mustard in human skin grafts on athymic nude mice. I. Gross and light microscopic changes, *J. Toxicol. Cut. Ocular Toxicol.* 3:371-391.

88. Papirmeister, B., Gross, C.L., Petrali, J.P., and Meier, H.L., 1984, Pathology produced by sulfur mustard in human skin grafts on athymic nude mice. II. Ultrastructural changes, *J. Toxicol. Cut. Ocular Toxicol.* 3:393-408.

89. McGown, E.L., van Ravenswaay, T., and Dumlao, C.R., 1987, Histologic changes in nude mouse skin and human skin xenografts following exposure to sulfhydryl reagents: Arsenicals, *Toxicol. Pathol.* 15:149-156.

90. Westrom, D.R., 1987, Animal models for vesicant-induced injury, in: *Proceedings of the Vesicant Workshop*, February, 1987. U.S. Army Medical Research Institute of Chemical Defense, Aberdeen Proving Ground, MD, 21010-5425. pp. 91-96.

91. Mitcheltree, L.W., Mershon, M.M., Wall, H.G., Pulliam, J.D., and Manthei, J.H., 1989, Microblister formation in vesicant-exposed pig skin, *J. Toxicol. Cut. Ocular Toxicol.* 8:309-319.

92. Mol, M.A.E., De Vries, R., and Kluivers, A.W., 1991, Effects of nicotinamide on biochemical changes and microblistering induced by sulfur mustard in human skin organ cultures, *Toxicol. Appl. Pharmacol.* 107:439-449.

93. King, J.R., and Monteiro-Riviere, N.A., 1990, Cutaneous toxicity of 2-chloroethyl methyl sulfide in isolated perfused porcine skin, *Toxicol. Appl. Pharmacol.* 104:167-179.

94. Riviere, J.E., Brooks, J.D., Williams, P.L., and Monteiro-Riviere, N.A., 1995, Toxicokinetics of topical sulfur mustard penetration, disposition and vascular toxicity in isolated perfused porcine skin, *Toxicol. Appl. Pharmacol.* 135:25-34.

95. Spoo, J.W., Monteiro-Riviere, N.A., and Riviere, J.E., 1995, Detection of sulfur mustard (bis-2-chloroethyl sulfide) and metabolites after topical application in the isolated perfused porcine skin flap, *Life Sci.* 56:1385-1394.

96. Monteiro-Riviere, N.A., and Inman, A.O., 1995, Indirect immunohistochemistry and immunoelectron microscopy distribution of eight epidermal-dermal junction epitopes in the pig and in isolated perfused porcine skin treated with bis (2-chloroethyl) sulfide. *Toxicol. Path.* 23:313-325.

97. Monteiro-Riviere, N.A., and Inman, A.O., 1996, Ultrastructural characterization of sulfur mustard-induced vesication in isolated perfused porcine skin, *Micro. Res. Tech.* In Press.

98. Yaoita, H., Gullino, M., and Katz, S.I., 1976, Herpes gestationis. Ultrastructure and ultrastructural localization of *in vivo*-bound complement: Modified tissue preparation and processing for horseradish peroxidase staining of skin, *J. Invest. Dermatol.* 66:383-388.

99. Monteiro-Riviere, N.A., and Riviere, J.E., 1991, Cutaneous toxicity of mustard and lewisite on the isolated perfused porcine skin flap. DAMD17-87-C-7139: NTIS, ADA254419 Final Report, pp. 1-140.

100. Papirmeister, B., Gross, C.L., Meier, H.L., Petrali, J.P., and Johnson, J.B., 1985, Molecular basis for mustard-induced vesication, *Fund. Appl. Toxicol.* 5:S134-S149.

101. Bernstein, I.A., Brabec, M.J., Conolly, R.C., Gray, R.H., and Kulkarn, A., 1987, Chemical blistering: Cellular and macromolecular components, Annual Summary Report AD-A190 313, pp. 1-34.

102. Gray, P.J., 1989, A literature review on the mechanism of action of sulfur and nitrogen mustard. Report No. MRL-TR-89-24.

103. Dannenberg, A.M., Jr., Pula, P.J., Liu, L.H., Harada, S., Tanaka, F., Vogt, R.F., Jr., Kajiki, A., and Higuchi, K., 1985, Inflammatory mediator and modulators released in organ culture from rabbit skin lesions produced in vivo by sulfur mustard. I. Quantitative histopathology, PMN, basophil, and mononuclear cell survival, and unbound (serum) protein content, *Am. J. Pathol.* 121:15-27.

104. Harada, S., Dannenberg, A.M., Jr., Kajiki, A., Higuchi, K., Tanaka, F., and Pula, P.J., 1985, Inflammatory mediators and modulators released in organ culture from rabbit skin lesions produced *in vivo* by sulfur mustard. II. Evans blue dye experiments that determined the rates of entry and turnover of serum protein in developing and healing lesions, *Am. J. Pathol.* 121:28-38.

105. Harada, S., Dannenberg, A.M., Jr., Vogt, R.F., Jr., Myrick, J.E., Tanaka, F., Redding, L.C., Merkhofer, R.M., Pula, P.J., and Scott, A.L., 1987, Inflammatory mediators and modulators released in organ culture from rabbit skin lesions produced *in vivo* by sulfur mustard. III. Electrophoretic protein fractions, trypsin-inhibitory capacity, 1-proteinase inhibitor, and 1- and 2- macroglobulin proteinase inhibitors of culture fluids and serum, *Am. J. Pathol.* 126:148-163.

106. Higuchi, K., Kajiki, A., Nakamura, M., Liu, L.H., Harada, S., Pula, P.J., Scott, A.L., and Dannenberg, A.M., Jr., 1988, Proteases released in organ culture by acute dermal inflammatory lesions produced in vivo in rabbit skin by sulfur mustard: Hydrolysis of synthetic peptide substrates for trypsin-like and chymotrypsin-like enzymes, *Inflammation* 12:311-334.
107. Stanley, J.R., Alvarez, O.M., Bere, E.W., Eaglstein, W.H., and Katz, S.I., 1981, Detection of basement membrane zone antigens during epidermal wound healing in pigs, *J. Invest. Dermatol.* 77:240-243.
108. Rigal, C., Pieraggi, M.T., Vincent, C., Prost, C., Bouissou, H., and Serre, G., 1991, Healing of full-thickness cutaneous wounds in the pig. I. Immunohistochemical study of epidermal-dermal junction regeneration, *J. Invest. Dermatol.* 96:777-785.
109. Monteiro-Riviere, N.A., Inman, A.O., Spoo, J.W., Rogers, R.A., and Riviere, J.E., 1993, Studies on the pathogenesis of bis (2-chloroethyl) sulfide (HD) induced vesication in porcine skin, in: *Proceedings of the Medical Defense Bioscience Review*. U.S. Army Medical Research Institute of Chemical Defense, Aberdeen Proving Ground, MD., pp. 31-40.
110. Woodley, D., Sauder, D., Talley, M.J., Silver, M., Grotendorst, G., and Qwarnstrom, E., 1983, Localization of basement membrane components after dermal-epidermal junction separation, *J. Invest. Dermatol.* 81:149-153.

AN *IN VITRO* PIG SKIN MODEL FOR PREDICTING HUMAN SKIN PENETRATION AND IRRITATION POTENTIAL

William G. Reifenrath, PhD,[1*] Barbara W. Kemppainen, PhD,[2] and Winifred G. Palmer, PhD[3]

[1] Reifenrath Consulting and Research
Richmond, California 94804
[2] Department of Physiology and Pharmacology
Auburn University
Auburn, Alabama 36849
[3] The U.S. Army CHPPM
Fort Detrick
Frederick, Maryland 21701

1. ABSTRACT

Percutaneous absorption is a complex process by which molecules traverse the stratum corneum or barrier layer of skin, pass through the viable layers and finally leave via the skin's microcirculation. When the penetrant is directly toxic to the skin, a more complicated series of events can lead to skin irritation. However, if a substance cannot reach the viable layers of the skin, irritation will not result. If we can identify which factors are the most important for predicting skin penetration potential, we will establish some of the necessary conditions for toxic events subsequent to penetration of the barrier layer.

In previous work, we firmly established pig skin as predictive of human skin absorption. Building on this data base, we developed a promising empirical relationship between penetration, molecular weight and vapor pressure for organic molecules applied to skin in volatile solvents.

Using benzoic acid as a model penetrant, we demonstrated decreases in barrier function of skin with exposure to a complex mixture known to cause skin irritation. Also, we were able to evaluate the rate of barrier function recovery subsequent to exposure to the mixture.

---

[*] Reprint requests to: Dr. William Reifenrath, Reifenrath Consulting and Research, 1315 So. 46th St., Richmond, CA 94804 (510) 231-9463.

Advances in Swine in Biomedical Research, edited by Tumbleson and Schook
Plenum Press, New York, 1996

Using a radiolabeled vesicant, we were able to show a correlation between the incidence of penetration of radiolabel through excised pig skin and the occurrence of skin irritation/inflammation *in vivo*. With the addition of data for other compounds and vehicles, it will be possible to develop a general model for assessing skin absorption and irritation potential.

2. INTRODUCTION

Insecticides and solvents are familiar examples of volatile substances which come in contact with the skin. Mechanical removal, evaporation and penetration are key processes which contribute to the removal of these substances from the skin surface. The interaction between these two processes is dependent on several factors.

2.1. Vapor Pressure

Vapor pressure is a thermodynamic measurement of the equilibrium between the vapor state and the liquid or solid state of a substance. A list of compounds whose vapor pressure spans approximately 9 orders of magnitude is depicted in Table 1. Circumstances will dictate whether these compounds are considered volatile or nonvolatile. For example, water may be regarded as nonvolatile compared with acetone. However when we consider exposure of the skin to μg amounts of chemical, even compounds with vapor pressures as low as that of DDT can have measurable evaporative loss. From this perspective, water would be considered an extremely volatile compound.

2.2. Dose

Consider evaporation of a chemical from a large dose that completely covers a skin surface of constant area, the so called "infinite dose". Under these conditions, loss by skin absorption will be small compared to the amount of chemical on the skin surface. The evaporation rate will be independent of the depth of the chemical layer and will be constant or "zero order". Due to frictional resistance, air movement is relatively nil close to the exposed surface of the applied chemical and transport through this stagnant air layer (boundary layer) will occur only by molecular diffusion and will be the slowest or rate limiting step in the overall evaporation process. Under these conditions, the evaporation rate of a chemical will depend on its vapor pressure (VP) and its molecular weight (M) as given by the following equation[1]:

Table 1. Vapor pressure of liquids and solids[1]

Compound	Vapor Pressure(mm Hg @ 20C)
Acetone	270 (@ 30C)
Hexane	120
Ethanol	43
Water	17
Ethylene Glycol	2×10^{-2}
Glycerol	3×10^{-3}
Lindane	9×10^{-6}
DDT	2×10^{-7}

$$\text{Evaporation rate (infinite dose)} = f(VP)(M)^{0.5} \qquad (1)$$

The proportionality constant (f) will depend on the factors that control the depth of the boundary layer, such as the geometry of the evaporation chamber, surface roughness and air speed.

Now consider finite dose conditions, when the chemical no longer covers the entire surface of the test area. The evaporation rate will be determined by many factors, including the rate of skin absorption. Compared to infinite dose conditions, the evaporation rate will begin to decrease with time (t) in proportion to the amount of chemical remaining on the skin surface. Ignoring skin absorption, the evaporation rate may be predicted by the following exponential equation:

$$\text{Evaporation Rate (finite dose)} = A\,(e^{-kt}) \qquad (2)$$

where A is the evaporation rate at t = 0 and k is a constant.

2.3. Air Flow

As the rate of air flow across the skin increases, the depth of the boundary layer will decrease and the overall rate of evaporation will increase. The Beaufort wind scale[2] gives a perspective of the magnitude of wind speeds to which the skin may be exposed outdoors. "Calm conditions", when smoke rises vertically, exist with wind speeds less then 1 mph (zero on the Beaufort scale). "Light air" exists with winds of 1 to 3 mph, when smoke drifts with air but weather vanes are inactive. "Light breeze" results from winds of 4 to 7 mph, when weather vanes are active, wind is felt on the face and leaves rustle. A "gentle breeze" comes from winds of 8 to 12 mph, when leaves and small twigs move and light flags extend. The Beaufort scale continues from nos. 4 to 7 for moderate to strong breezes (13 to 31 mph) through nos. 7 to 17 for gales, storms and hurricanes.

Indoors, the skin is still subjected to air flows from sources such as ventilation. Current standards for dwellings designed for human occupancy call for an air flow no greater than 30 ft/min or 0.34 mph in winter and no greater than 50 ft/min or 0.57 mph in summer (ANSI/ASHRAE Standard 55-1981, Thermal Environmental Conditions for Human Occupancy). Body motion also will generate an air flow across the skin, since an ordinary walking speed of approximately 4 to 5 mph will generate "light breeze" conditions in otherwise still air. Additionally, body temperature can produce air movement over the skin. Skin temperature, when measured by pressing the tip of a thermocouple on the skin of a resting individual at normal room temperature, is approximately 30 to 32C. The temperature differential between the skin and the surroundings (approximately 20C) will induce a convective air flow. Warm air surrounding the skin rises and cooler air moves in to take its place, thus setting up currents. It has been estimated that approximately 15% of the body's heat loss is accomplished by convective currents.[3] However, compared to air movement caused by ventilation, wind or body motion, the effect is small.

2.4. Skin Penetration

For compounds of intermediate volatility, penetration into the skin will compete with evaporative loss. Compounds can be characterized by a permeability coefficient (K_p), generally determined by dividing the steady state absorption rate from a saturated solution in water by the saturation concentration. While compounds with higher Kp values have greater potential for penetrating the skin, the conditions under which Kp measurements are made are artificial and no simple relationship exists between Kp and skin penetration resulting from actual exposure conditions.

The permeability coefficient has been measured for relatively few compounds. Values have been reported as low as 3×10^{-6} cm/hr for hydrocortisone to as high as 1.2 cm/hr for ethyl benzene, with most compounds in the range of 10^{-2} to 10^{-5} cm/hr. A method has been proposed[4] for estimating Kp values, based on the penetrants molecular weight (MW) and octanol/water partition coefficient (Koct):

$$\log Kp \ (cm/hr) = -2.72 + 0.71(\log Koct) - 0.0061(MW) \tag{3}$$

A similar expression was developed by Hostynek *et al.*[5] to calculate Kp from Koct and MR (molar refraction)

$$\log Kp = -2.71 + 0.801(\log Koct) - 0.0260(MR) \tag{4}$$

Values for Koct also can be estimated if measured values are unobtainable.[6] MR is defined by the following equation

$$MR = (M/\text{liquid density}) \ (n_D^2 - 1)/(n_D^2 + 2) \tag{5}$$

where M is the molecular weight (g/mole), liquid density is in g/ml, and n_D is the refractive index at the sodium D line wavelength. Methods for estimation of MR are available.[7]

3. METHODS

3.1. *In vitro* Skin Penetration/Evaporation Cells

Reifenrath and Hawkins developed cells (Fig. 1) to measure skin penetration and evaporation from the skin surface. These penetration/evaporation cells allow control of air temperature and flow over the outer skin surface and permit collection of volatilized chemical. Together with the apparatus shown in Fig. 2, these cells can simulate indoor air flows ranging from stagnant air (0 mph) to those resulting from recommended ventilation standards (0.3 to 0.5 mph). Liquid or solid chemicals can be suspended in the evaporation cell to enable the study of skin absorption from the vapor phase.

3.2. *In vivo* Skin Penetration

In vivo skin penetration studies can be conducted in animals or humans by application of test compound to a defined skin area. The percent of the dose excreted in the urine or feces is determined and divided by the fraction of an injected dose recovered in the excreta to correct for other routes of excretion. A patch (Fig. 3) is placed around the application area to protect it from abrasion or rubbing;[8] however, volatile compounds can evaporate freely from the application device. A device has been developed to collect volatiles arising from topical application (Fig. 4); however, air flow is restricted.[9] A number of recent reviews have been published regarding the many techniques for *in vivo* skin penetration studies.[10-12]

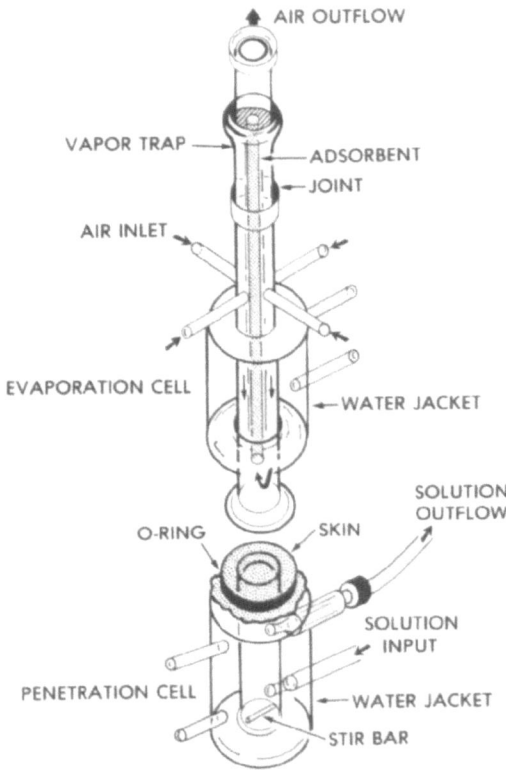

Figure 1. Skin penetration/evaporation cells.

4. DISCUSSION OF PREVIOUS RESULTS

4.1. The Effect of Air Flow on Evaporation and Skin Penetration

Compounds were applied (4 μg/cm^2) to excised pig skin mounted on skin penetration/evaporation cells. Air was caused to flow through the evaporation cell at 60 or 600 ml/min. These flows corresponded to 0.03 and 0.3 mph through the 0.8 cm^2 cross sectional area of the evaporation cell, with the higher air flow providing a better simulation of indoor air conditions. As expected, evaporative loss was greater at the higher air flow rate (Table 2). There was no simple correlation between vapor pressure and evaporative loss.

4.2. A Comparison of *in vitro* and *in vivo* Skin Penetration as a Function of *in vitro* Air Flow

A comparison was made between *in vitro* pig skin penetration, as measured by radioactivity recovered in the dermis plus receptor fluid,[13] and *in vivo* pig skin penetration for 9 compounds.[14] Each compound was applied at a dose of 4 μg/cm^2 in ethanol vehicle. *In vitro* values were obtained at an air flow of 60 or 600 ml/min through the evaporation cell. Three of the compounds were nonvolatile (progesterone, testosterone and fluocinolone acetonide), while the remainder (caffeine, DDT, benzoic acid, malathion, parathion, DEET

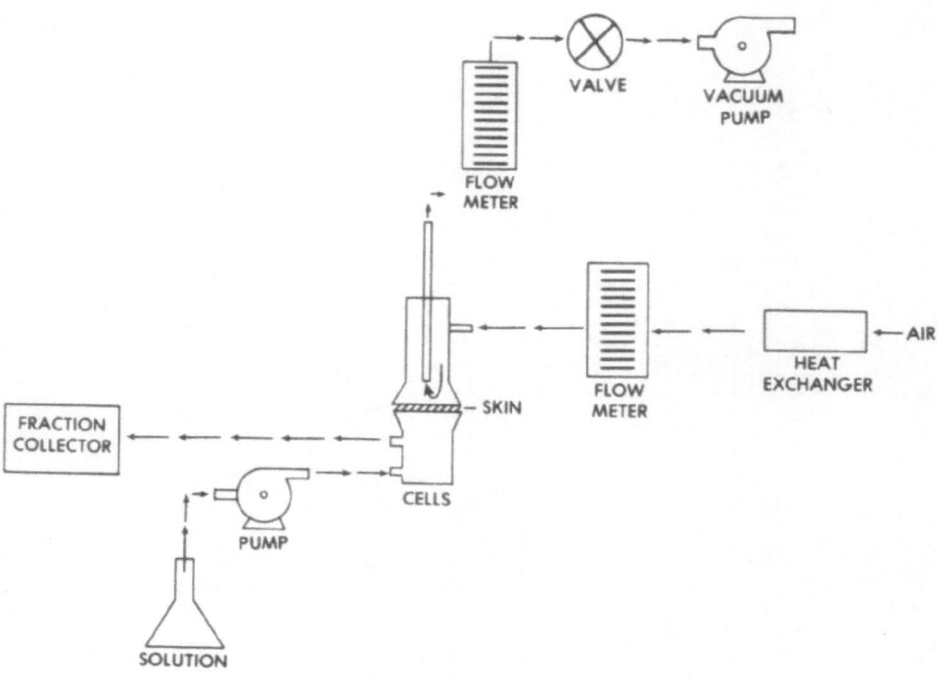

Figure 2. Diagram of apparatus for making skin penetration and evaporation measurements.

and lindane) had detectable evaporative loss from excised pig skin mounted in the skin penetration/evaporation cells (Table 3).

At the termination of the *in vitro* studies (48 hr), excised pig skin was separated into epidermis and dermis. Compound recovered from the epidermis (skin surface plus epidermis) consisted largely of residues remaining on the surface of the stratum corneum as well as lesser amounts in the epidermis itself. These values were compared to residues recovered

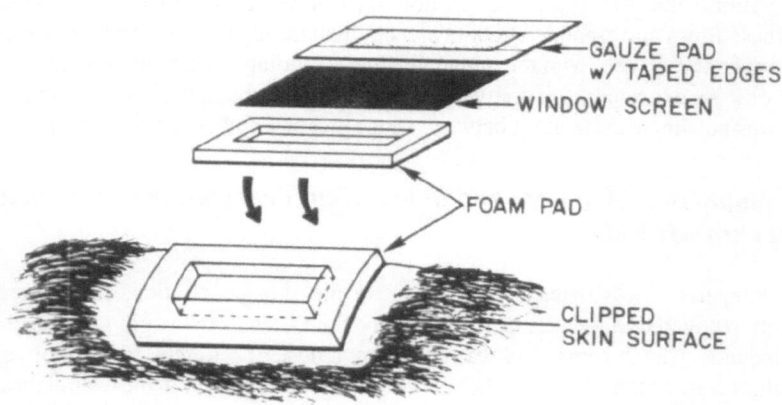

Figure 3. Patch for protection of the application site during *in vivo* skin penetration studies.

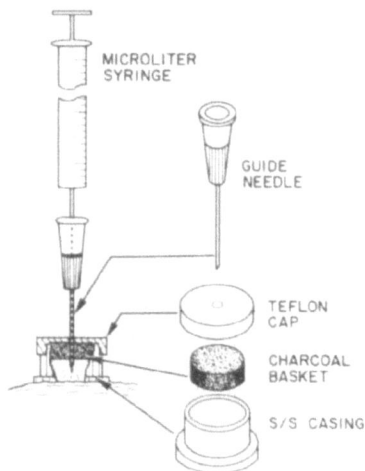

MICROLITER
SYRINGE

GUIDE
NEEDLE

TEFLON
CAP

CHARCOAL
BASKET

S/S CASING

Figure 4. Apparatus for collection of volatiles during *in vivo* skin penetration studies.

by wiping pig skin with an ethanol dampened cotton swab 48 hr after topical application *in vivo* (600 ml/min flow rate, Fig. 5). Significant correlations were obtained ($p < 0.05$) for either the 60 ml/min ($r = 0.88$) or 600 ml/min ($r = 0.82$) flow rates. However, when the nonvolatile steroids were removed from the comparison, a much better correlation was obtained for the higher flow rate data ($r = 0.95$) than for the lower flow rate data ($r = 0.63$).

Values for *in vivo* penetration correlated better with *in vitro* penetration obtained at the 600 ml/min air flow rate (Fig. 6, $r = 0.72$) as compared to the corresponding correlation at the 60 ml/min air flow rate ($r = 0.46$). The correlation is maintained when the nonvolatile compounds are removed ($r = 0.76$) from data collected at the high air flow rate; however, the correlation becomes much lower when the nonvolatile compounds are removed from the data collected at the low air flow rate ($r = 0.22$).

Thus, the ability of the *in vitro* model to predict *in vivo* skin penetration and skin surface residues is dependent on the simulation of air flows corresponding to indoor conditions. No correlation exists between calculated Kp values and the experimentally

Table 2. Evaporation of radiolabeled compounds 48 hr after application
(4 μg/cm^2) to excised pig skin[13]

Compound	Vapor Pressure x 10^{-5}mm Hg@20C	Evaporation (% of Appl. Dose)	
		Air Flow = 60 ml/min	Air Flow = 600 ml/min
Caffeine	-	4± 2	17± 10
DDT	0.015	5± 5	28± 5
Parathion	0.47	33± 7	63± 14
Malathion	0.55	24± 11	74± 5
Lindane	3.3	62± 14	77± 4
Benzoic Acid	38	25± 15	45± 10
N,N-Diethyl-m-toluamide	103	53± 12	73± 10

Table 3. Comparison of calculated Kp with experimentally derived maximum flux
and percent of absorbed dose after application of radiolabeled compounds
(4 $\mu g/cm^2$) to excised pig skin[a]

Compound	Calculated Kp x 10^{-3} cm/hr	Experimental Max. Flux $\mu g/cm^2$-hr x 10^{-3}	Experimental % absorbed mean ± S.D.
Caffeine	0.085	1.96	20± 4
Malathion	1.23	0.84	10± 6
N,N-Diethyl-m-Toluamide	4.26	2.28	6± 1
Benzoic Acid	8.9	9.16	15± 8
Parathion	29.5	3.28	10± 4
DDT	47	-	9± 5
Lindane	59	0.36	6±2

[a]Values of Kp were estimated according to the method of Magee[5] except for DDT, whose value was estimated by the method of Potts and Guy.[4] Percent absorbed was measured over 48 hr.[13] Values of maximum flux were derived from the raw data obtained for percent absorption.[13]

determined maximum flux or percent absorbed from the high air flow *in vitro* data (Table 3).

4.3. The Effect of Dose

The effect of dose on skin penetration and evaporation can be demonstrated with data (Table 4) for the insect repellent N,N-diethyl-m-toluamide (DEET). At a very low dose of 4 $\mu g/cm^2$ applied to excised pig skin in an acetone vehicle, evaporative loss follows Eq. 2, with the rate constant k = 1.90 (Fig. 7). At actual use doses of DEET (360 $\mu g/cm^2$, Fig. 8), evaporative loss data can still be fit with Eq. 2; however, the rate constant k is much smaller (k = 0.19). At very high doses (a pool of DEET on the skin surface), the evaporation rate is constant as predicted by Eq. 1 (Fig. 9).

As the skin dose of a pure compound is increased, the rate of absorption, expressed as mass/(area x time), will increase until the capacity of the skin is reached. However, when data are expressed as "percent of applied dose", percent absorption decreases when the dose becomes large (Table 4).

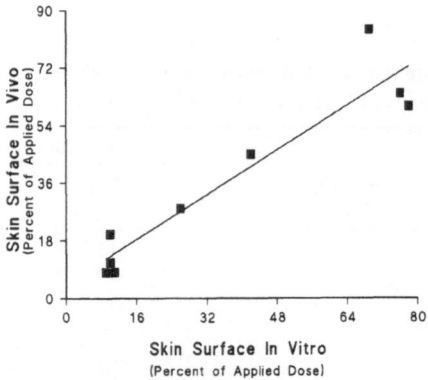

Figure 5. Comparison of skin surface residues 48 hours after *in vitro* and *in vivo* topical application of compounds to pig skin.

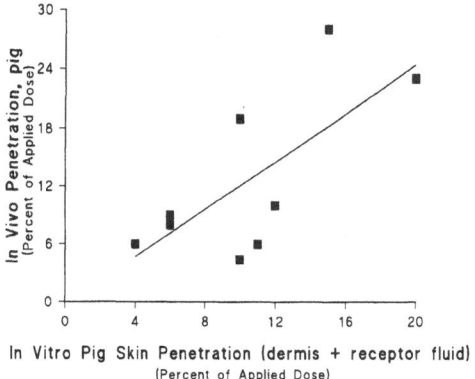

Figure 6. Comparison of percutaneous penetration determinations after *in vitro* and *in vivo* topical application of compounds to pig skin.

4.4. The Effect of Air Temperature and Humidity

An increase in air temperature would be expected to increase skin absorption of a compound, because its thermodynamic activity would be increased. However, in the case of a volatile compound, evaporation also would be expected to increase because its vapor pressure increases with temperature. In the case of the insect repellent DEET, evaporation and percutaneous penetration were studied at air temperatures of 24 and 32C. The 2 processes appear to nearly offset each other with elevated temperature, resulting in a small increase in percutaneous penetration (from 18 to 23%) and a slight decrease in evaporation (from 42 to 39%, Reifenrath, W.G., unpublished data).

The effect of air humidity on percutaneous penetration was examined by conducting the *in vitro* experiments in an environmental chamber. High relative humidity (70%) was found to increase preferentially the percutaneous penetration of relatively hydrophilic compounds (Log Poct < 3) and had no effect on more lipophilic compounds (Log Poct > 3, Fig. 10).

4.5. Occlusion

Occlusion of topical application sites generally results in increased absorption because the increased hydration state of the stratum corneum presumably makes it more permeable. In the case of volatile compounds, occlusion has the compounded effect of

Table 4. Effect of dose on evaporation and percutaneous penetration of N,N-diethyl-m-toluamide (DEET) on excised pig skin[a]

Dose (μg/cm²)	Percent Evaporation Mean ± S.D. (N)	Percent Penetration Mean ± S.D. (N)
4	70± 10 (9)	5 ± 2 (9)
360	73± 4 (9)	13 ± 2 (9)
125000	2.2± 0.6 (6)	0.26± 0.10 (6)

[a]Reifenrath, unpublished data

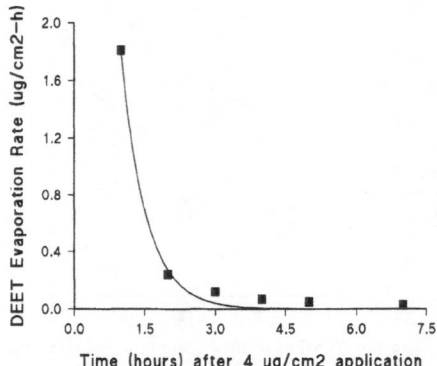

Figure 7. Evaporation rate of N,N-diethyl-m-toluamide (DEET) following topical application to excised pig skin at a dose of 4 μg/cm².

increased skin permeability and decreased evaporative loss, leading to large increases in percent absorption (Table 5).

4.6. Skin Irritation/Skin Damage

Many different forms of damage to the barrier properties of skin have been demonstrated to increase skin penetration of chemicals. Methods of skin damage have included 1) partial or complete removal of the stratum corneum by abrasion (using a hypodermic needle, scalpel or sandpaper), adhesive tape stripping,[15] or midinfrared laser,[16] 2) UV irradiation induced erythema and inflammation,[17] 3) irritant induced dermatitis[18] and 4) atopic dermatitis.[19] The general conclusions from these studies are as follows. Firstly, the greatest increase in skin permeability results from the largest amount of disruption or removal of the stratum corneum.[15] Secondly, disruption of the stratum corneum barrier results in greater enhancement of skin penetration by lipophilic drugs[18] and drugs poorly absorbed through intact skin.[17]

In most of these studies, the intent was to investigate the effect of skin damage on percutaneous penetration of test chemicals or drugs. An alternative approach is to measure

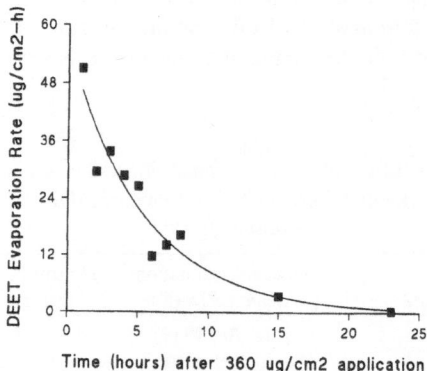

Figure 8. Evaporation rate of N,N-diethyl-m-toluamide (DEET) following topical application to excised pig skin at a dose of 360 μg/cm².

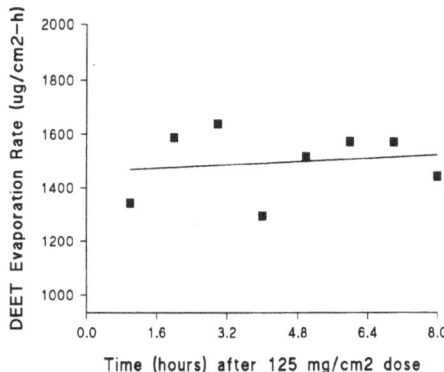

Figure 9. Evaporation rate of N,N-diethyl-m-toluamide (DEET) following topical application to excised pig skin at a dose of 125 mg/cm^2.

skin penetration of a model compound to assess the extent of damage to the barrier properties of skin induced by insult, e.g., irritation. We measured the *in vitro* skin penetration of benzoic acid to compare damage to the barrier properties of pig skin caused by *in vivo* and *in vitro* exposure to an irritant material composed of a complex mixture of chemicals.[20]

The irritant used in the study was liquid gun propellant (LP) which is being tested for use in a new artillery system. LP is composed of a highly irritating mixture of hydroxylammonium nitrate, triethanolammonium nitrate and water. There was concern that skin exposure to this irritating mixture of chemicals compromised the barrier properties of skin, resulting in enhancement of its own penetration.

When a penetrant is composed of a mixture of chemicals, it is much more difficult to assess changes in its skin penetration by using traditional chromatography methods to measure diffusion of the test article into and through skin layers. Consequently, the use of a model compound, e.g., carbon-14 labeled benzoic acid, facilitated the measurement of changes in skin barrier properties induced by the irritating mixture of chemicals in LP.

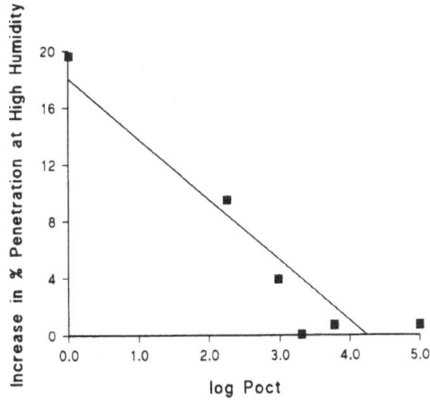

Figure 10. Increase in percutaneous penetration of several compounds as a function of the logarithm of the octanol-water partition coefficient (log Poct). The compounds and their log P values are as follows: caffeine, 0.01; N,N-diethyl-m-toluamide, 2.29; malathion, 2.98; testosterone, 3.31; progesterone, 3.78; DDT, 5.0.

Table 5. Percutaneous absorption of fragrances by human skin *in vitro*[22]

Compound	Dose	% Absorbed Non-occluded Skin (N)	% Absorbed occluded skin (N)
Cinnamyl Anthranilate	4 μg/cm^2	24± 5 (8)	53 ± 7 (7)
Safrole	4 μg/cm^2	15± 2 (6)	38 ± 2 (9)
Cinnamic alcohol	4 μg/cm^2	34± 7 (6)	66 ± 8 (7)
Cinnamic Acid	4 μg/cm^2	18± 5 (6)	61 ± 10 (7)

For the *in vivo* exposure study, 3 sites along the upper back of a group of weanling pigs were dosed with control (saline) and 3 sites were dosed with LP. Application sites were covered with nonocclusive patches for various exposure periods (1 to 5 da) to assess the time course of damage and recovery. At the end of the exposure period, pigs were euthanized and skin at application sites was removed and mounted in skin penetration cells. For the *in vitro* exposure study, skin was excised from another group of euthanized pigs, placed in skin penetration cells, and topically exposed to control (saline) or test chemical (LP) for 1 da. Barrier properties of skin disks exposed *in vivo* or *in vitro* to saline or LP were tested by measuring *in vitro* penetration of benzoic acid during a 24 hr exposure period.

Visual inspection of the pig skin after *in vivo* exposure to LP indicated the lesions (erythema and small pustules) were minimal after 1 and 5 da exposure, and most severe (ruptured pustules and crust formation on epidermal surface) after 3 da exposure. The penetration of benzoic acid through pig skin was increased maximally (8.2 fold increase) after *in vivo* exposure to LP for 1 da, with the penetration returning to normal by 4 da after application (Fig. 11). The difference between the time course of the gross lesions and enhanced skin penetration of benzoic acid could be due to different chemical and physiological processes. The erythema and pustule formation are due to an inflammatory reaction, while increased skin penetration may be due to a direct corrosive effect of LP on stratum corneum.

There was no difference between the increase in skin penetration of the model compound (benzoic acid) following 24 hr *in vivo* or *in vitro* exposures to LP (Fig. 12). *In*

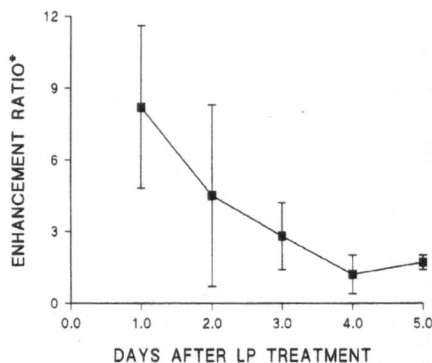

DAYS AFTER LP TREATMENT

*TREATED/CONTROL IN VITRO PERCUTANEOUS ABS. OF BENZOIC ACID

Figure 11. Enhancement of the percutaneous penetration of benzoic acid as a function of time after pretreatment with liquid propellant (LP). Skin was exposed *in vivo* to a single dose of LP and benzoic acid penetration was measured every da for 5 da. The enhancement ratio was determined as the penetration of benzoic acid in pretreated samples divided by the corresponding value for control skin samples. Values are mean ± 1 SD.

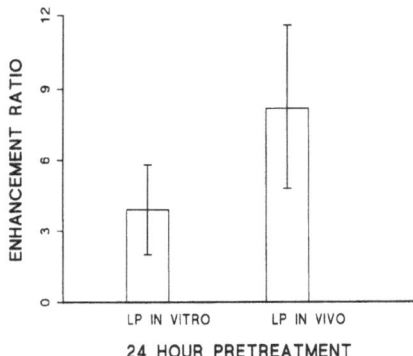

Figure 12. Comparison of the enhancement of ^{14}C-benzoic acid penetration at 24 hr following liquid propellent exposure *in vitro* vs *in vivo*. See Figure 11 for the definition of "enhancement ratio".

vivo and *in vitro* exposures to a potent irritant result in similar damage to barrier properties of pig skin.

In another study, we tested the validity of using changes in barrier function of excised pig skin to predict chemical induced irritation *in vivo*. In this study, several skin pretreatments were challenged with topical exposure to a vesicant [bis(2-chloroethyl)sulfide (HD)]. There was a good correlation (r = 0.97, Fig. 13) between (1) the occurrence of ^{14}C penetration through excised pig skin following *in vitro* topical application of radiolabeled HD and (2) the occurrence of *in vivo* irritation/inflammation induced in intact rabbits by topical exposure to HD (Reifenrath, W.G., unpublished data).

Wilhelm *et al.*[18] reported that *in vivo* exposure of hairless guinea pigs to an irritant (sodium lauryl sulfate) resulted in a much smaller enhancement of skin penetration of drugs compared to *in vitro* exposure. Bronaugh *et al.*[17] reported close agreement between enhanced skin penetration in rat skin following *in vivo* or *in vitro* skin damage caused by mechanical damage (abrasion or tape stripping), but enhancement of skin penetration was greater following *in vivo* exposure to a treatment that caused damage via a physiologic response (erythema and eschar formation caused by UV irradiation) than *in vitro* exposure. In most

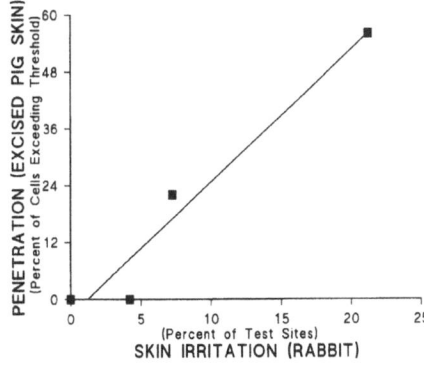

Figure 13. Correlation of *in vitro* percutaneous penetration of radiolabel following topical application of ^{14}C-labeled bis(2-chloroethylsulfide) with *in vivo* skin irritation/inflammation following *in vivo* application of bis(2-chloroethylsulfide) to the rabbit. The different points represent various pretreatments to protect the skin.

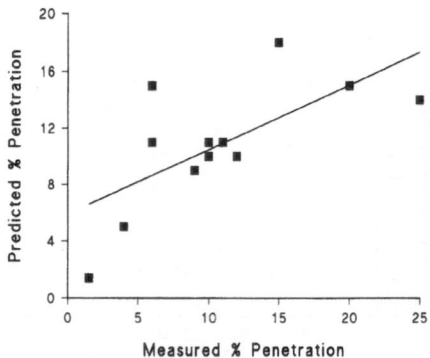

Figure 14. Comparison of experimentally determined percutaneous penetration values in the pig vs corresponding values derived from equation 6 (see text).

cases, when the treatment caused direct damage to stratum corneum (mechanical damage, corrosive or vesicant action), there was agreement between enhancement of skin penetration following *in vivo* or *in vitro* exposure. In conclusion, measurement of skin penetration by a model compound greatly facilitates assessment of changes in the barrier properties of skin caused by complex mixtures. This technique is particularly relevant to assessment of effects of occupational and environmental exposure to toxicants which are frequently complex chemical mixtures whose skin penetration is difficult to analyze by standard chromatographic methods. The pig skin model using the dermal penetration of radiolabeled benzoic acid as a secondary measure of altered skin permeability can be used effectively as a measure of the irritation potential of test chemicals.

5. SUMMARY

Potential evaporation of a chemical from the skin surface is related to its vapor pressure, but actual evaporative loss will depend on ambient conditions, e.g. temperature, humidity and air flow, and interaction with the skin surface, e.g. binding or chemical reaction, penetration. Once inside the skin, the chemical's fate may be dependent on skin metabolism or its stability in an aqueous environment. Uptake of chemical into the general circulation may depend on blood flow to the particular anatomical site. Thus, no single physical property will predict the disposition of topically applied compounds. Rather, a chemical's disposition is dependent on the complex interaction of the physical properties of the chemical, the ambient conditions, and the properties and functioning of the skin itself.

Based on the *in vitro* percent absorption data obtained with skin penetration/evaporation cells at 4 µg/cm^2 for 13 compounds (caffeine, malathion, N,N-Diethyl-m-toluamide, benzoic acid, parathion, DDT, lindane, progesterone, testosterone, fluocinolone acetonide, diethyl malonate, diisopropyl fluorophosphate and trinitrotoluene), we have made a preliminary attempt to correlate percent absorption with molecular weight, an index of diffusability of the compound in skin and with vapor pressure, an index of potential evaporative loss (Eq. 6).

$$\%\text{Penetration} = 22.2 - 0.238 \times 10^{-3} \text{ (VP)} - 0.038 \text{ (MW)} \qquad (6)$$

where VP is the vapor pressure in mm Hg x 10^{-5} mm Hg @ 20C and MW is the molecular weight. Predicted percent absorption vs measured percent absorption is shown in Fig. 14 (multiple $r = 0.66$, $p = 0.0567$). Although there is substantial variation not explained by Eq. 3, we are encouraged by the preliminary results from this rather simple approach. The inclusion of additional compounds in the data base, together with the refinement of Eq. 6, would provide a useful tool for toxicologists to estimate skin absorption from readily available physical properties.

6. REFERENCES

1. Spencer, W.F., and Farmer, W.J., 1980, Assessment of the vapor behavior of toxic organic chemicals, in: *Dynamics, Exposure and Hazard Assessment of Toxic Chemicals* (R. Hague, ed.), Ann Arbor Science, Ann Arbor, pp. 143-161.
2. Miller, J.E., 1968, Wind, *World Book Encyclopedia*, Volume 29, pp. 278-279.
3. Tokay, W., 1970, *Fundamentals of Physiology*, Barnes and Noble, New York, pp. 269-274.
4. Potts, R.O., and Guy, R.H., 1992, Predicting skin permeability, *Pharm. Res.* 9:663-669.
5. Hostynek, J.J., Reifenrath, W.G., Melendres, J.L., and Magee, P.S., 1995, Correlation of *in vivo* and *in vitro* percutaneous absorption with a mathematical model, in: *Exogenous Dermatology: Advances in Skin-Related Allergology, Bioengineering, Pharmacology, and Toxicology, Current Problems in Dermatology*, Volume 22 (C. Surber, P. Elsner, and A.J. Bircher, eds.), Karger, Basel, pp. 139-145.
6. Lyman, W.J., 1990, Octanol/water partition coefficient, in: *Handbook of Chemical Property Estimation Methods* (W.J. Lyman, W.F. Reehl, and D.H. Rosenblatt, eds.), American Chemical Society, Washington, DC, pp. 1-51.
7. Rechsteiner, C.E., 1990, Boiling point, in: *Handbook of Chemical Property Estimation Methods* (W.J. Lyman, W.F. Reehl, and D.H. Rosenblatt, eds.), American Chemical Society, Washington, DC, pp. 12.8-12.16.
8. Snodgrass, H.L., 1990, Physical resources needed for *in vivo* animal studies, in: *Methods for Skin Absorption* (B.W. Kemppainen and W.G. Reifenrath, eds.), CRC, Boca Raton, pp. 111-125.
9. Susten, A.S., Dames, B.L., Burg, J.R., and Niemeier, R.W., 1985, Percutaneous penetration of benzene in hairless mice: an estimate of dermal absorption during tire-building operations, *Am. J. Indust. Med.*, 7:323.
10. Wester, R.C., and Maibach, H.I., 1989, *In vivo* methods for percutaneous absorption measurements, in: *Percutaneous Absorption Mechanisms-Methodology-Drug Delivery*, 2nd ed. (R.L. Bronaugh, and H.I. Maibach, eds.), Dekker, New York, pp. 215-220.
11. Klain, G.J. and Reifenrath, W.G., 1991, *In vivo* assessment of dermal absorption, in: *Dermal and Ocular Toxicology* (D.W. Hobson, ed.), CRC, Boca Raton, pp. 247-266.
12. Hall, L.L., and Shah, P.V., 1990, *In vivo* methods for determining percutaneous absorption and metabolism of xenobiotics: indirect methods, in: *Methods for Skin Absorption* (B.W. Kemppainen and W.G. Reifenrath, eds.), CRC, Boca Raton, pp. 81-109.
13. Reifenrath, W.G., Hawkins, G.S., and Kurtz, M.S., 1991, Percutaneous penetration and skin retention of topically applied compounds: an *in vitro - in vivo* study. *J. Pharm. Sci.*, 80:526-532.
14. Reifenrath, W.G., Chellquist, E.M., Shipwash, E.A., Jederberg, W.W., and Krueger, G.G., 1984, Percutaneous penetration in the hairless dog, weanling pig and grafted athymic nude mouse: evaluation of models for predicting skin penetration in man. *Brit. J. Dermatol. III*, Suppl. 27:123-135.
15. Scott, R.C., Dugard, P.H., and Doss, A.W., 1986, Permeability of abnormal rat skin, *J. Invest. Dermatol.* 86:201-207.
16. Nelson, J.S., McCullough, J.L., Glenn, T.C., Wright, W.H., Liaw, L.-H.L., and Jacques, S.L., 1991, Mid-infrared laser ablation of stratum corneum enhances *in vitro* percutaneous transport of drugs, *J. Invest. Dermatol.* 97:874-879.
17. Bronaugh, R.L., and Stewart, R.F., 1985, Methods for *in vitro* percutaneous absorptions studies V: Permeation through damaged skin, *J. Pharm. Sci.* 74:1062-1066.
18. Wilhelm, K.P., Surber, C., and Maibach, H.I., 1991, Effect of sodium lauryl sulfate-induced skin irritation on *in vitro* percutaneous penetration of four drugs, *J. Invest. Dermatol.* 97:927-932.
19. Turpeinen, M., Mashkilleyson, N., Bjorksten, F., and Salo, O.P., 1988, Percutaneous absorption of hydrocortisone during exacerbation and remission of atopic dermatitis in adults, *Acta Derm. Venereol.* (Stockholm) 68:331-335.

20. Kemppainen, B.W., Terse, P., Madhyastha, M.S., Lenz, S.D., Palmer, W.G., and Reifenrath, W.G., 1993, *In vitro* assessment of in vivo damage to the barrier properties of pig skin caused by a complex mixture, *J. Toxicol.-Cut. Ocular Toxicol.* 12:239-248.

21. Grain, C.F., 1990, in: *Handbook of Chemical Property Estimation Methods* (W.J. Lyman, W.F. Reehl and D.H. Rosenblatt, eds.), American Chemical Society, Washington, DC, pp. 14-16.

22. Bronaugh, R.L. and Stewart, R.F., 1985, Comparison of percutaneous absorption of fragrances by humans and monkeys, *Fd. Chem. Toxic.*, 23:111-114.

40

SWINE LIVER USAGE IN EXTRACORPOREAL DETOXIFICATION

Vyacheslav E. Ryabinin, MD[*]

Department of Chemistry
Chelyabinsk Medical Institute
454092, Chelyabinsk
Vorovsky str., 64, Russia

1. THEORETICAL PRINCIPLES FOR DETOXIFICATION METHOD DEVELOPMENT USING THE LIVER SUPPORT SYSTEM

Such pathologic conditions as liver failure, thermal and radiation injuries, traumas and infections diseases may result in alterations of the liver functional state that are followed by pronounced metabolic disturbances and toxemic syndrome development. As has been shown in liver disease, biological oxidation, especially mitochondrial and microsomal electron transfer chain components are decreased considerably.[1-3]

1.1. Biomedical Significance of Mitochondrial Oxidation

One of the main mitochondrial functions is synthesis of ATP, energetic currency, which supports all metabolic system functioning. Priority of bioenergetic reactions is determined by the universal character of generated energy form, ATP. It would be interesting to know the mechanisms of adverse factor influence on the functional state of mitochondria as in this case energetic metabolism exceeds physiologic limits. Application of mathematical models enabled us to establish the general mechanism of stress factor activity and to suggest that any influence on the cell may result in oxidative phosphorylation dissociation, decreases of ATP and GTP as well as increases of AMP and ADP.[4]

Decrease of ATP level causes disturbances of ionic pump function, accumulation of Ca^{2+} in cytoplasm, subsequent activation of phospholipase A_2, hydrolysis of phospholipids and increased permeability of membrane. Mitochondria degradation is connected closely with excessive free radical oxidation and increased lipid peroxidation. Similar changes are observed in different hepatic diseases, e.g., traumatic shock, hemorrhage and toxic hazard. Thus, energetic balance support appears to be of primary importance in disease management.

[*]Reprint requests to: Dr. Vyacheslav Ryabinin, Vorovskystr, 64, Chelyabinsk Medical Institute, Chelyabinsk, Russia, 454092, 0073512349716.

Injection of ATP is ineffective as a method of energy supply. Therefore, usage of artificial organ and liver support systems may be essential.

1.2. Biomedical Importance of Microsomal Oxidation

Microsomal oxidation is fundamental protection of the human body from different toxigenic substances.[5] This system is a complex consisting of an electron transfer chain and a terminal cytochrome P-450 component implementing oxidation of different metabolites (hormones, fatty acids, bile acids, prostaglandins and other substrates) and various xenobiotics to the polar products with their subsequent removal from the body or blood, to the electrofilic substances possessing high reactivity. A fermentative system includes 4 components, 2 flavoproteins and 2 hemoproteins. One flavoprotein is oxidized NADPH, the other is NADH. The mechanism of electron transfer on cytochrome P-450 is not clear; however, in NADPH oxidation, one electron is transferred from the flavoprotein and the other from cytochrome b5. The mechanism of monooxygenesis reactions implies interaction of substrates with cytochrome P-450, reduction of this complex and interaction with molecular O_2, with formation of peroxicytochrome P-450 which disintegrates to the generated free radical oxygen and oxidated substrate. Metabolic transformations in microsomes are the reactions where the chemical compounds are undergoing a number of oxidations, reductions and hydrolytic transformations. As a rule, this may lead to the appearance of functional groups able to increase molecular polarity and to act as a center to initiate the second phase, conjugation.[6]

Microsomal oxidation processes in various types of hepatic disorders have been well studied. Decreased activity of liver microsomes without profound changes in biochemical blood indices is evidence of lipid dystrophia.[7] Decreased microsomal activity with simultaneous hyperbilirubinemia and increased activity of alkaline phosphatase is evidence of cholestasis of various origins. Where decreased mircosomal activity is correlated only with increased alanine transaminase (ALT) activity, we propose the presence of degenerative alteration of hepatocytes. Considerable changes of microsomal oxidation also are observed in radiation exposure, thermal trauma and combined radiation thermal injury.[1-3] Therefore, studies on management techniques contributing to the maintenance of optimal intensity of microsomal oxidation are of particular importance. Development of an artificial system based on principles of cytochrome P-450 functioning and other metabolic processes has relevance.

1.3. Basic Characteristics of Artificial Cells

An artificial cell, according to T. Chang, is not a specific structure but an idea to create microscopic size structures able to compensate for the deficiency of some cellular functions in the body.[8] At the same time, this structure may be used as a model to study various influences on the processes of detoxification and biotransformation and also for the analysis of xenobiotic metabolism. There may be indirect electrochemical oxidation using active oxygen carriers in simulation of liver tissue detoxificating functions.[9] As the development of the electrochemical model of liver tissue monooxygenase system encountered a serious problem of compatibility between the electrochemical unit and blood, a procedure of indirect electrochemical oxidation of blood was proposed where carriers of active oxygen were used. Blood did not contact with the electrochemical system and electrolysis occurred in solution of oxygen carrier (0.89% NaCl), which accumulated active oxygen in the form of sodium hypochlorite. Sodium hypochlorite, administered into patients, enabled a bypass of the effect of protein protection of toxic metabolites; it simulated both liver tissue monooxygenase functions and phagocytosis molecular mechanisms. Active oxygen generated by different hemoproteins play an important role in bacteriocidicity and mutagenicity. However, the mechanism of the bactericidal and mutagenic action of different active oxygen forms is discussed. The production of hypochlorite is caused by the bacteriocidal effect of myeloperoxidase. Cytochrome P-450 containing systems generate active oxygen that can cause bacteriocidity and mutagenicity. Bacteriocidal and mutagenic actions of human leukocyte myeloperoxidase and soluble reconstituted monooxygenase system were studied.[10] The main impact in the bacteriocidal action belonged to hypochlorite, formed during the myeloperoxidase reaction. The difference between bacteriocidal and mutagenic effects on myeloperoxidase and soluble reconstituted monooxygenase system can be explained by the fact that the bacteriocidal action is affected mainly by hypochlorite generated by myeloperoxidase, while the mutagenic effect of myeloperoxidase is the same as in reconstituted monooxygenase system. This can be caused by the production of different active oxygen which can damage DNA cells. The other and more reliable way consists of the creation of biocompatible membranes with specialized functions. One of the trends in this area is to elaborate athrombogenic artificial cells for hemocarboperfusion.

Application of anion exchange reacting with albumin to remove bilirubin from the blood, encapsulation of combined systems such as enzymeabsorbent and others are known.[11] There are examples of subcutaneous, intramuscular and intraabdominal implantation of artificial cells.[11,12] The drawback of all these methods is in the interaction of xenobiotics with the tissues and blood of a recipient, that may be accompanied by the development of unfavorable immunological and biochemical reactions.

2. PROBLEMS AND PROSPECTS TO DEVELOP NEW METHODS OF DETOXIFICATION

Most methods used for management of endo and exogenic intoxication in humans are based on the extracorporeal detoxification techniques, i.e., hemodialysis, hemosorption and plasmaphoresis. But their limitations and insufficient efficacy required improvement of the new ways of detoxification based on the pathogenetic considerations resulting in normal metabolic processes in cases of intoxication and exposure to adverse environmental factors. Special emphasis has been placed on the development of prompt, cheap and efficient methods of treatment in cases of natural disasters and situations caused by technogenic accidents (thermal trauma, radiation exposure and combined injuries).

There are known different detoxification methods by means of subcutaneous injection of hepatocytes or the latter are injected directly into the blood.[14,15] But it is difficult to obtain intact hepatocytes. The procedure requires liver perfusion *in situ* in animals under anesthesia, special operative techniques to cannulate blood vessels, selection of excretion media, blood oxygenation, pH control and application of collagenase. Thus, hepatocytes obtained in optimal conditions are intact only for 1 to 2 hr. After that, it is necessary to introduce supplementary components to stabilize their state. Therefore, usage of existing methods of treatment is limited, especially if cases are numerous.

2.1. Liver Usage in Extracorporeal Detoxification

To improve all known detoxification methods and to create a stable liver preparation which can be preserved for a long time, it was suggested to use a specifically made preparation of liver (PL). The preparation was derived by means of swine liver homogenization and subsequent lyophilization. Evaluation of PL functional ability showed high

Table 1. Activity of liver preparation enzymes (nmol/min/mg)

Time after preparation (da)	Experimental conditions		
	DDMA	NADPH-oxidase	NADPH: 2, 6-dichlorphenol-indophenolreductase
1	7.5 - 7.8	0.3 - 0.4	45 - 65
7	7.9 - 8.0	0.5 - 0.6	70 - 90
90	7.8 - 8.4	0.4 - 0.6	90 - 100
300	10.2 - 10.8	0.4 - 0.6	120 - 130
730	9.0 - 10.6	0.4 - 0.5	100 - 110

Note: DDMA = demethylase of dimethylaniline. Average findings of 4 to 5 estimations.

enzyme activity. When PL was stored for a long period, the activity was not changed (Table 1).

Analysis of the therapeutic effect was conducted on nonthoroughbred white rats weighing 200 to 220 g. Antipyrine pharmacokinetic analysis was used to study biotransformation and detoxification ability of the liver. The operation was performed under IV thiopentalsodium anesthesia (50 mg/kg). The carotid artery and jugular vein were isolated, an arteriovenous shunt was made and connected to a system consisting of 2 compartments separated from each other by a membrane. The liver fraction was pumped into the first compartment at a constant speed of 1.5 ml/min, blood from the carotid enters into the second compartment, and later is returned to the jugular vein. Five min after the arteriovenous shunt was established, antipyrine was introduced at a dose of 40 mg/kg; in 40 min, a blood sample and dialysate were taken to estimate antipyrine concentration. Three types of dialyzation solutions were used: tyrode solution, preparation of liver with 3 mM NADPH and without it (Table 2). PL, especially in combination with NADPH, enhances antipyrine metabolism decreasing its half life (T 1/2) and accelerating elimination.

Table 2. Analysis of efficiency of extracorporeal antipyrine biotransformation

Incubation media	Liver preparation	Liver preparation + NADPH	Tyrode solution
Dose of antipyrine (mg/kg)	43	43	46
Concentration of antipyrine in blood (mcg/kg)	initial 5.8	18.8	18.7
	after 40 min 13.2	12.2	15.0
Concentration of antipyrine in liver preparation (mcg/mg)	after 5 min 0.75	0.5	2.6
	after 40 min 0.18	0.12	3.5
T1/2 (min)	100	60	110
Clearance (Cl)(ml/min)	0.019	0.026	0.015
Volume distribution (Vd)(ml/g)	2.7	2.3	2.46
Constant elimination(K_el.)	0.007	0.0115	0.0063

3. EXPERIMENTAL AND CLINICAL BASIS TO USE THE EXTRACORPOREAL DETOXIFICATION METHOD

In various pathologic conditions, there occur functional disorders in the liver resulting in some metabolic disturbances and toxemic syndrome development. This is associated with the fact that liver is an unique organ able to synthesize the bulk of biologically active compounds, participate in the detoxification process, aid in biotransformation of chemical substances and implement systemic functional connections between different organs. Endoplasmic reticulum of hepatocytes plays a role in these processes implementing oxidation, hydrolysis, esterification and conjugation. The hazards of infectious diseases, thermal traumas, radiation thermal injuries and liver failures raised a question of developing simple and highly effective methods of extracorporeal detoxification. This is not due to the lack of urgency for liver support system. One way of using the biochemical approach for artificial liver support is the application of immobilized enzymes, especially artificial cells.[16,17]

Artificial cells have been made to contain multienzyme systems, including cofactor recycling systems. Hepatic microsomes have been immobilized for possible metabolic function.[18] It is nearly impossible to assess the effectiveness of liver support systems in the treatment of patients with liver failure. There are large variations in survival rates, depending on age, etiology and other factors. Analysis of accumulated results from different centers is hampered because of variations between centers, including the types of conservative treatment used.[16] A detailed comparative study of the numerous approaches to liver support systems can only be done in suitable animal systems.

3.1. Experimental Studies on the Extracorporeal Detoxification Method for Efficiency in Rats

We were able to carry out extracorporeal detoxification of the blood bringing it into contact with the liver preparation. Experimental studies were performed in 3 groups of animals: a) control animals were given hemodialysis with tyrode medium, b) animals with thermal burns of IIIa to IIIb degree (15 to 20% of the body surface) with burn toxemia (7 da after injury) were performed extracorporeal detoxification using LP and c) control animals with thermal burns (sham). Burn disease was taken as a model to assess the efficiency of this therapeutic method. This choice was stipulated by the fact the thermal trauma causes alterations in the liver functional state and corresponding metabolic disturbances. As has been shown in thermal trauma, monooxygenase activity is decreased, especially in the middle and terminal microsome electron transfer chain components.[19] There is an essential decrease of drug metabolizing microsomal system in burned animals: UDP-glucuronyl transferase activity has decreased, chloral hydrate prolonged soporific effect, cytochrome P-450 dependent oxidation system has been depressed and sleep prolongation after hexenal injection has been observed.[20] Studies of antipyrine pharmacokinetics revealed profound disturbances in metabolism of burned animals: observation during the early period of burn toxemic and 7 da after the burn showed an 80 to 40% increase of T1/2, 120 to 70% increase of area under the curve concentration time (S), 30% antipyrine clearance decrease and from 50 to 70% volume of distribution (Vd) decrease. Extracorporeal detoxification was performed on 13 Wistar rats under IV thiopental sodium anesthesia (50 mg/kg). To increase the efficiency of extracorporeal detoxification, there was elaborated an improved system consisting of thermoregulating container with LP into which a cell was submerged, the latter was filled with the mixture of heparin (1,000 U/kg), reopolyclucinum and tyrode medium (total volume 2.5 to 3.0 ml). The efficiency was assessed by the pharmacokinetic analysis of antipyrine, the concentration of glucose, urea and bilirubin, the infurosian toxicity test,

Table 3. Influence of extracorporeal detoxification on antipyrine biotransformation in experimental animals

Experimental conditions	Healthy animalsn=5	Control group(sham operation)n=5	Animals with thermal trauma treated by means of extracorporeal detoxificationn=5
Initial concentration of antipyrine (C°)(mcg/ml)	18.25 ± 0.02	25.2 ± 5.9	43.0 ± 9.8
T1/2 (min)	92.5 ± 37.00	45.2 ± 5.1	24.9 ± 1.5[a]
K_el.	0.009 ± 0.004	0.016 ± 0.002	0.028 ± 0.013
Vd (ml/g)	2.75 ± 0.05	1.16 ± 0.15	1.2 ± 0.2
Cl (ml/min)	0.024 ± 0.01	0.019 ± 0.005	0.034 ± 0.003[b]
S	2542 ± 1636	1568 ± 317	1528 ± 119

[a]t=3.8 p<0.01

[b]t=2.5 p<0.05

and the content of lipid peroxidation products. The application of this method contributed to antipyrine biotransformation velocity increase, and was accompanied by 100% decrease of T1/2, and 79% increase of antipyrine clearance in the treated animals as compared to rats which had undergone sham operations (Table 3). There was evidence of enhancement of monooxygenase cytochrome P-450 dependent reactions and substitution of the corresponding functions of the injured liver. Therapy decreased plasma toxicity as compared to animals which had pseudooperations, normalized glucose level, stabilized urea content and contributed to the decrease in lipid peroxidation to some degree. Thus, model experiments and animal studies demonstrated safety and high therapeutic efficiency of this detoxification method. On these grounds, clinical trials were undertaken.

3.3. Therapeutic Efficiency Assessment of Extracorporeal Detoxification in Patients

For clinical trials, LP was obtained from swine liver homogenate in a sterile setting. Efficiency assessment was demonstrated in the following cases.

Case 1. Patient P.A., age 49, was admitted to the infectious unit of the community hospital with the diagnosis of viral hepatitis B. In spite of intensive therapy, the state of the patient deteriorated and in 3 da became precomatose. Extracorporeal detoxification, using swine liver, was initiated. Subclavicular and femoral veins were isolated and catheterized to form a venovenostomy. Blood and LP velocity through the main vessels was estimated to be 80 ml/min. In 50 min, bilirubin level decreased 12%, urea level increased 14%, AST and ALT activity decreased 25 and 45 units, respectively, and the level of middle molecules decreased several fold. This was evidence of both biosynthetic liver function enhancement and normalization of other metabolic processes. While this procedure was being carried out, the state of the patient remained stable. Anaphylactic responses and other complications did not develop. Two da after the procedure, the state of the patient improved; he was alert, could answer questions and sit up. His blood analyses showed positive dynamics in some indices, i.e., AST and ALT activity decreased and bilirubin level decreased by 1.5 fold. Subsequently, he was discharged from the hospital.

Case 2. Patient P., age 72, was admitted to Chelyabinsk community hospital with diagnoses of obstructive jaundice and acute cholecystitis. Following cholecystectomy, the patient's condition remained poor; toxicosis signs aggravated, his consciousness was

Table 4. Pre and post operative bilirubin content and alaninaminotransferase (ALT) activity in the blood of patient P

Assessed indices	Preoperatively	Postoperatively		
		1 da	2 da	3 da
Total bilirubin (mcm/l)	162	123	97	43
Bilirubin, direct (mcm/l)	126	90	66	19
Bilirubin, indirect (mcm/l)	36	33	31	24
ALT (mmol/l/hr)	53	44	43	20

blurred, signs of hepatic failure became evident and a precomatose state developed. Extracorporeal detoxification was carried out in the same as in the first case. During the procedure, the patient's state remained stable. Postoperatively, in 40 min, blood glucose level decreased 14% (to normal), the number of middle molecules decreased 38% and the number of middle molecular peptides 9 to 12%. Changes of bilirubin content and ALT activity (Table 4) indicated hepatic function improvement after this procedure. Later a positive clinical course was observed.

Case 3. Patient S., age 70, was admitted to Chelyabinsk regional hospital with a diagnosis of acute hepatitis. Two da after admission, laparotomy and segmental electrocoagulation of the liver were performed. The state of the patient was poor. The diagnosis was active liver cirrhosis with signs of cellular failure. Subsequently, bilirubin level increased. The patient had vomiting and sickness, which was evidence of endogenic intoxication. She developed a comatose state and unconsciousness. In contrast to the previous cases, the dose of liver preparation was decreased 40%. During the procedure, the patient's condition remained stable. In 50 min, total bilirubin level decreased 9%, direct bilirubin 31% and midmolecular peptides 5 to 7%. For the next 2 da, bilirubin level continued to decrease: total bilirubin decreased 36% and indirect bilirubin 26%. Three da after the operation, there was an increase in bilirubin level and the patient's condition was aggravated again. In cases of severe hepatic failure, it is necessary to perform repeatedly the procedure; therefore, it should be done as soon as possible after admission.

Model experiments were carried out aimed at efficiency assessment of long term LP activity as well as its ability to adjust low molecular weight metabolites. To donor blood (2.5 l) was added NH_4CL to an ammonia concentration of 0.960 mmol/l; then it was connected to the liver support device. During the next 8 hr, blood samples and LP were taken hourly. Blood ammonia level decreased in an hour to 0.150 mmol/l; in 7 hr, its level could not be determined practically. During the same time, LP ammonia concentration increased from 0.219 to 1.95 mmol/l. As there was no balance of ammonia concentration on both sides of the membrane and urea concentration in LP increased from 2.63 to 3.30 mmol/l, one may consider the preparation able to bind NH_3 for a long period and to synthesize urea. Maintenance of DMA demethylase activity (5.26 to 4.21 nmol/min/mg of protein for 8 hr of incubation), insignificant fluctuation of cytochrome$_{B5}$ content (0.19 to 0.15 nmol/mg of protein for 6 hr of incubation) and pH (7.26 to 7.05 for 8 hr of incubation) all help define functional activity of the liver preparation. The level of cytochrome P-450 was changing more sharply, decreasing 2 and 20 fold in 2 and 6 hr, respectively. Model and experimental clinical studies on therapeutic efficiency of the extracorporeal detoxification method enabled us to draw a conclusion of advisability of its application in medical practice.

3.4. Potentials of the Extracorporeal Detoxification Method Application in Toxicology

The suggested method of extracorporeal detoxification may be applied in management of hepatic failure, thermal traumas and other conditions, including all kinds of poisoning. Metabolism of xenobiotics is known to result in the formation of less active substances, i.e., detoxification occurred or it may be accompanied by formation of highly toxic metabolites exerting carcinogenic, mutagenic and other effects. While doing the operation using LP, it is necessary to know the metabolic peculiarities of each xenobiotic which causes a toxic state and to take into account the time of its elimination from the body. With poisoning substances, it is advisable to use LP for detoxification during the first days but with care because of microsomal oxidation and the above substances may be transformed into more active compounds. Thus, chloroform may be transformed into phosgene, trichloroethylene into carbonic oxide, 3,4-benzopyrene into epoxide, etc. Microsomal oxidation activation by inductors such as phenobarbital may contribute to the decrease of impairment due to various xenobiotics. Possibly, this occurs due to conjugation activation. Nevertheless, it is safer to use LP detoxification at the late stages of intoxication with these substances. Some xenobiotics are known to cause intoxication and it is possible to use LP at the early stage of the disease as there is metabolism into less active compounds. This emphasizes the idea that the outcome of many diseases depends on timely procedures. For example, in thermal or radiation injuries, hemorrhagic shock, kidney or liver failure, poisoning the therapeutic effect will be greater if the treatment is pathogenic. The earlier the treatment is started, the more favorable the outcome. It should be taken into account when the primary management of such patients is carried out in the field hospital setting.

4. SWINE LIVER USAGE FOR XENOBIOTICS INDICATION AND MONITORING

Presently, there are 6 million substances with 25 to 30 thousand new compounds synthesized annually. Of more than 100,000 substances used by man, only 10% are tested for ecologic and toxicologic properties. Moreover, there are sanitation and hygiene standards for even fewer of the substances. There have been testing and determination of biological activity of only 30 chemical substances. A profound investigation of biochemical activity of one substance is a time consuming and expensive procedure; for instance, toxicological studies only will require 0.5 to 2 million dollars, more than 800 animals and not less than 3 yr of work. Finally, for a complete estimation of one substance, it is necessary to spend more than 2 million dollars.[21] The total number of mammals used in these types of experiments in the world exceeds 100 to 160 million. There are recommendations to use *in vitro* alternative testing methods, which do not require laboratory animals. Today, such alternatives are the approaches associated with the usage of cell cultures, cellular organelle and other model membrane systems. A microsomal system, particularly its central part, cytochrome P-450, in man and animals is the most sensitive to environmental deterioration. Some parts of the proposed methods are based on the investigation of microsomal and cytochrome P-450 functional activity because xenobiotic biotransformational processes occur mainly in liver and kidney microsomes with cytochrome P-450 and may be accompanied both by detoxification and toxification; that depends on the nature of the compounds. Thus, investigation of xenobiotic biotransformation mechanism provides the opportunity to evaluate potential biological activity of substances, their ecological risk and a basis to work out adequate methods of detoxification and diminishing the level of pollutants in the human organism. Suggested methods may be used to study xenobiotic metabolism dynamics and biotransformation, rate and coefficient of their diffusion, half life and clearance, to characterize enzymes metabolizing xenobiotics and to model biotechnological systems for environmental protection.

5. REFERENCES

1. Griffeth, L.K., Rosen, G.M., and Rauckman, E.J., 1984, Effects of model traumatic injury on hepatic drug metabolism in the rat, *Drug Metab. Dispos.* 12:588-595.
2. Fruncillo, R.J., and Di Gregorio, G.J., 1983, The effect of thermal injury on drug metabolism in the rat, *J. Trauma* 23:523-529.
3. Ryabinin, V.E., and Lifshits, R.I., 1992, Characteristics and correlation between processes of mitochondrial and microsomal oxidation at various intensities of lipid peroxidation, *Biochemistry* 56:1406-1411.
4. Bilenko, M.V., 1989, Ischaemic and Repurfusion Injury to Organs, Meditsina, Moscow.
5. Williams, R.T., 1963, Detoxication mechanisms in man, *Clin. Pharm. Ther.* 4:234-239.
6. La Du, B.N., Mandel, H.G., and Way, E.L. (eds.), 1971, Fundamentals of Drug Disposition, Baltimore.
7. Bluger, A.F., Gorshtein, E.S., and Majore, A.J., 1985, Microsomal monooxygenase activity in the main types of experimental liver damages, in: Cytochrome P-450 and Protection of Human Environment, Scientific Center of Biological Research of the Academy of Sciences of the USSR, Pushchino.
8. Chang, T.M.S., 1977, Biomedical Applications of Immobilized Enzymes and Proteins, Plenum Press, New York.
9. Sergienko, V.I., Martynov, A.K., and Vasilev, Yu.V., 1990, Indirectelectrochemical oxidation using active oxygen carriers in simulation of the liver tissue detoxicating function, *Vopr. Med. Chem.* 36:28-32.
10. Khailov, P.M., Sivov, I.G., and Adrianov, N.V., 1991, Bacteriocidal and mutagenic effect of myeloperoxidase and soluble reconstituted monooxygenase system, cytochrome P-450, Proc. 7th Intl. Conf. Biochem. Biophys. Cytochrome P-450, Moscow, Russia, pp. 543-545.
11. Chang, T.M.S. (ed.), 1978, Artificial Kidney, Artificial Liver and Artificial Cells, Plenum Press, New York.
12. Lamers, W.H., Been, W., and Charles, R., 1990, Hepatocytes explanted in the spleen, preferentially express carbomoylphosphate synthetase rather than glutamine synthetase, *Hepatology* 12:701-709.
13. Grundmann, R., Koebe, H.G., and Waters, W., 1986, Transplantation of cryopreserved hepatocytes or liver cytosol injection in the treatment of acute liver failure in rats, *Res. Exp. Med.* 186:141-149.
14. Eiseman, B., Van Wyk, J., and Griffen, W.O., 1966, Methods for extracorporeal hepatic assist, *Res. Exp. Med.* 123:522-530.
15. Touraine, J.L., 1991, In utero transplantation of fetal liver stem cells in humans, *Blood Cells* 17:379-386.
16. Chang, T.M.S., 1964, Semipermeable microcapsules, *Science* 146:524.
17. Chang, T.M.S., 1984, Liver support systems, in: Life Support Systems in Intensive Care, INC., Chicago, pp. 461-485.
18. Sofer, S.S., and McCallum, R.E., 1978, Effect of bacterial endotoxin on an immobilized microsomal hepatic detoxification system, *Artif. Organs* 2(Suppl.):289.
19. Ryabinin, V.E., and Lifshits, R.I., 1986, The functional state of liver microsomes and lipid peroxidation at the early period of burns toxemia, *Vopr. Med. Chem.* 32:115-118.
20. Ryabinin, V.E., and Lifshits, R.I., 1991, Cytochrome P-450 inactivation in pathology state, Proc. 7th Intl. Conf. Biochem. Biophys. Cytochrome P-450, Moscow, Russia, p. 57.
21. Calleja, M.C., Persoone G., and Geladi, P., 1993, The predictive potential of a battery of ecotoxicological tests for human acute toxicity, as evaluated with the first 50 MEIC chemicals, *ALTA* 21:330-349.

41

PIG BEHAVIOR AND BIOMEDICAL RESEARCH

Suitable Subjects and Experimental Models

Harold W. Gonyou, PhD[*]

Prairie Swine Centre, Inc.
Saskatoon, SK, Canada, S7H 5N9

1. ABSTRACT

Ethology (the study of animal behavior) relates to the interaction of animals with their environment, both animate and inanimate. As such, ethology has three roles to play in relation to animals used in biomedical research. The first relates to providing an appropriate environment for the animal used as a biomedical model. The second role is to determine if the species behavioral capabilities can be used to facilitate answering nonbehavioral research questions. The final area is to use the animal as a model for studying behaviors which are of interest beyond the species itself. The behavior of pigs is examined and suggestions made on their care as experimental subjects and use as behavioral models.

2. INTRODUCTION

The domestic pig has an established position within biomedical research. It serves as a model in cardiovascular research, nutritional studies, organ transplantation, endocrinology and surgical training.[35] Animal models must be managed in such a way that the results of the study are reliable and free from the effects of nonexperimental stressors. Appropriate guidelines have been developed for the care of pigs as laboratory animals and in agricultural research.[6,29] Guidelines are best prepared, interpreted and applied by those with an understanding of their scientific basis. Many aspects of the guidelines relate to the behavior of the animal. Issues such as social groupings, handling techniques, space allowances and environmental enrichment can be understood only through a knowledge of behavior.

Behavior is the primary means of communication among animals and it represents a means for an experimental model to convey the effects of a manipulation to the researcher.

---

[*] Reprint requests to: Dr. Harold Gonyou, Box 21057, 2105 8th Street East, Saskatoon, SK, Canada, S7H 5N9, (306) 477-7452.

Advances in Swine in Biomedical Research, edited by Tumbleson and Schook
Plenum Press, New York, 1996

Some behaviors, such as aggression, are normal in pigs and are affected by physiological or pharmacological treatments. In other cases, animals may be trained to respond in a specific way to demonstrate changes in their physiological status. Operant conditioning techniques are useful when asking an animal if its environmental needs have changed.

Behavioral strategies for coping with stress and improving fitness are shared by many species. Reactions to stress include behavioral adaptation as well as changes in physiological function. Stress induced behaviors such as stereotypies exist in both humans and nonhuman species. Exaggerated social and reproductive strategies in one species also may make them favored as a biomedical model.

The purpose of this article is to examine the behavior of pigs and how it relates to their use in biomedical research. I hope to demonstrate that pig behavior is amenable to management in a research setting and also is of interest in itself. In terms of their behavior, pigs are both suitable research subjects and appropriate models.

3. BEHAVIOR AND MANAGEMENT

Pigs are social animals. The most common social grouping observed in feral pigs is 2 to 4 sows with their offspring. Adult males are solitary except during mating season.[14] Females separate from their group prior to parturition,[19] but return with their litter approximately 7 to 10 days after delivery.[28] Social groups are related closely and interactions with nonrelatives, except to mate, are rare. The Belding's ground squirrel, with a similar matriarchal society, is capable of recognizing close relatives even if separated at birth.[18] When pigs are cross fostered and later regrouped with estranged littermates, aggression is just as intense as that among unrelated and unfamiliar animals.[34] Although group housing or social contact is recommended for experimental pigs, regrouping must be accomplished in advance or the resulting conflict imposes a significant stressor on the animals. Conflict may continue even if animals are penned adjacent to each other but with no opportunity to resolve their social status. Sows housed in adjacent stalls with horizontal bars apparently fail to resolve social conflict and maintain elevated levels of free cortisol compared to sows in stalls with vertical bars.[4] Some implications of the social behavior of pigs on their management are that mature boars should be individually penned, pigs should be thoroughly familiarized before the study begins and penning should allow resolution of conflict between adjacent animals.

Although there is little doubt that severe handling of animals is stressful and detrimental to most biomedical studies, little has been known until recently about the effect of subtle handling methods. Prompted by reports of differential reproductive performance being related to apparent fearfulness toward humans,[16] we determined the reaction of pigs to specific signals associated with handling by humans. Pigs were less frightened if humans squatted rather than stood in their pens, and if pigs were allowed to approach the human rather than vice versa.[17] When these aversive signals were incorporated into the daily management routine, pigs demonstrated an adrenal response and reduced growth compared to nonhandled controls or pleasantly handled animals.[11] Recently, researchers confirmed that handlers can have a negative effect on their animals and that a positive attitude towards animals is a necessity for good human/animal relationships.[15] Restraint of any kind is likely to be stressful to pigs. Although castration without the use of anesthetic is understandably painful to the pig, the vocalizations of the animals suggest that the restraint involved is nearly as stressful as the surgery.[37] Fortunately, pigs may be habituated or conditioned to react neutrally to handling. The Panepinto sling is an effective restraint device,[30] and pigs have been trained using sweet rewards to stand without restraint for blood collection (T. Hartsock, personal communication).

The amount of space required by pigs is dependent upon their behavior and size. Although guidelines for space allowances usually recommend a single value for a wide range of body weights, a continuous function based on body weight.[667] is most appropriate.[31] It only remains to determine the appropriate constant for each behavioral type. Growing pigs use space for lying, eating, drinking, elimination and social interactions. The amount of space required for lying varies with temperature above or below the point of thermoneutrality.[5] In general, recommendations for groups of growing pigs in agricultural conditions vary from 0.030 to 0.038 m^2 per kg[667]. The shape of pen also affects how pigs use space[38] as well as group size. To accommodate such variation, recommendations for pigs used as laboratory animals are more spacious.

Environmental enrichment is a poorly understood concept in animal care.[27] How to enrich a pen often is left to the caretaker. Research has shown that objects on the floor of pig pens are poor means of environmental enrichment as pigs lose interest in them once they are soiled.[13] Pigs prefer to chew on malleable and destructible objects, such as rope or straw, rather than a chain.[9] Although such objects need to be replaced frequently, their value as enrichment devices is much greater than that of heavy floor objects frequently used.

4. BEHAVIORAL ASSESSMENT OF EXPERIMENTAL MANIPULATIONS

A change in behavior is indicative of a change in the motivational state or physiological condition of an animal. Some behaviors are affected directly by physiological or pharmacological interventions. The social aggression discussed above, although quite different from that observed in humans, may be used to study tranquilizers and anxiolytic compounds.[7] Treatment of pigs with azaperone or amperozide results in a delay or reduction of aggression associated with regrouping.[12]

Operant conditioning techniques are used to determine environmental preferences and motivational levels. Pigs readily learn to activate infrared or microwave heaters to alleviate cool conditions.[26,32] Operant methods have been used to determine the effect of diet or manipulation of the hypothalamus on thermal preferences of pigs.[3] Hunger also is measurable using operant methods, with a progressive ratio technique preferred to using a fixed ratio.[20] An example of the use of such methodology is the assessment of the effect of dietary bulk on hunger.[22] Pigs learn quickly and the usefulness of operant conditioning to monitor the effects of physiological and pharmacological interventions is limited only by the inventiveness of the researcher.

5. PIG BEHAVIOR AS A MODEL

Several areas of behavioral research on pigs may have application in biomedical research. Environment induced stereotypies have been studied extensively in sows. A second area related to welfare research in pigs is the development of methods to assess the importance of differing motivations. Finally, the large litter size of pigs makes them useful models for studying the physiology of nursing, intrauterine effects on behavior and social effects on sex ratio. These areas certainly are of importance in other species, and may be relevant to humans.

Stereotypies generally are defined as unvarying, repetitive behavior which have no obvious function.[23] They are considered indicators of stress in both humans and pigs. Physical restriction and boredom commonly were believed to be the cause of stereotypies in gestating sows. However, after it was observed that sows fed less than the normal gestation ration had higher levels of stereotypies,[2] the search for a cause was redirected towards feeding motivation. A factorial study in which 2 levels of feeding, as well as 2 degrees of physical restriction, were imposed demonstrated clearly that the more important causative factor was feed restriction. Additional studies led to the expansion of a motivational model for stereotypy development to include arousal, sensitization and channeling of behavior.[21] Research on stereotypy development in sows serves as an example for studying behavior related to stress and welfare. Similar environment induced stereotypies occur in humans and there may be some value in using pigs as an experimental model.

Another research area related to welfare in pigs is the comparison of motivation for various behaviors. Consumer demand theory was developed originally for application to humans, using price as the cost of a resource. The extension of its use to animals required substituting another form of cost.[8] One way in which this has been done is to use operant conditioning. This technique was applied to determine the relative ranking of food, social contact and environmental novelty as necessities for pigs.[24] The demand for food was inelastic, while that for social contact was less elastic than for novelty. The application of this technique using nonmonetary costs requires further development, but could be used to assess biological requirements in humans.

Nursing and suckling involve an interaction of behavior and physiology. The large litters and synchronized nursings in pigs makes them a potential model for biomedical research. Nursing and suckling in pigs involves a series of vocal signals, as well as massage both before and after milk flow, all of which is synchronized around a 15 sec release of milk.[10] Blood sampling in relation to behavior and known physiological responses can be very precise. The degree of teat stimulation is related not only to prolactin, but also to somatostatin, glucagon and vasoactive intestinal peptide.[1]

The in utero hormonal environment can affect the anatomy, physiology and behavior of the resulting offspring. Species with large litters represent an opportunity to study subtle differences in these effects. While in utero, piglets may be spaced between 2 males, adjacent to 1 male, or between 2 females. Females which developed between 2 males displayed shorter periods of receptivity during estrus, but mounted other females more often than did females which developed between females.[32] Males which developed between 2 brothers were more successful than other males in competitive situations.[33] Determining the mechanism for these changes in pigs may direct research into the possibility of similar perinatal hormonal effects in humans.

Sociobiology relates social behavior to reproductive success. Current theories predict that the social status of a female may skew the sex ratio of her offspring, favoring males in some situations and females in others. Dominant sows produce a higher proportion of males than do subordinate females.[25] The physiological mechanism for this effect is not known. Although it may not be possible to identify a similar strategy in humans, once the mechanism is known, it may provide insight to the physiology of gender determination.

6. CONCLUSIONS

With proper facility design, good animal handling, and appropriate use of habituation or conditioning to stressful procedures, pigs make excellent experimental subjects. In addition the behavior of pigs, particularly their versatility in operant studies, can be used to assess the effects of physiological and pharmacological treatments. Pigs also exhibit a number of stress and reproductive behaviors which make them appropriate models for biomedical studies. Greater interaction among biomedical scientists and ethologists could be beneficial to both research communities.

7. REFERENCES

1. Algers, B., Madej, A., Rojanasthien, S., and Uvnas-Moberg, K., 1991, Quantitative relationships between suckling-induced teat stimulation and the release of prolactin, gastrin, somatostatin, insulin, glucagon and vasoactive intestinal polypeptide in sows, *Vet. Res. Comm.* 15:395-407.
2. Appleby, M.C., and Lawrence, A.B., 1987, Food restriction as a cause of stereotypic behaviour in tethered gilts, *Anim. Prod.* 45:103-110.
3. Baldwin, B.A., 1979, Operant studies on the behavior of pigs and sheep in relation to the physical environment, *J. Anim. Sci.* 49:1125-1134.
4. Barnett, J.L., Hemsworth, P.H., Cronin, G.M., Newman, E.A., and McCallum, T.H., 1991, Effects of design of individual cage-stalls on the behavioural and physiological responses related to the welfare of pregnant pigs, *Appl. Anim. Behav. Sci.* 32:23-33.
5. Boon, C.R., 1981, The effect of departures from lower critical temperature on the group postural behaviour of pigs, *Anim. Prod.* 33:71-79.
6. Consortium for Developing a Guide for the Care and Use of Agricultural Animals in Agricultural Research and Teaching. 1988, Guide for the Care and Use of Agricultural Animals in Agricultural Research and Teaching. Association Headquarters, Champaign, IL.
7. Dantzer, R., and Mormede, P., 1979, Effects of lithium on aggressive behaviour in domestic pigs, *J. Vet. Pharm. Therap.* 2:299-303.
8. Dawkins, M.S., 1983, Battery hens name their price: Consumer demand theory and the measurement of ethological "needs", *Anim. Behav.* 31:1195-1205.
9. Feddes, J.J.R., and Fraser, D., 1993, Non-nutritive chewing by pigs: Implications for tail-biting and behavioural enrichment. Livestock Environment IV, Amer. Soc. Agric. Engin., St. Joseph, MI. pp. 521-527.
10. Fraser, D., 1980, A review of the behavioural mechanisms of milk ejection of the domestic pig, *Appl. Anim. Ethol.* 6:247-255.
11. Gonyou, H.W., Hemsworth, P.H., and Barnett, J.L., 1986, Effects of frequent interactions with humans on growing pigs, *Appl. Anim. Behav. Sci.* 16:269-278.
12. Gonyou, H.W., Rohde Parfet, K.A., Anderson, D.B., and Olson, R.D., 1988, Effects of amperozide and azaperone on aggression and productivity of growing-finishing pigs, *J. Anim. Sci.* 66:2856-2864.
13. Grandin, T., 1989, Effect of rearing environment and environmental enrichment on behavior and neural development in young pigs, PhD Thesis, University of Illinois, Urbana, IL.
14. Graves, H.B., 1984, Behavior and ecology of wild and feral swine, *J. Anim. Sci.* 58:482-492.
15. Hemsworth, P.H., Barnett, J.L., and Coleman, G.J., 1993, The human-animal relationship in agriculture and its consequences for the animal, *Anim. Welf.* 2:33-51.
16. Hemsworth, P.H., Brand, A., and Willms, P., 1981, The behavioural response of sows to the presence of humans and its relation to productivity, *Livest. Prod. Sci.* 8:67-74.
17. Hemsworth, P.H., Gonyou, H.W., and Dziuk, P.J., 1986, Human communication with pigs: The behavioural response of pigs to specific human signals, *Appl. Anim. Behav. Sci.* 15:45-54.
18. Holmes
19. Jensen, P., 1986, Observations on the maternal behaviour of free-ranging domestic pigs, *Appl. Anim. Behav. Sci.* 16:131-142.
20. Lawrence, A.B., and Illius, A.W., 1989, Methodology for measuring hunger and food needs using operant conditioning in the pig, *Appl. Anim. Behav. Sci.* 24:273-285.
21. Lawrence, A.B., and Terlouw, E.M.C., 1993, A review of behavioral factors involved in the development and continued performance of stereotypic behaviors in pigs, *J. Anim. Sci.* 71:2815-2825.
22. Lawrence, A.B., Appleby, M.C., Illius, A.W., and McLeod, H.A., 1989, Measuring hunger in the pig using operant conditioning: The effect of dietary bulk, *Anim. Prod.* 48:213-220.
23. Mason, G.J., 1993, Forms of stereotypic behaviour, In: Stereotypic Animal Behaviour: Fundamentals and Applications to Welfare, (A.B. Lawrence and J. Rushen, eds.), CAB International, Wallingford, UK. pp. 7-40.
24. Matthews, L.R., and Ladewig, J., 1994, Environmental requirements of pigs measured by behavioural demand functions, *Anim. Behav.* 47:713-719.

25. Meikle, D.B., Drickamer, L.C., Vessey, S.H., Rosenthal, T.L., and Fitzgerald, K., 1993, Maternal dominance rank and secondary sex ratio in domestic swine, *Anim. Behav.* 46:79-85.
26. Morrison, W.D., Amyot, E., McMillan, I., Otten, L., and Pei, D.C.T., 1987, Effect of duration of reward upon operant heat demand of piglets receiving microwave or infrared heat, *Can. J. Anim. Sci.* 67:903-907.
27. Newberry, R., 1995, Environmental enrichment: Increasing the biological relevance of captive environments. *Appl. Anim. Behav. Sci.* 44:229-243.
28. Newberry, R.C., and Wood-Gush, D.G.M., 1986, Social relationships of piglets in a semi-natural environment, *Anim. Behav.* 34:1311-1318.
29. NRC (National Research Council), 1985, Guide for the Care and Use of Laboratory Animals, U.S. Department of Health and Human Services, Washington, D.C.
30. Panepinto, L.M., Phillips, R.W., Norden, S.E., Pryor, T.C., and Cox, R., 1983, A comfortable, minimum stress method of restraint for Yucatan miniature swine, *Lab. Anim. Sci.* 33:95-97. 31. Petherick, J.C., 1983, A biological basis for the design of space in livestock housing, In: Farm Animal Housing and Welfare, (Baxter, S.E., eds.) pp 103-120.
32. Rohde Parfet, K.A., Ganjam, V.K., Lamberson, W.R., Rieke, A.R., vom Saal, F.S., and Day, B.N., 1990, Intrauterine position effects in female swine: Subsequent reproductive performance, and social and sexual behavior, *Appl. Anim. Behav. Sci.* 26:349-362.
33. Rohde Parfet, K.A., Lamberson, W.R., Rieke, A.R, Cantley, T.C., Ganjam, V.K., vom Saal, F.S., and Day, B.N., 1990, Intrauterine position effects in male and female swine: Subsequent survivability, growth rate, morphology and semen characteristics, *J. Anim. Sci.* 68:179-185.
34. Stookey, J.M., 1991, Factors Influencing Aggression in Newly Mixed Pigs, Ph.D. Thesis. University of Illinois, Urbana. pp.147.
35. Swindle, M.M., Smith, A.C., Laber-Laird, K., and Dungan, L., 1994, Swine in biomedical research: Management and models, ILAR News 36(1):1-5.
36. Verstegen, M.W.A., Siegerink, A., Van der hel, W., Geers, R., and Brandsma, C., 1987, Operant supplementary heating in groups of growing pigs in relation to air velocity, *J. Therm. Biol.* 12:257-261.
37. Weary, D.M., Braithwaite, L.A., Lalonde, J.L.M., and Buckley, D., 1995, Measuring pig calls: Grunts, squeals and digital processing, Proc. Intern. Soc. Appl. Ethol., Exeter, UK, paper 97.
38. Wiegand, R.M., Gonyou, H.W., and Curtis, S.E., 1994, Pen shape and size: Effects on pig behavior and performance, *Appl. Anim. Behav. Sci.* 39:49-61.

42

CURRENT STATUS OF *IN VITRO* PRODUCTION OF PORCINE EMBRYOS

Hiroaki Funahashi, PhD and Billy N. Day,[*] PhD

Department of Animal Sciences
University of Missouri-Columbia
Columbia, Missouri 65211

1. INTRODUCTION

Approximately 210,000 primordial follicles exist in a porcine ovary.[1] The establishment of an efficient system to produce porcine embryos *in vitro* from follicular oocytes of slaughterhouse ovaries would reduce the cost for procurement of embryos for various research purposes. Further, it would accelerate the use of embryos in the study of embryo micromanipulation directed toward production of transgenic pigs, embryological development and cryopreservation of gametes. As reviewed by the authors[2,3] and others,[4-7] research on *in vitro* maturation (IVM) and *in vitro* fertilization (IVF) has been conducted recently in an effort to resolve serious problems that prevented the in vitro production of porcine embryos, such as a high incidence of polyspermy[8-10] and a reduced incidence of male pronuclear formation.[8,10] In turn, *in vitro* techniques for production of embryos have been developed that result in acceptable rates of blastocyst development[11-14] and have produced live piglets.[11,15,16] Here we review these technological advancements.

2. OOCYTE CUMULUS COMPLEXES AS MATERIALS

2.1. Oocyte Growth *in vitro*

As described above, there are large number of primordial follicles in a porcine ovary (210,000), as compared with other species such as humans (151,000), cattle (105,000) and mice (2135).[1] Utilization of primordial oocytes for *in vitro* embryo production would bring about a major reduction in the cost and effort required to produce large numbers of embryos. However, it has been known that when subjected to IVM conditions, oocytes in the growing stage do not have the competence to progress beyond the metaphase I stage during meiosis,[17]

---

[*] Reprint requests to: Dr. Billy Day, Department of Animal Sciences, University of Missouri-Columbia, Columbia, MO 65211, (573) 882-7555.

and the pattern of histone H1 kinase activity of growing oocytes differs from that of fully grown oocytes during the maturation period.[18] To utilize growing oocytes for *in vitro* embryo production, more successful oocyte growth will need to be achieved *in vitro*. This will require an increase in the volume of oocyte cytoplasm as well as the number of cumulus cells.[17] Further, increased activation of cytoplasmic functions are needed (see reviews by Motlik and Fulka[19] and Moor *et al.*[20]). This research area has recently become more active, and some physiological requirements for oocyte growth have been identified.[18,21-24] Hirao *et al.*[25] obtained oocyte maturation and sperm penetration of porcine oocytes following culture of preantral follicles. However, the incidence of oocytes grown and matured *in vitro* is still low (40%).[25] Developmental ability of this type of oocyte following *in vitro* fertilization has not been reported. It is expected that further research with oocytes derived from primordial and growing follicles will need to be undertaken to increase the number of oocytes available from a single ovary for IVM/IVF.

2.2. Grown Oocytes

Currently, the materials used for *in vitro* embryo production largely have been limited to fully grown oocytes. Extensive use has been made of oocyte cumulus complexes (OCCs) with uniform ooplasm and a compact cumulus cell mass collected from antral follicles of slaughtered prepubertal gilts. However, only limited information is available on quality of OCCs used in these studies. The size of follicles that are selected for collection of OCCs has varied within investigators[3]. Further, even when OCCs are obtained from follicles which are selected by size, there is a large variation in the morphology of the germinal vesicle (Table 1). In contrast, the nuclear stage of oocytes is synchronized closely in gilts 72 hr after injection with equine chorionic gonadotropin (H. Funahashi, T.C. Cantley and B.N. Day, unpublished results).

There are dramatic changes in number and size of follicles in gilts during da 16 to 21 of the estrous cycle.[26-28] Although follicular heterogeneity in steroid concentration and gonadotropin binding ability has been observed in both cyclic[26,29] and gonadotropin treated gilts,[30] steroid concentrations and gonadotropin binding ability both increase until da 20 as the follicle grows and then decrease on da 21. Also the number of granulosa cells per follicle increases between da 20 and da 21.[26] These changes may be required to reduce the variation in the dictyate stage of the first meiotic prophase. OCCs used for IVM and IVF should be synchronized meiotically and of high quality. Further investigations which lead to a reduction in the heterogeneity of OCCs may be expected to improve the developmental rate of IVM/IVF embryos.

3. *IN VITRO* MATURATION OF PORCINE OOCYTES

3.1. Nuclear Maturation

Porcine oocytes remain at the germinal vesicle stage until 18 hr after hCG injection, and achieve first meiosis between 35 and 37 hr.[31] Recently it has been shown, using ultrasonography, that the interval from the onset of the LH surge to ovulation is on average 44 ± 3 hr.[32] To achieve oocyte maturation, many investigators have cultured porcine OCCs in modified Medium 199 containing various hormonal supplements for 36 to 48 hr.[33-39] However, exposure of OCCs to gonadotropins for only the first 2 hr of IVM is adequate to achieve nuclear maturation in 75% of the oocytes.[40] Removing cumulus cells 24 hr after the start of culture does not affect nuclear maturation.[41] Inhibitor studies indicate that heterogenous nuclear RNA synthesis

Table 1. Germinal vesicle stages in porcine oocytes collected from follicles of various sizes

Diameter of follicles(mm)	No. of oocytes examined	Percentage of germinal vesicle oocytes in the category of:				
		GV-X	GV-I	GV-II	GV-III	GV-IV
d<2	53	69.7	18.9	3.8	5.7	1.9
2≤d<4	48	16.7	56.3	8.3	8.3	10.4
4≤d<6	42	4.7	40.5	31.0	2.4	21.4

Germinal vesicle stage was determined by the categories GV-I to GV-IV as defined by Motlik and Fulka.[132] Oocytes regarded to be in a stage earlier than GV-I are shown as GV-X, Funahashi and Day (unpublished data).

and protein phosphorylation required to induce nuclear maturation occur during the first 12 hr of IVM.[42,43] Protein kinase A, protein kinase C and calmodulin pathways modulate protein synthesis and phosphorylation of oocytes and cumulus cells which, consequently, appear to affect nuclear maturation.[44] A transient increase in cAMP levels has been observed *in vivo* around 12 hr following hCG injection and has been induced by LH during the first 12 hr of IVM (in the presence of follicular cells, but not in the absence of follicular cells[45]). A stimulated protein kinase A pathway is known to increase the activity of the plasminogen activator[46] and the degree of intercellular coupling in porcine OCCs.[47] Further, external calcium influx may be required by porcine oocytes for nuclear maturation past metaphase I.[48]

The role of growth factors in nuclear maturation is presently under investigation. EGF,[49-52] which is present in porcine follicular fluid,[53,54] and TGF-alpha,[55] which bind to the EGF receptor, are known to enhance nuclear maturation of porcine oocytes. Gonadotropins[50], especially FSH,[51] increase the stimulating effect of EGF because specific binding of EGF to granulosa cells is increased by FSH.[56] On the other hand, TGF-beta, which has been detected in follicular fluid from small size porcine follicles,[57] is reported to inhibit porcine oocyte maturation.[55] TGF-beta also has been found to produce dose and time dependent effects on FSH induced progesterone production in porcine granulosa cells.[58] The concentration of IGF-I in porcine follicular fluid increases in a time dependent manner for 72 hr after an equine chorionic gonadotropin injection.[59] FSH appears to be an effective stimulator of IGF-I secretion by porcine granulosa cells, and LH and estradiol are facilitators.[60] The addition of IGF-I to the maturation medium does not affect nuclear maturation of porcine oocytes.[49]

3.2. Cytoplasmic Maturation

An abnormally low incidence of male pronuclear formation was observed frequently in earlier studies when IVM oocytes were fertilized *in vitro*.[8,34,37] The frequency of this failure in development has been reduced by modifications of IVM conditions.[2-6] The acquisition of factors which are needed for male pronuclear formation (and occasionally early embryonic development) frequently is defined to be cytoplasmic maturation and is distinguished from nuclear maturation, which may measure only a continuation of the meiotic processes of oocyte chromatin. The ability of porcine oocytes to form a male pronucleus has been known to be affected by hormonal levels,[34,40,61] follicular secretions,[10,37,38,62-67] intracellular ionic strength[11,68,69] and supplementation of media with cysteine.[70-72] In addition, recent studies[68,70] have indicated that the incidence of male pronuclear formation is associated with oocyte glutathione content at the end of maturation. Glutathione appears to be associated with a required reduction in the disulfide bond cross linking of sperm protamine to initiate sperm decondensation.[73] Oocyte glutathione content is increased during oocyte maturation when cysteine is added to the maturation medium; whereas, the content is decreased without added cysteine.[70] Although the presence of cumulus cells surrounding the oocyte is required throughout IVM to maintain a high oocyte glutathione level,[71] male pronuclear formation is prevented by glutathione synthesis inhibitor only when exposed during the first 24 hr of IVM.[74] Although the mechanism of how oocyte glutathione content is regulated during maturation is not clear, increasing oocyte glutathione content by adding cysteine[70,74] and antioxidants (Funahashi and Day, unpublished observation) resulted in a high rate of male pronuclear formation of IVM oocytes following IVF (> 85%).

3.3. Sodium Concentration in Maturation Medium

Elevated NaCl concentrations in maturation medium has a detrimental effect on cytoplasmic maturation as reflected not only by histone H1 kinase activity,[75] microfilament organization[11] and glutathione content of porcine oocytes at the end of IVM, but also by the incidence of male pronuclear formation[68] and *in vitro* development[69] following IVM and IVF. Supplementation of maturation media containing relatively high NaCl levels with organic osmolytes, such as taurine and sorbitol, reduces the severity of these detrimental effects[11]. Therefore, a low concentration of NaCl or the presence of organic osmolytes in maturation medium appears to be a requirement

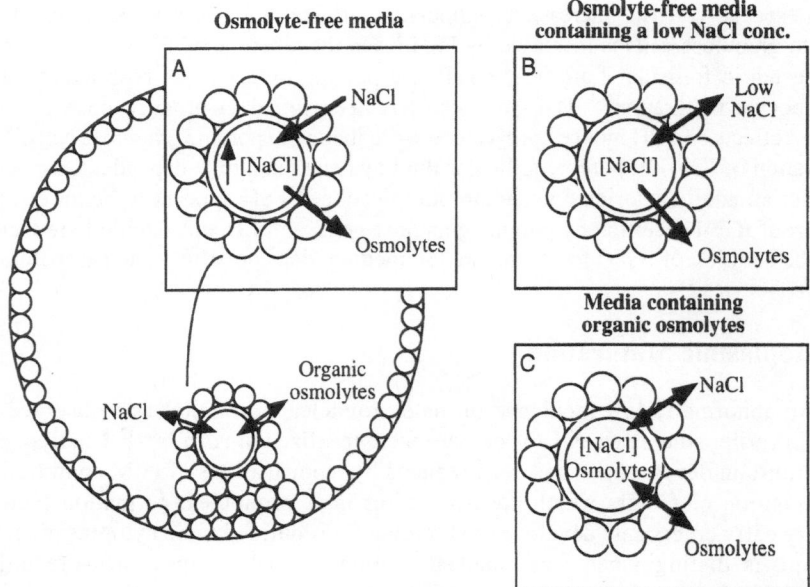

Figure 1. Proposed interactive effect of NaCl and organic osmolytes on intracellular ionic strength. When oocyte-cumulus complexes (in which intracellular ionic strength has been balanced with extracellular NaCl and organic osmolytes in follicles) are transferred to osmolyte free media (A), intracellular ionic strength is increased by escape of the osmolytes and by influx of salts. However, the increase of intracellular ionic strength may not occur in the oocytes when OCCs are transferred to osmolyte free media containing a low NaCl concentration (B) or when transferred to media containing organic osmolytes (C).

for normal oocyte metabolism, and consequently for achievement of cytoplasmic maturation in porcine oocytes (Fig. 1).

4. IVF CONDITIONS AND SPERM PENETRATION

A high incidence of polyspermic penetration was observed when porcine oocytes were inseminated *in vitro*;[8-10,76] whereas, monospermic penetration occurs in over 95% of the oocytes fertilized *in vivo*.[77,78] Polyspermic fertilization appears to be prevented by two major mechanisms: (1) the female reproductive tract restricts the number of sperm arriving simultaneously at the surface of oocytes[79,80] and (2) the direct response of oocytes to sperm penetration, including cortical and zona reactions.[81-83]

4.1. Number of Spermatozoa Reaching Oocytes

At the time of insemination *in vitro*, an increased number of spermatozoa per oocyte is known to be associated with a higher incidence of polyspermic penetration.[84,85] In the current IVF conditions, oocytes may be exposed to several waves of capacitated spermatozoa after insemination.[6] A reduced number of capacitated spermatozoa that can attach to the zona pellucida of oocytes appears to be needed at the time of IVF to prevent polyspermy. The presence of macromolecules secreted from oviductal epithelial cells[86,87] and a low concentration of follicular fluid[88] in the fertilization medium reduce the incidence of polyspermic penetration in oocytes. Oviductal and follicular sulfate proteoglycans in preincubation and/or fertilization media likely stimulate the rate of sperm capacitation and acrosome reaction, compete with sperm receptors for binding to zona ligands, and consequently reduce the number of spermatozoa attaching to the surface of oocytes.[3] The duration of coincubation of porcine oocytes matured *in vivo* with spermatozoa is reported to affect not only the rate of sperm penetration but also the frequency of polyspermic penetration.[89]

4.2. Oocyte Reaction to Sperm Penetration

During fertilization in mammals, exocytosis of the cortical granule contents containing hydrolytic enzymes and saccharide components to the perivitelline space leads to the formation of a cortical granule envelope,[90] modification of zona proteins and the inactivation of sperm receptors.[81,91,92] However, the cortical reaction mechanism may not be sufficiently effective in porcine IVF systems to prevent polyspermy. Sperm penetration is not blocked after electrical activation of oocytes.[76,93] The number of spermatozoa in a penetrated oocyte at 12 hr after IVF is lower when ovulated oocytes are used as compared to IVM oocytes.[76] On the other hand, the sperm receptor activity of the zona pellucida of IVM oocytes is reduced during IVF,[94] and porcine IVM oocytes release all cortical granules following electrical stimulation[95] or by microinjection of a G-protein stimulator, guanosine-5'-O-(3'-thiotriphosphate).[96] Cortical granule constituents in the perivitelline space of porcine IVM-IVF ova disperse with time following sperm penetration.[97] It has been suggested that an increase in the width of the perivitelline space reduces the incidence of polyspermic penetration.[68] Further, oviductal secretory proteins have been detected within the zona pellucida, perivitelline space and plasma membrane of oviductal porcine oocytes and embryos[98] which may affect sperm penetration. Also, it has been demonstrated that the proteolytic digestion of zona pellucida is delayed after oocytes and embryos have been exposed to the oviduct.[99] Recently, we showed an effect of oviductal fluid on the exocytosis of cortical granules following IVF.[100]

5. CULTURE OF PORCINE EMBRYOS

5.1. Culture of *in vivo* Porcine Embryos

An *in vitro* block in development of swine embryos occurs at the 4 cell stage in some media, which also is the time of transition from maternal to embryonic control of early development.[101,102] As reviewed by Davis[103] and then Petters and Wells,[104] recent technological progress in culture of 1 cell porcine embryos matured and fertilized *in vivo* has been achieved by the use of simple media.[105,106] Successful blastocyst formation from 1 and 2 cell porcine embryos occurs when cultured in Whitten's medium containing a high BSA concentration (1.5%) in an atomosphere of 5% O_2, 5% CO_2 and 90% N_2[105,107,108] or 5% CO_2 in air,[109] and in NCSU23 and NCSU37 media in an atomosphere of 5% CO_2 in air.[104,106] Major differences in formulation between Whitten's medium and NCSU media, except for BSA concentration, is the presence or absence of organic osmolytes, lactate and pyruvate, and the concentration of sodium. Detrimental effects of high sodium concentration in culture media and beneficial effects of organic osmolytes also have been observed in systems used for culture of mouse embryos.[110-114] Available methods for the successful culture of porcine embryos to the blastocyst stage have made it possible to examine the developmental ability of IVM/IVF porcine embryos.

5.2. Culture of IVM/IVF Embryos

IVM/IVF embryos develop to the blastocyst stage when cultured in Whitten's medium and NCSU23 medium,[11,12,14,69,115] amniotic fluid from chick embryos[13] or a coculture system.[116] The ability of IVM/IVF embryos to develop to the blastocyst stage *in vitro* has been improved mainly by modification of IVM conditions.[11,12,69] However, the incidence (3 to 12%) is still low compared with that of *in vivo* produced embryos.[104] Since a high blastocyst rate has been achieved in the culture of *in vivo* embryos, inadequacies existing in IVM systems rather than embryo culture systems appear to be the major cause of the reduced rate of early embryonic development of IVM/IVF embryos. Recently, Funahashi and Day[14] demonstrated that preincubation of OCCs for 12 hr before exposing to gonadotropins enhances the development of IVM/IVF embryos to the blastocyst stage (23%). Further investigation on preparation of OCCs and on the IVM system will be required to improve the developmental ability of IVM/IVF embryos.

There is also a lower number of cells in the trophectoderm of IVF embryos as compared to those produced *in vivo*.[117] Recently, adding IGF-I to culture medium has been shown to increase the cleavage rate of IVM/IVF embryos.[118] Exogenous IGFs, which are known to bind to trophectderm,[119] have a stimulating effect on metabolism and cellular proliferation in porcine blastocysts.[120] Modification of culture conditions for IVM/IVF embryos likely will be improved with an increased understanding of the regulation of early embryonic development by growth factors.

6. EMBRYO TRANSFER AND STORAGE

6.1. Transfer of Porcine Embryos

Traditionally, embryo transfer in pigs has been accomplished surgically with a high pregnancy rate.[121] As indicated previously, piglets from IVM/IVF oocytes have been produced in a few laboratories.[11,15,16] In our laboratory, transfer of IVM/IVF porcine embryos frequently has resulted in extended estrous cycles with apparent failure of pregnancy during later stages of gestation. Inappropriate concentration or timing of administration of hormones during IVM may be major causes of embryonic and fetal mortality. This hypothesis is supported by *in vivo* results which indicate that embryo survival is related to the interval between the peak concentration of estradiol and LH.[32,122,123] Recently, a set of instruments for nonsurgical embryo transfer has been developed to deliver porcine embryos into a single uterine horn.[124]

6.2. Storage of Porcine Embryos

Porcine embryos at the cleavage to blastocyst stages are known to be sensitive to temperatures below 15C;[121] whereas, blastocysts at the perihatching stages are more resistant to chilling.[125,126]

The birth of piglets from frozen and thawed embryos has been reported from the perihatching stages.[127-129] Recently, Nagashima *et al.*[130,131] reported that the sensitivity of porcine embryos to lowered temperatures is related to their cytoplasmic lipid content, and they obtained piglets from 2 to 4 cell porcine embryos by removing the cytoplasmic lipid and then freeze thawing. Further improvement in the survival rate of porcine embryos following freeze thawing is needed for commercial use of IVM/IVF embryos.

7. SUMMARY

Successful development of systems for culture of swine embryos to the blastocyst stage has stimulated research on the utilization of follicular oocytes for *in vitro* production of embryos. Recent progress in overcoming problems associated with oocyte maturation and fertilization *in vitro* has resulted in acceptable rates for production of blastocysts that may be used in biomedical research. However, further research is needed to obtain enhanced efficiency of *in vitro* embryo production. Establishing the culture system for preantral follicles will increase the number of oocytes available from individual animals for *in vitro* production of embryos. Reducing the heterogeneity of oocyte cumulus complexes by resolving the required interactions among oocytes, companion cells and culture environment is needed to improve the quality of oocytes selected for culture. Further, improved culture conditions during IVM and reduction of the incidence of polyspermic penetration is needed to increase the developmental rate of IVM/IVF embryos.

8. ACKNOWLEDGMENTS

This manuscript was prepared while supported by NRI Competitive Grants Program/USDA 93-37203-9186 and 94-37203-1087 and is a contribution from the Missouri Agriculture Experiment Station. Journal Series Number 12,419. We thank Mrs. Betty Nichols for secretarial assistance with the preparation of this manuscript.

9. REFERENCES

1. Gosden, R.G., and Telfer, E., 1987, Number of follicles and oocytes in mammalian ovaries and their allometric relationships, *J. Zool., Lond.* 211: 169-175.
2. Funahashi, H., and Day, B.N., 1994, In vitro maturation/in vitro fertilization of porcine oocytes, *Proceeding of the Annual Conference of the Society for Theriogenology* 206-214.

3. Day, B.N., and Funahashi, H., 1996. In vitro maturation and fertilization of pig oocytes, Beltsville Symposium in Agriculture Research XX, Biotechnology's Role in the Genetic Improvement of Farm Animals (Miller, R.H., Pursel, V.G., and Normal, H.D., eds.), American Society of Animal Science, IL, pp. 125-144.

4. Niwa, K., 1993, Effectiveness of *in vitro* maturation and *in vitro* fertilization techniques in the pig, *J. Reprod. Fertil. Suppl.* 48: 49-59.

5. Nagai, T., 1994, Current status and perspectives in IVM-IVF of porcine oocytes, *Theriogenology* 41: 73-78.

6. Mattioli, M., 1994, Recent acquisitions in pig oocyte maturation and fertilization in vitro, *Reprod. Dom. Anim.* 29: 346-348.

7. Sirard, M.A., Dubuc, A., Bolamba, D., Zheng, Y., and Coenen, K., 1993, Follicle-oocyte-sperm interactions in vivo and in vitro in pigs, *J. Reprod. Fertil. Suppl.* 48: 3-16.

8. Nagai, T., Niwa, K., and Iritani, A., 1984, Effects of sperm concentration during preincubation in a defined medium on fertilization *in vitro* of pig follicular oocytes, *J. Reprod. Fertil.* 70: 271-275.

9. Cheng, W.T.K., Polge, C., and Moor, R.M., 1986, In vitro fertilization of pig and sheep oocytes, *Theriogenology* 25: 146 (abstr.).

10. Mattioli, M., Galeati, G., and Seren, E., 1988, Effect of follicle somatic cells during pig oocyte maturation on egg penetrability and male pronucleus formation, *Gamete Res.* 20: 177-183.

11. Funahashi, H., Kim, N.-H., Stumpf, T.T., Cantley, T.C., and Day, B.N., 1995, Presence of organic osmolytes in maturation medium enhances cytoplasmic maturation of porcine oocytes, *Biol. Reprod.* 54: 1412-1419.

12. Grupen, C.G., Nagashima, H., and Nottle, M.B., 1995, Cysteamine enhances in vitro development of porcine oocytes matured and fertilized in vitro, *Biol. Reprod.* 53: 173-178.

13. Ocampo, M.B., Ocampo, L.C., Mori, T., Ueda, J., Shimizu, H., and Kanagawa, H., 1994, Blastocyst formation of pig embryos derived from in vitro fertilization of in vitro matured pig oocytes in the amniotic fluid of a developing chick embryo, *Anim. Reprod. Sci.* 37: 65-73.

14. Funahashi, H., Cantley, T.C., and Day, B.N., 1996, Preincubation of oocyte-cumulus complexes before expose to gonadotropins improves the developmental ability of porcine embryos matured and fertilized *in vitro*. Theriogenology (In Press).

15. Mattioli, M., Bacci, M.L., Galeati, G., and Seren, E., 1989, Developmental competence of pig oocytes matured and fertilized *in vitro*, *Theriogenology* 31: 1201-1207.

16. Yoshida, M., Mizoguchi, Y., Ishigaki, K., Kojima, T., and Nagai, T., 1993, Birth of piglets derived from *in vitro* fertilization of pig oocytes matured *in vitro*, *Theriogenology* 39: 1303-1311.

17. Motlik, J., Crozet, N., and Fulka, J., 1984, Meiotic competence in vitro of pig oocytes isolated from early antral follicles, *J. Reprod. Fertil.* 72: 323-328.

18. Christmann, L., Jung, T., and Moor, R.M., 1994, MPF components and meiotic competence in growing pig oocytes, *Mol. Reprod. Dev.* 38: 85-90.

19. Motlik, J., and Fulka, J., 1986, Factors affecting meiotic competence in pig oocytes, *Theriogenology* 25: 87-97.

20. Moor, R.M., Mattioli, M., Ding, J., and Nagai, T., 1990, Maturation of pig oocytes *in vivo* and *in vitro*, *J. Reprod. Fertil. Suppl.* 40: 197-210.

21. Lazzari, G., Galli, C., and Moor, R.M., 1994, Functional changes in the somatic and germinal compartments during follicle growth in pigs, *Anim. Reprod. Sci.* 35: 119-130.

22. Petr, J., Tepla, O., Grocholova, R., and Jilek, F., 1994, Inhibition of meiotic maturation in growing pig oocytes by factor(s) from cumulus cells, *Reprod. Nurt. Dev.* 34: 149-156.

23. Petr, J., Rozinek, J., Fulka, J.J., and Jilek, F., 1994, Influence of cytoplasmic microinjection on meiotic competence in growing pig oocytes, *Reprod. Nurt. Dev.* 34: 81-87.

24. Price, C.A., Carriere, P.D., Bhatia, B., and Groome, N.P., 1995, Comparison of hormonal and histological changes during follicular growth, as measured by ultrasonography, in cattle, *J. Reprod. Fertil.* 103: 63-68.

25. Hirao, Y., Nagai, T., Kubo, M., Miyano, T., Miyake, M., and Kato, S., 1994, *In vitro* growth and maturation of pig oocytes, *J. Reprod. Fertil.* 100: 333-339.

26. Grant, S.A., Hunter, M.G., and Foxcroft, G.R., 1989, Morphological and biochemical characteristics during ovarian follicular development in the pig, *J. Reprod. Fertil.* 86: 171-183.

27. Hunter, M.G., and Wiesak, T., 1990, Evidence for and implications of follicular heterogeneity in pigs, *J. Reprod. Fertil. Suppl.* 40: 163-177.

28. Hunter, R.H.F., and Baker, T.G., 1975, Development and fate of porcine graafian follicles identified at different stages of the oestrous cycle, *J. Reprod. Fertil.* 43: 193-196.

29. Hunter, M.G., Grant, S.A., and Foxcroft, G.R., 1989, Histological evidence for heterogeneity in the development of preovulatory pig follicles, *J. Reprod. Fertil.* 86: 165-170.

30. Wiesak, T., Hunter, M.G., and Foxcroft, G.R., 1990, Differences in follicular morphology, steroidogenesis and oocyte maturation in naturally cyclic and PMSG/hCG-treated prepubertal gilts, *J. Reprod. Fertil.* 89: 633-641.

31. Hunter, R.H.F., and Polge, C., 1966, Maturation of follicular oocytes in the pig after injection of human chorionic gonadotrophin, *J. Reprod. Fertil.* 12: 525-531.

32. Soede, N.M., Helmond, F.A., and Kemp, B., 1994, Periovulatory profiles of oestradiol, LH and progesterone in relation to oestrus and embryo mortality in multiparous sows using transrectal ultrasonography to detect ovulation, *J. Reprod. Fertil.* 101: 633-641.

33. Nagai, T., Takahashi, T., Masuda, H., Shioya, Y., Kuwayama, M., Fukushima, M., Iwasaki, S., and Hanada, A., 1988, In-vitro fertilization of pig oocytes by frozen boar spermatozoa, *J. Reprod. Fertil.* 84: 585-591.

34. Mattioli, M., Galeati, G., Bacci, M.L., and Seren, E., 1988, Follicular factors influence oocyte fertilizability by modulating the intercellular cooperation between cumulus cells and oocyte, *Gamete Res.* 21: 223-232.

35. Wang, W.H., Niwa, K., and Okuda, K., 1991, In-vitro penetration of pig oocytes matured in culture by frozen-thawed ejaculated spermatozoa, *J. Reprod. Fertil.* 93: 491-496.

36. Yoshida, M., Ishizaki, Y., and Kawagishi, H., 1990, Blastocyst formation by pig embryos resulting from in-vitro fertilization of oocytes matured *in vitro*, *J. Reprod. Fertil.* 88: 1-8.

37. Funahashi, H., and Day, B.N., 1993, Effects of different serum supplements in maturation medium on meiotic and cytoplasmic maturation of pig oocytes, *Theriogenology* 39: 965-973.

38. Zheng, Y.S., and Sirard, M.A., 1992, The effect of sera, bovine serum albumin and follicular cells on *in vitro* maturation and fertilization of porcine oocytes, *Theriogenology* 37: 779-790.

39. Ding, J., Moor, R.M., and Foxcroft, G.R., 1992, Effects of protein synthesis on maturation, sperm penetration, and pronuclear development in porcine oocytes, *Mol. Reprod. Dev.* 33: 59-66.

40. Funahashi, H., and Day, B.N., 1993, Effects of the duration of exposure to supplemental hormones on cytoplasmic maturation of pig oocytes *in vitro*, *J. Reprod. Fertil.* 98: 179-185.

41. Kameyama, Y., and Ishijima, Y., 1994, Effect of cumulus cells on in vitro maturation of pig oocytes, *Jpn. J. Fertil. Steril.* 39: 66-69.

42. Meinecke, B., and Meinecke-Tillmann, S., 1993, Effects of α-amanitin on nuclear maturation of porcine oocytes in vitro, *J. Reprod. Fertil.* 98: 195-201.

43. Jung, T., Fulka, J.J., Lee, C., and Moor, R.M., 1993, Effects of the protein phosphorylation inhibitor genistein on maturation pig oocytes in vitro, *J. Reprod. Fertil.* 98: 529-535.

44. Jung, T., Lee, C., and Moor, R.M., 1992, Effects of protein kinase inhibitors on pig oocyte maturation in vitro, *Reprod. Nurt. Dev.* 32: 461-473.

45. Mattioli, M., Galeati, G., Barboni, B., and Seren, E., 1994, Concentration of cyclic AMP during the maturation of pig oocytes *in vivo* and *in vitro*, *J. Reprod. Fertil.* 100: 403-409.

46. Kim, N.-H., and Menino, A.R.J., 1995, Effects of stimulators of protein kinase A and C and modulators of phosphorylation on plasminogen activator activity in porcine oocyte-cumulus cell complexes during in vitro maturation, *Mol. Reprod. Dev.* 40: 364-370.

47. Racowsky, C., 1985, Effect of forskolin on maintenance of meiotic arrest and stimulation of cumulus expansion, progesterone and cyclic AMP production by pig oocyte-cumulus complexes, *J. Reprod. Fertil.* 74: 9-21.

48. Kaufman, M.L., and Homa, S.T., 1993, Defining a role for calcium in the resumption and progression of meiosis in the pig oocyte, *J. Exp. Zool.* 265: 69-76.

49. Reed, M.L., Estrada, J.L., Illera, M.J., and Petters, R.M., 1993, Effects of epidermal growth factor, insulin-like growth factor-I, and dialyzed porcine follicular fluid on porcine oocyte maturation in vitro, *J. Exp. Zool.* 266: 74-78.

50. Ding, J., and Foxcroft, G.R., 1994, Epidermal growth factor enhances oocyte maturation in pigs, *Mol. Reprod. Dev.* 39: 30-40.

51. Singh, B., Barbe, G.J., and Armstrong, D.T., 1993, Factors influencing resumption of meiotic maturation and cumulus expansion of porcine oocyte-cumulus cell complexes in vitro, *Mol. Reprod. Dev.* 36: 113-119.

52. Sommer, P., Rath, D., and Niemann, H., 1992, In vitro maturation of porcine oocytes in the presence of follicular granulosa cells, FSH and/or EGF, Proceedings of the 12th International Congress on Animal Reproduction, pp. 378-380.

53. Hsu, C.-J., Holmes, S.D., and Hammond, J.M., 1987, Ovarian epidermal growth factor-like activity. Concentrations in porcine follicular fluid during follicular enlargement, *Biochem. Biophys. Res. Commun.* 47: 242-247.

54. Feng, P., Knecht, M., and Catt, K., 1987, Hormonal control of epidermal growth factor receptors by gonadotropins during granulosa cell differentiation, *Endocrinology* 120: 1121-1126.

55. Coskun, S., and Lin, Y.C., 1994, Effects of transforming growth factors and activin-A on in vitro porcine oocyte maturation, *Mol. Reprod. Dev.* 38: 153-159.

56. Fujinaga, H., Yamoto, M., Nakano, R., and Shima, K., 1992, Epidermal growth factor binding site in porcine granulosa cells and their regulation by follicle-stimulating hormone, *Biol. Reprod.* 46: 705-709.

57. Gangrade, B.K., and May, J.V., 1990, The production of transforming growth factor-ß in the porcine ovary and its secretion in vitro, *Endocrinology* 127: 2372-2380.

58. Chang, W.Y., Ohmura, H., Kulp, S.K., and Lin, Y.C., 1993, Transforming growth factor-ß1 regulates differentiation of porcine granulosa cells in vitro, *Theriogenology* 40: 699-712.

59. Hammond, J.M., Hsu, C.-J., Klindt, J., Tsang, B.K., and Downey, B.R., 1988, Gonadotropins increase concentrations of immunoreactive insulin-like growth factor-I in porcine follicular fluid in vivo, *Biol. Reprod.* 38: 304-308.

60. Hsu, C.-J., and Hammond, J.M., 1986, Gonadotropins and estradiol stimulate immunoreactive IGF-I production by porcine granulosa cells in vitro, *Endocrinology* 120: 198-207.

61. Funahashi, H., Cantley, T.C., and Day, B.N., 1994, Different hormonal requirement of porcine oocyte-complexes during maturation *in vitro*, *J. Reprod. Fertil.* 101: 159-165.

62. Naito, K., Fukuda, Y., and Toyoda, Y., 1988, Effects of porcine follicular fluid on male pronucleus formation in porcine oocytes matured *in vitro*, *Gamete Res.* 21: 289-295.

63. Mattioli, M., Bacci, M.L., Galeati, G., and Seren, E., 1991, Effects of LH and FSH on the maturation of pig oocytes in vitro, *Theriogenology* 36: 95-105.

64. Yoshida, M., Ishizaki, Y., Kawagishi, H., Bamba, K., and Kojima, Y., 1992, Effects of pig follicular fluid on maturation of pig oocytes *in vitro* and on their subsequent fertilizing and developmental capacity *in vitro*, *J. Reprod. Fertil.* 95: 481-488.

65. Nagai, T., Ding, J., and Moor, R.M., 1993, Effect of follicle cells and steroidogenesis on maturation and fertilization *in vitro* of pig oocytes, *J. Exp. Zool.* 266: 146-151.

66. Ding, J., and Foxcroft, G.R., 1994, FSH-stimulated follicular secretions enhanced oocyte maturation in pigs, *Theriogenology* 41: 1473-1481.

67. Ding, J., and Foxcroft, G.R., 1992, Follicular heterogeneity and oocyte maturation *in vitro* in pigs, *Biol. Reprod.* 47: 648-655.

68. Funahashi, H., Cantley, T.C., Stumpf, T.T., Terlouw, S.L., and Day, B.N., 1994, Use of low salt culture medium for in vitro maturation of porcine oocytes is associated with elevated oocyte glutathione levels and enhanced male pronuclear formation after in vitro fertilization, *Biol. Reprod.* 51: 633-639.

69. Funahashi, H., Cantley, T.C., Stumpf, T.T., Terlouw, S.L., and Day, B.N., 1994, *In vitro* development of *in vitro* matured porcine oocytes following chemical activation or *in vitro* fertilization, *Biol. Reprod.* 50: 1072-1077.

70. Yoshida, M., Ishigaki, K., Nagai, T., Chikyu, M., and Pursel, V.G., 1993, Glutathione concentration during maturation and after fertilization in pig oocytes: relevance to the ability of oocytes to form male pronucleus, *Biol. Reprod.* 49: 89-94.

71. Funahashi, H., and Day, B.N., 1995, Effects of cumulus cells on glutathione content of porcine oocyte during in vitro maturation, *J. Anim. Sci.* 73 (Suppl. 1): 90 (abstr.).

72. Yoshida, M., Ishigaki, K., and Pursel, V.G., 1992, Effect of maturation media on male pronucleus formation in pig oocytes matured *in vitro*, *Mol. Reprod. Dev.* 31: 68-71.

73. Perreault, S.D., 1990, Regulation of sperm nuclear reactivation during fertilization, in: *Fertilization in mammals* (Bavister, B.D., Cummins, J., and Roldan, E.R.S., eds.), Serono Symposia USA, Norwell, pp. 285-296.

74. Yoshida, M., 1993, Role of glutathione in the maturation and fertilization of pig oocytes *in vitro*, *Mol. Reprod. Dev.* 35: 76-81.

75. Funahashi, H., Stumpf, T.T., Kim, N.H., and Day, B.N., 1994, Effects of sodium chloride concentration in maturation media on histone H1 kinase (H1K) activity and cytoplasmic maturation of porcine oocytes, *J. Reprod. Fertil.* 38 (abstr.).

76. Funahashi, H., Stumpf, T.T., Terlouw, S.L., and Day, B.N., 1993, Effects of electrical stimulation before or after *in vitro* fertilization on sperm penetration and pronuclear formation in pig oocytes, *Mol. Reprod. Dev.* 36: 361-367.

77. Hunter, R.H.F., 1967, The effects of delayed insemination on fertilization and cleavage in the pig, *J. Reprod. Fertil.* 13: 133-147.

78. Hunter, R.H.F., 1972, Local action of progesterone leading to polyspermic fertilization in pigs, *J. Reprod. Fertil.* 31: 433-444.

79. Hunter, R.H.F., 1990, Fertilization of pig eggs *in vitro* and *in vivo*, *J. Reprod. Fertil. Suppl.* 40: 211-226.

80. Hunter, R.H.F., 1991, Oviduct function in pigs, with particular reference to the pathological condition of polyspermy, *Mol. Reprod. Dev.* 29: 385-391.

81. Wassarman, P.M., 1988, Zona pellucida glycoproteins, *Ann. Rev. Biochem.* 57: 415-442.

82. Jones, R., 1990, Identification and functions of mammalian sperm-egg recognition molecules during fertilization, *J. Reprod. Fertil. Suppl.* 42: 89-105.

83. Cran, D.G., and Esper, C.R., 1990, Cortical granules and the cortical reaction in mammals, *J. Reprod. Fertil. Suppl.* 42: 177-188.

84. Rath, D., 1992, Experiments to improve in vitro fertilization techniques for in vivo-matured porcine oocytes, *Theriogenology* 37: 885-896.

85. Coy, P., Martinez, E., Ruiz, S., Vazquez, J.M., Roca, J., and Gadea, J., 1993, Environment and medium volume influence in vitro fertilization of pig oocytes, *Zygote* 1: 209-213.

86. Kano, K., Miyano, T., and Kato, S., 1994, Effect of oviductal epithelial cells on fertilization of pig oocytes in vitro, *Theriogenology* 42: 1061-1068.

87. Nagai, T., and Moor, R.M., 1990, Effect of oviduct cells on the incidence of polyspermy in pig eggs fertilized *in vitro*, *Mol. Reprod. Dev.* 26: 377-382.

88. Funahashi, H., and Day, B.N., 1993, Effects of follicular fluid at fertilization *in vitro* on sperm penetration in pig oocytes, *J. Reprod. Fertil.* 99: 97-103.

89. Coy, P., Martinez, E., Ruiz, B., Vazquez, J.M., Roca, J., Matas, C., and Pellicer, M.T., 1993, In vitro fertilization of pig oocytes after different coincubation intervals, *Theriogenology* 39: 1201-1208.

90. Dandekar, P., and Talbot, P., 1992, Perivitelline space of mammalian oocytes: Extracellular matrix of unfertilized oocytes and formation of a cortical granule envelope following fertilization, *Mol. Reprod. Dev.* 31: 135-143.

91. Gulyas, B.J., 1980, Cortical granules of mammalian eggs, *Int. Rev. Cytol.* 63: 357-392.

92. Wassarman, P.M., 1990, Regulation of mammalian fertilization by zona pellucida glycoproteins, *J. Reprod. Fertil. Suppl.* 42: 79-87.

93. Funahashi, H., Stumpf, T.T., Cantley, T.C., Kim, N.-H., and Day, B.N., 1995, Pronuclear formation and intracellular glutathione content of in vitro-matured porcine oocytes following in vitro fertilization and/or electrical activation, *Zygote* 3: 273-281.

94. Hatanaka, Y., Nagai, T., Tobita, T., and Nakano, M., 1992, Changes in the properties and composition of zona pellucida of pigs during fertilization *in vitro*, *J. Reprod. Fertil.* 95: 341-440.

95. Sun, F.Z., Hoyland, J., Huang, X., Mason, W., and Moor, R.M., 1992, A comparison of intercellular changes in porcine eggs after fertilization and electroactivation. *Development* 115: 947-956.

96. Machaty, Z., Mayes, M.A., and Prather, R.S., 1995, Parthenogenetic activation of porcine oocytes with guanosine-5'-O-(3'-thiotriphosphate), *Biol. Reprod.* 52: 753-758.

97. Yoshida, M., Cran, D.G., and Pursel, V.G., 1993, Confocal and fluorescence microscopic study using lectins of the distribution of cortical granules during the maturation and fertilization of pig oocytes, *Mol. Reprod. Dev.* 36: 462-468.

98. Buhi, W.C., O'Brien, B., Alvarez, I.M., Erdos, G., and Dubois, D., 1993, Immunogold localization of porcine oviductal secretory proteins within the zona pellucida, perivitelline space, and plasma membrane of oviductal and uterine oocytes and early embryos, *Biol. Reprod.* 48: 1274-1283.

99. Broermann, D.M., Xie, S., Nephew, K.P., and Pope, W.F., 1989, Effects of the oviduct and wheat germ agglutinin on enzymatic digestion of porcine zona pellucidae, *J. Anim. Sci.* 67: 1324-1329.

100. Kim, N.-H., Funahashi, H., Abeydeera, L.R., Moon, S.J., Prather, R.S., and Day, B.N., 1996, Effects of oviductal fluid on sperm penetration and cortical granule exocytosis during in vitro fertilization of porcine oocytes, *J. Reprod. Fertil.* 107: 79-86.

101. Polge, C., and Frederick, C.L., 1968, Culture and strage of fertilized pig eggs, Proc. 6th Int. Congr. Anim. Reprod. & A.I., Paris, pp. 211 (abstr.).

102. Rundell, J.M., and Vincent, C.K., 1968, In vitro culture of swine ova, *J. Anim. Sci.* 27: 1196 (abstr.).

103. Davis, D., 1985, Culture and strage of pig embryos, *J. Reprod. Fertil. Suppl.* 38: 115-124.

104. Petters, R.M., and Wells, K.D., 1993, Culture of pig embryos, *J. Reprod. Fertil. Suppl.* 48: 61-73.

105. Beckmann, L.S., and Day, B.N., 1993, Effect of media NaCl concentration and osmolarity on culture of the early stage porcine embryo and viability of embryos cultured in a selected superior medium, *Theriogenology* 39: 611-622.

106. Petters, R.M., and Reed, M.L., 1991, Addition of taurine or hypotaurine to culture medium improves development of one- and two-cell pig embryos, *Theriogenology* 35: 253 (abstr.).

107. Lindner, G.M., and Wright, R.W.J., 1978, Morphological and quantitative aspects of the development of swine embryos in vitro, *J. Anim. Sci.* 46: 711-718.

108. Menino, A.R.J., and Wright, R.W.J., 1982, Development of one-cell porcine embryos in two culture systems, *J. Anim. Sci.* 54: 583-588.

109. Galvin, J.M., Stewart, A.N.V., and Meredith, S., 1993, Higher sodium chloride concentration can induce a four-cell block in porcine embryos, *Theriogenology* 39: 224 (Abstr.).

110. Biggers, J.D., Lawitts, J.A., and Lechene, C.P., 1993, The protective action of betaine on the deleterious effects of NaCl on preimplantation mouse embryos *in vitro*, *Mol. Reprod. Dev.* 34: 380-390.

111. Lawitts, J.A., and Biggers, J.D., 1991, Overcoming the 2-cell block by modifying components in a mouse embryo culture medium, *Biol. Reprod.* 45: 245-251.

112. Lawitts, J.A., and Biggers, J.D., 1992, Joint effects of sodium chloride, glutamine, and glucose in mouse preimplantation embryo culture media, *Mol. Reprod. Dev.* 31: 189-194.

113. Anbari, K., and Schultz, R.M., 1993, Effect of sodium and betaine in culture media on development and relative rates of protein synthesis in preimplantation mouse embryos *in vitro*, *Mol. Reprod. Dev.* 35: 24-28.

114. Ho, Y., Doherty, A.S., and Schultz, R.M., 1994, Mouse preimplantation embryo development in vitro: Effect of sodium concentration in culture media on RNA synthesis and accumulation and gene expression, *Mol. Reprod. Dev.* 38: 131-141.

115. Funahashi, H., Stumpf, T.T., Terlouw, S.L., Cantley, T.C., Rieke, A., and Day, B.N., 1994, Developmental ability of porcine oocytes matured and fertilized in vitro, *Theriogenology* 41: 1425-1433.

116. Nagai, T., and Takahashi, M., 1992, Culture of in vitro matured and fertilized pig oocytes, *12th Int. Cong. Anim. Reprod.* 3: 1324-1326.

117. Rath, D., Niemann, H., Tao, T., and Boerjan, M., 1995, Ratio and number of inner cell mass and trophoblast cells of in vitro and in vivo produced porcine embryos, *Theriogenology* 43: 304 (abstr.).

118. Xia, P., Tekpetey, F.R., and Armstrong, D.T., 1994, Effect of IGF-I on pig oocyte maturation, fertilization, and early embryonic development in vitro, and on granulosa and cumulus cell biosynthetic activity, *Mol. Reprod. Dev.* 38: 373-379.

119. Corps, A.N., Brigstock, D.R., Littlewood, C.J., and Brown, K.D., 1990, Receptors for epidermal growth factor and insulin-like growth factor-I on preimplantation trophectoderm, *Development* 110: 221-227.

120. Lewis, A.M., Kaye, P.L., Lising, R., and Cameron, R.D.A., 1992, Stimulation of protein synthesis and expansion of pig blastocysts by insulin in vitro, *Reprod. Fertil. Dev.* 4: 119-123.

121. Polge, C., 1982, Embryo transplantation and preservation, in: *Control of Pig Reproduction* (Cole, D.J.A., and Foxcroft, G.R., eds.), Butterworth Scientific, London, pp. 277-291.

122. Blair, R.M., Coughlin, C.M., Minton, J.E., and Davis, D.L., 1993, Embryonic survival and variation in embryonic development in gilts and primiparous sows on day 11 of gestation, Proceedings of the 4th International Conference on Pig Reproduction, Columbia, Missouri, pp. 19 (abstr.).

123. Hunter, M.G., and Picton, H.M., 1995, Effect of hCG administration at the onset of oestrus on early embryo survival and development in Meishan gilts, *Anim. Reprod. Sci.* 38: 231-238.

124. Li, J., Rieke, A., Day, B.N., and Prather, R.S., 1995, Technical breakthrough in non-surgical porcine embryo transfer, *Biol. Reprod.* 52 (Suppl. 1): 127.

125. Nagashima, H., Yamakawa, H., and Niemann, H., 1992, Freezability of porcine blastocysts at different peri-hatching stages, *Theriogenology* 37: 839-850.

126. Niemann, H., and Reichelt, B., 1993, Manipulating early pig embryos, *J. Reprod. Fertil. Suppl.* 48: 75-94.

127. Kashiwazaki, N., Ohtani, S., Nagashima, H., Yamakawa, H., Cheng, W.T.K., Lin, A.-T., Ma, R.C.-S., and Ogawa, S., 1991, Production of normal piglets from hatched blastocysts frozen at -196 C, *Theriogenology* 35: 221 (abstr).

128. Hayashi, S., Kobayashi, K., Mizuno, J., Saitoh, K., and Hirano, S., 1989, Birth of piglets from frozen embryos, *Vet. Rec.* 125: 43-44.

129. Fujino, Y., Ujisato, Y., Endo, K., Tomizuka, T., Kojima, T., and Oguri, N., 1993, Cryoprotective effect of egg york in cryopreservation of porcine embryos, *Cryobiology* 30: 299-305.

130. Nagashima, H., Kashiwazaki, N., Ashman, R.J., Grupen, C.G., Seamark, R.F., and Nottle, M.B., 1994, Removal of cytoplasmic lipid enhances the tolerance of porcine embryos to chilling, *Biol. Reprod.* 51: 618-622.

131. Nagashima, H., Kashiwazaki, N., Ashman, R.J., Grupen, C.G., and Nottle, M.B., 1995, Cryopreservation of porcine embryos, *Nature* 374: 416.

132. Motlik, J., and Fulka, J., 1976, Breakdown of the germinal vesicle in pig oocytes in vivo and in vitro, J. Exp. Zool. 198:155.

43

REPRODUCTIVE PHYSIOLOGY IN CHINESE MEISHAN PIGS

A University of Illinois Perspective

Brett R. White, MS, JoElla Barnes, BS, and Matthew B. Wheeler, PhD[*]

Laboratory of Molecular Embryology
Department of Animal Sciences
University of Illinois at Urbana-Champaign
Urbana, Illinois 61801

1. ABSTRACT

Experiments were designed to examine: 1) age at puberty in Chinese Meishan (Ms) and Yorkshire (Y) gilts, 2) litter size in Ms and Y gilts and 3) ejaculate characteristics in Ms and Y boars. Ms gilts reached sexual maturity 105 da earlier than Y gilts and Ms gilts were in estrus almost 1 da longer than Y gilts for their first, second and third estrus. No differences were detected between breeds for cycle length on the first or second estrous cycle. Validation of the Coat A Count Progesterone RIA procedure resulted in an accurate, rapid and efficient method for quantification of porcine serum progesterone and has extended to swine a reliable, repeatable methodology to confirm visual estrous detection which allows for accurate prediction of first ovulation at puberty. Luteal phase concentrations of progesterone are greater in Meishan gilts than Y control gilts during the first 2 estrous cycles. Reciprocal cross females (Ms x Y and Y x Ms) had higher ovulation rates and numbers of fetuses than Ms and Y females, possibly due to hybrid vigor. However, length of the uterus was not influenced by their larger litter sizes suggesting that uterine capacity was not challenged at this reproductive age. It appears there is a key point for Ms females where physiological and reproductive age must coincide to maintain improved prolificacy over domestic breeds of females. Further, the mechanism of fetal survival may be different in crossbred Ms females than in purebred Ms females. Therefore, of ovulation rate, uterine and embryonic interac-

---

[*] Reprint requests to: Dr. Matthew B. Wheeler, 366 Animal Sciences Laboratory, University of Illinois, Department of Animal Sciences, 1207 West Gregory Drive, Urbana, Illinois, USA, 61801 (217) 333-2239.

Advances in Swine in Biomedical Research, edited by Tumbleson and Schook
Plenum Press, New York, 1996

tions, including uterine capacity, and reproductive age are all important in the greater prolificacy of Ms females over occidental breeds of females. There were higher concentrations of estradiol-17β (E_2) and testosterone in the seminal plasma from Ms boars compared to seminal plasma from Y boars.

2. INTRODUCTION

Chinese pigs were imported into the U.S. due to their large litter sizes (4 to 5 more piglets per litter) and early age at puberty (3 mo in Ms gilts vs 5 to 6 mo in occidental breed gilts). The University of Illinois Imported Swine Research Laboratory received 21 Ms females in July 1989 as part of a joint University of Illinois, Iowa State University, USDA/ARS importation of 65 Ms gilts and 30 Ms, 24 Fengjing (F) and 21 Minzhu boars from the Peoples' Republic of China.[1] The Ms pigs represented 10 distinct families which were unrelated back to their grandparents. A group of Y gilts was obtained from 5 purebred producers in Illinois as a control population.

By using Chinese pigs as a model, we might better understand the factors limiting prolificacy of occidental pigs. Often overshadowed in importance by litter size, the early age at puberty of Ms gilts could provide insight into mechanisms involved with puberty attainment. Much has been done to assess the role that the Ms female plays in swine reproduction; however, there are few findings as to the boar's importance in this regard. In addition, swine offer some distinct advantages over laboratory animal species such as mice and hamsters because they are immunologically and physiologically closer to humans and thus serve as a better human research model.[2]

Characterization of age at puberty in Ms and Y gilts[3] instigated experiments to investigate the functionality of early estrous cycles of Ms gilts. Previously we characterized ovulation rate, uterine length and fetal spacing parameters for Ms and Y second parity sows.[3] We initiated an experiment to investigate different genotypes of fetuses in different uterine environments as well as using Ms and Y gilts which were at a standardized reproductive age. Further, we reported semen characteristics and seminal plasma composition for Ms, F and Y boars[4]. There were also differences between breeds for levels of E_2. Therefore, we investigated levels of E_2 and testosterone in repeated collections from Ms and Y boars.

The goal of our research at the University of Illinois was to examine some of the mechanisms involved with the early age at puberty of Ms gilts and the role of ovulation rate and uterine capacity in the mechanism of prolificacy of the Ms female. We also examined the components of seminal plasma from Ms boars that may influence the female reproductive tract to increase litter size.

3. MATERIALS AND METHODS

3.1. Age at Puberty

3.1.1. Characterization of Age at Puberty. Daily estrous detection on 50 Ms and 34 Y gilts began at 60 and 120 da of age (SD = 10), respectively, to determine age at first estrus. Yorkshire gilts were not exposed to boars at the same age as Ms gilts to avoid delayed attainment of puberty, which would not be representative of the Y population. Gilts exposed too early to boars could be delayed in sexual development.[5-9] Meishan and Y gilts were housed in adjacent pens in the same environmentally controlled building and were fed the same diet. Both breeds were moved from the nursery at 50 da of age (SD=10) into the grower unit so they could adjust to a new environment and both were moved at the same age to the finishing unit. Daily boar exposure for 15 to 20 min continued until each gilt expressed 3 estrous periods.

3.1.2. Progesterone Profiles at Puberty. Six gilts from each breed (Ms and Y) were selected randomly to be bled weekly.[4] As controls for weekly blood sampling, one gilt of each breed was sampled twice daily. Blood samples were collected weekly via jugular venipuncture starting at 60 and 120 da of age for Ms and Y gilts, respectively. Blood sampling continued until each gilt had been observed in estrus during 2 successive estrous cycles. Twice daily blood samples were collected from indwelling catheters for a complete estrous cycle. Blood samples were collected in 10 ml silicone coated vacutainer tubes and allowed to clot at 4C for 1 hr. Samples were centrifuged (900 g, 15 min, 4C) and serum was collected in 5 ml plastic scintillation vials and stored at -20C. Serum concentrations of progesterone were determined by a radioimmunoassay (RIA; Coat-A-Count Progesterone, Diagnostic Products Corp., Los Angeles, CA) with the following modifications. This assay is a no extraction, solid phase ^{125}I RIA designed to measure quantitatively progesterone in human serum or plasma. Serum from 8 age matched barrows of different breeds was collected separately and stored at -20C. Progesterone (Sigma Chemical Co., St. Louis, MO) standards were prepared in serum from each barrow to contain the following amounts: 0, 0.14, 0.5, 1.0, 5.0, 10.0, 20.0, 40.0 or 50.0 ng/ml. Porcine serum progesterone standards (8 sets) were compared with human serum based progesterone standards which accompanied the kit. Standard curves which resulted from the porcine serum progesterone standards (8 sets) were identical and parallel to the standard curve generated by the human standards. Recovery of progesterone from porcine standards was 95%. Serum from the 8 barrows subsequently was pooled and a progesterone standard stock solution (100 ng/ml) was prepared. Porcine serum progesterone standards, to be used in the assay, were prepared from this standard stock solution. Serial dilutions (25, 50 and 100 µl) of a standard sera pool collected from 48 da gestating gilts were parallel to the standard curve generated by porcine serum progesterone standards. Minimum detectable concentration of progesterone was 0.14 ng/ml. Serum samples were analyzed in triplicate following Coat-A-Count Progesterone RIA procedures with the exception of porcine serum standards. Serum samples were analyzed in 3 assays with inter and intra assay coefficients of variation of 3.65 and 2.61%, respectively.

Criteria, similar to those in cattle,[10] were established to confirm age at first ovulation from progesterone hormone profiles. Gilts were considered to have ovulated and formed CL: 1 to 2 da previously if progesterone concentrations were ≥ 2 ng/ml and ≤ 6 ng/ml, 3 ± 1 da previously if progesterone concentrations were ≥ 8 ng/ml and < 10 ng/ml, approximately 5 ± 2 da previously if progesterone concentrations were > 20 ng/ml with the previous sample at less than basal concentrations (≤ 2 ng/ml), 12 ± 2 da previously if progesterone concentrations were >20 ng/ml with the previous sample at greater than basal concentrations (> 2 ng/ml), and 19 ± 2 da previously if progesterone concentrations were at basal levels (≥ 2 ng/ml) with the previous sample > 20 ng/ml. These guidelines are based on the hormonal profile of the porcine estrous cycle.[11] Analysis of variance (ANOVA) was performed on puberty traits using the general linear models (GLM) procedure of the statistical analysis system (SAS).[12] Means for puberty traits were adjusted for the effects of month farrowed. Progesterone profiles were analyzed using Cluster analysis[13] to determine width between maxima, progesterone concentration maxima and area under the curve for individual gilts; means for each breed were determined using GLM.

3.2. Mechanism of Prolificacy

3.2.1. Experiment 1. To determine the role of the uterus in development of fetuses with the same genetic background, we produced 1/2 Ms and 1/2 Y fetuses in 4 different maternal genotypes. This experimental design allowed us to examine the role of the uterus, after standardization of the fetus, in the relationship of number of fetuses and uterine length. In addition, we standardized the estrous cycle at time of breeding in both experiments to the third estrous cycle to compare number of ovulations between Ms and Y females with the same number of previous heats.

Gilts were housed in adjacent pens in the same environmentally controlled building and fed the same diet. All breeds were moved from the nursery at 50 da of age (SD = 10) into the grower unit so they could adjust to a new environment and were moved at the same age (110 da of age) to the finishing unit. Daily observations for detection of estrus for 10 Ms, 10 Ms x Y, 10 Y x Ms and 10 Y gilts began at 60, 100, 100 and 180 da of age (SD = 10), respectively, to determine age at puberty. Gilts represented 6 Ms, 6 Ms x Y, 7 Y x Ms and 5 Y families. Because sexual development can be delayed in prepubertal gilts when exposed to boars too early,[5-9] Y and crossbred gilts were not exposed to boars at the same age as Ms gilts to avoid delayed attainment of puberty which would not be representative of either population. Daily boar exposure for 15 to 20 min continued until each gilt had 3 heats. Estrus was determined when gilts stood to be mounted by the boar and exhibited the lordosis response.

At third estrus, Ms, Ms x Y, Y x Ms and Y gilts were bred by artificial insemination to Y, Ms x Y, Y x Ms and Ms sires, respectively, at 12 hr intervals each da they exhibited estrus beginning 12 hr after the onset of estrus in Ms x Y, Y x Ms and Y gilts and 24 hr after the onset of estrus in Ms gilts. Meishan gilts have been reported to ovulate approximately 12 to 14 hr later than Large White (LW) gilts relative to the onset of estrus.[14,15] Each mating was to a different boar of the appropriate breed. Gilts were slaughtered at 51 da of gestation (SD = 2); blood samples were collected in 10 ml silicone coated vacutainer tubes and allowed to clot at 4C for 1 hr. Samples were centrifuged (900 g, 15 min, 4C); serum was collected in 5 ml plastic scintillation vials and stored at -20C. Reproductive tracts (broad ligament removed) were examined 6 hr after removal from the animal to standardize conditions for measurement.[16] Ovulation rate was determined by dissecting and counting each CL. Uterine length was determined using a uterometer with measurements recorded in 1 cm increments.[16] Fetuses were counted and spacing among fetuses was measured. Space per fetus was calculated as the sum of half the distance from neighboring fetuses. For fetuses at the tip of the uterine horn, the distance from the tip and half the distance from the neighboring fetus were summed. Head-tail orientation of fetuses was determined by palpation and sex of fetuses was recorded. A 1.5 ml sample was taken from each allantoic and amnionic fluid sac with a 3 ml syringe and 18 ga needle. These samples were placed in 1.7 ml microcentrifuge tubes and stored at -70C. Allantoic and amnionic fluid volumes from each fetoplacental unit were measured separately in a 100 ml graduated cylinder. Fetal crown-rump length and weight were recorded. Finally, 2 amnionic and 2 allantoic samples were selected randomly from each uterine horn of each female to be analyzed for progesterone concentration. Serum and allantoic and amnionic fluid concentrations of progesterone were determined by RIA.

ANOVA was conducted using the GLM procedure of SAS.[12] Differences between means for each trait were compared using paired *t*-test procedures. All traits were analyzed with breed as the main effect. Gestation length was added as a covariate for the analyses of number of fetuses and number of fetuses/number of CL. Gestation length and number of fetuses nested within breed were added as covariates for the analysis of uterine length. For analysis of fetal traits, gilt nested within breed was added as a random effect and was used as the error term in calculating standard errors and probabilities. Chi square analysis was used to examine the effect of breed on sex ratio and head-tail orientation of fetuses.

3.2.2. Experiment 2. Next, we produced purebred fetuses in Ms (Ms-P) and Y (Y-P) gilts and compared these results to those determined for crossbred fetuses in Ms (Ms-X) and Y (Y-X) gilts in the first experiment. This experimental design allowed us to examine effects of different fetal genotypes in a similar uterine environment on number of fetuses and uterine length.

To determine age at puberty, daily observations for detection of estrus for 10 Ms and 10 Y gilts began when gilts reached 60 and 180 da of age (SD = 10), respectively. Gilts represented 6 Ms and 5 Y families and were treated as explained previously. At third estrus, Ms and Y gilts were bred to Ms and Y sires, respectively, as described above. Gilts were slaughtered at 51 da of gestation (SD = 2) and blood samples were taken and prepared for analysis as described above. Reproductive tracts were handled and traits were measured as explained previously. One amnionic and one allantoic sample were selected randomly from each female to be analyzed for progesterone concentration. Serum and allantoic and amnionic fluid concentrations of progesterone were determined by RIA.

ANOVA was conducted using the GLM procedure of SAS.[12] All traits were analyzed similarly to Experiment 1 with the following modifications. Breed of gilt, genotype of fetus and their interaction were included as main effects. In addition, gilt nested within the interaction between breed of gilt and genotype of fetus was added as a random effect and was used as the error term in calculating standard errors and probabilities for analysis of fetal traits.

3.3. Seminal Plasma

3.3.1. Experiment 1. Six sexually mature F and 12 Ms boars ranging in age from 6 to 34 mo and 6 Y boars at 10 mo of age were used for semen collection. All boars were housed in an environmentally controlled building in individual sow gestation crates under artificial lighting. The boars were limit fed a balanced diet and water was available *ad libitum*.

Boars were collected using the gloved hand technique either on a sow in standing estrus or on a phantom mount. All boars were allowed a minimum of 3 da sexual rest between collections. All fractions of the semen were collected into clean styrofoam cups and only complete collections were analyzed. The whole semen was transported to the laboratory in an insulated thermos. The gel fraction was separated from the semen by filtration through cheese cloth. The gel and gel free semen volumes were measured in a 1000 ml graduated cylinder. A 1 ml aliquot of gel free semen was frozen at -20C for E_2 analysis.

Seminal plasma concentrations of E_2 were determined by Coat-A-Count RIA (Diagnostic Products Corporation, Los Angeles, CA) with the following modifications. Coat-A-Count E_2 is a no extraction, solid phase ^{125}I RIA designed to measure quantitatively E_2 in human serum or plasma. The antibody is immobilized to the wall of polypropylene tubes and the antibody bound fraction of ^{125}I labelled E_2 is separated from the free fraction by decanting the supernatant. The tubes are then counted for 1 min in a gamma counter (Gamma 5500; Beckman Instruments, Inc. Fullerton, CA). Estradiol (1,3,5 [10]-estratriene-3, 17β-diol; Sigma Chemical Co., St. Louis, MO) stock solutions were prepared in 0.1% gelatin phosphate buffered saline (PBSG, 0.1M, pH 7.5). An E_2 standard stock solution (1pg/10μl) was prepared in PBSG, Dulbecco's modified Eagle's medium (DMEM) and a barrow sera pool. Serial dilutions (6.25, 12.5, 25, 50 and 100 μl) of the PBSG, DMEM and barrow sera E_2 standard stock solutions were identical to and parallel to the standard curve generated by human standards. Recovery of E_2 from the standard stock solutions was 95%. Seminal plasma samples were analyzed in duplicate and standard Coat-A-Count E_2 RIA procedures were followed. All seminal plasma samples were analyzed in one assay. Minimum detectable concentration of E_2 was 2.16 pg/ml and intraassay coefficient of variation was 2.76%. The antibody used was specific for E_2 with a cross reactivity to other estrogens of < 1%. Crossreactivity of the antibody to androgens was < 0.01%. This system maintains an interassay variation of < 10% and an intraassay variation of < 5%.

The E_2 concentration data were transformed logarithmically to reduce inequality of variances between breeds.[17] ANOVA was conducted using the GLM procedure of SAS[12]. Trait means for E_2 were analyzed with a model that included breed and boar nested within breed as fixed effects and age of boar and collection date as covariates. Differences were determined using the t-test for pairwise significant differences.

3.3.2. Experiment 2. We wished to determine if repeated collections influenced E_2 concentrations. Prior to selection of boars, all Ms and Y boars were screened for high and low E_2 concentrations. Once these parameters were determined, 4 mature boars from each breed were selected, 2 with high concentrations and 2 with low concentrations of E_2. The 4 Ms boars were an average age of 469 da with a range of 350 to 560 da of age. The 4 Y boars averaged 398 da of age with a range of 278 to 560 da of age. Boars were housed in the same environment during and for 6 mo prior to the experiment. Animals were limit fed a balanced diet with water available *ad libitum*. Semen was collected by gloved hand technique from each boar over 2 wk with at least 48 hr sexual rest. Boars had been trained previously to allow semen collection from a phantom. The ejaculate was collected into a styrofoam cup and filtered through cheesecloth, separating the gel from the gel free fraction. Measurement of gel free volume was performed using a 1000 ml graduated cylinder.

Next, a drop of semen was put on a slide and examined under a microscope to determine progressive motility. The number of forwardly motile cells divided by the number of total cells was defined as the percent progressively motile. Aliquots of whole semen were frozen at -20C. One of the aliquots of semen was thawed at room temperature and diluted 1:100 with physiological saline prior to counting sperm on the hemocytometer. The second aliquot of semen was thawed and sperm were pelleted by centrifugation. Seminal plasma was used for determination of protein, E_2 and testosterone concentrations. Seminal plasma was diluted 1:50 and protein was determined spectrophotometrically with a bicinchoninic assay (Pierce Chem. Co., Rockford, IL). Estradiol-17ß concentrations were determined by RIA as described above.

Testosterone concentrations were determined by a solid phase RIA (Coat-A-Count Testosterone; Diagnostic Products Corporation, Los Angeles, CA) previously validated in swine.[18] Determinations were made on 100 µl aliquots of seminal plasma. Assay sensitivity was 0.04 ng testosterone/ml of seminal plasma. The antibody used was specific for testosterone with a cross reactivity to estrogens of < 1% and to androgens of < 0.3%. Seminal plasma samples were diluted to fall within the linear portion of the standard curve and assayed in duplicate. Inter and intraassay variations were < 13 and < 5%, respectively.

E_2 concentration data were transformed logarithmically to reduce inequality of variances among breeds.[17] ANOVA was conducted using the GLM procedure of SAS.[12] The effects of sample were analyzed by ANOVA with repeated measures. Means for all traits were analyzed with a model that included breed and boar nested within breed as main effects.[1]

4. RESULTS AND DISCUSSION

4.1. Age at Puberty

4.1.1. Characterization of Age at Puberty. Forty-five of 50 Ms gilts and 29 of 34 Y gilts were in estrus by 280 da of age (Table 1). Ms gilts reached sexual maturity at 95 da of

Table 1. Puberty trait least squares means for Meishan and Yorkshire gilts[a]

Trait	Meishan			Yorkshire		
	n	Mean	SEM	n	Mean	SEM
Age at puberty (da)	46	95.5[**] ±	5.8	29	201.2 ±	5.5
Length of estrus[b] (da)	46	2.7[**] ±	0.2	29	1.9 ±	0.2
Estrous cycle length[c] (da)	45	20.3 ±	0.6	26	20.3 ±	0.6
Length of estrus[d] (da)	46	2.6[*] ±	0.2	27	1.9 ±	0.2
Estrous cycle length[e] (da)	42	20.2 ±	0.3	24	20.0 ±	0.3
Length of estrus[f] (da)	42	2.9[***] ±	0.2	24	2.1 ±	0.2

[a] From White et al.[3].
[b] Days exhibiting estrus at first estrus.
[c] Cycle length of first estrous cycle.
[d] Days exhibiting estrus at second estrus.
[e] Cycle length of second estrous cycle.
[f] Days exhibiting estrus at third estrus.
[***] $P < 0.001$; [**] $P < 0.01$; [*] $P < 0.05$.

age vs 201 da of age for Y gilts ($P < 0.01$). Meishan gilts were in estrus almost 1 da longer than the Y gilts for their first (2.7 vs 1.9; $P < 0.01$), second (2.6 vs 1.9; $P < 0.05$) and third (2.9 vs 2.1; $P < 0.001$) estrus. No differences were detected between breeds for cycle length on the first ($P > 0.95$) or second ($P > 0.63$) estrous cycle.

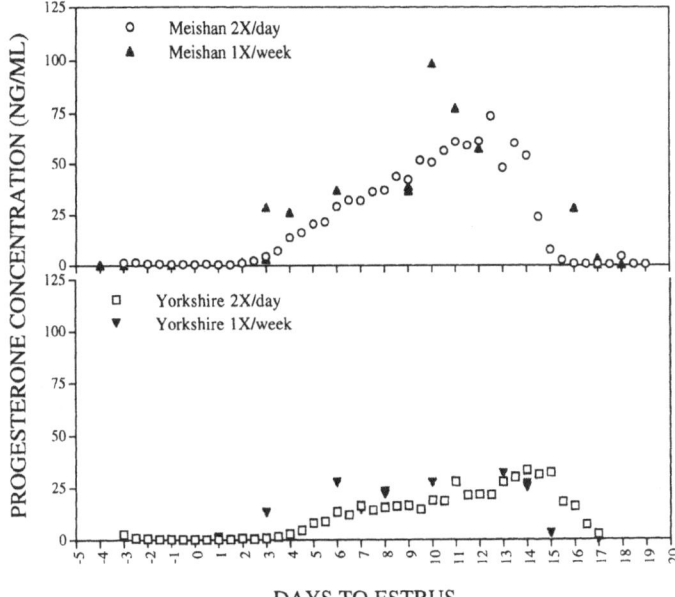

Figure 1. Comparison of progesterone concentrations (ng/ml) collected either once weekly or twice daily for Meishan and Yorkshire gilts. Progesterone concentrations are plotted against days to onset of estrus (Day 0).

4.1.2. Progesterone Profiles at Puberty. Examination of serum progesterone was an efficient method for confirmation of first ovulation in Ms and Y gilts. Further, weekly blood samples showed a hormone profile similar to twice daily samples (Fig. 1). Increases in progesterone concentrations were consistent with behavioral estrus in 10 of 12 gilts (83%). One Ms gilt had a silent estrous cycle as indicated by an increase in progesterone prior to the date of first observed estrous behavior. Progesterone concentrations of one Y gilt did not increase and estrous detection records indicated 3 questionable periods of estrus. Therefore, it was concluded that this gilt had not been in estrus nor ovulated. Age at first estrus and first predicted ovulation were 94 ± 24 and 87 ± 25 da, respectively, for Ms gilts and 188 ± 17 and 192 ± 19 da, respectively, for Y gilts (Table 2). Means for age at puberty and age at first predicted ovulation were not different for either breed (P > .85). However, age at first predicted ovulation for Ms gilts was slightly less than age at puberty because 1 gilt had a silent estrous cycle which caused age at first predicted ovulation to be approximately 21 da earlier than age at puberty for this gilt. Results for age at first estrus and age at first ovulation suggest that the early estrous cycles of the Ms gilts are functional. Meishan gilts had higher concentrations of serum progesterone (P < .01) and areas under the curve (P < .05) than did Y gilts. Serum progesterone concentration and area under the curve during the first 2 estrous cycles were higher in Ms than in Y gilts. In addition, differences between breeds for width between maxima of progesterone were not detected (P > .25). Width between progesterone maxima was similar to the result for estrous cycle length (P > .10).

Validation of the Coat-A-Count Progesterone RIA procedure resulted in an accurate, rapid and efficient method for quantification of porcine serum progesterone. Concentrations of progesterone in Y gilts reported from this assay procedure were consistent with those previously reported for occidental breeds.[11] Finally, this study has extended to swine a

Table 2. Puberty trait least squares means for Meishan and Yorkshire gilts

Trait	No.	Meishan Mean		SEM	No.	Yorkshire Mean		SEM
Age at puberty[a] (da)	6	93.9**	±	24.0	6	187.6	±	17.4
Age at first ovulation[b] (da)	6	86.5**	±	24.9	5	192.2	±	18.6
Estrous cycle length[c] (da)	6	20.3	±	1.1	6	21.0	±	0.8
Estrous cycle length[d] (da)	6	20.3	±	0.5	6	19.4	±	0.4
Width between maxima (da)	6	20.6	±	0.6	5	19.6	±	0.7
Progesterone conc. (ng/ml)	6	62.8**	±	5.7	5	32.8	±	6.3
Area[e] (ng*d/ml)	6	421.9*	±	43.4	5	229.1	±	47.5

[a] Age at puberty did not differ from age at first ovulation for either breed (*P* > .85).

[b] Age at first ovulation as predicted by progesterone hormone profiles.

[c] Cycle length of first estrous cycle.

[d] Cycle length of second estrous cycle.

[e] Area under curve (in excess of baseline).

** *P* < .01 compared to control (Yorkshire) gilts; * *P* < .05 compared to control (Yorkshire) gilts.

reliable, repeatable methodology, previously described in cattle,[10] to confirm visual estrous detection and allows for accurate prediction of first ovulation at puberty. In addition, luteal phase concentrations of progesterone were greater in Ms gilts than Y control gilts during the first 2 estrous cycles.

4.2. Mechanism of Prolificacy

4.2.1. Experiment 1. Large differences existed among Ms, Y and reciprocal cross females for age at slaughter and slaughter weight ($P < .05$; Table 3). Age at puberty differed among groups; Ms gilts were youngest (103 da), Y gilts were oldest (223 da) and reciprocal cross gilts were intermediate (148 and 150 da; $P < .05$). Serum progesterone concentration was highest in Ms x Y gilts ($P < .05$). Gestation length differences were not detected.

Meishan gilts had 14.2 ovulations, Y gilts had 12.5 ovulations and each group of crossbred gilts had 17.3 ovulations ($P < .05$; Table 3). Crossbred females had a greater number of ovulations than either Ms or Y females, which may be the effect of hybrid vigor. Wilmut *et al.*[19] reported that ovulation rates for crossbred Ms x Large White (LW) sows were intermediate to purebreds. Ovulation rates increase with increased age, number of estrous cycles and parities to a greater extent in Ms females than in occidental females.[20,21] Second parity Ms females had approximately 6 more ovulations than did Y females in our previous study;[3] whereas, Ms gilts bred at the third estrus had approximately 2 more ovulations than Y gilts in the current study. Ovulation rates for Ms gilts bred at fifth or sixth estrus[3] (17 CL) were higher than those for Ms gilts bred at third estrus in the present studies (14 CL for both experiments). Thus, this supports the suggestion that breed differences for ovulation rate

Table 3. Reproductive trait least squares means for Meishan, Yorkshire and reciprocal cross gilts[a]

Trait	Breed of Gilt				
	Ms	Ms x Y	Y x Ms	Y	SEM
Number of gilts	10	10	10	10	
Age at puberty (da)	103.0[c]	148.0[d]	150.0[d]	223.0[e]	14.7
Age at slaughter (da)	196.4[c]	238.1[d]	240.7[d]	313.6[e]	15.0
Slaughter weight (kg)	86.3[c]	128.6[d]	130.4[d]	146.3[e]	13.2
Gestation length (da)	50.9	51.1	51.2	51.1	2.8
Number of CL[b]	14.2[c]	17.3[d]	17.3[d]	12.5[c]	2.3
Number of fetuses	9.2[c]	12.9[de]	14.7[e]	10.9[cd]	2.4
Number of fetuses/no. of CL x 100	65.3[c]	74.9[cd]	85.9[de]	87.3[e]	12.3
Uterine length, cm	325.1	342.6	320.5	333.4	36.4
Progesterone concentration (ng/ml)	2591.8[c]	2045.9[d]	2177.8[c]	1927.4[c]	324.2

[a] Ms = Meishan; Y = Yorkshire; From White and Wheeler.[46]

[b] Number of CL determined by dissection.

[c,d,e] Means in same row with different superscripts differ ($P < .05$).

diverge, in favor of Ms females, as estrous cycles and parities increase. However, a definitive study has not yet been reported on repetitive determinations of ovulation rate on the same animals as estrous cycles and parities increase.

Number of fetuses tended to be lower in Ms and Y gilts than in crossbred gilts (P < .10) as observed previously[14,22,23] but number of fetuses in Y and Ms females did not differ (P > .05; Table 3). Although both groups of crossbred gilts had similar numbers of fetuses, there was a trend for the Y x Ms crossbred females to have larger litters than Ms x Y gilts (14.7 vs 12.9; P = .10) indicating a larger maternal effect for number of fetuses in Ms than Y gilts. A large Ms maternal effect has been demonstrated by larger studies examining crossbreeding parameters.[24]

Fetal survival was higher for Y gilts than Ms x Y gilts in Experiment 1 (P < .05); whereas, the reciprocal cross gilts did not differ for fetal survival (P > .05; Table 3). Further, Ms and Ms x Y gilts did not differ for fetal survival (P > .05). The proportion of CL represented as fetuses was higher for Y and Y x Ms females (87 and 86%, respectively) than for Ms females (65%; P < .05). This result was unusual since Chinese Ms gilts have been reported to have a greater embryonic survival per CL than occidental gilts.[21,25] However, very few studies have examined fetal survival[3,26] and none have examined fetal survival at 50 da of gestation as in the present study.

After standardization of the genotype of fetuses in Experiment 1, uterine length did not differ among breeds (Table 3). Therefore, the genotype of the fetus has an important role in determining uterine length because uterine length was not different among the 4 groups of gilts despite differences in numbers of fetuses. Although, Wu et al.[27] determined that in occidental breeds of pigs ovulation rate was correlated positively with number of fetuses up to 18 CL and 13 fetuses. Above this point, ovulation rate was no longer correlated with number of fetuses but number of fetuses correlated positively with uterine length. In Experiment 1, 7 Y x Ms females had litter sizes of 13 or more fetuses; whereas, females from the other groups had only 5 or fewer gilts with litter sizes of 13 or more fetuses. The uteri of Y x Ms were probably the only uteri that could have been challenged at this stage of gestation although uteri from these females tended to be the shortest indicating that more than 13 fetuses are required to challenge the uterus in Y x Ms gilts. Therefore, at this reproductive age, the differences between crossbred gilts and purebred gilts for number of fetuses at 50 da of gestation probably is determined by ovulation rate and not uterine capacity. The relationship among ovulation rate, number of fetuses and uterine length may be inherently different among breeds and may differ as the number of estrous cycles and parities increases. Although all females were treated similarly, factors that could have influenced these results are artificial insemination technician, semen handling, boar usage and nutrition. Boar or fertilization effects were unlikely to be important because gilts were bred to a different boar at each mating and semen was examined for motility and not used if stored over 24 hr. Although Ms gilts reached puberty earlier, Y and crossbred Ms gilts may have been closer to their mature size than Ms gilts and therefore nutrient partitioning and utilization to the uterus may have been more advanced in Y and crossbred Ms gilts. Therefore, a key point where physiological and reproductive age coincide must be reached for Ms females to have improved prolificacy over occidental breeds of females. This emphasizes that Ms gilts should not be bred until 5 to 6 mo of age.

Fetuses within Ms uteri occupied more space than fetuses in other groups in Experiment 1 (P < .05; Table 4). Fetuses in Y and Ms x Y uteri had a similar amount of space (P > .05); whereas, fetuses in Y x Ms gilts occupied the least amount of space per fetus (P < .05). Allantoic volumes and allantoic and amnionic progesterone concentrations were highest in Ms females (P < .05; Table 4). Allantoic and amnionic progesterone concentrations were lowest for Y gilts while levels in reciprocal cross gilts were intermediate to Ms and Y gilts (P < .05). Amnionic volumes per fetus, crown-rump length and fetus weight were highest in

Table 4. Fetal development trait least squares means for Meishan, Yorkshire and reciprocal cross gilts[a]

Trait	Maternal Breed				SEM
	Ms	Ms x Y	Y x Ms	Y	
Total number of fetuses	92	129	147	109	
Space/fetus (cm)	34.0[b]	28.8[c]	25.6[d]	30.5[c]	10.0
Amnionic volume/fetus (ml)	45.5[b]	37.9[c]	45.2[b]	52.1[d]	13.5
Allantoic volume/fetus (ml)	176.9[b]	134.5[cd]	152.0[c]	122.0[d]	78.4
Crown rump length (mm)	94.2[b]	90.6[c]	94.4[b]	97.0[d]	3.5
Weight/fetus (g)	50.6[b]	45.2[c]	49.9[b]	53.6[d]	5.6
Amnionic P_4 concentration (ng/ml)	9409.4[b]	8746.4[c]	8534.0[c]	8061.9[d]	1028.9
Allantoic P_4 concentration (ng/ml)	7590.9[b]	6751.7[c]	6535.3[c]	5780.1[d]	871.4

[a] Ms = Meishan; Y = Yorkshire; From White and Wheeler.[46]

[b,c,d] Means in same row with different superscripts differ ($P < .05$).

Y gilts, lowest in Ms x Y females, and the similar values for Ms and Y x Ms gilts were intermediate ($P < .05$). No differences existed among groups for head-tail orientation or sex ratio of fetuses. Fetuses within Y females had more amnionic fluid, greater crown-rump lengths and were heavier than in the other groups even though the genotype of the fetus was similar among all groups. Lower values for these measurements in reciprocal cross females probably can be attributed to the larger number of fetuses within uteri of similar lengths. Even though number of fetuses was similar for Y and Ms females, Y females had fetuses with more amnionic fluid, greater crown-rump lengths and heavier fetal weights ($P < .05$). Reasons for this remain to be elucidated but this does suggest a uterine effect for these traits. Allantoic volumes were highest for fetuses within Ms females, lowest for fetuses from Y females and intermediate for reciprocal crossbred females. It is unclear why the differences among breeds existed for this trait.

Others have suggested that the Ms uterus has an inhibitory effect on embryonic growth rate[28,29] and fetal growth rate.[26] In Experiment 1 of the present study, standardized fetuses were 2.8 mm longer and 3.0 g heavier in Y uteri than in Ms uteri which might indicate that Ms uteri had an inhibitory effect. However, results from Experiment 2 showed genotype of fetus effects for both of these traits and a breed of gilt x genotype of fetus interaction for crown-rump length. Thus, the communication between the uterus and fetuses appears to be important for fetal traits.

4.2.2. Experiment 2. Breed of gilt effects (Table 5) were detected for age at puberty, age at slaughter and slaughter weight; whereas, effects of fetus genotype were detected for age at puberty ($P < .05$). A breed of gilt x genotype of fetus interaction was detected for serum progesterone concentration ($P < .05$) with the lowest progesterone concentration in Y-X females. Large differences existed between the Ms and Y groups for age at puberty, age at slaughter and slaughter weight ($P < .05$; Table 6). Gestation length differences were not detected.

Table 5. Observed significance levels (probabilities) from analyses of variance of reproductive traits in Meishan and Yorkshire gilts carrying purebred or crossbred fetuses[a]

Trait	Breed	Genof	Breed x Genof
Age at puberty (da)	.000	.009	.790
Age at slaughter (da)	.000	.095	.505
Slaughter weight (kg)	.000	.159	.652
Gestation length (da)	.884	.064	.884
Number of CL[b]	.026	.834	.944
Number of fetuses	.651	.846	.087
Number of fetuses/no of CL (%)	.008	.752	.029
Uterine length (cm)	.999	.175	.994
Progesterone concentration (ng/ml)	.707	.154	.006
Space/fetus (cm)	.006	.061	.000
Amnionic volume/fetus (ml)	.000	.000	.876
Allantoic volume/fetus (ml)	.000	.000	.071
Crown rump length (mm)	.000	.000	.000
Weight/fetus (g)	.418	.004	.917
Amnionic P_4 concentration (ng/ml)	.016	.213	.004
Allantoic P_4 concentration (ng/ml)	.000	.000	.004

[a] Breed = Breed of Gilt; Genof = Genotype of fetus; From White and Wheeler.[46]
[b] Number of CL determined by dissection.

Breed of gilt effects (P < .05; Table 5) were detected for number of ovulations as evidenced by 2 more ovulations for Ms than Y females. Bazer et al.[30] reported that after the third estrus, LW gilts had a higher ovulation rate than Ms gilts; whereas, Biggs et al.[31] found that Ms and LW x Landrace females had similar ovulation rates at third or fourth estrus. Differences between these studies probably can be attributed to differences between Ms and control populations of animals as well as different management procedures. No genotype of fetus or breed of gilt x genotype of fetus interaction effects were detected (P > .05).

Number of fetuses was unaffected by the fetal genotype (purebred or crossbred), maternal genotype (Ms or Y) or their interaction (P > .05; Table 5). However, Ms-P gilts had a similar number of fetuses to the other groups despite having the shortest uteri. The role of a crossbred fetus in a Y uterus may be different from that in a Ms uterus. Therefore, the interaction between the uterus and fetuses in Ms and Y gilts appears to be important at this reproductive age.

Yorkshire gilts carrying crossbred fetuses had the highest fetal survival (P < .05); whereas, the other 3 groups were not different (P > .05; Table 6). Fetal survival was quite high for the Y gilts in these 2 experiments (87% in Experiment 1; 75 and 88% in Experiment

Table 6. Reproductive trait least squares means for Meishan and Yorkshire gilts carrying purebred or crossbred fetuses[a]

	Breed of Gilt				
	Meishan		Yorkshire		
	Genotype of Fetus				
Trait	Pure	Cross	Pure	Cross	SEM
Number of gilts	10	10	10	10	
Age at puberty (da)	123.0^c	103.0^d	239.0^e	223.0^e	21.2
Age at slaughter (da)	202.9^c	196.4^c	328.5^d	313.6^d	19.7
Slaughter weight (kg)	82.5^c	86.3^c	139.0^d	146.3^d	12.2
Gestation length (da)	49.7	50.9	49.7	51.1	2.1
Number of CL[b]	14.0	14.2	12.4	12.5	2.3
Number of fetuses	10.4	9.2	9.4	10.9	2.4
Number of fetuses/no. of CL x 100	73.1^c	66.5^c	75.3^c	88.3^d	13.3
Uterine length (cm)	239.1^c	303.2^d	336.3^e	314.3^{de}	33.9
Progesterone concentration (ng/ml)	2357.6^c	2543.7^c	2568.6^c	1994.4^d	343.3

a   Pure = Purebred; Cross = Crossbred; From White and Wheeler.[46]

b   Number of CL determined by dissection.

c,d,e   Means in same row with different superscripts differ ($P < .05$).

2) suggesting that differences between control populations from our study and previous reports with embryonic survival may have influenced contrasting results. Again, the mechanism may be altered somewhat as the females undergo increasing numbers of estrous cycles and parities. In addition, an interaction between breed of gilt x genotype of fetus was detected ($P < .05$; Table 5) for fetal survival with Y-X gilts having higher fetal survival rates than Y-P gilts; whereas, Ms-P and Ms-X gilts did not differ ($P > .05$). Therefore, a crossbred fetus has a different role in a Y uterus than in a Ms uterus. We also detected a significant breed of gilt effect for fetal survival (Table 5) with Y gilts having higher fetal survival rates than Ms gilts. In a reciprocal embryo transfer study, Rivera et al.[26] found similar results with a tendency for higher survival rates for Y fetuses (66%) than Ms fetuses (45%) regardless of recipient breed at 90 da of gestation. It is possible the communication between a Ms uterus and Ms fetuses may be different from the communication between a Y uterus and Y fetuses.

Meishan females carrying purebred fetuses had the shortest uterine lengths ($P < .05$; Table 6). Gilts carrying purebred fetuses had longer uteri than Ms-X gilts. However, Ms-X and Y-X females had similar uterine lengths and both groups of Y females had similar uterine lengths ($P > .05$). Uteri from Ms-X females were longer than Ms-P females despite having similar numbers of fetuses; whereas, Y-X and Y-P females had similar uterine lengths and numbers of fetuses. Although no breed of gilt x genotype of fetus interaction (Table 5) was detected, this was an interesting result. The results of uterine length for the Ms females certainly indicates the smaller space requirement of purebred Ms fetuses compared to crossbred fetuses. However, uteri from Ms-X females were longer. This indicates the Ms uterus has the ability to expand to meet space requirements of the fetal genotype. Bazer *et al.*[30] determined that LW gilts had longer uteri from 8 to 12 da after the onset of estrus regardless of pregnancy status (bred or unbred). Therefore, Ms uteri had the ability to expand with crossbred fetuses to uterine lengths similar to those of Y-X females and still maintain a similar number of fetuses at 50 da of gestation. Meishan uteri were challenged at this reproductive age; whereas, Y uteri may not have been challenged due to fewer embryos. If Ms uteri can maintain the ability to expand and meet the space requirements of the fetuses as they undergo more estrous cycles and parities, i.e., ovulation rate increases and uterine length increases, this may account for part of the improved prolificacy of the Ms breed. Because ovulation rates increase to a greater extent in Ms females than in occidental females, Ms females should have a greater ability to produce larger litters with reproductive age as long as uterine capacity increases at similar rates in both breeds. In addition, uterine capacity may increase to a greater extent in Ms females than in occidental breeds of females. More investigation into uterine capacity at later estrous cycles and parities needs to be done.

Breed of gilt and breed of gilt x genotype of fetus interaction effects were detected for space per fetus (P < .05; Table 5). Fetuses from Ms-X females and fetuses from Y-P females required the most uterine space (34 and 35 cm, respectively; Table 7), fetuses from Y-X females were intermediate (30 cm) and fetuses from Ms-P females required the least uterine space (25 cm; P < .05). Space per fetus requirements were much less for Ms fetuses than for Y fetuses (25.4 vs 35.0 cm; Experiment 2). This had been suggested previously.[3,32] Therefore, more fetuses should be able to survive in a uterus of a given absolute length. However, Ms-P females had shorter uteri than Ms-X females in Experiment 2. Also, Ms gilts

Table 7. Fetal development trait least squares means for Meishan and Yorkshire gilts carrying purebred or crossbred fetuses[a]

Trait	Maternal Breed				
	Meishan		Yorkshire		
	Genotype of Fetus				
	Pure	Cross	Pure	Cross	SEM
Total number of fetuses	104	92	94	109	
Space/fetus (cm)	25.4[b]	34.0[c]	35.0[c]	30.5[d]	10.5
Amnionic volume/fetus (ml)	27.8[b]	45.5[c]	34.8[d]	52.1[e]	12.9
Allantoic volume/fetus (ml)	99.1[b]	176.9[c]	69.2[d]	122.0[e]	65.8
Crown-rump length (mm)	79.2[b]	94.2[c]	90.1[d]	97.0[e]	2.6
Weight/fetus (g)	37.9[b]	50.6[cd]	41.7[c]	53.6[d]	41.0
Amnionic P_4 concentration (ng/ml)	8976.6[b]	9409.4[b]	9110.1[b]	8061.9[c]	984.8
Allantoic P_4 concentration (ng/ml)	5781.3[b]	7590.9[c]	5228.3[b]	5780.1[b]	845.4

[a]   Pure = Purebred; Cross = Crossbred; From White and Wheeler.[46]

[b,c,d,e]   Means in same row with different superscripts differ (P < .05).

in Experiment 1 had the greatest uterine space available to fetuses. In Experiment 2, we detected a breed of gilt x genotype of fetus interaction effect for space per fetus. Space was created for the larger crossbred fetuses from Ms-X females to maintain the same number of fetuses with an identical ovulation rate as Ms-P females. However, fetuses from Y-P females had more uterine space than fetuses from Y-X females despite much larger size of crossbred fetuses than purebred Y fetuses. Yorkshire females, at this reproductive age, had uteri with enough space to provide for all the fetuses even if they had an embryonic survival rate of 100% because they only ovulated 12.5 eggs. Communication between the uterus and fetuses is very important for uterine space available to fetuses in Ms females at this reproductive age and may not be as important in Y females because of fewer embryos.

Breed of gilt effects (Table 5) were detected for allantoic and amnionic volumes, allantoic and amnionic progesterone concentrations and crown-rump length of fetuses. Effects of genotype of fetus were detected for allantoic and amnionic volumes, allantoic progesterone concentration, crown-rump length and fetal weight ($P < .05$). Breed of gilt x genotype of fetus interaction effects were detected for allantoic and amnionic progesterone concentrations. Allantoic fluid volumes for fetuses from Ms-X, Y-X, Ms-P and Y-P were 177, 122, 99 and 69 ml, respectively ($P < .05$; Table 7). Differences among all groups were determined for amnionic volume and crown-rump length ($P < .05$). The highest allantoic progesterone concentration was present in Ms-X females ($P < .05$). Fetuses from Y-X females were the longest, heaviest and had the most amnionic fluid and the highest amnionic progesterone concentration followed by fetuses from Ms-X, Y-P and Ms-P females, respectively. Fetuses from Ms-P gilts weighed less than fetuses from any other group ($P < .05$). Fetal weights from Y-X females were heavier than fetal weights from Y-P females. However, no differences were determined between fetuses from Y-X and Ms-X or fetuses from Ms-X and Y-P ($P > .05$). No effects of breed of gilt or genotype of fetus were detected among groups for head-tail orientation or sex ratio of fetuses. Other researchers reported uterine environment and fetal genotype effects for fetal size[26,33] and uterine environment effects for placental fluid volume,[26] fetal weight[28] and allantoic fluid volume.[33]

A breed of gilt x genotype of fetus interaction was detected for crown-rump length ($P < .05$). In a reciprocal embryo transfer study, Ashworth et al.[33] found no interactions between recipient and donor breeds for allantoic fluid volume, crown-rump length and fetal weight at 30 da of gestation. In another reciprocal embryo transfer experiment, Ford et al.[28] detected a fetal genotype x uterine genotype interaction for amnionic fluid volume; whereas, no interaction was detected for this trait in our study. Thus, the communication between fetuses and the uterus could be different between the 2 breeds.

4.3. Seminal Plasma

4.3.1. Experiment 1. Estradiol-17ß concentrations in Ms seminal plasma were higher ($P < .05$) than concentrations in F and Y seminal plasma, while F boars had higher E_2 concentrations in seminal plasma ($P < .05$) than Y boars. Mean E_2 concentrations for Ms, F and Y were 3.49, 0.75 and 0.19 ng/ml, respectively. Not only were there breed differences regarding seminal characteristics, but within breed variation also has been observed.[4] Within breed variation and between boar effects were detected as indicated by the wide range of E_2 concentrations. Mean E_2 concentrations ranged from 2.48 to 4.54 ng/ml in Ms boars, 2.40 to 3.52 ng/ml in F boars and 1.00 to 3.62 ng/ml in Y boars. The highest concentration of E_2 observed in Ms boars was 34.6 ng/ml compared to the lowest concentration of E_2 observed in Y boars of 0.009 ng/ml.

Table 8. Summary of least squares means for semen and seminal plasma traits in Chinese
Meishan and Yorkshire boars

Trait	Meishan			Yorkshire		
	n	Mean	SEM	n	Mean	SEM
Age (da)	16	469		16	398	
Gel free volume (ml)	16	89.3	± 23.2	16	217.2*	± 23.2
Progressive motility (%)	16	72.5	± 3.2	16	84.0*	± 3.2
Sperm concentration (10^6/ml)	16	2.28	± .18	16	3.07*	± .18
E_2-17β concentration (ng/ml)	16	4.66†	± 2.0	15	.48	± 2.0
Protein concentration (mg/ml)	16	75.82	± .26	16	75.88	± .26
Testosterone concentration (ng/ml)	16	14.18†	± 3.96	16	2.40	± 2.89

* $P < .05$; † $P < .10$.

4.3.2. Experiment 2. Examination of E_2 in repeated collections from each boar
indicated the mean E_2 concentration (Table 8) was 4.66 ng/ml in Ms boars which was higher
than that of Y boars (0.48 ng/ml; $P < .10$). This difference possibly is affected by between
and within boar effects which is evidenced by the range of 0.627 ng/ml to 69.4 ng/ml for Ms
boars and 0.142 ng/ml to 3.03 ng/ml for Y boars. Considerations must be made for within
boar sample effects as were seen in this study. An infrequent collection schedule, as opposed
to a repeated collection schedule, could have an effect on E_2 concentrations. This was
observed in Experiment 1 where there was a minimum of 3 da and a maximum of 6 mo
sexual rest between semen collections. Without a regular collection schedule, there is the
possibility that E_2 could accumulate in the epididymis. To examine this aspect, Experiment
2 was designed to standardize collections by allowing only 48 hr sexual rest between
collections. With repeated collections, E_2 did not increase but remained either constant or
decreased. Season also affects estrogen concentration in domestic boar ejaculates[34]. All boars
were housed in a daylight controlled environment and semen collections occurred inside this
environment during the last 2 wk of August. A higher concentration of E_2 has been seen in
the ejaculate of domestic boars during October to December.[35] Collections would have to be
made during this time period to determine whether this is also the situation for Chinese boars.

An important component of the ejaculate, E_2 comprises 73% of seminal estrogens.[36]
Seminal estrogens also may influence uterine and behavioral characteristics of the female[37,38]
as well as the male sex drive to copulate.[39] Estradiol-17β has been shown to influence
myometrial contractions in the female.[37,38] A 2 fold increase in E_2 contractions has been
observed when females were infused with saline containing this estrogen.[37] An increase in
myometrial contractions possibly could aid spermatozoon in reaching the site of fertilization.[40] In castrated males, E_2 administration initiated mounting and combined with testosterone caused copulation, ejaculation and increased salivation.[39] Testosterone concentrations
were higher in ejaculates from Ms boars than Y boars in this study.

Another possible effect of E_2 is on the timing and duration of ovulation.[41] Prostaglandin F-2α influences contraction of smooth muscle tissue, i.e., the uterine tract, thereby
delivering spermatozoa to the uterotubal junction. Claus[37,38] showed that intrauterine applications of 10 μg of estrogen lead to a sudden rise in prostaglandin F-2α in the uterine vein.

Coincidentally, an increase in uterine contractile waves in the tubal direction was observed. Sows infused transcervically with prostaglandin F-2α in saline ovulated 12 hr earlier than those infused with saline alone. Therefore, seminal estrogens influence the timing of ovulation under the control of prostaglandin F-2α. In addition to increased levels of estrogen, FSH concentrations appear to be greater in Ms boar blood plasma.[42] In females, when E_2 rises, FSH levels decrease.[43] Insemination of females with seminal plasma containing dead sperm prior to insemination with a live ejaculate resulted in 1.3 more piglets per litter than those not infused with seminal plasma containing dead sperm before breeding.[44] Possibly, hormonal factors of the ejaculate are beneficial in reducing embryonic mortality.

Gel free volume was higher in Y than Ms boars. These results support the study of Gerfen et al.[4] who concluded that gel free volume was different between the Chinese breeds F and Ms and the occidental breed Y. Volume of semen introduced into the uterus has been shown to influence uterine contractions.[45] The minimum amount required to stimulate uterine contraction is > 50 ml. Both breeds used in this study are well above this threshold volume, with the Y exceeding that volume by more than 150 ml. However, there was variation within and between boars. The highest single ejaculate volume for each breed was 130 ml for Ms and 350 ml for Y.

Sperm concentration was higher in semen from Y than in Ms boars. This is in disagreement with Gerfen et al.[4] who reported a higher concentration of spermatozoa in Ms than Y boars. Again, the collection schedule differences between experiments could have influenced results. Analysis of progressive motility showed that Y boars had more motile spermatozoa ($P < .05$) than Ms boars. Gerfen et al.[4] reported no differences between Ms and Y ejaculates for progressive motility. Variations within the experimental design of Gerfen et al.[4] could account for the observed differences.

Seminal plasma protein concentrations were not different among breeds. However, total seminal protein concentration per ejaculate was higher ($P < .05$) in Y than in Ms boars. This coincides with the higher volume of semen in Y boars as compared to Ms boars. Gerfen et al.[4] found similar concentrations of seminal plasma protein in ejaculates from F, Ms and Y boars.

In conclusion, much is still unknown as to the prolificacy of the Chinese breeds. A difference was observed with regard to E_2 and testosterone concentrations in seminal plasma of Chinese and occidental breeds. It is possible the high levels of E_2 in the seminal plasma of Ms boars affects myometrial contractions in the female and aids in transport of sperm to the site of fertilization, thus improving litter size.

5. REFERENCES

1. Rothschild, M.F., McLaren, D.G., Young, L.D., Christian, L.L., Hsieh, C.-Y., and White, B.R., 1990, Preliminary reproductive results from Meishan gilts imported from the Peoples Republic of China (PRC) to the United States, J. Anim. Sci. 68 (Suppl. 1): 228 abstr.
2. Phillips, R.W., and Tumbleson, M.E., 1986, Models, Swine in Biomedical Research 1: 437-440.
3. White, B.R., McLaren, D.G., Dziuk, P.J., and Wheeler, M.B., 1993, Age at puberty, ovulation rate, uterine length, prenatal survival and litter size in Chinese Meishan and Yorkshire females, Theriogenology 40: 85-97.
4. Gerfen, R.W., White, B.R., Cotta, M.A., and Wheeler, M.B., 1994, Comparison of the semen characteristics of Fengjing, Meishan and Yorkshire boars, Theriogenology 41: 461-469.
5. Zimmerman, D.R., Carlson, R., and Nippert, L., 1969, Age at puberty in gilts as affected by daily heat checks with a boar, J. Anim. Sci. 29 (Suppl. 1): 203 abstr.
6. Brooks, P.H., and Cole, D.J.A., 1976, Effect of boar presence on the age at puberty of gilts, Report from the School of Agriculture, University of Nottingham. 74 abstr.
7. Doroshkov, V.B., 1974, Stimulatory effect of males on reproductive performance of female pigs, Anim. Breed. Abstr. 42: 3830 abstr.

8. Hughes, P.E., and Cole, D.J.A., 1976, Reproduction in the gilt. 1. Influence of age and weight at puberty on ovulation rate and embryo survival in the gilt, *Anim. Prod.* 21: 183-189.

9. Hughes, P.E., 1982, Factors affecting the natural attainment of puberty in the gilt, in: *Control of Pig Reproduction*, (D.J.A. Cole, and G.R. Foxcroft, eds.), Butterworth Scientific, London, pp. 117-138.

10. Wheeler, M.B., Anderson, G.B., BonDurant, R.H., and Stabenfeldt, G.H., 1982, Postpartum ovarian function and fertility in beef cattle that produce twins, *J. Anim. Sci.* 54: 589-593.

11. Hansel, W., Concannon, P.W., and Lukaszewska, J.H., 1973, Corpora lutea of the large domestic animals, *Biol. Reprod.* 8: 222-245.

12. SAS, 1985, *SAS User's Guide: Statistics*, 5th ed., SAS Inst., Inc., Cary, NC.

13. Veldhuis, J.D., and Johnson, M.L., 1986, Cluster analysis: A simple, versatile, and robust algorithm for endocrine pulse detection, *Am. J. Physiol.* 250: E486-E493.

14. Martinat-Botte, F., Bazer, F.W., and Terqui, M., 1989, Embryonic survival mechanisms in Chinese Meishan (MS) and hyperprolific Large White (LWh) gilts, *Proc. 3rd Int. Conf. Pig Reprod.* 45 abstr.

15. Haley, C.S., Ashworth, C.J., Lee, G.J., Wilmut, I., Aitken, R.P., and Ritchie, W., 1990, British studies of the genetics of prolificacy in the Meishan pig, in: *Symp. Sur Le Porc Chinois* (M. Molenat, and C. Legault, eds.), Tolouse, France, pp. 86-97.

16. Wu, M.C., and Dziuk, P.J., 1988, Procedures for measuring length of the pig uterus, *J. Anim. Sci.* 66: 1712-1720.

17. Scheffe, H., 1959, *The Analysis of Variance*, John Wiley and Sons, Inc., New York, pp. 365-367.

18. Almond, G.W., Esbenshade, K.L., and Smith, C.A., 1992, Effects of chronic gonadotropin-releasing hormone agonist treatment on serum luteinizing hormone and testosterone concentrations in boars, *Am. J. Vet. Res.* 53: 22-25.

19. Wilmut, I., Ritchie, W.A., Haley, C.S., Ashworth, C.J., and Aitken, R.P., 1992, A comparison of rate and uniformity of embryo development in Meishan and European white pigs, *J. Reprod. Fertil.* 95: 45-56.

20. Christenson, R.K., 1993, Ovulation rate and embryonic survival in Chinese Meishan and white crossbred pigs, *J. Anim. Sci.* 71: 3060-3066.

21. Haley, C.S., and Lee, G.J., 1993, Genetic basis of prolificacy in the Meishan pig, *J. Reprod. Fertil.* Suppl. 48: 247-259.

22. Legault, C., Caritez, J.C., Gruand, J., and Bidanel, J.P., 1984, Le point de l'expérimentation sur les races chinoises en France: 'reproduction' et 'production', *J. Rech. Porcine Fr.* 16: 481-494.

23. Bidanel, J.P., Caritez, J.C., and Lagant, H., 1990, Ovulation rate and embryonic survival in gilts and sows with variable proportions of Meishan (MS) and Large White (LW) genes. Mean performance and crossbreeding parameters between MS and LW breeds, in: *Symp. Sur Le Porc Chinois* (M. Molenat, and C. Legault, eds.), Tolouse, France, p. 110. abstr.

24. Bidanel, J.P., Caritez, J.C., and Legault, C., 1989, Estimation of crossbreeding parameters between Large White and Meishan porcine breeds. I. Reproductive performance, *Genet. Sel. Evol.* 21: 507-526.

25. Galvin, J.M., Wilmut, I., Day, B.N., Ritchie, M., Thomson, M., and Haley, C.S, 1993, Reproductive performance in relation to uterine and embryonic traits during early gestation in Meishan, Large White and crossbred sows, *J. Reprod. Fertil.* 98: 377-384.

26. Rivera, R.M., Christenson, L.K., Youngs, C.R., and Ford, S.P., 1994, Competitive survival, growth rate and estrogen secretory activity of Meishan and Yorkshire fetuses, *J. Anim. Sci.* 72 (Suppl. 1): 78 abstr.

27. Wu, M.C., Hentzel, M.D., and Dziuk, P.J., 1987, Relationships between uterine length and number of fetuses and prenatal mortality in pigs, *J. Anim. Sci.* 65: 762-770.

28. Ford, S.P., Christenson, L.K., Rivera, R.M., and Youngs, C.R., 1994, Inhibitory effects of the Meishan uterus on growth rate and estradiol-17β (E_2) secretion of day 30 conceptuses, *Biol. Reprod.* 50 (Suppl. 1): 175 abstr.

29. Youngs, C.R., Christenson, L.K., and Ford, S.P., 1994, Investigations into the control of litter size in swine: III. A reciprocal embryo transfer study of early conceptus development, *J. Anim. Sci.* 72: 725-731.

30. Bazer, F.W., Thatcher, W.W., Martinat-Botte, F., and Terqui, M., 1988, Sexual maturation and morphological development of the reproductive tract in Large White and prolific Chinese Meishan pigs, *J. Reprod. Fertil.* 83: 723-728.

31. Biggs, C., Tilton, J.E., Craigon, J., Foxcroft, G.R., Ashworth, C.J., and Hunter, M.G., 1993, Comparison of follicular heterogeneity and ovarian characteristics in Meishan and Large-White hybrid pigs, *J. Reprod. Fertil.* 97: 263-269.

32. Bidanel, J.P., Caritez, J.C., and Lagant, H., 1990, Characteristics of the reproductive organs of gilts and sows with variable proportions of Meishan (MS) and Large White (LW) genes, in: *Symp. Sur Le Porc Chinois* (M. Molenat, and C. Legault, eds.), Tolouse, France, p. 108. abstr.

33. Ashworth, C.J., Haley, C.S., Aitken, R.P., and Wilmut, I., 1990, Embryo survival and conceptus growth after reciprocal embryo transfer between Chinese Meishan and Landrace x Large White gilts, *J. Reprod. Fertil.* 90: 595-603.

34. Borg, K.E., Lunstra, D.D., and Christenson, R.K., 1993, Semen characteristics, testicular size, and reproductive hormone concentrations in mature Duroc, Meishan, Fengjing, and Minzhu boars, *Biol. Reprod.* 49: 515-521.

35. Claus, R., Schopper, D., and Wagner, H.G., 1983, Seasonal effect on steriods in blood plasma and seminal plasma of boars, *Steriod Biochem.* 19: 725-729.

36. Eiler, H., and Graves, C.N., 1977, Oestrogen content of semen and the effect of exogenous oestradiol-17ß on the oestrogen and androgen concentration in semen and blood plasma of bulls, *J. Reprod. Fertil.* 50: 17-21.

37. Claus, R., 1990, Physiological role of seminal components in the reproductive tract of the female pig, *J. Reprod. Fertil.* Suppl. 40: 117-131.

38. Claus, R., Hoang-Vue, C., Ellendorff, F., Meyer, H.D., Schopper, D., and Weiler, U., 1987, Seminal oestrogens in the boar: Origin and functions in the sow, *Steriod Biochem.* 27: 331-335.

39. Parrott, R.F., and Booth, W.D., 1984, Behavioral and morphological effects of 5α-dihydrotestosterone and oestradiol-17ß in the prepubertally castrated boar, *J. Reprod. Fertil.* 71: 453-461.

40. Willmen, T., Rabeler, J., Everwand, A., Waberski, D., and Weitze, K.F., 1991, Influence of seminal plasma and oestrogenens in the inseminate on fertilization rate, sperm transport and ovulation time, in: *Boar Semen Preservation II* (L.A. Johnson, and D. Rath, eds.), Paul Parey Scientific Publishers, Berlin, Germany, pp. 379-383.

41. Weiler, U., and Claus, R., 1991, Endocrine aspects of testicular function, especially hormones in the seminal plasma and their fate in the female reproductive tract: Testicular steriods and their relevance for male and female reproductive functions, in: *Boar Semen Preservation II* (L.A. Johnson, and D. Rath, eds.), Paul Parey Scientific Publishers, Berlin, Germany, pp. 41-61.

42. Ford, J.J., Wise, T., Lunstra, D.D., and Klindt, J., 1994, Follicle-stimulating hormone secretion in purebred Meishan (MS) and white crossbred (WC) boars, *J. Anim. Sci.* 77 (Suppl 1): 163 abstr.

43. Van De Wiel, D.F.M., Erkens, J., Koops, W., Vos, E., and Van Landeghem, A.A.J., 1981, Periestrous and midluteal time courses of circulating LH, FSH, Prolactin, Estradiol-17ß and Progesterone in the domestic pig, *Biol. Reprod.* 24: 223-233.

44. Murray, F.A., and Grifo, A.P., Jr., 1986, Intrauterine infusion of killed semen to increase litter size in gilts, *J. Anim. Sci.* 62: 187-190.

45. Stratman, F.W., and Self, H.L., 1960, Effect of semen volume and number of sperm on fertility and embryo survival in artificially inseminated gilts, *J. Anim. Sci.* 19: 1081-1088.

46. White, B.R., and Wheeler, M.B., 1995, Examination of ovulation rate, uterine and fetal interactions, and reproductive age in Chinese Meishan, Yorkshire, and reciprocal cross gilts: effects of fetal and maternal genotypes, *Anim. Reprod. Sci.* 39: 147-158.

44

SELECTION FOR EXTREMES IN SERUM FSH CONCENTRATIONS RESULTS IN REDUCED TESTIS SIZE AND FERTILITY IN MEISHAN AND WHITE COMPOSITE BOARS

Donald D. Lunstra, PhD, J. Joe Ford, PhD, and Thomas H. Wise, PhD[*]

U.S. Department of Agriculture
Agricultural Research Service
Roman L. Hruska U.S. Meat Animal Research Center
Clay Center, Nebraska 68933-0166

1. ABSTRACT

The Meishan breed of swine was imported from China for its high ovulation rate and as a model to increase fecundity in the swine industry. Concentrations of follicle stimulating hormone (FSH) are increased in Meishan boars (5 to 10 ×) as compared to European breeds utilized for meat production in the United States. In human reproduction, elevated FSH concentrations in men are associated with impaired gonadal function and infertility. In the current studies, concentrations of serum FSH were related negatively to testis size and weights in Meishan boars and White composite boars of European origin. Selection for high FSH concentrations in boars was correlated with reduced testis weight and fertility, all common characteristics of hypergonadotropism associated infertility in men. Meishan boars that have extremes in FSH concentrations and testis weights may be a useful model for the study of hypergonadotropic hypogonadal function in men.[†]

2. INTRODUCTION

A number of scenarios produce elevated follicle stimulating hormone (FSH) concentrations with associated decreased fertility and Sertoli cell function: these include Klinefel-

---

[*] Reprint requests to: Dr. T. Wise, USDA, ARS, RLH US Meat Animal Research Center, Clay Center, NE 68933-0166. (402) 762-4185.

[†] Mention of trade names are necessary to report factually on available data; however, the USDA neither guarantees nor warrants the standard of the product, and the use of the name by USDA implies no approval of the product to the exclusion of others that also may be suitable.

ter's syndrome[1,2] and chemotherapy[3] but are minor in comparison to the numerous reports of subfertility and increased FSH concentrations of unknown cause in man.[4-9]

Experimental analysis of this reproductive problem with human subjects has its limits due to numbers available and invasive techniques. In contrast, genetic swine lines (which can be selected for various characteristics) that show high FSH concentrations in conjunction with decreased testis weight and spermatogenetic function are available as a possible model for this aspect of infertility in man. Objectives of the present studies were to investigate relationships between serum concentrations of FSH and testis weight and function in Chinese Meishan and European White composite lines of swine that are divergent in FSH concentrations.

3. METHODS

Boars utilized in this study were Chinese Meishan and a White composite line of European breeds ($\frac{1}{4}$ Chester White, $\frac{1}{4}$ Large White, $\frac{1}{4}$ Landrace, $\frac{1}{4}$ Yorkshire). Crossbreds of the above 2 lines were utilized to produce maximum variation in FSH concentrations (3/4 Meishan:$\frac{1}{4}$ White composite, 3/4 White composite:$\frac{1}{4}$ Meishan and $\frac{1}{2}$ Meishan:$\frac{1}{2}$ White composite). All boars were sexually mature. Plasma FSH and testosterone concentrations were analyzed by RIA (FSH[10,11], testosterone[12]). Hormonal values generally were from 3 or more blood evaluations per animal.

The *in vivo* measurement of testicular volume was by method of Young and coworkers[13] (volume = 4/3 $\pi \times$ ($\frac{1}{2}$ length) $\times$ ($\frac{1}{2}$ width)$^2 \times 2$). Testicular weights were acquired at slaughter or castration. Right testes weight was utilized as an index of testicular weight. Daily and total sperm productions were calculated by the method of Amann.[14] Briefly, 3 samples of testicular parenchyma (proximal, mid and distal thirds; 1 to 2 g) were placed into 30 ml of cold Dulbecco phosphate buffer (0.15 M, 7.4 pH) with 0.1% Triton X-100 and homogenized. A portion of the homogenized suspension was removed (0.5 ml) and mixed with 0.5 ml of Trypan Blue (0.4% in saline solution). Elongated spermatid nuclei were counted in duplicate via hemacytometer. Daily sperm production/g testis was calculated by averaging the sperm/g results and dividing by the species specific constant number of da. The constant was based on the proportion of the spermatogenic cycle that homogenized resistant spermatid nuclei were present (swine = 4.37 da). Total testicular sperm production was calculated from the daily sperm production/g of tissue multiplied by the paired testicular weight of the animal.

Fertility analysis was conducted with 3/4 Meishan:$\frac{1}{4}$ White composite (n = 6) and 3/4 White composite:$\frac{1}{4}$ Meishan (n = 7) boars selected for extremes in FSH concentrations. Each boar was mated to approximately 15 estrous females (total n = 206 females exposed to males). Females were mated twice to the same boar approximately 20 hr apart.

4. RESULTS

Analysis of FSH concentrations and testis weights in boars (n = 68) indicated a negative relationship in White composite (Fig. 1a; n = 24), Meishan (Fig. 1b; n = 27) and crossbreds (Fig. 1c; n = 17) of these 2 breeds (overall correlation for 3 breed groups; r = -0.75). Animals with high FSH concentrations (>1000 ng/ml) were associated with lower testis weights and extremes were in Meishan and Meishan crossbred boars. Testis weight was greater in White composite boars as compared to Meishan boars (289.4 ± 21.0 vs 153.2 ± 16.9 g, respectively) and testis weight of crossbreds of the 2 breeds was intermediate (198.6 ± 20.7 g). Concentrations of FSH were elevated in the Meishan breed as compared to White

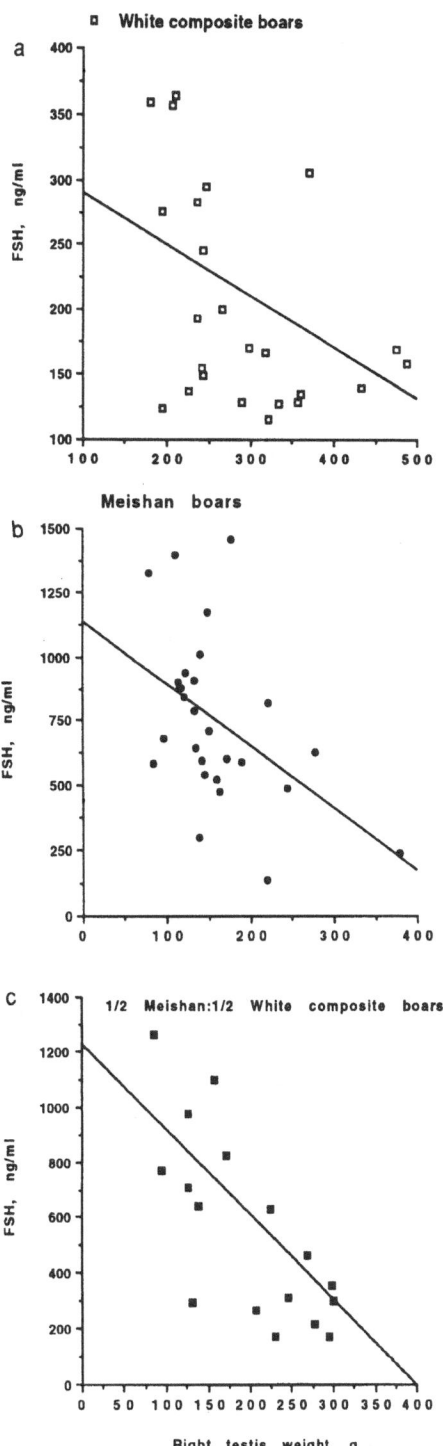

Figure 1. Inverse relationships of serum FSH and right testis weight (p<0.01) in White composite (a), Meishan (b) and White composite:Meishan crossbred boars (c). Crossbred boars were selected for extremes in FSH concentrations.

composite boars (775 ± 73 vs 202 ± 19 ng/ml, respectively; and $\frac{1}{2}$ White composite:$\frac{1}{2}$ Meishan crossbreds were intermediate (557 ± 102 ng/ml).

Meishan boars had higher circulating testosterone concentrations than did White composite boars (Meishan = 16.8 ± 2.6 vs White composite = 3.8 ± 1.3 ng/ml). Testosterone concentrations were not related to total sperm production in either breed (Figs. 2a, 2c). There was no relationship between testicular weight and circulating testosterone concentrations in the Meishan or White composite boars (Fig. 2b, 2d).

In vivo measurement of testicular volume for comparison with testicular weights of crossbred Meishan × White composite boars showed similar relationships with FSH concentrations (Fig. 3). Correlation between estimated testicular volume and actual testicular weight was 0.67.

Daily sperm production per gram of testis did not differ in Meishan (n = 22) and White composite boars (n = 22). However, total daily sperm production was greater in White composite boars, reflecting the increased testis weight in this swine genetic line (Table 1). Total daily sperm production was related negatively to FSH concentrations with each of the 3 breed groups (Fig. 4).

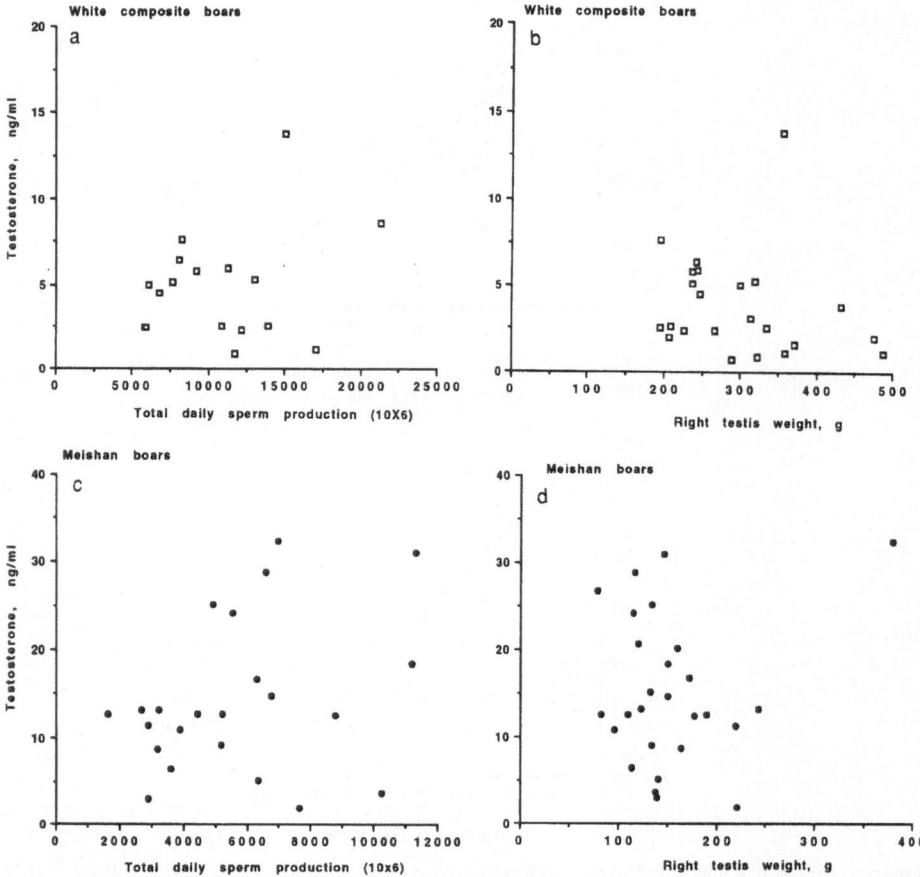

Figure 2. Relationships between testosterone and total daily sperm production and the right testis weight of White composite (a,b) and Meishan boars (c,d) were not significant.

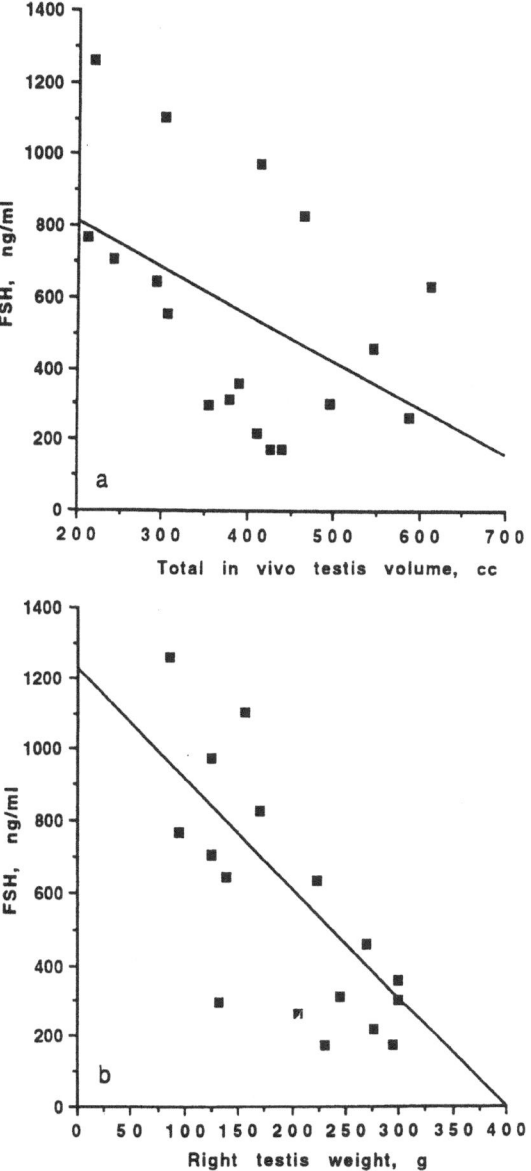

Figure 3. *In vivo* measurements of testicular volume (a) and testis weight (b) and FSH concentrations were related inversely (p<0.01). Correlation of the 2 methods of measuring the testis was 0.67.

Relationships of FSH and testis weight (Fig. 5a) and sperm production (Fig. 5b) were correlated negatively in 3/4 Meishan and 3/4 White composite sires. Fertility of Meishan and White composite sires (n = 13) also demonstrated that as FSH concentrations increased in boars, conception declined (206 females exposed, Fig. 5c). Fertility ranged from 27 to 93%, and analysis of FSH concentrations and conception rates showed a negative relationship between percent conception and FSH (r = -0.74).

Table 1. Differences in testis weight, daily sperm production and total daily
sperm production in White composite and Meishan boars

Breed	Testis weight	Daily spermproduction[a]	Total daily sperm production[b]
White composite (n = 22)	272.4 ± 13.7[c]	21.7 ± 1.3[d]	10823 ± 821[cd]
Meishan (n = 22)	149.7 ± 12.4	20.4 ± 1.5	5585 ± 579

[a]Sperm production/g testis.
[b]Total testicular sperm production.
[c]Mean ± SEM, p<0.01.
[d]10^6 cells.

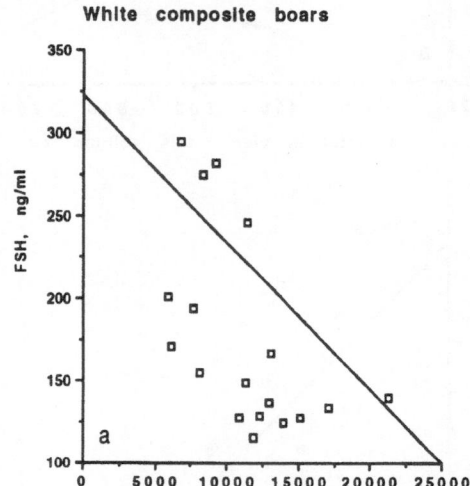

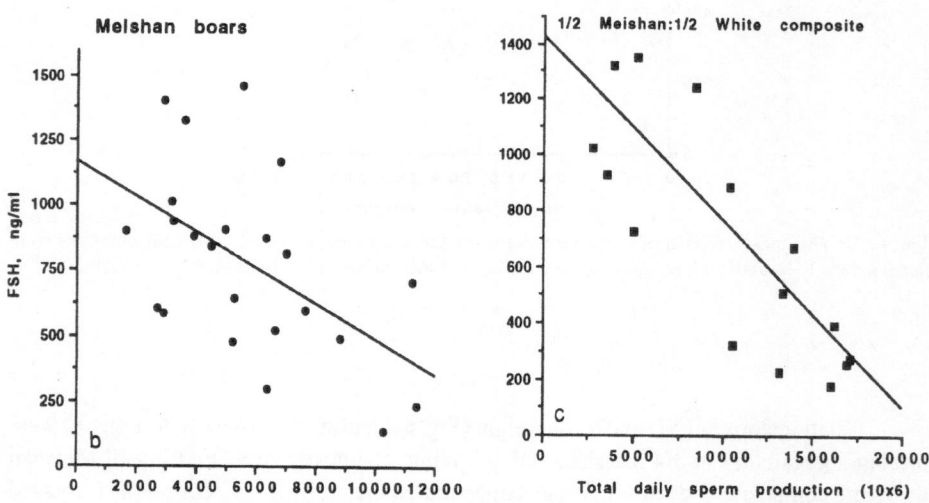

Figure 4. Relationships between serum FSH concentrations and total sperm production in White composite (a), Meishan (b) and White composite:Meishan crossbred boars (c). Concentrations of FSH and total daily sperm production were negatively related (p<0.01).

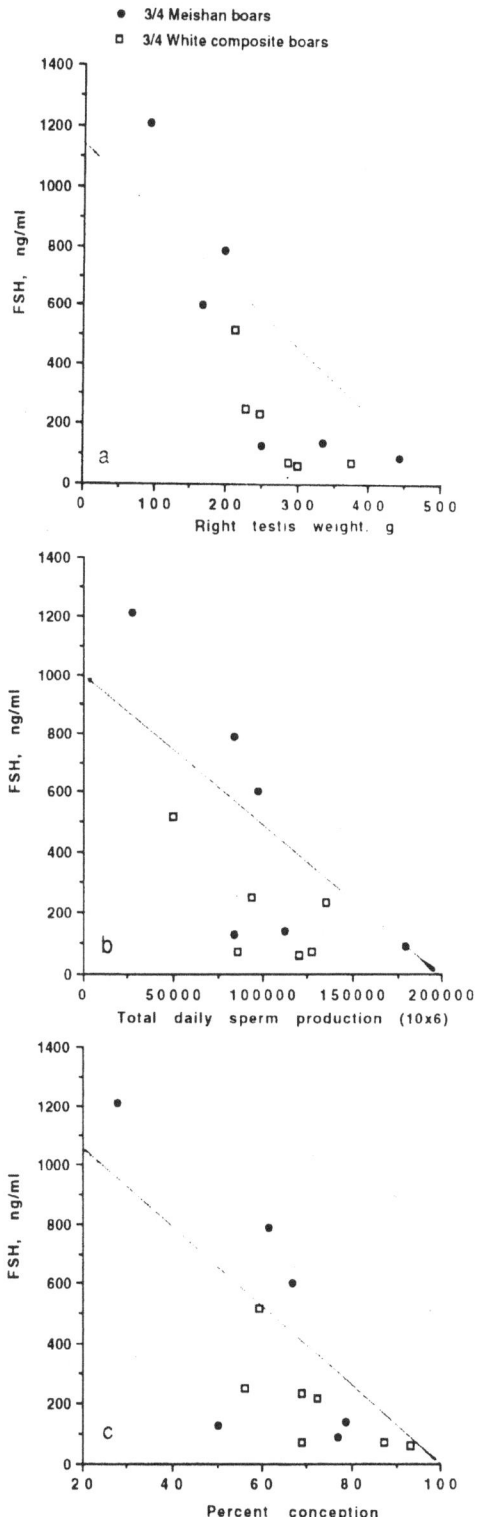

Figure 5. Relationships of serum FSH and right testis weight (a), total daily sperm production (b) and percent conception (c) in 3/4 White composite and 3/4 Meishan sires. Testis weight, total daily sperm production and percent conception were related negatively to serum FSH concentrations (p<0.01).

5. DISCUSSION

In mammals, FSH plays a major role in the regulation and maintenance of spermatogenesis.[15] Ongoing selection for high FSH concentrations in boars has made the authors recognizant of the relationships of testis size/weight, elevated FSH concentrations and possible infertility, a problem in men. Concentrations of FSH can be elevated in men with dysfunction of the seminiferous tubules and infertility.[1,4,7-9] This seems paradoxical as increased FSH should stimulate Sertoli cell function and provide for a better environment for sperm maturation.

The mechanism associated with elevated FSH and infertility in men is unknown. Increases in circulating FSH concentrations were hypothesized to result from decreased inhibin due to the damaged Sertoli cells.[16,17] The decreases in testicular weights with increases in FSH concentrations are much more prominent in the Meishan breed with higher circulating gonadotropin concentrations (Fig. 1). No differences in inhibin concentrations have been detected in White composite and Meishan boars or between Meishan boars selected for high or low circulating FSH concentrations,[18] thus the mechanism of elevated FSH concentrations in Meishan boars is unknown. Testosterone concentrations may be an indicator of Leydig cell function but were not related to testis weight or sperm production in these boar models (Fig. 2). Similar to White composite boars, most hypergonadotropic hypogonadism in man exhibits normal testosterone and luteinizing hormone concentrations suggesting Leydig functions were unaffected.[1,4,8]

Another type of hypergonadotropic hypogonadism in which both LH and FSH are elevated produces Leydig cell hypoplasia and elevated FSH associated infertility.[19] In Meishan boars in which both LH and FSH are elevated as compared to White composite boars,[18,20] Leydig cells in the testis of Meishan boars are increased in size 30%.[21] Increases in testosterone concentrations in the Meishan boars (Fig. 2) may be due to both increased LH secretion and the larger Leydig size, which would provide greater synthetic capability. Although most investigators report Sertoli cell damage in men with infertility and elevated FSH concentrations,[2,7,9] spermatogenesis can be impaired in hypergonadotropic men with increased FSH levels associated with germinal cell damage.[4,8,9] Leonard et al.[5] reported no relationship with FSH levels and sperm count, but others[7,9] noted a relationship between low sperm counts and FSH. As Kula[9] noted, decreased Sertoli function and increased circulating FSH concentrations do not necessarily alter the ability to maintain complete spermatogenesis. In fact, increased FSH may stimulate spermatogonia development as FSH receptors are also found on spermatogonia.[22]

Sires ($\frac{3}{4}$ Meishan and $\frac{3}{4}$ White composite) had inverse relationships between FSH concentrations and testicular weights, total daily sperm production and percent conception. Fertility data indicated that in these swine models a negative relationship existed between elevated FSH concentrations and testicular function. The swine models utilized in these studies are similar to FSH related alterations in spermatogenic function and infertility relationships noted in man and offer considerable advantages for clinical analysis, i.e., numbers of animals, fertility data and homogenous or heterogeneous genetic populations.

6. REFERENCES

1. Rosen, S.W., and Weintraub, D.B., 1971, Monotropic increase in serum FSH correlated with low sperm count in young men with idiopathic oligospermia and aspermia, *J. Clin. Endocrinol. Metabol.* 32:410-416.
2. de Kretser, D.M., Burger, H.G., Fortune, D., Hudson, B., Long, A.R., Paulsen, C.A., and Taft, H.P., 1972, Hormonal, histological and chromosomal studies in adult males with testicular disorders, *J. Clin. Endocrinol. Metabol.* 35:392-401.

3. van Theil, D.H., Sherins, R.J., Myers, G.H., and De Vita, V.T., 1972, Evidence of a specific seminiferous tubule factor affecting follicle-stimulating hormone secretion in man, *J. Clin. Invest.* 51:1009-1024.

4. Franchimont, P., Millet, D., Vendrely, E., Letawe, J., Legros, J.J., and Netter, A., 1972, Relationship between spermatogenesis and serum gonadotropin levels in azoospermia and oligozoospermia, *J. Clin. Endocrinol. Metabol.* 34:1003-1008.

5. Leonard, J.M., Leach, R.B., Couture, M., and Paulsen, C.A., 1972, Plasma and urinary follicle-stimulating hormone levels on oligozoospermia, *J. Clin. Endocrinol. Metabol.* 34:209-214.

6. de Kretser, D.M., Burger, H.G., and Hudson, B., 1974, The relationship between germinal cells and serum FSH levels in males with infertility, *J. Clin. Endocrinol. Metabol.* 38:787-793.

7. Hunter, W.M., Edmond, P., Watson, G.S., and McLean, N., 1974, Plasma LH and FSH levels in subfertile men, *J. Clin. Endocrinol. Metabol.* 39:740-749.

8. Wang, C., Dahl, K.D., Leung, A., Chan, S.Y.W., and Hseuh, A.J.W., 1987, Serum bioactive follicle-stimulating hormone in men with idiopathic azoospermia and oligozoospermia, *J. Clin. Endocrinol. Metabol.* 65:629-633.

9. Kula, K., 1991, Hyperactivation of early steps of spermatogenesis compromises meiotic insufficiency in men with hypergonadotropism. A possible quantitative assay for high FSH/low testosterone availabilities, *Andrologica* 23:127-133.

10. Krystek, S.R., Jr, Diaz, J.A., Reichert, L.E., Jr, and Andersen, T.T., 1985, Prediction of antigenic site in follicle stimulating hormones. Differences profiles enhance antigenicity prediction methods, *Endocrinology* 117:1125-1134.

11. Borg, K.E., Lunstra, D.D., and Christenson, R.K., 1993, Semen characteristics, testicular size and reproductive hormone concentrations in mature Duroc, Meishan, Fengjing and Minzhu boars, *Biol. Reprod.* 49:515-521.

12. Ford, J.J., Christenson, R.K., and Maurer, R.R., 1980, Testosterone concentrations in embryonic and fetal pigs during sexual differentiation, *Biol. Reprod.* 23:1033-1038.

13. Young, L.D., Leymaster, K.A., and Lunstra, D.D., 1986, Genetic variation in testicular development and its relationship to female reproductive traits in swine, *J. Anim. Sci.* 63:17-26.

14. Amann, R.P., 1970, Sperm production rates, in: *The Testis,* Volume 1 (A.D. Johnson, W.R. Gomes, and N.L. VanDemark, eds.), Academic Press, New York, pp. 443-482.

15. Marshall, G,R,, Zorub, D.S., and Plant, T.M., 1995, Follicle-stimulating hormone amplifies the population of differentiated spermatogonia in the hypophysectomized testosterone-replaced adult Rhesus monkey (Macaca mulatta), *Endocrinology* 136:3504-3511.

16. Baker, H.W.G., Bremmner, W.J., Burger, H.G., de Kretser, D.M., Dulmanis, A., Eddie, L.W., Hudson, B., Keogh, E.J., Lee, W.K., and Rennie, G.C., 1976, Testicular control of follicle stimulating hormone secretion, *Rec. Progr. Horm. Res.* 32:429-437.

17. de Jong, F.H., and Robertson, D.M., 1985, Inhibin: update on action and purification, *Mol. Cell. Endocrinol.* 35:95-103.

18. Wise, T., Lunstra, D.D., and Ford, J.J., 1996, Differential pituitary and gonadal function of Chinese Meishan and European White composite boars: Effect of GnRH stimulation, castration and steroidal feedback, *Biol. Reprod.* 54: 146-153.

19. Toledo, S.P.A., Arnhold, I.J.P., Luthold, W., Russo, E.M.K., and Saldanha, P.H., 1985, Leydig cell hypoplasia determining familial hypergonadotropic hypogonadism, in: *Endocrine Genetics and Genetics of Growth: Fourth International Clinical Genetics Seminar* (C.J. Pappadatos, and C.S. Bartsocas, eds.), Alan R Liss Inc, New York, pp. 311-314.

20. Wise, T., Lunstra, D.D., and Ford, J.J., 1994, Changes in serum testosterone and FSH in Cross-Bred (CX) and Meishan (MS) boars after castration and in response to steroidal challenge, *Biol. Reprod.* 50(Suppl 1:50.

21. Lunstra, D.D., Borg, K.E., and Klindt, J., 1993, Changes in porcine testicular structure during development in Meishan boars, *Biol. Reprod.* 48(Suppl 1):126.

22. Orth, J., and Christensen, A.K., 1978, Autoradiographic localization of specifically bound ^{125}I-labelled follicle stimulating hormone on spermatogenesis of the rat testis, *Endocrinology* 103:1944-1951.

45

REPRODUCTIVE CHARACTERISTICS IN SINCLAIR MINIATURE SWINE

Guy F. Bouchard, DVM, MS,[1,2*] Ronald M. McLaughlin, DVM, MS,[3]
Mark R. Ellersieck, PhD,[4] Gary F. Krause, PhD,[4]
Craig Franklin, DVM, MS,[5] and Chada S. Reddy, DVM, PhD[6]

[1] Sinclair Research Center
Columbia, Missouri 65203
[2] Veterinary Medicine and Surgery
[3] Office of Laboratory Animal Medicine
[4] Agricultural Experimental Station
[5] Veterinary Pathology
[6] Veterinary Biomedical Sciences
University of Missouri
Columbia, Missouri 65211

1. ABSTRACT

Three hundred seventy one litter records collected between 1985 and 1993 from 156 Sinclair(S-1) miniature sows, a Hormel derived strain of miniature swine, were analyzed retrospectively and compared with previously published records representing 1950 to 1952 and 1963 to 1965. Effects of several variables, such as season and month of parturition, age of sow, parity and litter size, on reproductive parameters of Sinclair miniature swine were evaluated. Preovulatory follicular and ovulation rates were evaluated in 230 virgin Sinclair miniature gilts. Means $\pm$ SEM for litter size, number of liveborn, number of stillborn and litter size at weaning of Sinclair(S-1) miniature swine were 7.20 ± 0.12, 6.57 ± 0.12, 0.63 ± 0.06 and 5.75 ± 0.12 piglets, respectively. From a total of 2436 liveborn piglets, 2133 (87.6%) were weaned. Litter size at birth was similar to that reported previously for this strain of swine; whereas, mean litter size at weaning increased from 4.7 piglets during 1963 to 1965 to 5.8 piglets in our study. Average birth weight decreased from 0.90 kg in 1950 to 1952 and 0.72 kg in 1963 to 1965 to 0.59 kg in our study. Reproductive variables that affected miniswine reproduction included month of parturition, age of sow, parity and litter size. Primiparous sows had the smallest litter size and lowest number of weaned piglets. Sows during their second and third parity or sows between 2 and 4 yr old had the best reproductive

* Reprint requests to: Dr. Guy Bouchard, Sinclair Research Center, Inc., 5701 South Sinclair Rd, Columbia, Missouri 65203, (314) 446-6464.

Advances in Swine in Biomedical Research, edited by Tumbleson and Schook
Plenum Press, New York, 1996

performances. Litter size and number of stillborn increased with parity and age of sow resulting in fewer piglets weaned in older sows. Litter size had a curvilinear effect on preweaning mortality with the highest mortality occurring in small and large litters. The overall preovulatory follicular and ovulation rates were 9.0 ± 0.4 and 7.5 ± 0.1, respectively. At 10.7, 20.5 and 27.0 mo, ovulation rates were 6.6, 7.9 and 8.1, respectively. Ovulation rate paralleled litter size. Overall, the production efficiency of Sinclair miniswine has been stable or improved since 1950, all reproductive variables evaluated, except season, were found to have an effect on miniswine reproduction; ovulation rate of Sinclair(S-1) miniature swine increases until about 20 mo of age and levels off thereafter; the small litter size of Sinclair(S-1) miniature swine is related to the low preovulatory follicular rate.

2. INTRODUCTION

In response to the problem of the large size of domestic pigs, several stocks of miniature swine were developed that reached a mature size small enough to be handled easily, maintained under laboratory conditions and yet possessed anatomy and physiology similar to that of domestic swine.[1] The Hormel miniature pig was developed specifically for research by the Hormel Institute in 1949 and became the first strain of miniature pig available to scientists.[2-4] The Hormel strain was developed from the crossbreeding of 4 American feral stocks of pigs.[5] A Yorkshire boar was added later to the cross for the introduction of a white hair coat.[3] Sinclair miniature pigs, direct descendants of the Hormel strain, were acquired from the Hormel Institute and raised at the University of Missouri since 1965.

The American feral pig used in the development of the Hormel strain of miniature swine represents domesticated pigs that escaped to the wild and reverted to some of the characteristics of wild pigs. Domesticated pigs from Europe were introduced to the American continent by explorers and missionaries during the colonization.[6,7] In contrast, European wild pigs have somewhat more primitive characteristics and were introduced during the later part of the nineteenth and the early part of the twentieth centuries. In some cases there has been cross breeding of European wild pigs and American feral pigs.[8] There are several differences between the 2 stocks of pigs. For instance, the European wild pig is a seasonal reproductor, while the American feral pig can breed and farrow piglets year around much like the domestic pig.[6,9,10]

There is little information about miniature pig reproduction. Additionally, the information available is often out of date, incomplete and does not mention the effect of several variables known to influence domestic swine reproduction. The objectives of the present study were to compile the recent reproductive information collected at the University of Missouri, compare this reproductive information to data previously published for the Hormel strain, evaluate preovulatory follicular and ovulation rates and evaluate any effect of variables such as season, month of parturition, age of sow, parity and litter size on reproductive parameters of Sinclair miniature pigs.

3. MATERIALS AND METHODS

Reproductive records of Sinclair(S-1) miniature, which farrowed between January, 1985 and July, 1993 at the Sinclair Research Farm (SRF) of the University of Missouri, were analyzed retrospectively. The SRF herd of miniature swine, referred to as Sinclair miniature swine, was kept as a closed herd in conventional housing. The breeding herd was maintained in large outside lots with access to shelters. All breedings were natural and breeding dates were recorded when observed.

Sows were dewormed and vaccinated during prenancy and transferred into farrowing pens 1 to 2 wk prior to the expected parturition. The ambient temperature was maintained between 72 and 78F. Supplemental heat was provided to the piglets in hutches by heating lamps to maintain their environment at about 80 to 85F. Piglets were examined during the first 24 hr of life for any congenital defects and administered 100 mg of iron dextran IM during the first da of life and 1 wk later. Piglets were vaccinated prior to weaning and a booster vaccine was administered after weaning.

The following information was collected for each litter: sow and boar identification; breeding date; parity (litter number); previous farrowing date; abortion date; current farrowing date; litter size (total number of liveborn and stillborn); liveborn males and females; stillborn (recorded as small or normal size stillborn or mummified offspring); congenital defects; litter weight at birth; weaning date; and sex, identification, weight and color of weaned piglets. Clinical data for any mortalities or sicknesses of the dam or piglets also were recorded.

The total number of liveborn and stillborn, preweaning mortality (total number of liveborn piglets that died before weaning), percent of total born weaned (weanlings/total number of piglets born) and average birth weight of liveborn were computed for each sow record. Piglets which died before or after parturition were considered as either stillborn or preweaning mortality, respectively. Physical characteristics such as size, apparent degree of maturity and lung flotation also were used to classify piglets as stillborn or preweaning mortality.

To determine preovulatory follicular rate and ovulation rate, ovaries from Sinclair miniature gilts were collected between January, 1991 and March, 1995 during standard ovariectomy or necropsy procedures on Sinclair miniature swine. At the time of ovary extraction, age of each miniature female was recorded. Once extracted, ovaries were dissected using a scalpel blade; ovarian structures were identified and counted. Graafian follicles, corpora hemorrhagicum (CH) and corpora luteum (CL) were distinguished and measured in millimeters. Graafian follicles were defined as fluid filled, smooth, convex, thin walled and fluctuating structures at palpation. Only follicles larger than 3 to 4 mm were recorded. CH were defined as well circumscribed hemorrhagic ovarian structures with a protrusion or crown of varying size. CL were defined as well circumscribed ovarian structures, liver like consistency, yellow to dark orange or brown in color and with a prominent protrusion or crown. The presence of numerous large follicles with the presence of regressing CL was indicative of the proestrous or estrus period of the estrus cycle. Preovulatory follicular rate was defined as the number of graafian follicles present during proestrus or estrus.

3.1. Statistical Analysis

Historical means, representing 1950 to 1952 and 1963 to 1965, were compared to the means representing 1985 to 1993 using a 2 tailed student-t test with the assumption that the unknown historical variance held. The effect of month and season of parturition, parity, age of sow and litter size on different reproductive parameters was evaluated by analysis of variance (ANOVA). Mean differences were determined using Fisher's least significant difference (LSD) when the overall F value was significant. For the purpose of ANOVA, sow ages were rounded to the closest unit. Effects of parity, age of sow and litter size on different reproductive parameters was evaluated further by simple regression analysis testing for linear and curvilinear (quadratic) effects. P values for goodness of fit of the model were calculated. Multiple R^2 values also were calculated. The denominator of sum of squares (model)/sum of squares (total) includes variation from many sources. R^2 values, in this case, do not indicate goodness of fit due to model after adjustment for all other identifiable sources of variation. All differences were considered significant if P was equal to or less than 0.05. Ovulation rate, as defined by the number of luteal structures on the ovaries, was compared between age classes using ANOVA. When significantly different, means were separated using the least square difference. For proestrus and estrus, the preovulatory number of graafian follicles and the number of regressing CL were recorded as 2 different observations for statistical analyses. For percent of ovulation and estimated embryonic survival, we pooled preovulatory follicular rate and ovulation rate for the 2 older groups of miniature swine, namely 20.5 and 27.0 mo, and compared this data to litter size of similar age groups. Estimated percent ovulation was calculated by dividing the ovulation rate by the preovulatory follicular rate. Estimated ovulation yielding successful conception was calculated by dividing litter size by ovulation rate.

4. RESULTS

A total of 371 gestations resulting from 156 Sinclair miniature sows between January 1985 and July 1993, excluding 5 abortions, were included in the statistical analyses. In Table 1 are summarized means, standard deviations and ranges for all reproductive parameters evaluated. A total of 2671 piglets were born, of which 2436 (91.2%) were liveborn and 235 (8.8%) were stillborn. A total of 2133 piglets were weaned for a preweaning mortality percentage (percent of liveborn piglets that died before weaning) of 12.4% or a liveborn weaning percentage of 87.6%.

Table 1. Reproductive parameters evaluated for Sinclair miniature swine during the period of January, 1985 to July, 1993

Reproductive parameters	Mean	SEM	Range Low	Range High
Sow age at parturition (yr)	2.38	0.05	0.93	6.7
Parity at parturition	2.20	0.07	1	8
Litter Size	7.20	0.12	1	16
Liveborn				
Males	3.32	0.09	0	9
Females	3.25	0.09	0	10
Total	6.57	0.12	1	13
Stillborn				
Mummified	0.03	0.01	0	1
Small Size	0.14	0.03	0	6
Normal Size	0.46	0.05	0	8
Total	0.63	0.06	0	8
Weaned				
Males	2.87	0.09	0	8
Females	2.88	0.09	0	8
Total	5.75	0.12	0	11
Percent Total Born Weaned	79.78	1.22	0	100
Preweaning Mortality	0.82	0.06	0	6
Litter Weight (kg)	3.98	0.07	0.55	8.00
Average Birth Weight (kg)	0.59	0.01	0.26	0.90
Gestation Length (da)	115.17	0.07	112	118

¶From: Bouchard et al.[11]

Table 2. Comparison of reproductive data collected between 1950 to 1952, 1963 to 1965 and 1985 to 1993

Reproductive parameter	1950-1952[§]		1963-1965[§]		1985-1993	
	N[†]	Mean[‡]	N[†]	Mean[‡]	N[†]	Mean[‡]
Litter Size (liveborn)	67	6.8	172	6.4	371	6.6
Piglets Weaned	N/A	N/A	172	4.7[a]	380	5.8[a]
Average Birth Weight (kg)	67	0.90[b]	172	0.72[c]	320	0.59[b,c]

[¶]From: Bouchard et al.[11]
[§]From: Dettmers et al.[2,4]
[†]Number of litters
[‡]Mean refers to the average number of piglets per litter or average birth weight. Means with similar superscripts in the same row differ (P<0.05).

For the Hormel miniature pig, liveborn litter size did not change between 1950 to 1952, 1963 to 1965 and 1985 to 1993 (Table 2). However, the number of piglets weaned increased between 1963 to 1965 and 1985 to 1993, while average birth weight decreased between 1950 to 1952 and 1985 to 1993, and between 1963 to 1965 and 1985 to 1993.

Month of parturition affected litter size, males liveborn, total number of liveborn piglets, males weaned, total number of weanlings, litter weight and average birth weight (Table 3). Lowest and highest litter sizes, liveborn and weanling piglets were seen in sows that farrowed in September and June or July, respectively. Differences seen between months were significant. No differences, however, were seen in number of stillborn, preweaning mortality or percent total born piglets weaned. Highest litter weight and average birth weight were seen in litters delivered in June or July. In contrast, season of parturition had no effect on any evaluated reproductive parameters.

Parity had an effect on litter size, males and females liveborn, total number of liveborn piglets, mummified offspring, small size stillborn, total number of stillborn, males and females weaned, total number of weanlings, percent total born weaned, litter weight and average birth weight (Table 4). Primiparous sows yielded the smallest litter size, lowest litter

Table 3. Effect of farrowing month on reproductive parameters of Sinclair miniature swine (Mean ± SEM)

Farrowing month	N[§]	Litter size[†]	Liveborn[†]	Weaned piglets[†]	N[§,‡]	Litter weight[†](kg)	Average birth weight[†](kg)
Jan	28	7.1±0.4[bc]	6.3±0.4[b]	5.5±0.4[bc]	25	4.0±0.3[ab]	0.60±0.02[abc]
Feb	26	7.1±0.4[bc]	6.4±0.4[b]	5.5±0.4[bc]	21	3.7±0.3[a]	0.56±0.02[ab]
Mar	27	6.6±0.4[ab]	6.3±0.4[b]	6.0±0.4[bcd]	25	3.7±0.3[a]	0.58±0.02[abc]
Apr	33	7.4±0.4[bc]	6.5±0.4[b]	5.7±0.4[bc]	29	3.7±0.3[a]	0.58±0.02[c]
May	41	7.3±0.3[bc]	7.1±0.3[bc]	6.0±0.4[bcd]	40	4.1±0.2[ab]	0.57±0.02[ab]
Jun	30	8.0±0.4[c]	7.7±0.4[c]	6.9±0.4[d]	29	4.9±0.3[c]	0.63±0.02[c]
Jul	30	8.1±0.4[c]	7.6±0.4[c]	6.6±0.4[cd]	23	4.7±0.3[bc]	0.63±0.02[c]
Aug	19	6.6±0.5[ab]	6.1±0.5[ab]	5.0±0.5[ab]	11	3.7±0.4[a]	0.63±0.03[bc]
Sep	15	5.7±0.6[a]	4.9±0.6[a]	4.1±0.6[a]	4	3.6±0.7[abc]	0.66±0.06[bc]
Oct	49	7.4±0.3[bc]	6.6±0.3[b]	5.8±0.3[ab]	48	4.0±0.2[ab]	0.61±0.02[bc]
Nov	32	6.9±0.4[ab]	6.1±0.4[ab]	5.4±0.4[ab]	27	3.5±0.3[a]	0.54±0.02[a]
Dec	41	6.9±0.3[ab]	6.2±0.3[b]	5.3±0.4[ab]	38	3.7±0.2[a]	0.60±0.02[bc]

[¶]From: Bouchard et al.[11] [§]Number of litters applies to the columns on the right. [†]Means bearing different superscripts within a column are different (P<0.05). [‡]Under certain circumstances, some litters were not weighed.

Table 4. Effect of parity on reproductive parameters of Sinclair miniature swine (Mean ± SEM)

Parity	N[§]	Litter size[†]	Liveborn[†]	Stillborn[†]	Weaned piglets[†]	Percent total born weaned[†]	N[§,‡]	Litter weight[†](kg)
1	150	6.5±0.2[a]	5.8±0.2[a]	0.64±.09[ab]	5.0±0.2[a]	77.1±1.9[a]	124	3.5±0.1[a]
2	100	7.6±0.2[bc]	7.2±0.2[b]	0.47±0.11[ab]	6.4±0.2[bc]	84.5±2.3[b]	85	4.4±0.2[bc]
3	60	7.9±0.3[bc]	7.4±0.3[b]	0.52±0.14[ab]	6.7±0.3[c]	85.2±3.0[b]	55	4.5±0.2[c]
4	35	7.2±0.4[ab]	6.3±0.4[a]	0.91±0.18[b]	5.4±0.4[a]	76.2±3.9[ab]	34	3.9±0.2[ab]
5	16	7.5±0.5[abc]	6.9±0.5[ab]	0.63±0.27[ab]	5.5±0.6[ab]	71.3±5.8[a]	13	4.0±0.4[abc]
>6	10	8.9±0.7[c]	7.0±0.7[ab]	1.90±0.34[c]	5.8±0.7[abc]	64.8±7.3[a]	9	4.3±0.5[abc]

¶From: Bouchard et al.[11]
§Number of litters applies to the columns on the right.
†Means with different superscripts within a column are different (P<0.05).
‡Under certain circumstances, some litters were not weighed.

weight, and lowest number of liveborn and weanling piglets. Second and third parity sows had the highest number of liveborn, weanling piglets, percent total born piglets weaned and litter weight and the lowest number of stillborn. Older sows with parity of 6 or higher produced litters with the greatest number of piglets, but with the highest incidence of stillbirths, and lowest percent total born piglets weaned. Normal size stillborn, preweaning mortality and average birth weight were unaffected by parity (Table 4).

Effect of sow age on reproductive parameters was similar to that of parity. Litter size, small and normal size stillborn, total number of stillborn, total number of weanlings, preweaning mortality, percent total born weaned, litter weight and average birth weight were all affected (Table 5). Yearling sows and sows older than 5 yr of age had the lowest number of weanling pigs, percent total born weaned and litter weight. Sows between the ages of 2 and 4 yr had the largest litter size and the highest number of weanling pigs, percent total born weaned and litter weight. The number of stillborn increased steadily with the sow age.

Changes in litter size, when considered as a reproductive variable, were associated with changes in all reproductive parameters evaluated except for percent total born weaned and average birth weight (Table 6). Litter size positively influenced the number of liveborn, stillborn, weanlings and litter weight. Litter size had a curvilinear effect on preweaning

Table 5. Effect of the age of the sow on reproductive parameters of Sinclair miniature swine (Mean ± SEM)

Sow age	N[§]	Litter size[†]	Stillborn[†]	Weaned piglets[†]	Percent total born weaned[†]	N[§,‡]	Litter weight[†](kg)	Average birth weight[†](kg)
1	72	6.4±0.3[a]	0.39±0.13[a]	5.2±0.3[a]	78.8±2.7[a]	69	3.4±0.2[a]	0.56±0.01[a]
2	164	7.4±0.2[bc]	0.54±0.08[ab]	6.2±0.2[b]	84.1±1.8[ab]	142	4.2±0.1[b]	0.61±0.01[b]
3	88	7.2±0.2[bc]	0.72±0.12[bc]	5.7±0.2[ab]	78.7±2.4[ab]	68	4.1±0.2[b]	0.59±0.01[ab]
4	34	7.7±0.4[bc]	1.09±0.19[c]	5.6±0.4[ab]	72.1±3.9[ab]	30	4.1±0.3[b]	0.62±0.02[b]
5	6	6.0±0.9[ab]	0.67±0.44[abc]	4.5±0.9[ab]	69.7±9.3[abc]	5	3.8±0.6[ab]	0.62±0.05[ab]
>6	7	8.6±0.8[c]	2.00±0.41[d]	4.4±0.9[a]	49.0±8.7[c]	6	3.3±0.6[ab]	0.51±0.05[a]

¶From: Bouchard et al.[11]
§Number of litters applies to the columns on the right.
†Means bearing different superscripts within a column are different (P<0.05).
‡Under certain circumstances, some litters were not weighed.

Table 6. Effect of litter size on reproductive parameters of Sinclair miniature swine
(Mean ± SEM)

Litter size	N[§]	Liveborn[†]	Stillborn[†]	Weaned piglets[†]	Preweaning mortality[†]	N[§,‡]	Litter weight[†](kg)
2	6	1.7±0.4[a]	0.33±0.44[a,b]	1.0±0.6[a]	0.7±0.4[a,b,c]	5	1.0±0.4[a]
3	12	2.7±0.3[a]	0.33±0.31[a,b]	2.3±0.4[a,b]	0.3±0.3[a,b]	10	1.6±0.3[a]
4	25	3.7±0.2[b]	0.32±0.22[a]	3.3±0.3[b,c]	0.4±0.2[a]	19	2.4±0.2[b]
5	39	4.5±0.2[c]	0.49±0.17[a]	4.0±0.2[c]	0.5±0.2[a,b]	30	2.8±0.2[b]
6	55	5.5±0.1[d]	0.51±0.15[a]	4.9±0.2[d]	0.6±0.1[a,b]	48	3.6±0.1[c]
7	58	6.4±0.1[e]	0.60±0.14[a,b]	5.6±0.2[e]	0.8±0.1[a,b,c]	51	3.7±0.1[c]
8	67	7.4±0.1[f]	0.55±0.13[a]	6.3±0.2[f]	1.2±0.1[c]	60	4.4±0.1[d]
9	54	8.0±0.1[g]	0.96±0.15[b,c]	7.3±0.2[g]	0.8±0.1[a,b]	51	4.9±0.1[e]
10	35	9.4±0.2[h]	0.57±0.18[a,b]	8.4±0.3[h]	1.0±0.2[b,c]	32	5.5±0.2[f]
11	16	9.6±0.3[h]	1.44±0.27[c]	8.6±0.4[i]	0.9±0.3[a,b,c]	13	5.4±0.3[e,f]

[¶]From: Bouchard et al.[11]
[§]Number of litters applies to the columns on the right.
[†]Means bearing different superscript within a column are different (P<0.05).
[‡]Under certain circumstances, some litters were not weighed.

mortality with the lowest incidence of preweaning mortality occurring in litter sizes varying between 3 and 6 piglets and the highest prevalence in smaller and larger litters.

A total of 230 virgin miniature swine resulting in 246 observations were evaluated in an effort to determine preovulatory follicular and ovulation rates. Animal age ranged from 9.1 to 27.3 mo. They were grouped, based on their age distribution at the time of the ovary collection, as follows: groups 1, 2 and 3 consisted of 51, 157 and 22 miniature gilts of 10.7 ± 0.04, 20.5 ± 0.02 and 27.0 ± 0.04 mo, respectively.

Based on structures present on ovaries, 48 miniature swine were either in proestrus or estrus periods of the estrus cycle. The number of preovulatory follicles on the left, right and combined ovaries were 4.9 ± 0.3, 4.1 ± 0.3, and 9.0 ± 0.4, respectively. Preovulatory follicle size ranged between 4 and 12 mm. Overall ovulation rates on the left, right and combined ovaries were 3.7 ± 0.1, 3.8 ± 0.1 and 7.5 ± 0.1, respectively. Mature CL sizes ranged between 5 and 11 mm. Estimated percents of ovulation for 10.7 and 21.4 mo old miniature gilts were 92 and 83%, respectively. Similarly, estimated ovulation yielding successful conception rates for 10.7 and 21.4 mo old miniature gilts were 92 and 89%, respectively.

5. DISCUSSION

Reproductive data collected from Sinclair miniature swine in the past 10 yr compare well with data collected 30 and 43 yr ago from their ancestors. Sinclair miniature swine still are reproductively healthy 44 yr after inception. The number of weaned piglets per sow increased in the last 44 yr which might be the result of improved management. In addition, the average birth weight of piglets decreased during the 1970's and 1980's despite no attempt to decrease the size of Sinclair miniature swine. The decrease in birth weight of piglets in the absence of an increase in the litter size could be due to management practices, such as using different criteria to classify stillborn piglets, or a continuing adjustment of the Sinclair miniature pig body weight. Dettmers *et al.*[2] did not find a correlation between birth weight and 140 da body weight, thus the meaning of the lower birth weight of the piglets remains elusive.

In temperate climates, season is recognized to have a considerable impact on domestic swine productivity. In fact, most of the seasonal effect on domestic swine productivity appears to be linked to extreme temperature and humidity fluctuations rather than nonspecific seasonal effects.[12-14] For instance, severe heat stress can delay puberty or return to estrus in postweaned sows, decrease conception rate, cause early embryonic death or abortion and even decrease litter size at weaning.[15-19] Primiparous sows appear more susceptible to the seasonal postpartum anestrus.[19] This poor productivity syndrome often is referred as seasonal or summer infertility.[20,21] Additionally, infectious disease incidence, such as diarrhea, pneumonia and septicemia, in piglets is reported to increase during fall and winter, when ambient temperature and ventilation are most difficult to control.[12-14] We did not detect any effect of season on reproductive parameters evaluated. This is peculiar since primiparous sows constituted nearly half of the population of miniature swine evaluated in this study. Lack of seasonal effects as observed in our study is not a unique finding as other investigators[15,20,21] also reported similar findings in domestic swine.

Smallest and largest litters were observed in September and June or July, respectively. In domestic swine, there is no evidence of a monthly effect on litter size. However, there is a trend toward smaller litter size resulting from breeding in late winter and hot summer months.[15] The effect of month of parturition on litter size in our study remains unexplained.

Influences of parity and sow age on reproductive parameters of Sinclair miniature swine are similar. Primiparous sows had the smallest litter size resulting in the lowest number of weaned piglets. Meanwhile, litter size increased with parity or sow age but the number of stillborn and preweaning mortality (effect of sow age only) increased as well, resulting also in fewer piglets weaned and a lower percent total born weaned in older sows. Average birth weight was heavier in 2 to 5 yr old sows. Overall, litters with the highest viability occurred during the second and third parity or in 2 to 4 yr old sows. A similar trend has been observed in domestic swine.[13,24-26] Similar to our findings, Vaillancourt[13] and Nielsen[22] reported mortality lowest in piglets born from second litter sows and highest among piglets born to sows with parity of 6 or higher.

In domestic swine, the relationship between litter size and preweaning mortality is curvilinear, with the highest prevalence of mortality occurring in small and large litters.[14] The larger the litter size, the greater the chance of placental insufficiency due to lack of uterine space.[27,28] In such cases, stressed piglets are disturbed physiologically and maladjusted at birth. Enlarged adrenal glands, abnormal cortisol levels, lower levels of hemoglobin and higher glucose levels are common findings in hypoxic stressed piglets. The fate of these piglets is seriously jeopardized since they cannot compete efficiently for a nipple.[29] These piglets often are presented as stillborn intrapartal, weak or undersized, splay legged or traumatized.[14,30] This problem is compounded when litter size exceeds the number of functional nipples.[14] Piglets do not ingest an adequate amount of colostrum and are less resistant to gastrointestinal conditions, septicemia, polyarthritis or other infectious diseases.[22,23] Similarly, litter size had a curvilinear effect on preweaning mortality in Sinclair miniature swine, with the highest prevalence of mortality occurring in small and large litters. Litter size also directly influenced litter weight but did not have an effect on average birth weight of individual piglets in our study. Contrary to conventional knowledge, lower birth weight of piglets observed in our study, in relation to historical data, did not contribute to higher preweaning mortality. This may reflect different management practices, such as using different criteria to classify stillborn, and/or a normal continuing downward adjustment of the body weight of Sinclair miniature swine. No adverse environmental or nutritional factors contributed to lower birth weight.

Ovulation rate increases from 10.7 to 20.5 mo and remains stable until 27.0 mo of age. A similar pattern of ovulation rate increase with age was seen in domestic swine. However, the increase in ovulation rate in domestic European swine appears to plateau sooner after puberty.[31] The pattern of increase of the ovulation rate in the highly prolific Chinese Meishan is closer to Sinclair miniature swine. Chinese Meishan swine ovulation rate increase is still marked at 785 da of age[31,32] (25.6 mo). Ovulation rate of gilts reaching puberty earlier (5 to 7 mo) will increase with each estrous period; whereas, in older gilts, increases in ovulation rate with successive estrous periods is reduced.[31] This certainly applies to the Chinese Meishan since it reaches puberty around 118 da of age.[31] The age at which Sinclair swine reach puberty has not been determined. However, successful accidental breedings have been known to occur in Sinclair gilts as early as 3 mo of age.

The ovulation rate of Sinclair swine appears to match very closely the litter size. The overall esitimated ovulation yielding successful conception was 91%. Embryonic survival at 30 da of gestation for the Chinese Meishan and white crossbred pigs are 73 and 81%, respectively.[32] Therefore, Sinclair miniature swine are well equipped to support the number of embryos produced and the ovulation rate appears to be the limiting factor to litter size. Similarly, the overall percent of ovulation was 83%. In other words, for an average of 9.0 preovulatory follicles, 7.5 CL were recorded. In domestic swine, during proestrus and estrus, 15 to 40 follicles approach maturity of which only 10 to 20 ovulate.[33] Therefore, Sinclair miniature swine are efficient ovulators and limited follicular growth may be responsible for the lower ovulation rate than domestic swine.

We established a database of important reproductive parameters for Sinclair miniature swine. These reproductive parameters have remained stable during the past 44 yr and are similar to those of domestic and American feral swine. Such an extensive evaluation of the reproductive traits of miniature swine should be useful to scientists involved in teratology, pediatrics or any other research fields either directly or indirectly related to reproduction.

6. ACKNOWLEDGMENTS

Funded in part by the Canadian Medical Research Council. The authors wish to thank all personnel of the Sinclair Comparative Medicine Research Farm and particularly, Dr. Joe Safron, Mr. Darrell Hale, Mr. Ray Glendening and Mrs. Barbara Rodgers for their invaluable help in data collection.

7. REFERENCES

1. Bustad, L.K., and McClellan, R.O., 1968, Miniature swine: development, management and utilization, *Lab. Anim. Care* 18:280-287.
2. Dettmers, A.E., Rempel, W.E., and Comstock, R.E., 1965, Selection for small size in swine, *J. Anim. Sci.* 24:216-220.
3. Dettmers, A.E., and Rempel, W.E., 1968, Minnesota's miniature pigs, *Lab. Anim. Care* 18:104-109.
4. Dettmers, A.E., Rempel, W.E., and Hacker, D.E., 1971, Response to recurrent mass selection for small size in swine, *J. Anim. Sci.* 33:1212-1215.
5. England, D.C., and Panepinto, L.M., 1986, Conceptual and operational history of the development of miniature swine, in: *Swine in Biomedical Research* (M.E. Tumbleson, ed.), Plenum Press, New York, pp. 17-22.

6. Belden, R.C., and Frankenberger, W.B., 1989, History and biology of feral swine, *Proc. Feral Pig Symp.* (April, 1989), pp. 3-10.

7. Hanson, R.P., and Karstak, L., 1959, Feral swine in the southeastern United States, *J. Wildl. Manage.* 23:64-74.

8. Mersman, H.J., 1986, The pig: a concise source of information, in: *Swine in Cardiovascular Research*, Volume 1 (H.C. Stanton, and H.J. Mersmann, eds.), CRC Press, Boca Raton, pp. 1-9.

9. Delcroix, I., Mauget, R., and Signoret, J.P., 1989, Existence d'une synchronisation de la reproduction au sein d'un groupe social chez le sanglier, *J. Rech. Porcine en France* 21:133-138.

10. Sweeney, J.M., Sweeney, J.R., and Provost, E.E., 1979, Reproductive biology of a feral hog population, *J. Wildl. Manage.* 43:555-559.

11. Bouchard, G.F., McLaughlin, R.M., Ellersieck, M.R., Krause, G.F., Franklin, C., and Reddy, C.S., 1995, Retrospective evaluation of production characteristics in Sinclair miniature swine — 44 years later, *Lab. Anim. Sci.* 45:408-414.

12. Nielsen, N.C., Bille, N., Larsen, J.L., and Svendsen, J., 1975, Preweaning mortality in pigs - 7. Polyarthritis, *Nord. Vet-Med.* 27:529-543.

13. Gracey, J.F., 1955, Survey of pig losses, *Vet. Rec.* 67:984-990.

14. Vaillancourt, J.P., Dial, G.D., Marsh, W.E., and Tubbs, R.C., 1991, Enzootic mortality among piglets between birth and weaning, *Compend. Cont. Ed.* 13:1642-1650.

15. Clark, L.K., 1986, Factors influencing live litter size, in: *Current Therapy in Theriogenology* (D.A. Morrow, ed.), W.B. Saunders, New York, pp. 928-930.

16. Wildt, D.E., Riegle, G.D., and Dukelow, W.R., 1975, Physiological temperature response and embryonic mortality in stressed swine, *Am. J. Physiol.* 229:1471-1475.

17. Omtvedt, I.T., Nelson, R.E., Edwards, R.L., Stephens, D.F., and Turman E.J., 1971, Influence of heat stress during early, mid and late pregnancy of gilts, *J. Anim. Sci.* 32:312-317.

18. Mavrogenis, A.P., and Robison, O.W., 1976, Factors affecting puberty in swine, *J. Anim. Sci.* 42:1251-1255.

19. Hurtgen, J.P., 1986, Noninfectious infertility in swine, in: *Current Therapy in Theriogenology* (D.A. Morrow, ed), W.B. Saunders, New York, pp. 962-966.

20. Love, R.J., 1978, Definition of a seasonal infertility problem in pigs, *Vet. Rec.* 103:443-446.

21. Hennessy, D.P., and Williamson, P.E., 1984, Stress and summer infertility in pigs, *Aust. Vet. J.* 61:212-215.

22. Nielsen, N.C., Christensen, K., Bille, N., and Larson, J.L., 1974, Preweaning mortality in pigs - 1. Herd investigations, *Nord. Vet-Med.* 26:137-150.

23. Svensmark, B., Jorsal, S.E., Nielsen, K., and Willeberg, P., 1989, Epidemiological studies of piglet diarrhoea in intensively managed Danish sow herds I. Pre-weaning diarrhoea, *Acta Vet. Scand.* 30:43-53.

24. Rasajski, M., 1990, The investigation of sow fertility in connection with the age of boar and sow at fertilization, *World Rev. Anim. Prod.* 25:23-38.

25. Leman, A.D., Knudson, C., Rodeffer, H.E., and Mueller, A.G., 1972, Reproductive performance of swine on 76 Illinois farms, *J. Am. Vet. Med. Ass.* 161:1248-1250.

26. Kernkamp, H.C.H., 1965, Birth and death statistics on pigs of preweaning age, *J. Am. Vet. Med. Ass.* 146:337-340.

27. Christenson, R.K., Leymaster, K.A., and Young, L.D., 1987, Justification of unilateral hysterectomy-ovariectomy as a model to evaluate uterine capacity in swine, *J. Anim. Sci.* 65:738-744.

28. Knight, J.W., Bazer, F.B., Tatcher, W.W., Frank D.E., and Wallace, H.D., 1977, Conceptus development in intact and unilaterally hysterectomized-ovariectomized gilts: interrelationships among hormonal status, placental development, fetal fluids and fetal growth, *J. Anim. Sci.* 44:620-637.

29. Bereskin, B., Shelby, C.E., and Cox, D.F., 1973, Some factors affecting pig survival, *J. Anim. Sci.* 36:821-827.

30. Svendsen, J., 1992, Perinatal mortality in pigs, *Anim. Reprod. Sci.* 28:59-67.

31. Christenson, R.K., 1993, Ovulation rate and embryonic survival in Chinese Meishan and white crossbred pigs, *J. Anim. Sci.* 71:3060-3066.

32. Faillace L.S., Biggs, C., and Hunter, M.G., 1994, Factors affecting the age at onset of puberty, ovulation rate and time of ovulation in Chinese Meishan gilts, *J. Reprod. Fertil.* 100:353-357.

33. Roberts, S.J., 1986, *Veterinary Obstetrics and Genital Diseases*, Edwards Brothers, Ann Arbor, pp. 636-653.

46

A COMPARISON OF PHYSIOLOGICAL AND BIOCHEMICAL PARAMETERS IN FULLY CONSCIOUS HORMEL-HANFORD AND YUCATAN STRAINS OF MINIATURE PIG

George J. Ikeda, PhD,[*] Theodore C. Michel, BS, Dennis W. Gaines, MS, Vira L. Olivito, Philip P. Sapienza, BS, Leonard Friedman, PhD, Curtis N. Barton, PhD, and Michael W. O'Donnell, MS

U.S. Food and Drug Administration
Center for Food Safety and Applied Nutrition
Beltsville Research Facility
8501 Muirkirk Road
Laurel, Maryland 20708

1. INTRODUCTION

The Yucatan miniature pig was introduced to FDA's Beltsville Research Facility in the early 1980s. We were impressed by the docile, calm temperament of this strain of miniature pig. The existing strain of miniature pig at our facility was a cross between the Hormel and Hanford strains, bred to be a uniform, compact sized animal with a known pedigree. Because of similarities in the cardiovascular systems of humans and swine, swine often are used in cardiovascular risk assessment studies. The objective of these experiments was to determine whether the docile, calm temperament of the Yucatan miniature pig would make this strain more useful or advantageous than our resident strain in the study of cardiovascular responses to cardioactive agents, particularly while the animals were conscious and restrained. These experiments were designed to measure a spectrum of physiological and biochemical responses to restraint, intravenous intervention and cardioactive agent administration in groups of both Hormel-Hanford and Yucatan miniature pigs.

---

[*] Reprint requests to G.J. Ikeda, U.S. Food and Drug Administration, HFS-506, Beltsville Research Facility, 8501 Muirkirk Rd, Laurel, MD 20708 (301) 594-1516.

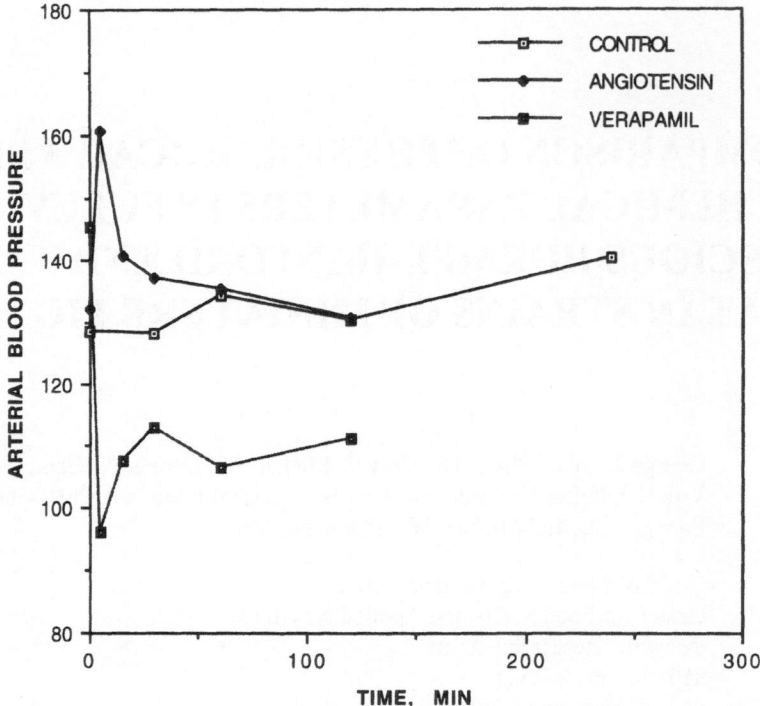

Figure 1. Arterial blood pressure, Hormel strain.

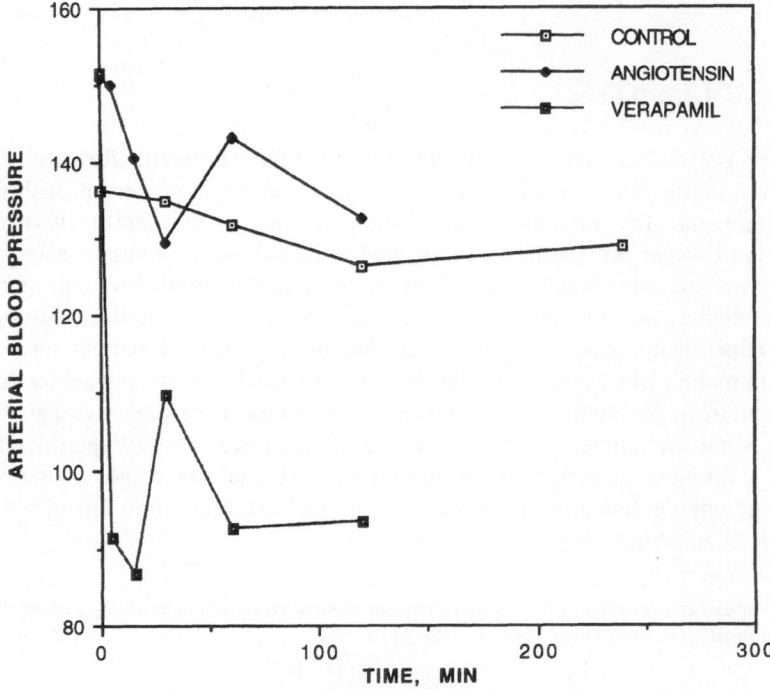

Figure 2. Arterial blood pressure, Yucatan strain.

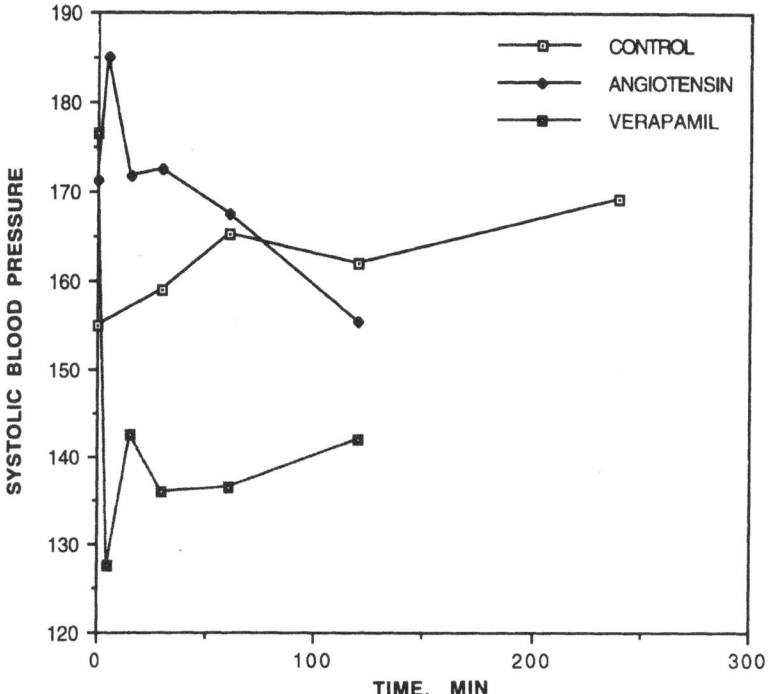

Figure 3. Systolic blood pressure, Hormel strain.

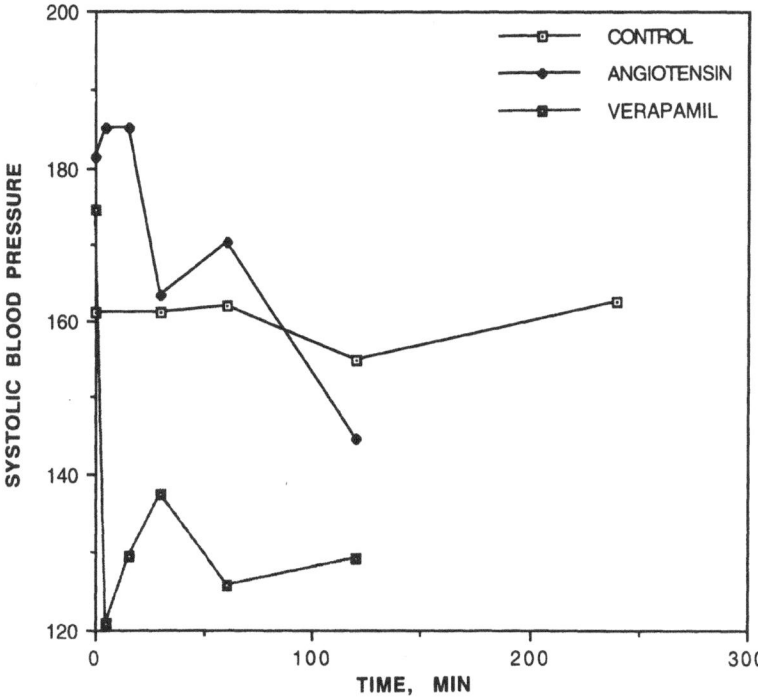

Figure 4. Systolic blood pressure, Yucatan strain.

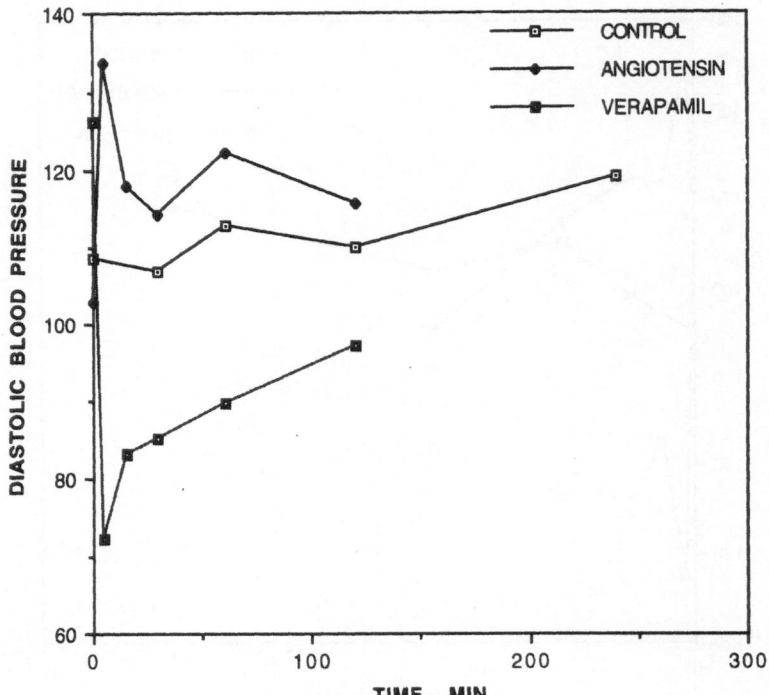

Figure 5. Diastolic blood pressure, Hormel strain.

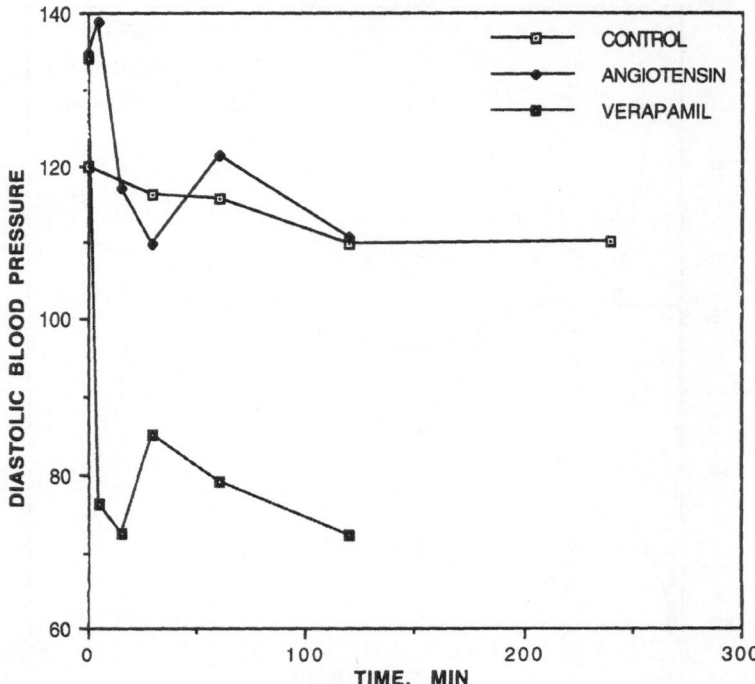

Figure 6. Diastolic blood pressure, Yucatan strain.

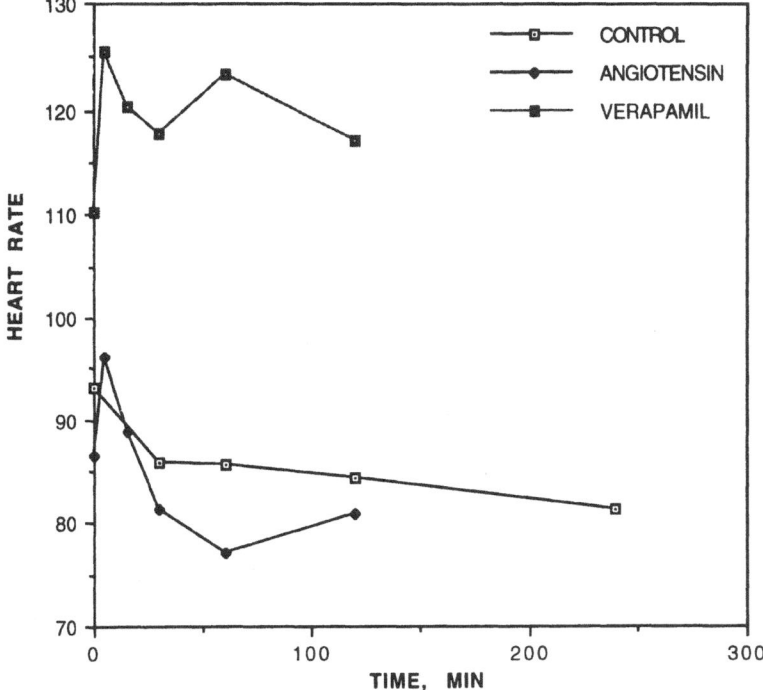

Figure 7. Heart rate, Hormel strain.

2. MATERIALS AND METHODS

2.1. Animals

The Hormel-Hanford pigs used in these experiments were bred by a contractor, Environmental Consultants, Inc., Suffolk, VA. The Yucatan miniature pigs were purchased from Colorado State University, Fort Collins, CO. The animals were housed in individual kennels and fed SR 1160 diet (obtained from the Granary, U.S. Department of Agriculture, Beltsville, MD) and tap water *ad libitum*. None of the animals had been used previously in any experiment.

2.2. Equipment

The Critikon neonatal cuff, Size No. 1; Dinamap research monitor, Model 1255; and Dinamap trend recorder were purchased from Critikon, Inc., Tampa, FL. The Panepinto sling was purchased from Colorado State University, Fort Collins, CO. PE-90 polyethylene tubing can be obtained from any scientific supply company.

2.3. Chemicals and Reagents

Angiotensin II was obtained from Sigma Chemical Co., St. Louis, MO; verapamil was obtained from Knoll Pharmaceutical, Whippany, NJ. The Cat-A-Kit (catecholamine [³H] radioenzymatic assay) was obtained from AMERSHAM, Arlington Heights, IL; the Cortisol

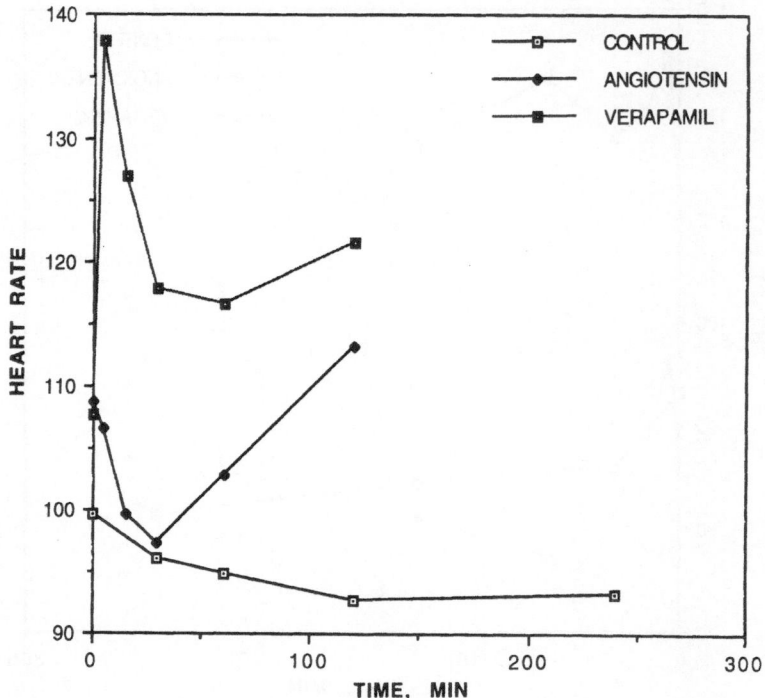

Figure 8. Heart rate, Yucatan strain.

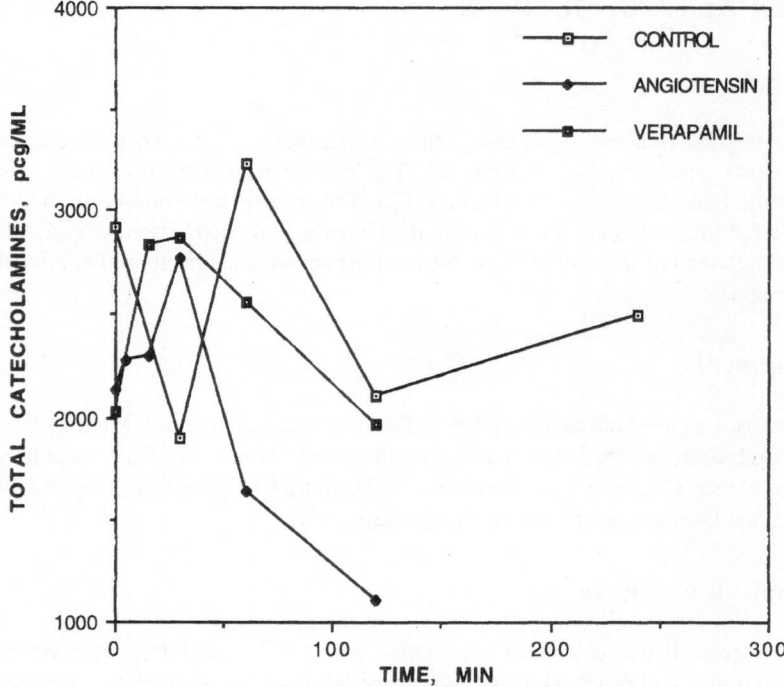

Figure 9. Blood catecholamines, Hormel strain.

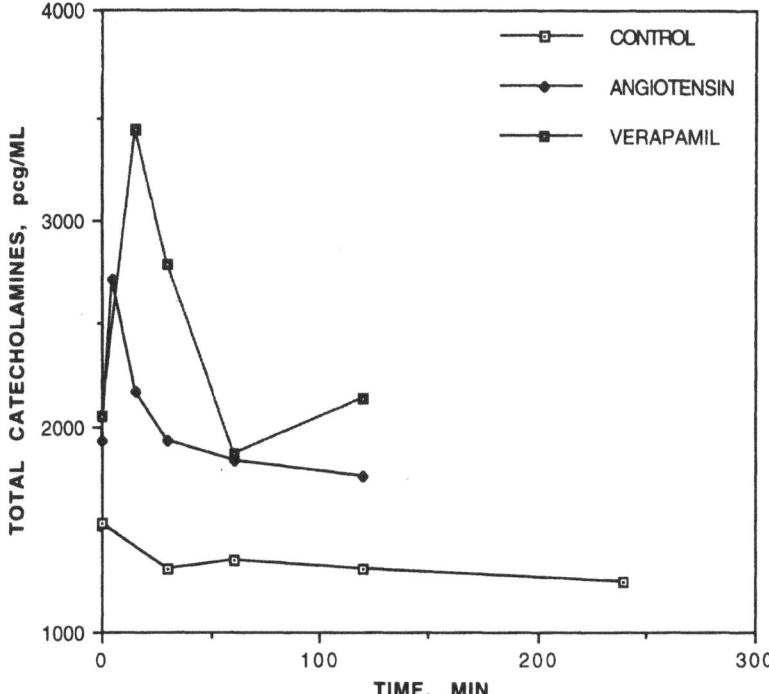

Figure 10. Blood catecholamines, Yucatan strain.

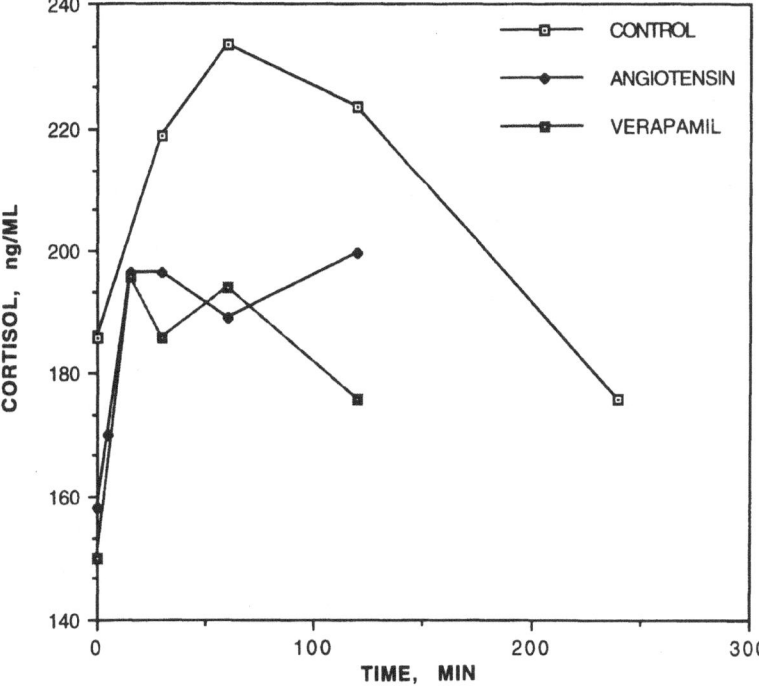

Figure 11. Blood cortisol levels, Hormel strain.

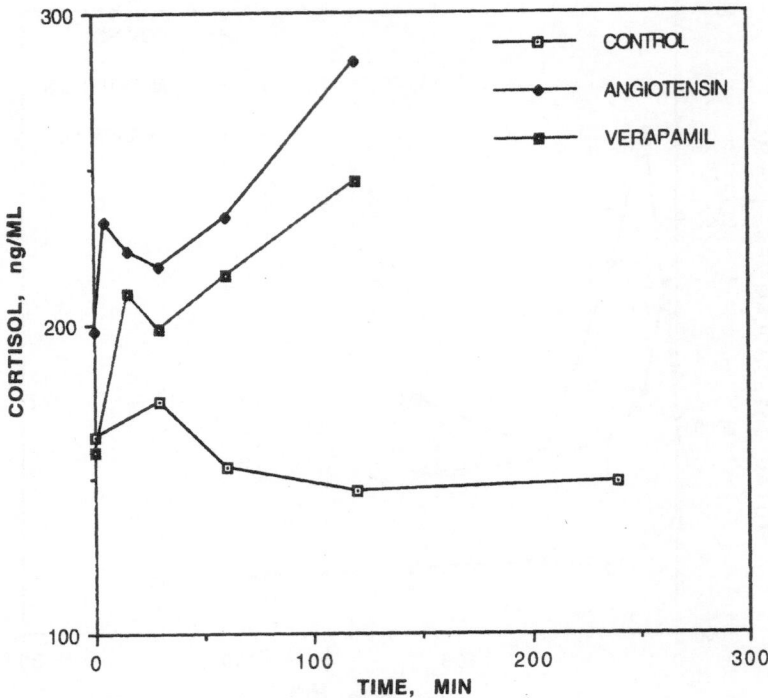

Figure 12. Blood cortisol levels, Yucatan strain.

(H³) Kit (steroid radioimmunoassay kit) was obtained from Radioassay Systems Laboratories, Inc., Carson, CA.

2.4. Experimental Procedures

Two groups of 5 adult female Yucatan and 5 adult female Hormel-Hanford miniature pigs were used in the study. Animals in both groups were housed individually in pens and maintained under similar conditions. Both groups were untrained with respect to exposure to restraint systems. After 3 acclimation sessions, the pigs were used in the experiments. In an acclimation session, the pigs were placed in a Panepinto sling for a period of 1 hr, but no other test procedures were implemented during the session. Each Yucatan minipig was paired with a Hormel-Hanford minipig.

On the day of experiment, each pig was placed in a V-trough. A 16 ga needle was directed caudodorsally toward the brachiophalic vein, a small amount of blood was withdrawn and the syringe was removed. Intramedic PE-90 polyethylene tubing was inserted through the needle into the vein, the needle was removed and the cannula was sutured to the pig's neck. The cannula was filled with heparinized saline, and the pig was placed in the Panepinto sling and moved to a quiet room. A Critikon neonatal cuff was placed snugly around the pig's tail and taped in place. The cuff was attached to a Dinamap research monitor (Model 1255). A Dinamap trend recorder was used to record the heart rate, mean arterial blood pressure, and systolic and diastolic blood pressures.

Each minipig participated in 5 experimental sessions. The first 3 were control sessions, in which the pig was placed in the Panepinto sling and fitted with IV cannula and Dinamap monitor. Measurements of heart rate and systolic, diastolic and mean arterial blood pressures were obtained at various times after the pig was placed in the sling; blood samples were taken via the cannula for determination of total catecholamine and cortisol levels. In each session, the pig was allowed a period of time to "settle down" in the sling. In most cases, this required 10 to 15 min after attachment of the monitoring equipment (readily determined by monitoring heart rate and blood pressure). Following this, physiological measurements and blood samples were taken at 0, 1/2, 1, 2 and 4 hr. In the fourth session, after attachment of the cannula, placement in the sling and attachment of the physiological monitoring equipment, the pig was administered IV 10 μg/kg angiotensin II in saline (via an ear vein) to elevate blood pressure. During the fourth session, responses were measured at 0, 5, 15, 30, 60 and 120 min; blood specimens were taken at 0, 15, 30, 60 and 120 min. After a suitable recovery period (at least 1 wk), the pig was used in a fifth session. 0.5 mg/kg of verapamil was administered IV (via an ear vein) to lower blood pressure. Physiological responses were measured at 0, 5, 15, 30, 60 and 120 min; blood specimens were taken at 0, 15, 30, 60 and 120 min.

2.5. Chemical Analyses

Blood specimens were analyzed according to the directions on Cat-A-Kit and the Cortisol (H[3]) Kit.

2.6. Statistical Analyses

Because the purpose of the study was to compare species under different conditions, separate analyses were performed for the control, angiotensin and verapamil data. For the control data, the mean for the three control sessions was used.

A repeated measures analysis of variance (ANOVA) was used for each response measure.[1] Strain was a between subjects effect, min (after start of the experiment) and the min by strain interaction were within subject effects. Tests of within subject effects were conducted using the Greenhouse-Geisser adjusted F-test.[2] All tests were conducted with a level of 0.05 for statistical significance.

3. RESULTS AND DISCUSSION

Both species of miniature pig responded in similar fashion to the administration of angiotensin or verapamil: as expected, administration of a single dose of angiotensin II elevated the blood pressure, with an immediate effect that was evident for less than 1 hour. The administration of a single dose of verapamil resulted in an immediate decrease in blood pressure, with an immediate effect that was evident more than 2 hr after the dose. (Figures 1-6). Verapamil increased heart rate in both species and the effect was evident longer than 2 hr after dosing; a single dose of angiotensin II elevated heart rate slightly in the Yucatan strain of minipig, but had little effect on heart rate in the Hormel strain. (Figures 7 and 8). Mean values for the various measurements, along with the standard errors of the mean (SEM), are presented in Tables 1-6. One strain does not appear to be more advantageous than the other if the goal of the experiment is to measure physiological parameters such as blood pressure or heart rate.

Measurement of blood catecholamines showed a marked difference between the strains of miniature pig. Catecholamines were elevated by treatment with angiotensin II and verapamil, but the control values for catecholamines fluctuated so widely in the Hormel-Hanford strain that it was difficult to determine whether or not there was an effect in this strain (Figure 9). By contrast, the control (or baseline) levels of catecholamine were

Table 1

Time, Min	Control	S E M	Angiotensin	S E M	Verapamil	S E M
			Hormel strain			
0	128.5	5.3	132.0	10.7	145.4	8.2
5	-	-	160.8	8.3	96.2	5.6
15	-	-	140.8	13.7	107.6	3.5
30	128.3	3.6	137.2	13.9	112.8	7.8
60	134.3	3.7	135.4	13.5	106.6	9.4
120	130.4	4.7	130.6	18.1	111.2	7.5
240	140.2	3.9	-	-	-	-
			Yucatan			
0	136.3	7.4	150.8	7.1	151.8	7.1
5	-	-	150.0	3.7	91.5	12.5
15	-	-	140.6	8.0	87.0	18.7
30	134.9	5.7	129.4	11.2	109.6	15.9
60	131.8	5.7	143.2	11.2	92.6	16.6
120	126.3	4.5	132.6	3.6	93.4	12.8
240	128.9	3.8	-	-	-	-

N = 15 animals for controls and N=5 for treated animals

* Units are millimeters of mercury

Table 2. Systolic blood pressure*, Hormel and Yucatan strain

Time, Min	Control	S E M	Angiotensin	S E M	Verapamil	S E M
			Hormel			
0	154.9	5.9	171.2	8.7	176.4	5.0
5	-	-	185.0	7.3	127.6	6.0
15	-	-	171.8	7.0	142.4	9.5
30	159.0	4.6	172.4	10.6	136.0	9.9
60	165.2	3.8	167.4	9.2	136.6	10.3
120	161.9	3.3	155.6	15.4	142.0	8.8
240	169.2	4.4	-	-	-	-
			Yucatan			
0	161.1	8.2	181.4	6.8	174.6	11.3
5	-	-	185.2	10.7	120.8	14.6
15	-	-	185.2	9.5	129.4	14.3
30	161.1	6.5	163.4	16.5	137.4	14.3
60	161.9	6.2	170.2	13.1	125.6	14.9
120	154.9	5.3	144.6	12.4	129.2	14.5
240	162.5	2.4	-	-	-	-

N = 15 animals for controls and N=5 for treated animals

* Units are millimeters of mercury.

Table 3. Diastolic blood pressure[*], Hormel and Yucatan strain

Time, Min	Control	S E M	Angiotensin	S E M	Verapamil	S E M
			Hormel			
0	108.5	5.5	102.8	12.7	126.4	12.2
5	-	-	133.8	13.7	72.4	4.3
15	-	-	118.0	16.7	83.2	5.7
30	106.9	3.7	114.2	15.7	85.2	6.8
60	112.9	4.3	122.2	16.9	89.8	7.5
120	110.1	6.1	115.6	21.2	97.2	6.2
240	119.2	4.4	-	-	-	-
			Yucatan			
0	120.1	6.6	134.8	5.6	134.2	8.3
5	-	-	139.0	5.9	76.3	10.7
15	-	-	117.0	7.9	72.6	14.5
30	116.3	5.3	109.8	8.5	85.2	10.4
60	115.7	5.0	121.4	10.4	79.2	10.6
120	109.6	4.4	110.6	6.6	72.2	10.2
240	109.9	4.2	-	-	-	-

N = 15 animals for controls and N=5 for treated animals
[*] Units are millimeters of mercury.

Table 4. Heart rate[*], Hormel and Yucatan strain

Time, Min	Control	S E M	Angiotensin	S E M	Verapamil	S E M
			Hormel			
0	93.1	3.6	86.4	4.7	110.2	9.8
5	-	-	96.2	2.7	125.6	13.7
15	-	-	88.8	8.0	120.4	12.9
30	85.9	3.4	81.4	5.4	117.8	12.3
60	85.7	2.8	77.0	2.9	123.4	5.3
120	84.3	4.3	81.0	3.7	117.2	4.7
240	81.4	3.0	-	-	-	-
			Yucatan			
0	99.7	4.6	108.8	15.1	107.6	10.6
5	-	-	106.6	14.6	137.8	16.4
15	-	-	99.6	12.0	127.0	8.6
30	96.1	5.8	97.4	11.6	117.8	10.2
60	94.9	6.7	102.8	12.6	116.6	11.0
120	92.7	5.8	113.2	13.7	121.6	9.6
240	93.3	6.8	-	-	-	-

N = 15 animals for controls and N=5 for treated animals
[*] Units are millimeters of mercury.

Table 5. Blood catecholamines[*], Hormel and Yucatan strain

Time, Min	Control	S E M	Angiotensin	S E M	Verapamil	S E M
			Hormel			
0	2917	530	2131	502	2030	332
5	-	-	2273	511	-	-
15	-	-	2299	428	2831	320
30	1903	309	2773	566	2869	192
60	3224	521	1639	320	2556	439
120	2102	222	1111	55	1962	376
240	2485	560	-	-	-	-
			Yucatan			
0	1531	295	1935	302	2053	383
5	-	-	2719	414	-	-
15	-	-	2172	558	3447	663
30	1312	171	1928	288	2786	648
60	1350	238	1840	251	1863	626
120	1307	158	1764	265	2136	367
240	1242	196	-	-	-	-

N = 15 animals for controls and N=5 for treated animals
[*] Units are picograms per milliliter.

consistently low in the Yucatan strain, so that effects of angiotensin II and verapamil on blood catecholamines were seen (Figure 10). Blood cortisol levels in the Hormel-Hanford control group were so elevated that any effects from angiotensin II or verapamil were obscured (Figure 11). However, in the Yucatan strain, blood cortisol levels in the control group were sufficiently low and consistent, so that the effects of angiotensin II and verapamil were evident (Figure 12). If experiments involve measurement of blood catecholamine or cortisol levels, there appears to be a good reason to select the Yucatan strain over the Hormel-Hanford strain of miniature pig, as the control levels of catecholamines and cortisol were lower and more consistent in the Yucatan strain.

Table 6. Blood cortisol[*], Hormel strain

Time, Min	Control	S E M	Angiotensin	S E M	Verapamil	S E M
			Hormel			
0	186	19	158	13	150	12
5	-	-	170	14	-	-
15	-	-	196	24	196	16
30	219	26	196	22	186	17
60	233	28	189	33	194	17
120	223	19	200	23	176	20
240	176	14	-	-	-	-
			Yucatan			
0	164	8	198	33	159	12
5	-	-	233	37	-	-
15	-	-	224	35	210	19
30	175	14	219	44	198	25
60	154	15	234	48	215	25
120	146	18	284	37	246	19
240	149	21	-	-	-	-

N = 15 animals for controls and N=5 for treated animals
[*] Units are nanograms per milliliter plasma.

4. CONCLUSIONS

There does not appear to be any advantage to the use of one strain of miniature pig over the other if the design of the experiment involves measurement of physiological parameters such as blood pressure or heart rate. Whichever strain the investigator finds more convenient can be used. On the other hand, if the experiments involve measurement of biochemical parameters such as catecholamines or cortisol, the stable and lower control values for these parameters shown by the Yucatan strain would make it preferable.

5. ACKNOWLEDGMENTS

The authors thank Messrs. Roger Matthews, Widmark Johnson, Terry Gaither and Stewart Long for their help in husbandry and manipulation of the miniature swine during the experiments. We also thank Ms. Carolyn Jeletic for editing the manuscript.

6. REFERENCES

1. Kirk, R.E., 1968, *Experimental Design: Procedures for the Behavioral Sciences*, Wadsworth Publishing Company, Belmont, California.
2. Greenhouse, S.W. and Geisser, S., 1959, On methods in the analysis of profile data. *Psychometrika* 24: 95-112.

47

BIOCHEMICAL INDICES OF STRESS ASSOCIATED WITH SHORT-TERM RESTRAINT IN HORMEL AND YUCATAN MINIATURE SWINE

Leonard Friedman, PhD,[1][*] Linda Panepinto, BS,[2] Dennis W. Gaines, MS,[1] Ruey Chi, PhD,[1] Robert C. Braunberg, MS[1] and James Terris, PhD[3]

[1] U.S. Food and Drug Administration
Division of Toxicological Research
8501 Muirkirk Road
Laurel, Maryland 20708
[2] P&S Associates
Box 192, Redstone Canyon
Masonville, Colorado 80541
[3] Uniformed Services University of the Health Sciences
F. Edward Hebert School of Medicine
4301 Jones Bridge Road
Bethesda, Maryland 20814-4799

1. INTRODUCTION

One difficulty in conducting animal experiments and attempting to assess the effect of treatment on various physiological, biochemical and immunological functions is the frequently unavoidable imposition of stress factors (known or unknown) on the results of the measurements. One of these factors, which is especially a problem when working with larger animals, is the stress of restraint. Restraint usually is necessary for obtaining serial blood or other body fluid specimens and for monitoring bodily functions under relatively controlled conditions. Restraint related stress has been shown to influence a number of biological phenomena,[1-3] including those related to immune function[4] or *in utero* development,[5] which in turn could be related to changes in endocrine or neuroendocrine organ activity.[6,7]

[*] Reprint requests to Dr. Leonard Friedman, U.S. Food and Drug Administration, Division of Toxicological Research, 8501 Muikirk Road, Laurel, MD 20708 (301) 594-3062.

At the Beltsville Research Facility (BRF), miniature pigs are one of our animal models and occasionally must be placed under restraint when certain procedures are conducted. In this study, 2 strains of miniature swine, subjected to 3 different commonly used methods of restraint, were assessed for their relative degrees of stress by measuring levels of cortisol, ß-endorphin and lactic acid dehydrogenase (LDH) in blood sera. Cortisol, which is secreted from the adrenal cortex in response to adrenocorticotrophic hormone, and ß-endorphin, which is synthesized and stored within cells of the pars distalis and pars intermedia of the pituitary,[8,9] are released into the blood after stressful stimuli of various sorts.[10,11] Blood LDH was measured because in preliminary studies it was found to rise occasionally in miniature pigs maintained under restraint conditions for varying periods of time.

2. MATERIALS AND METHODS

2.1. Animals

Hormel-Hanford (FDA colony) and Yucatan (Colorado State University) male and female young adult miniature swine (48 to 75 kg; 19 to 23 months) were housed in individual concrete floor pens at ambient temperatures and fed on a standard diet (diet #1160, obtained from the U.S. Department of Agriculture Research Center, Beltsville, MD). Tap water was available *ad libitum*. Each piglet received 1 cc of iron dextran IM within 24 hr after birth; shortly after weaning at 8 wk, pigs were wormed and certified to be free of brucellosis and pseudorabies. Animals were grouped by sex and breed; all groups were maintained under similar environmental conditions, including pen size, feeding schedule, cleaning and contact with handlers and personnel. Pigs were tested in temperature controlled rooms (22 to 25C), although pens were at ambient temperatures. All animals and their maintenance for this study, which was conducted during the spring and summer of 1983, were under the care and supervision of a facility veterinarian. All procedures were conducted with utmost considera-tion given to humane treatment.

2.2. Experimental Design

Pigs were restrained under five different conditions using three types of restraint devices. Blood was removed by venipuncture of the anterior vena cava.

In Restraint Condition A, after 2 wk of maintenance in the assigned pens, animals were placed individually in the Panepinto sling[12] and bled 3 to 5 min later. In Restraint Condition B, 2 wk after Restraint Condition A, animals were place in a V-trough device and bled within 3 to 5 min. In Restraint Condition C, 2 wk after Restraint Condition B, animals were restrained with a snare and bled in approximately 3 to 5 min. In Restraint Condition D-1, 2 wk after Restraint Condition C, all pigs were trained, including 3 experiences in the sling of 5 min each. At the third slinging, blood was removed by venipuncture 3 to 5 min and 10 min after slinging; while this phase was being conducted, the area experienced a heat wave (temperatures of about 95°F). In Restraint Condition D-2, 2 wk after Restraint Condition D-1, pigs were restrained and tested again as for D-1 except that normal seasonal temperatures were present. Vena cavas of a small number of pigs of both sexes and strains were catheterized in the sling. After 2 hr, blood was removed from pigs loose in the pen without any restraint; these animals were categorized as "controls".

Another series of experiments was designed to test the effects of longer periods of restraint in the sling on some of the same pigs used for the other experiments described above. Three of the Hormel pigs and one of the Yucatan pigs were placed in the sling after catheterization of their vena cavas, and blood specimens were taken at intervals from 3 to 5 min up to 6 hr. Observations about behavior of the pigs were also noted. Blood was drawn into serum collection EDTA Vacutainer tubes (Becton, Dickinson & Co., Rutherford, NJ) and processed for plasma by centrifugation under refrigeration; the serum was stored at -75C.

2.3. Serum Analyses

All specimens were assayed at least in duplicate. Cortisol was measured by a radioimmunoassay procedure using a cortisol (^{3}H) kit (Radioassay Systems Laboratories, Inc., Carson, CA). The antiserum used cross reacted substantially with prednisolone (85%), 21-desoxycortisol (18.9%), 11-desoxycortisol (8.2%) and corticosterone (2.25%). Minimum detection level of cortisol was 0.05 ng/ml.

The ß-endorphin was determined by a radioimmunoassay using rabbit antiserum[13] and I-125 ß-endorphin. This assay can measure as little as 10 to 15 pg ß-endorphin. The antibody does not cross react with α-endorphin. LDH was measured by the method of Kornberg[14] in which the rate of NADPH oxidation is followed in the presence of pyruvate at pH 7.3. Units were expressed arbitrarily as change in absorbance at 340 nm/min/ml of sera.

2.4. Statistics

A one or two way ANOVA (as appropriate) was carried out, followed by an LSD test for multiple comparison among the means. In cases where the normality assumption was not met, results were verified with a Kruskal-Wallis one way ANOVA by ranks procedure. A p-value of ≤0.05 was accepted as an indication of difference.

3. RESULTS

3.1. Restraint Effects on Cortisol Levels

The effects of the various methods of restraint on cortisol levels are compared in Table 1. The only differences among restraints *per se* were seen with male Hormel pigs in which values were higher for the sling than for the V-trough or the snare. These results were reflected in the other groups tested. Remaining in the sling for another 5 to 7 min appeared to result in an increase in cortisol levels, although a difference was seen only for males. The only sex related difference was for the V-trough and Yucatan pigs; in this case, the value for females was almost twice that of males. There were no strain related differences. Because of a high degree of variability, especially in the no restraint group, no generalization can be made regarding the effects of restraint vs no restraint (control) on this parameter.

When all data were combined by not treating sex or strain as separate subgroups, replacing means with ranks, and testing the resulting data by a nonparametric technique (the Kruskal-Wallis procedure), the V-trough and snare restraint procedures produced significant (p = 0.009) lower cortisol values than were obtained by the sling procedure. Also, combining the data from both sexes yielded no difference between strains for any of the parameters or conditions of restraint. These results reflect short term restraint and may not apply to conditions of extended restraint.

Table 1. Effect of 3 methods of restraint on plasma cortisol levels (ng/ml) of miniature swine

Restraint	Yucatan		Hormel	
	Female	Male	Female	Male
Sling	97[a]	64[b]	64	75[cd]
	± 26	12	± 28	± 14
Sling*	142	105[b]	126	118
10 min	± 17	± 14	± 18	± 16
V-trough	80	44*	49	42[c]
	± 14	± 5.3	± 16	± 9.6
Snare	63	51	46	38[d]
	± 14	± 3.8	± 7.5	± 8.3
Nonrestrained	42	61, 148	160	19

[a]Mean ± SEM (n = 5).
[b]Borderline significance (p = 0.058)
[c,d]Same letter indicates p < 0.05 in comparisons between types or times of restraint;
*indicates p < 0.05 in comparisons between sexes. All bleeding was conducted at 3 to 5 min after restraint or for a total of 10 min as indicated.
[e]Nonrestrained "controls" (1 to 2 animals per group) were catheterized in sling and then bled, using no restraint, approximately 2 hr after being loose in the pen. Individual values are shown; for all values, mean ± SEM = 86 ± 29

3.2. Restraint Effects on Endorphin Levels

There were no differences among the methods of restraint (Table 2). The only sex related effect was for the sling after 10 min of restraint; levels of ß-endorphin were lower for males than for females. The mean value for all the restraint groups was higher than any of the 5 individual values obtained for the no restraint (control) group. When the various control animals were treated as a group, a mean ± SEM of 191 ± 41 was obtained; this value is different from each of the means for the individual treatment groups.

Table 2. Effect of 3 methods of restraint on plasma ß-endorphin levels (pg/ml) of miniature swine

Restraint	Yucatan		Hormel	
	Female	Male	Female	Male
Sling	590[a]	561	479	599
	± 132	± 117	± 75	± 37
Sling*	936	552*	652	994
10 min	± 204	± 158	± 160	± 72
V-trough	405	406	554	701
	± 56	± 56	± 158	± 207
Snare	658	644	667	440
	± 144	± 145	± 237	± 71
Nonrestrained[b]	122	122, 136	327	246

[a]mean ± SEM (n=5). No differences among methods of restraint; *indicates p<0.05 in comparisons between sexes. Bleeding times as indicated in Table 1.
[b]Nonrestrained "controls" (1 to 2 animals per group) were catheterized in sling and then bled, using no restraint, aproximately 2 hr after being loose in the pen. Individual values shown; for all values, mean ± SEM=191± 41.

Table 3. Effect of training and temperature stress on cortisol levels (ng/ml) of sling restrained miniature swine

	Yucatan		Hormel	
	Female	Male	Female	Male
No training	$97^a \pm 26$	$64^b \pm 12$	$64^e \pm 28$	75 ± 14
No training + 10 min	142 ± 17	$105^b \pm 14$	$126^e \pm 18$	118 ± 16
Training	96 ± 21	$74^c \pm 11$	$80^f \pm 16$	$66^g \pm 8.5$
Training + 10 min	125 ± 19	$105^c \pm 5$	$130^f \pm 14$	$103^g \pm 15$
Training + stress	87 ± 34	$61^d \pm 5.9$	82 ± 14	$73^h \pm 11$
Training + 10 min + stress	168 ± 24	$109^d \pm 5.9$	192	$104^h \pm 6.1^*$

[a]Mean $\pm$ SEM (n = 3 to 5).
[b,c,d,e,f]Same letter indicates $p < 0.05$ in comparisons between conditions of restraint.
[g,h]Same letter indicates $p < 0.1$ (borderline significance) in comparisons between conditions of restraint.
[*]Indicates $p < 0.05$ in comparisons between sexes.

3.3. Effect of Training and Temperature Stress on Cortisol Levels

Neither training nor the increased temperature due to the heat wave had an effect on the cortisol levels of sling restrained pigs (Table 3). As for the nontrained and nontemperature stress group, 5 to 7 min of additional time in the sling produced increases in cortisol levels. The only apparent sex related difference was for Hormel pigs in the group receiving training with an additional 5 to 7 min in the sling following the stress of increased air temperature (but only one female was subjected to this measurement).

3.4. Effect of Training and Temperature Stress on ß-Endorphin Levels

Female Hormel pigs were the only group in which a training related effect was demonstrated. Surprisingly, the effect was to increase the ß-endorphin blood level (Table 4). The effect also was seen in other groups, although in these cases the differences were nonsignificant. Temperature stress produced an increased level of endorphin in the Yucatan males and in the Hormel strain of pigs, although the latter effect was only for females.

A sex related effect was apparent only for the untrained Yucatan pigs that were restrained for an additional 5 to 7 min. However, even arithmetically there was no consistent effect of sex on endorphin levels.

Table 4. Effect of training and temperature stress on plasma endorphin levels (pg/ml) of sling restrained miniature swine

	Yucatan		Hormel	
	Female	Male	Female	Male
No training	$590^a \pm 132$	561 ± 117	$479^c \pm 75$	$599^b \pm 37$
No training + 10 min	936 ± 204	$552^* \pm 158$	652 ± 160	$994^d \pm 72$
Training	773 ± 346	765 ± 68	$734^c \pm 103$	662 ± 189
Training + 10 min	616 ± 245	1166 ± 439	982 ± 153	1568 ± 550
Training + stress	350 ± 46	910 ± 288	$1818^c \pm 1043$	781 ± 240
Training + 10 min + stress	735 ± 314	723 ± 136	905^b	1090 ± 317

[a]Mean $\pm$ SEM (n = 3 to 5).
[b]One animal.
[c,d]Same letter indicates $p < 0.05$ in comparisons among conditions of restraint; [*]indicates $p < 0.05$ in comparisons among sexes.

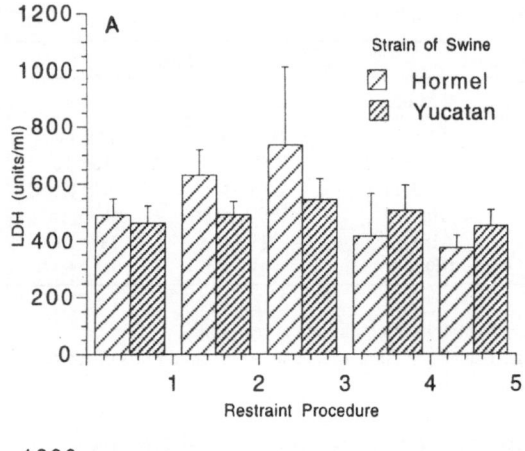

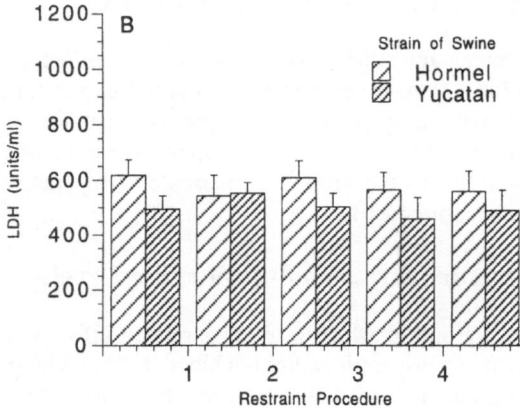

Figure 1. Comparison of plasma LDH values among groups (4 to 5 per group) of male (A) or female (B) miniature Hormel and Yucatan pigs exposed to various conditions of restraint. (1 = sling, no prior exposure; 2 = BRF V-trough; 3 = snare; 4 = sling, heat wave, prior exposure; 5 = sling, normal temperature, prior exposure). Measurements were made after 3 to 5 min of restraint. There was no difference among conditions of restraint or between the two strains of pigs.

3.5. LDH

No effects on plasma LDH related to restraint device, restraint condition, sex or strain resulted from these short term restraint trials (Figure 1). An additional 5 to 7 min in the sling did not result in an increase in LDH nor differences among sexes or strains of pigs (data not shown).

3.6. Effect of Prolonged Sling Restraint on Cortisol and LDH Levels

Figures 2 and 3 illustrate the type of results obtained for plasma cortisol by allowing Hormel miniature pigs to remain in the sling for periods of time of up to 7 hr. These pigs were placed in the sling and catheterized via the vena cava; blood specimens were removed at the time intervals indicated. In Figure 2A, cortisol levels increased initially for both pigs. After about 1 hr, however, they gradually decreased to below baseline levels for one pig that appeared calm during the study but increased and remained well above baseline levels for the other pig during the entire 6 hr. The latter pig appeared excitable and hyperactive during the same time period. A similar pattern was seen for this same pig at another day (Figure 2B); another pig that showed a calm disposition and was tested simultaneously had cortisol levels increase temporarily but then decline and remain close to baseline levels. In either case, levels of cortisol

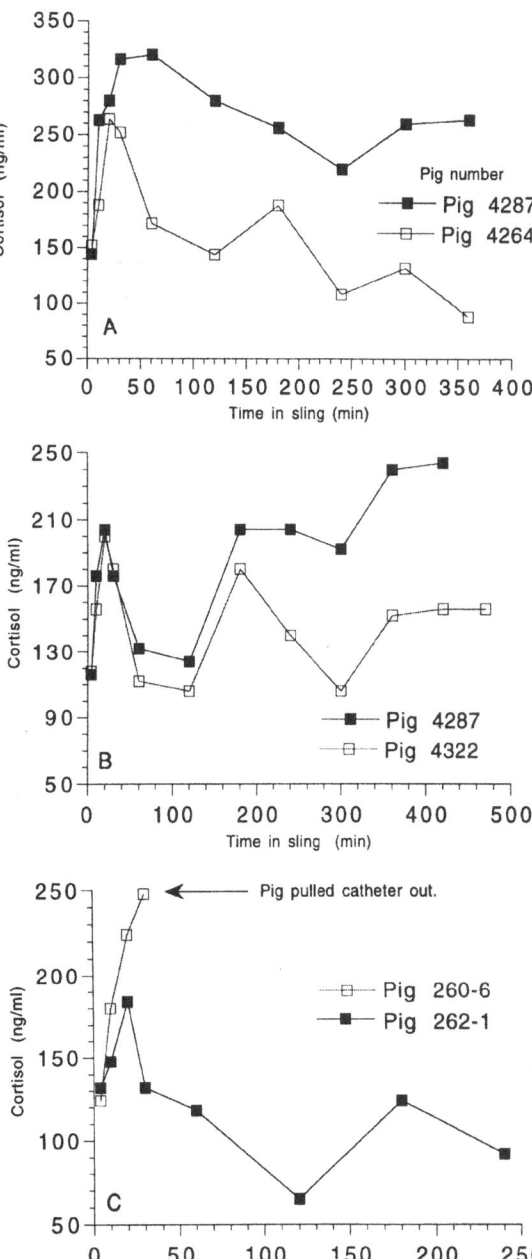

Figure 2. (A) Plasma cortisol levels in 2 female Hormel miniature pigs restrained in the Panepinto sling over a 6 hr period. Pig #4287 appeared very excitable; whereas, pig #4264 appeared relatively calm. Note that by 4 hr, cortisol values attained were lower than at initial bleeding (3 to 5 min after catheterization). (B) Plasma cortisol levels in 2 female Hormel miniature pigs restrained in the Panepinto sling over a 7 to 8 hr period. Again, pig #4287 appeared excitable, resulting in some trouble in bleeding. Pig #4322 appeared calm. (C) Plasma cortisol levels in 2 female Yucatan miniature pigs restrained in the Panepinto sling over a 1 to 4 hr period. One pig, #260s-6, was excitable, resulting in the catheter being pulled out at 1 h. Note that cortisol values for pig #260-6 at 1 hr were below those at initial bleeding.

at the baseline (time at initial bleeding) were not attained until the animals were allowed to remain in the sling for a minimum of 2 hr.

A similar situation was found in one experiment carried out with Yucatan pigs, although in this case baseline values occurred after 30 min (Figure 2C). In observing the corresponding LDH values from a sedate pig and the excitable pig after they had

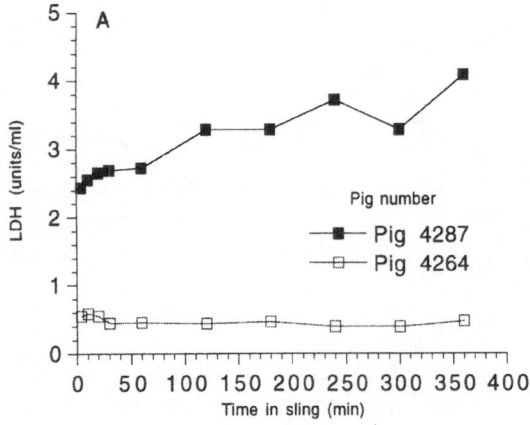

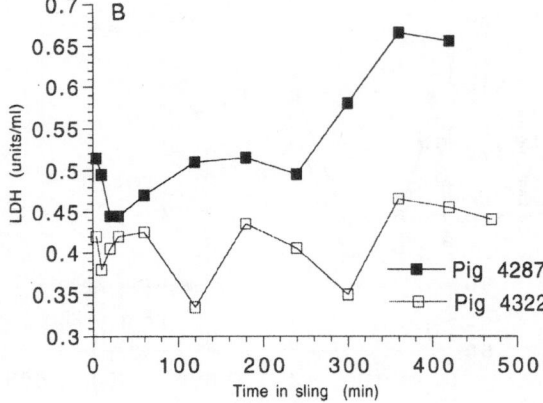

Figure 3. (A) Plasma LDH values in the same specimens analyzed for cortisol in Figure 2A. Note steady rise in LDH values for the excitable pig, #4287. (B) Plasma LDH values in the same specimens analyzed for cortisol in Figure 2B. Note again the almost steady rise in LDH for the excitable pig, #4287, after about 30 min in the sling. Variability in the LDH values was seen for the calm pig, #4322, but values never attained those present for the more excitable pig.

been in the sling for a prolonged period of time, there was a slow, but gradual, increase in blood LDH of the latter pig; whereas, no change was apparent in the former (Figure 3A). Repeating the test another day (Figure 3B) again demonstrated a steady rise in LDH values for this excitable pig, but not for another pig, which was calm throughout the period of measurements.

4. DISCUSSION

The studies described here were designed to assess the degree of stress endured by the FDA strain of Hormel and Yucatan miniature swine subjected to 3 commonly used methods of restraint for very short periods of time. Stress was ascertained by measuring plasma levels of cortisol and ß-endorphin. In a few experiments these measurements were made on pigs restrained in a sling for several hr; in this case, biochemical measurements were compared with observations of animal behavior during the period of restraint. The literature is replete with studies describing the effects of restraint on various species of animals. Most of the studies were conducted with rodents,[15-19] nonhuman primates,[20-23] cattle,[24-26] sheep,[27-30] cats and dogs,[31-33] and pigs.[1,34-38] Of the swine studies, only one specifically involved miniature pigs,[12] and this was primarily a description and application of the Panepinto sling.

In comparing the 3 methods of restraint, the 2 strains of miniature pigs or the sexes with regard to stress response using biochemical parameters of measurement, no generalizations could be derived. Under specified conditions, however, some variable related differences were discerned. First, plasma cortisol levels increased soon after restraint was imposed, not an unexpected finding based on studies carried out with rodents[16,19,39-43] and pigs.[1,36-38] Somewhat surprising, however, was the finding that when the data from strains and both sexes were combined, somewhat lower levels of cortisol resulted from using the snare and trough restraints than from using the sling, indicating that the pigs were experiencing less trauma in the first 2 devices, at least for measurements following short periods of restraint.

Another possible interpretation is that the few min of restraint did not give enough time for signs of biochemical stress to be manifested. To settle this question it will be necessary to carry out these measurements after restraining the pigs in the snare and trough for longer time periods. Wade *et al.*[1] reported apparent increases in cortisol levels in pigs in as little as 2.5 min after sling restraint. In the present study, measurements were initiated only at about this time (zero time measurements were not made). Measurements of a few animals catheterized and allowed to run around free with little restraint before they were bled produced values with such high variability they were considered of little use as possible reference points for nontreatment controls (although, as will be discussed later, ß-endorphin values from these same nonrestrained animals were useful).

Few studies assessing the effects of restraint stress on behavioral, physiological or biochemical parameters compared types of restraint. Two such studies involving nonhuman primates were reported, one comparing a chair restraint and tether on male rhesus monkeys of varying ages with growth hormone and cortisol as indicators,[23] and another comparing physical and chemical restraints on baboons.[21] One study with rats was concerned with how the type of restraint affected body temperature.[44] Some authors described differences in behavior or hormonal levels resulting from 2 or more methods of restraint or confinement on calves[26] and cattle.[45] Cortisol levels in cattle depended on the type of restraint. To the best of our knowledge, no studies of this nature have been carried out with swine.

Overall, in the present study few differences were attributable to sex, although a difference was observed for a particular strain and restraint device, e.g., Yucatan males produced lower levels of cortisol than did females when restrained by the V-trough. Again, few studies were reported in regard to the influence of strain on the effects of restraint. Although a trend appeared to exist in which cortisol values for female Hormel pigs were lower than those for female Yucatan pigs, this trend did not hold true for males, and there were no differences between strains regardless of whether the sexes were analyzed individually or the sex data were combined.

However, in another study carried out in this laboratory, the same 2 strains were compared by allowing them to be restrained for several hr in the sling after drug or vehicle (control) treatment. Although cortisol levels did not differ between strains, the Hormel strain showed a higher level for control plasma catecholamines.[46] This finding was not necessarily unexpected in light of the observations by several investigators that different hormones respond in different ways to the stress of restraint,[2,11,47] and this in turn is sometimes a function of the duration of the restraint period.[39,48]

Analysis of plasma for a number of hormones in pigs during restraint in a sling device showed increases in all the hormones studied (norepinephrine, epinephrine, renin, vasopressin, cortisol and aldosterone), indicating a commonality in the control mechanism relating stress to stimulation of the secretion and/or synthesis of these hormones of varied function and from various sites.[1] However, temporal relationships showed differences. For example, steady increases over time were seen for aldosterone and cortisol; whereas, for norepinephrine and epinephrine, peaks in activity occurred early after the beginning of restraint, followed by steady declines thereafter.

The small amount of training in the sling given the animals before the actual testing in the present study appeared to have no effect on the status of stress as reflected by cortisol levels. In some reported studies, training appeared to acclimatize the animals, although commonly more frequent or protracted periods of training were used.[49-51]

Cortisol levels in swine indicated that acclimatization occurred during their maintenance in tether stalls but required up to 4 da of restraint in these stalls.[35]

Heat stress superimposed on restraint stress had no effect on cortisol levels in the miniature swine, whether they received training in the sling or not. Although the literature contains many examples of the effects of cold temperatures on the hormonal response of animals, particularly rodents (reviewed by Dantzer and Mormede).[52] One is hard pressed to find reports on the effects of above ambient temperatures on these responses. In general, cold temperatures acted like other stresses in producing rapid increases in plasma corticosteroids and/or catecholamines. Kattesh et al.[53] found decreases in plasma glucocorticoid concentration of sows and gilts after heat stress was applied for 50 da during midgestation. In the present study, the particular short duration (2 da) of heat stress experienced by the miniature swine before testing may have been a factor in their apparent lack of response to this stress as reflected by cortisol levels. Also, ß-endorphin levels in this study generally were increased after heat stress, and there was evidence of a depressive action of this endogenous opioid on secretion of cortisol[54] as well as other hormones.[55-57]

When the miniature swine were maintained in the Panepinto sling for periods up to 6 hr, there were 3 general patterns of response as measured by cortisol levels. Each pig showed an initial surge in cortisol, followed either by a continuation of this surge, by a drop and second peak of activity as a biphasic response, or by a continuous decline to levels present before the initial and early measurement. Although the number of animals tested in this fashion was small, there seemed to be, not unexpectedly, a correlation between the type of pattern and the apparent excitability of the pig. Wade et al.,[1] carrying out similar studies with Duroc pigs, did not report such variability in response, but these pigs did have a continuous increase in cortisol levels for at least 1 hr. The important conclusion to be drawn from both pig studies is that at least some pigs undergo a prolonged state of excitation which may hinder or at least confound other measurements made in such animals. However, with the pigs and restraint device used in the present study, a relative state of calm can be achieved in these animals if care and selectivity are exercised in choosing pigs for the study.

The LDH levels measured during sling restraint increased only in those pigs that appeared excitable and that had corresponding increased levels of cortisol in their blood. LDH levels did not appear to increase until 1 to 2 hr after restraint; measurements taken early after restraint indicated no difference among strains or types of restraint or conditions associated with the restraint. In fact, other studies have shown that levels of some plasma enzymes, reflecting tissue damage, do indeed increase in some animals subjected to restraint. In one of these studies involving rhesus monkeys and chair restraint, plasma creatine phosphokinase increased early, but other enzymes such as LDH, GOT and GPT increased only several hr after the onset of restraint and remained high even after release from restraint.[22] In the other study with rhesus monkeys, these elevations of enzyme levels disappeared along with the rise in cortisol levels after several repetitions of the restraint procedures over a number of days. The authors concluded that the elevated enzyme levels at the onset of chair restraint were due to high sympathetic tone and the muscular activity expended in trying to avoid the restraint.[50]

As with cortisol, no consistent effects of any one variable on endorphin levels resulted from the restraint procedure used here. Temporal relationships were not studied to the extent that they were with cortisol. However, to some degree, endorphins appeared to be more responsive to restraint stress than was cortisol. Mean levels of ß-endorphin of nonrestrained animals were lower than the levels of all groups of restrained pigs; levels of plasma endorphin were higher in pigs after a brief increase in ambient temperatures due to the heat wave.

Although regulatory mechanisms, areas of synthesis, and receptor sites differ for the two hormones studied, one common characteristic is that the systems controlling their release in animals respond somewhat similarly to stress situations.[10,58-63] For example, in rats, increased levels of plasma corticosterone resulting from handling or cold stress were seen as early as 7 da of age, while at this stage of development, ß-endorphin levels were not elevated by these stresses.[64,65] Also, the complexity of interactions between these 2 classes of hormones is reflected in findings that glucocorticoids increase during restraint stress in rats, providing a negative feedback for the response of endogenous opioids to subsequent stress,[62] and that catecholamines, released in response to some stress stimuli (including restraint) but not others, activate the release of ß-endorphin from the intermediate lobe of the pituitary gland.[66]

Similar to the findings for plasma cortisol, there were few effects of sex, strain or prior training on levels of plasma ß-endorphin in the miniature swine after the animals in the present study were restrained. Unlike cortisol, however, there were no differences in ß-endorphin levels among methods of restraint. Although not as consistent as cortisol levels, there was a tendency toward higher levels of ß-endorphin in the pigs after an additional 10 min of restraint in the sling. This divergence between response of plasma cortisol levels and ß-endorphins may be at least partially attributed to the complex interactions between both types of hormones and their regulatory mechanisms.

Although it has been demonstrated that adrenergic agonists like norepinephrine stimulate ß-endorphin release in rats, the reverse may be true in humans experiencing stress where an inhibitory relationship may exist.[67] Unfortunately little work has been carried out to explore this phenomenon in pigs. In fact, not a single study could be found in the literature dealing with ß-endorphin levels in pigs with restraint stress. As for cortisol, the various stress parameters (type of restraint, strain, sex and prior training) appeared to have little influence on the general effect of restraint stress on ß-endorphin levels in the present study. Various types of stress situations that were applied to sheep, including immobilization, produced increases in ß-endorphin and cortisol levels, but the effects appeared to be similar regardless of the type of stress. The lack of effect of training on the response of endorphin levels to restraint in the sling was somewhat curious, since studies with rats have shown that, after repeated periods of immobilization, increases in plasma ß-endorphin levels due to acute immobilization stress were abolished.[68]

5. REFERENCES

1. Wade, C.E., Hannon, J.P., and Bossone, C.A., 1986, Cardiovascular and hormonal responses of conscious pigs during physical restraint, in: *Swine in Biomedical Research* (M.E. Tumbleson, ed.), Plenum Press, New York, pp. 1395-1404.
2. Shutt, D.A., Smith, A.I., Wallace, C.A., Connel, R., and Fell, L.R., 1988, Effect of myiasis and acute restraint stress on plasma levels of immunoreactive ß-endorphin, adrenocorticotrophin (ACTH) and cortisol in the sheep. *Aust. J. Biol. Sci.* 41: 297-301.
3. Girgis, M., 1980, Modification of rabbit head holder to facilitate stereotaxic brain surgery. *Brain Res. Bull.* 5: 769-770.
4. Steplewski, Z., Vogel, W.H., Ehya, H., Poropatich, C., and Smith, J.M., 1985, Effects of restraint stress on inoculated tumor growth and immune response in rats. *Cancer Res.* 45: 5128-5133.
5. Hood, R.D., and Rasco, J.F., 1993, Effect of restraint procedure on restraint stress-induced rib fusion in mice. *Toxicologist* 23: 254 (Abstract #958).
6. Kelley, K.W., Mertsching, H.J., and Salmon, H., 1984, Immunity changes in confined animals: a route to disease? *Ann. Rech. Vet.* 15: 201-204.

7. Kiss, A., 1983, Effect of acute and repeated immobilization stress on the ultrastructure of neurons of the rat hypothalamic ventromedial nucleus. *Z. Mikrosk. Anat. Forsch.* 97: 49-64.

8. Bloom, F.E., Rossier, J., Battenberg, E.L., Bayon, A., French, E., Henriksen, S.J., Siggins, G.R., Segal, D., Browne, R., Ling, N., and Guillemin, R., 1978, β-Endorphin cellular localization, electrophysiological, and behavioral effects. *Adv. Biochem. Psychopharmacol.* 18:89-109.

9. Li, C.H., Chung, D., and Doheen, B., 1976, Isolation, characterization and opiate activity of β-endorphin from human pituitary glands. *Biochem. Biophys. Res. Commun.* 72: 1542.

10. Flores, C.M., Hernandez, M.C., Hargreaves, K.M., Bayer, and B.M., 1990, Restraint stress-induced elevations in plasma corticosterone and ß-endorphin are not accompanied by alterations in immune function. *J. Neuroimmunol.* 28: 219-225.

11. Recher, H., Willis, G.L., Smith, G.C., and Copolov, D.L., 1988, ß-endorphin, corticosterone, cholesterol and triglyceride concentrations in rat plasma after stress, cingulotomy or both. *Pharmacol. Biochem. Behav.* 31: 75-79.

12. Panepinto, L.M., Phillips, R.W., Norden, S., Pryor, P.C., and Cox, R., 1983, A comfortable, minimum stress method of restraint for Yucatan miniature swine. *Lab. Anim. Sci.* 33: 95-97.

13. Pettibone, D.J., and Mueller, G.P., 1982, Evidence for independent secretion of β-endorphin immunoreactivity from rat pars distalis *in vivo. Endocrinology* 110: 469-473.

14. Kornberg, A., 1955, Lactic dehydrogenase of muscle, in: *Methods in Enzymology* (S.P. Colowick and N.O. Kaplan, eds.), Academic Press, New York, pp. 441-443.

15. Kelsey, J.E., Hoerman, W.A. IV, Kimball, L.D. III, Radack, L.S., and Carter, M.V., 1986, Arcuate nucleus lesions reduce opioid stress-induced analgesia (SIA) and enhance non-opioid SIA in rats. *Brain Res.* 382: 278-290.

16. Bakke, H.K., and Murison, R., 1989, Plasma corticosterone and restraint induced gastric pathology: age-related differences after administration of corticotropin releasing factor. *Life Sci.* 45: 907-916.

17. Berridge, C.W., and Dunn, A.J., 1989, CRF and restraint-stress decrease exploratory behavior in hypophysectomized mice. *Pharmacol. Biochem. Behav.* 34: 517-519.

18. Martin, G.E., and Papp, N.L., 1979, Effect on core temperature of restraint after peripherally and centrally injected morphine in the Sprague-Dawley rat. *Pharmacol. Biochem. Behav.* 10: 313-315.

19. Rattner, B.A., Michael, S.D., and Altland, P.D., 1983, Age-related responses to mild restraint in the rat. *J. Appl. Physiol.* 55: 1408-1412.

20. Bouyer, J.J., Dedet, L., Debray, O., and Eougeul, A., 1978, Restraint in primate chair may cause unusual behaviour in baboons; electrocorticographic correlates and corrective effects of diazepam. *Electroencephalogr. Clin. Neurophysiol.* 44: 562-567.

21. Goosen, D.J., Davies, J.H., Maree, M., and Dormeh., I.C., 1984, The influence of physical and chemical restraint on the physiology of the chacma baboon (*Papio ursinus*). *J. Med. Primatol.* 13: 339-351.

22. Tatsumi, T., Komatsu, H., and Adachi, J., 1990, Effects of chair restraint on plasma enzyme values in the rhesus monkey (*Macaca mulatta*). *Jikken Dobutsu* 39: 353-359.

23. Wheeler, M.D., Schutzengel, R.E., Barry, S., and Styne, D.M., 1990, Changes in basal and stimulated growth hormone secretion in the aging rhesus monkey: a comparison of chair restraint and tether and vest sampling. *J. Clin. Endocrinol. Metab.* 71: 1501-1507.

24. Klein, L., and Fisher, N., 1988, Cardiopulmonary effects of restraint in dorsal recumbency on awake cattle. *Am. J. Vet. Res.* 49: 1605-1608.

25. Dubois, M., Pickar, D., Cohen, M., Gershon, E., and Bunney, W.E., Jr., 1982, Effects of fentanyl on the response of plasma ß-endorphin immunoreactivity to surgery. *Anesthesiology* 57: 468-472.

26. Friend, T.H., Dellmeier, G.R., and Gbur, E.E., 1985, Comparison of four methods of calf confinement. I. Physiology. *J. Anim. Sci.* 60: 1095-1101.

27. Baxter, J.R., 1987, Response of sheep to short term restraint by electro-immobilisation. *Aust. Vet. J.* 64: 195.

28. Frey, M.J., and Moberg, G.P., 1980, Effect of intraventricular serotonin on the plasma cortisol response to restraint stress in unanesthetized sheep. *J. Anim. Sci.* 51: 380-385.

29. Faler, K., and Faler, K., 1987, Restraint of sheep. *Mod. Vet. Prac.* 68: 562-563.

30. Fenwick, D.C., Blackshaw, J.K., and Green, D.J., 1986, The effects of delays between restraint and sampling on some blood parameters in sheep. *Vet. Res. Commun.* 10: 309-315.

31. Barsanti, J.A., Mahaffey, M.B., Crowell, W.A., and Barber, D.L., 1984, Cystometry in dogs under oxymorphone and acepromazine restraint. *Am. J. Vet. Res.* 45: 2152-2153.

32. Malathi, S., and Batmanabane, M., 1983, Effects of varying periods of immobilization of a limb on the morphology of a peripheral nerve. *Acta Morphol. Neerl. Scand.* 21: 185-198.

33. Bowersox, S.S., Siegel, J.M., and Sterman, M.B., 1978, Effects of restraint on electroencephalographic variables and monomethylhydrazine-induced seizures in the cat. *Exp. Neurol.* 61: 154-164.

34. Wade, C.E., and Hannon, J.P., 1988, Confounding factors in the hemorrhage of conscious swine: a retrospective study of physical restraint, splenectomy, and hyperthermia. *Circ. Shock* 24: 175-182.

35. Becker, B., Christenson, R., Ford, J., Manak, R., Nienaber, J., Hahn, G., and Deshazer, J. 1984, Serum cortisol concentrations in gilts and sows housed in tether stalls, gestation stalls and individual pens. *Ann. Rech. Vet.* 15: 237-242.

36. Becker, B.A., Ford, J.J., Christenson, R.K., Manak, R.C., Hahn, G.L., and DeShazer, J.A., Cortisol response of gilts in tether stalls. *J. Anim. Sci.* 60: 264-270.

37. Rampacek, G.B., Kraeling, R.R., Fonda, E.S., and Barb, C.R., 1984, Comparison of physiological indicators of chronic stress in confined and nonconfined gilts. *J. Anim. Sci.* 58: 401-408.

38. Barnette, J., and Hemsworth, P.H., 1986, The impact of handling and environmental factors on the stress response and its consequences in swine. *Lab. Anim. Sci.* 36: 351-365.

39. Livezey, G.T., Miller, J.M., and Vogel, W.H., 1985, Plasma norepinephrine, epinephrine and corticosterone stress responses to restraint in individual male and female rats, and their correlations. *Neurosci. Lett.* 62: 51-56.

40. McMurtry, J.P., and Wexler, B.C., 1981, Hypersensitivity of spontaneously hypertensive rats (SHR) to heat, ether, and immobilization. *Endocrinology* 108: 1730-1736.

41. Groza, P., Bordeianu, A., Boca, A., Petrescu, A., and Cananau, S., 1979, Digestive structural modifications in rats submitted to 15 days of hypokinetical conditions. *Physiologie* 16: 243-247.

42. Groza, P., Cananau, S., Bordeianu, A., Boca, A., and Cananau, S., 1980, The digestive tract after 7 days of restraining conditions in normal and adrenalectomized rats. *Physiologie* 17: 257-259.

43. Soncrant, T.T., Holloway, H.W., Stipetic, M., and Rapoport, S.I., 1988, Cerebral glucose utilization in rats is not altered by hindlimb restraint or by femoral artery and vein cannulation. *J. Cereb. Blood Flow Metab.* 8: 720-726.

44. Martin, G.E., Pryzbylik, A.T., and Spector, N.H., 1977, Restraint alters the effects of morphine and heroin on core temperature in the rat. *Pharmacol. Biochem. Behav.* 7: 463-469.

45. Dunn, C.S., 1990, Stress reactions of cattle undergoing ritual slaughter using two methods of restraint. *Vet. Rec.* 126: 522-525.

46. Ikeda, G.J., T.C. Michel, D.W. Gaines, Olivito, V.L., Sapienza, P.P., Friedman, L., Barton, C.N., and O'Donnell, M.W., 1989, A comparison of physiological and biochemical parameters in two species of minipig. *Toxicologist* 9: 262.

47. Collu, R., Du Rusisseau, P., and Tache, Y., 1979, Role of putative neurotransmitters in prolactin, GH and LH response to acute immobilization stress in male rats. *Neuroendocrinology* 28: 178-186.

48. Ruisseau, P.D., Tache, Y., Brazeau, P., and Collu, R., 1978, Pattern of adenohypophyseal hormone changes induced by various stressors in female and male rats. *Neuroendocrinology* 27: 257-271.

49. Pierzchala, K., and Van Loon, G.R., 1990, Plasma native and peptidase-derivable Met-enkephalin responses to restraint stress in rats. Adaptation to repeated restraint. *J. Clin. Invest.* 85: 861-873.

50. Tatsumi, T., Koto, M., Komatsu, H., and Adachi, J., 1990, Effects of repeated chair restraint on physiological values in the rhesus monkey (*Macaca mulatta*). *Jikken Dobutsu* 39: 361-369.

51. Porro, C.A., and Carli, G., 1988, Immobilization and restraint effects on pain reactions in animals. *Pain* 32: 289-307.

52. Dantzer, R., and Mormede, P., 1983, Stress in farm animals: a need for reevaluation. *J. Anim. Sci.* 57: 6-17.

53. Kattesh, H.G., Kornegay, E.T., Knight, J.W., Gwazdauskas, F.G., Thomas, H.R., and Notter, D.R., 1980, Glucocorticoid concentrations, corticosteroid binding protein characteristics and reproduction performance of sows and gilts subjected to applied stress during mid-gestation. *J. Anim. Sci.* 50: 897-905.

54. Przekop, F., Mateusiak, K., Stukpnicka, E., Romanowicz, K., and Domanski, E., 1990, Suppressive effect of ß-endorphin and naloxone on the secretion of cortisol under stress conditions in sheep. *Exp. Clin. Endocrinol.* 95: 210-216.

55. Knepel, W., Nutto, D., and Anhut, H., 1983, ß-Endorphin controls vasopressin release during foot shock-induced stress in the rat. *Regul. Pept.* 7: 9-19.

56. Amir, S., Brown, Z.W., and Amit, Z., 1980, The role of endorphins in stress: evidence and speculations. *Neurosci. Biobehav. Rev.* 4: 77-86.

57. Grossman, A., 1985, Endorphins: "opiates for the masses." *Med. Sci. Sports Exerc.* 17: 101-105.

58. Armario, A., Marti, O., Gavalda, A., and Jolin, T., 1990, Blockade of opioid receptors with naltrexone inhibits thyrotropin increase after noise stress but does not prevent the decrease caused by immobilization. *Brain Res. Bull.* 25: 347-349.

59. Bruni, J.F., Hawkins, R.L., Yen, S.S., 1982, Serotonergic mechanism in the control of ß-endorphin and ACTH release in male rats. *Life Sci.* 30: 1247-1254.

60. Kant, G.J., Mougey, E.H., and Meyerhoff, J.L., 1986, Diurnal variation in neuroendocrine response to stress in rats: plasma ACTH, ß-endorphin, ß-LPH, corticosterone, prolactin and pituitary cyclic AMP responses. *Neuroendocrinology* 1986: 383-390.

61. Giagnoni, G., Santagostino, A., Senini, R., Fumagalli, P., and Gori, E., 1983, Cold stress in the rat induces parallel changes in plasma and pituitary levels of endorphin and ACTH. *Pharmacol. Res. Commun.* 15: 15-21.

62. De Souza, E.B., and Van Loon, G.R., 1989, Rate-sensitive glucocorticoid feedback inhibition of adreno-corticotropin and ß-endorphin/ß-lipotropin secretion in rats. *Endocrinology* 125: 2927-2934.

63. Kiiatkin, E.A., Polyntsev, I.U.V., Kushlinskii, N.E., and Amiragova, M.G., 1985, ACTH, corticosterone and ß-endorphin in the blood plasma of rats under prolonged immobilization stress. *Biull. Eksp. Biol. Med.* 100: 157-160.

64. Angelogianni, P., and Gianoulakis, C., 1989, Ontogeny of the ß-endorphin response to stress in the rat: role of the pituitary and the hypothalamus. *Neuroendocrinology* 50: 372-381.

65. Iny, L.J., Gianoulakis, C., Palmour, R.M., and Meaney, M.J., 1987, The ß-endorphin response to stress during postnatal development in the rat. *Dev. Brain Res.* 31: 177-181.

66. Berkenbosch, F., Tilders, F.J., and Vermes, I., 1983, ß-adrenoceptor activation mediates stress-induced secretion of ß-endorphin-related peptides from intermediate but not anterior pituitary. *Nature* 305: 237-239.

67. Troullos, E.S., Hargreaves, K.M., Goldstein, D.S., Stull, R., and Dionne, R.A., 1989, Epinephrine suppresses stress-induced increases in plasma immunoreactive ß-endorphin in humans. *J. Clin. Endocrinol. Metab.* 69: 546-551.

68. Vakulina, O.P., Tigranian, R.A., and Brusov, O.S., 1984, Opioid peptide content of the brain and blood of rats under immobilization stress. *Biull. Eksp. Biol. Med.* 98: 537-539.

48

THE MINIPIG AS A MODEL FOR THE STUDY OF AGING IN HUMANS

Selective Responses of Hormones Involved in Carbohydrate and Lipid Metabolism in Different Sexes

Sam J. Bhathena, PhD,[1]* Elliott Berlin, PhD,[1] and
Wesley A. Johnson,[2] DVM

[1] Metabolism and Nutrient Interactions Laboratory
Beltsville Human Nutrition Research Center
Agricultural Research Service
US Department of Agriculture
Beltsville, Maryland 20705
[2] Center for Food Safety and Applied Nutrition
Division of Toxicological Studies
Food and Drug Administration
US Department of Health and Human Services
Laurel, Maryland 20708

1. ABSTRACT

Both age and sex have been shown to affect lipid and carbohydrate metabolism in animals and humans. There is a gradual loss of glucose homeostasis and concomitant increase in insulin resistance with age. Carbohydrate and lipid metabolism in different metabolic conditions including obesity and noninsulin dependent diabetes mellitus (NIDDM), both of which are age dependent, are controlled by pancreatic, pituitary and adrenal hormones. Therefore, we assessed age related changes in plasma hormones in male and female Hormel miniature swine. Animals were fed an energy restricted cholesterol free, low fat stock diet postweaning. Animals were sacrificed at various ages, ranging from prepubertal, <0.5 yr; young, 0.5 to 2.0 yr; mature, 2 to 10 yr and old, >10 years. Levels of insulin, glucagon, adrenocorticotropic hormone (ACTH), cortisol and androstenedione were measured by radioimmunoassay in plasma from both sexes. In addition, dehydroepiandrosterone-sulfate (DHEA-S), aldosterone and testosterone were measured in male swine. No effect of age nor

---

* Reprint requests to Dr. Sam J. Bhathena, MNIL, BHNRC, ARS, Room 324 Building 307, BARC East, Beltsville, MD 20705, (301) 504-8422.

Advances in Swine in Biomedical Research, edited by Tumbleson and Schook
Plenum Press, New York, 1996

sex was observed on plasma insulin and ACTH levels though an interaction was observed between age and sex for ACTH. Thus, with age, levels of ACTH tended to increase in males but decrease in females. Plasma glucagon was higher in males than females and showed age dependent increases in males but not in females. Plasma androstenedione was 3 to 5 fold higher in males than in females and the level was lower in mature male swine than either prepubertal or older swine. There were sex and age dependent changes in plasma cortisol levels with an increase with age in males and a decrease with age in females. There were no age dependent changes in plasma DHEA-S and aldosterone levels in the males. Plasma testosterone levels were higher in prepubertal swine than in mature or older animals. In swine, both age and sex affect hormonal changes, which then impact on metabolic parameters controlled by hormones, notably carbohydrate and lipid metabolism. Energy restriction appears to have a beneficial effect in retarding the aging process and miniature swine can be used as a model for the study of aging in humans.

2. INTRODUCTION

Miniature swine have been recognized and used as models to study metabolic processes in humans.[1,2] Because of their similarities to humans in anatomic and physiologic characteristics and nutritional requirements,[2-4] they have been used to study nutritionally induced changes in humans. Other areas where minipigs are used as models for humans include immune function, cardiovascular disease, hypertension, arteriosclerosis, kidney disease, toxicity studies, gastric ulcer, obesity and ingestive behavior, alcoholism, osteoporosis, certain types of cancer, skin disorders and protein energy malnutrition.[3-6] Most of the studies carried out in minipigs are in neonatal or very young animals, generally before they reach the age of 1 yr. Limited studies have been carried out using older minipigs. Miniature swine have been used as models for studying human aging in a limited number of studies.

Aging is a continuous, time dependent process where anabolic processes predominate early in life with catabolic processes assuming predominance during the later phase of life. Though the aging process cannot be reversed, it can be slowed. Nutrition plays an important role in aging in that energy restriction, either by reducing food intake or by consuming foods with lower energy density, appears to be beneficial in reducing the detrimental effects of aging.[7-16] An energy intake that is more than required for sustaining metabolic balance and energy expenditure has been reported to lead to degenerative disorders such as obesity, diabetes, hypertension, cardiovascular diseases, kidney failure and cancer.[14,17] These overnutrition related disorders exacerbate the degenerative effects of aging.[18] The aging process, including age associated phenomena, the cause leading to it and the underlying mechanisms thereof have been studied extensively in rodent models, including mice, rats and hamsters.[19] Several epidemiological studies have been attempted to define the aging process in humans.[20] Age as well as sex have been reported to alter metabolic processes in humans and rodents.[21-23] Metabolic processes are under hormonal control. Since there is ample evidence to indicate that metabolic processes in swine are similar, if not identical, to humans, we conducted the present study with miniature swine to test the hypothesis that age related endocrine changes are responsible for the alteration in metabolic processes.

The role of the endocrine system involved in the growth of fetal and neonatal pigs, such as growth hormone, insulin like growth factor (IGF)-I, IGF-II, testosterone, luteinizing hormone, estrogen and prolactin, has been described in detail[24-33] but studies on the role of hormones involved in metabolic processes during development and aging in pigs are lacking. In pigs genetically selected for leanness or obesity of back fat,[34] higher levels of circulating growth hormone are observed in lean than obese lines of pigs.[35] However, no differences were observed in plasma concentrations of carbohydrates and lipids such as glucose, triacylglycerol and cholesterol or in levels of insulin[36] which controls the metabolism of these substrates.

3. MATERIALS AND METHODS

Twenty male and 16 female Hormel miniature swine and 1 male Hormel-Hanford swine were used for the study. Ages ranged from 5 mo to 13.9 yr for males and from 5 mo to 8 yr for females. The experimental protocol was approved by the Institutional Animal Care and Use Committee of the Food and Drug Administration, Laurel, MD. Animals were housed in individual pens with indoor and outdoor runs and provided with ad libitum water. Pens were heated during winter. After weaning at the age of 6 wk, animals were fed low fat, cholesterol free stock diet (USDA 1160) until sacrificed, except during pregnancy and lactation when they were fed USDA 1180. Piglets, from 6 wk to 6 mo, were fed USDA 1160 ad libitum. After 6 mo, the ration was restricted to weight maintenance (680 to 900 g/da). Females during gestation and lactation and suckling piglets up to 6 wk were fed ad libitum the high fat (USDA 1180) diet. Diet compositions are given in Table 1. Major nutrient contents of the USDA 1160 diet were 16% protein, 75% carbohydrate and 2.7% fat with a polyunsaturated to saturated fatty acid ratio of 3.8. Animals were divided into 4 groups depending on their age at sacrifice, prepubertal (less than 6 mo), young (6 mo to 2 yr), mature (2 to 10 yr) and old (>10 yr)(Table 2). Animals were sacrificed by electrocution and blood was collected in tubes containing EDTA (1.5 mg/ml of blood) and trasylol (aprotinin, 1000 U/ml of blood). Plasma was separated, portioned and stored at -70C until analysis. Plasma insulin, glucagon, cortisol, adrenocorticotropic hormone (ACTH), androstenedione, dehydroepiandrosterone-sulfate (DHEA-S), aldosterone and testosterone were measured using kits from Diagnostic Products Corporation, CA.

Data were analyzed statistically by analysis of variance (ANOVA) using a general linear models procedure to study the effect of age and its interaction with sex.[37] Values of $p<0.05$ were considered statistically significant.

Table 1. Diet composition (g/kg)

Ingredients	USDA 1160*	USDA 1180**
Corn	739	500
Cal Phos	21	22
Limestone	3.5	3.5
Mineral mix	1	2
Salt	5	5
Selenium	0.5	0.5
Vitamin mix	1	1.5
Alfalfa	50	—
Whey (dried)	—	150
Sugar	—	50
Soybean meal 49%	179	245
Soybean oil	—	20

*Fed ad libitum to piglets postweaning for 6 wk. Fed at 680 to 900 g/da for weight maintenance to adult pigs.
**Fed at 1150 to 1800 g/da to pregnant and lactating sows.

Table 2. Experimental design

Groups	Age(Yr)	Number of Animals		Weight at sacrifice (Kg)	
		Males	Females	Males	Females
I. Prepubertal	<0.5	6	6	36	29
II. Young	0.5 to 2	-	5	-	55
III. Mature	2 to 10	10	5	83	80
IV. Old	>10	5	-	79	-

4. RESULTS

Plasma levels of insulin, glucagon and ACTH of male and female minipigs of different ages are reported in Table 3. Neither sex nor age had effects on plasma insulin and ACTH levels. There was an interaction between age and sex for ACTH levels. Thus, in males, plasma levels increased with age while in females levels were lower in young and mature pigs compared to prepubertal animals. Plasma glucagon was higher in males than females. There was an age dependent increase in plasma glucagon levels in males but not in females.

There was an effect of age and sex on plasma cortisol (Table 4). Also, there was an interaction between age and sex. Plasma cortisol level increased with age in male animals but decreased with age in female animals. Though age had no effect on plasma levels of androstenedione in minipigs, levels were lower in female than male minipigs.

Plasma levels of DHEA-S, aldosterone and testosterone in male minipigs are shown in Table 5. In females, levels of these steroids were very low, i.e., below the detection limit. Plasma testosterone was lower in mature and older pigs than in prepubertal animals. The decrease in DHEA-S with age and increase in aldosterone with age were nonsignificant.

5. DISCUSSION

In the present study, we described hormonal changes that occur with aging in male and female minipigs. This study is unique because the animals were followed until ages of more than 10 yr (group IV) with the oldest animal at 13.9 yr. All animals in group IV were males. By food restriction, body weight was maintained around 80 kg once they reached the age of 2 yr.

Table 3. Effects of age and sex on plasma levels of peptide hormones on minipigs

Sex	Age group	Insulin pmol/l	Glucagon pmol/l	ACTH ng/l
Male	I	232 ± 41	16.9 ± 0.8	21.2 ± 2.6
	III	198 ± 13	18.4 ± 1.0	37.8 ± 6.7
	IV	187 ± 13	23.3 ± 1.6	45.0 ± 10.7
Female	I	217 ± 26	13.0 ± 1.2	32.6 ± 6.5
	II	200 ± 12	14.5 ± 1.7	13.4 ± 1.7
	III	209 ± 25	13.1 ± 1.5	16.3 ± 5.4
			ANOVA	
Age		ns	0.01	ns
Sex		ns	0.001	ns
Age x sex		ns	ns	0.02

Table 4. Effect of age and sex on plasma levels of cortisol and androstenedione in minipigs

Sex	Age Group	Cortisol nmol/l	Androstenedione nmol/l
Male	I	209 ± 29	1.68 ± 0.33
	III	199 ± 29	0.79 ± 0.10
	IV	404 ± 34	1.45 ± 0.34
Female	I	379 ± 55	0.30 ± 0.06
	II	284 ± 87	0.26 ± 0.08
	III	204 ± 23	0.34 ± 0.09
		ANOVA	
Age		0.05	ns
Sex		0.005	0.0001
Age x sex		0.07	0.01

In humans and rodents, aging produces gradual changes in metabolic processes; most notable is deterioration in glucose homeostasis leading to hyperglycemia and diabetes.[38-40] There is also an increase in plasma lipids, especially triacylglycerol and cholesterol levels and increased lipid peroxidation due to increased free radical generation.[14, 41-45] These lead to increases in blood pressure, development of atherosclerosis and increased incidence of either tumor formation or tumor growth.[14] High incidence of osteoporosis is common in aging populations.[46] The primary cause for these metabolic changes appears to be insulin resistance leading to hyperinsulinemia and decreased responsiveness of insulin by target tissues, notably muscle, adipose and liver.[47-50] Metabolic changes are slowed considerably by energy restriction, either by decreasing food intake[51,52] or by consuming diets with lower energy density.[53] The primary reason appears to be improvement in insulin resistance.[42] Conversely, excess energy intake and lack of exercise leads to insulin resistance and the associated metabolic processes described above leading to accelerated aging and decreased longevity.[42] Effects of energy intake on metabolic processes, antioxidant enzymes and insulin resistance have been demonstrated in rodents.[15, 54-57] Definitive studies in humans on the relationship between energy intake and life expectancy are lacking. Studies in swine will provide relevant data for aging processes in humans. In animals reported on in the present study, there were no differences in plasma glucose or triglycerides in food restricted minipigs aged 0.5 to 13.9 yr (unpublished observations, Johnson *et al.*), and plasma glucose levels were lower in older animals than in prepubertal animals. There also was an age dependent decrease in plasma cholesterol concentration.

Glucose homeostasis and perturbations in lipid metabolism are under hormonal control. Insulin tends to lower postprandial glucose by decreasing gluconeogenesis and increasing oxidation, and increases plasma lipid levels via increased lipogenesis.[58, 59] Counter regulatory hormones, glucagon, ACTH, growth hormone and cortisol, have the

Table 5. Effect of age on plasma levels of steroid hormones in male minipigs

Age Group	DHEA-S μmol/l	Aldosterone pmol/l	Testosterone nmol/l
I	0.13 ± 0.03	38.6 ± 10.2	29.2 ± 3.2
III	0.11 ± 0.04	41.6 ± 6.7	7.8 ± 1.2
IV	0.07 ± 0.04	51.2 ± 7.1	6.6 ± 2.0
	Anova		
Age	ns	ns	0.0001

opposite effect; they tend to increase glucose levels by increasing gluconeogenesis and glycogenolysis and decrease plasma lipids primarily by increasing lipolysis.[58, 59] In the present study, we did not observe any age dependent changes in insulin levels though there were age dependent changes in plasma glucagon and cortisol. Age dependent changes in these hormones were greater in males than females. There was a small decrease in plasma insulin with age and a concomitant increase in glucagon, ACTH and cortisol. Thus, as observed in rodents, energy restriction has beneficial effects on glucose homeostasis and lipid metabolism in minipigs which are due, in part, to alterations in the endocrine system. Since metabolic processes are similar in humans and pigs, this study tends to support the hypothesis that energy restriction will retard the deterioration in glucose homeostasis and lipid metabolism in humans and therefore may prolong life expectancy.

Though there was no age dependent deterioration in lipid peroxidation, as measured by free radical generation, both basal and inducible lipid peroxidation were higher in older animals than in prepubertal, young or mature animals.[60] Failure of an increase in free radical generation with age by food restriction may be one of the reasons for the absence of age dependent deterioration in lipid metabolism. What role, if any, the endocrine system plays in lipid peroxidation and free radical generation is not clear. Abnormal and excessive glycation of proteins also is involved in aging.[61] Glycation of hemoglobin and collegen are of primary importance. Again, it is not clear whether the endocrine system plays any role in glycation.

Another hormone involved in the aging process is DHEA, levels of which decrease with aging.[62, 63] In rats, DHEA decreases food intake and is used to treat obesity.[63, 64] DHEA treatment also prevents hyperglycemia in diabetic mice and lowers insulin levels in obese mice.[65-67] It may play a similar role in humans and swine. In humans, DHEA is present as a sulphate salt and is the principal steroid hormone. In the present study, we measured DHEA-S. The concentration was below detectable limit in females and there was nonsignificant decrease in males with aging. This may be due to the feeding of energy restricted diets. It is possible the level may have been lower in older pigs if they were fed ad libitum.

Wichmann et al.[68] measured several steroid hormones in plasma of boars of unknown origin and age[68]. DHEA-S levels in the present study were 10 to 20 fold higher than reported by Wichmann et al.[68] This may have been due to the fact that we measured the sulfated form (DHEA-S) while they reported the levels of DHEA. In humans, dehydroepiandrosterone is present in the sulfate form while in rats it is present predominantly in free form (DHEA). Based on the levels in our study and that of Wichmann et al.,[68] it is possible that in swine this steroid is present as DHEA-S. Stone and Seamark[69] reported 3 to 6 fold higher levels of DHEA-S compared to DHEA in nonpregnant and pregnant gilts. Further DHEA-S levels in our study in male swine were 100 fold higher than in females reported by Stone and Seamark.[69]

In female minipigs, plasma aldosterone was below the detectible limit. The reason for this is not clear as aldosterone plays an important role in electrolyte balance. The level of testosterone was lower in mature and older animals than in prepubertal minipigs, indicating that testosterone may play an important role during the early growth of male animals. Levels were similar to those observed by Wichmann et al.[68] in boars and 10 fold higher than reported in gilts by Stone and Seamark.[69] Others[31,32] have shown decreases in testosterone at 3 to 5 wk of age, compared to neonates. The lower levels observed in the present study could have been due to the fact that growth (weight gain) was restricted in mature and older minipigs by food restriction and hence less requirement for testosterone. Elevated growth hormone concentrations[28,70] and decreased IGF-I levels[70] have been reported during nutritional restriction in neonatal pigs. Depending on the method of assay, decreased and increased levels were observed for IGF binding protein-2 after fasting in neonatal pigs.[26] Prepubertal male minipigs were heavier than prepubertal female minipigs

(Table 2). Plasma levels of another steroid hormone, androstenedione, also were higher in male than female animals. In males, levels were higher in prepubertal and older animals compared to mature minipigs. Levels were of the same magnitude as reported previously for male[68] and female[69] animals. The effect of sex on plasma cortisol levels is complex; it depends on the age of the animals. It was higher in prepubertal females than in prepubertal males, but there was no sex related difference in mature animals. In males, older animals had higher levels of cortisol than prepubertal or mature adults. Since there were no female minipigs older than 10 yr, it is difficult to predict whether older females would have lower levels than younger or mature animals. Thus, steroid hormones in general were lower in females than males.

In conclusion, both age and sex affect hormonal changes which then impact on metabolic parameters controlled by the hormones, notably carbohydrate and lipid metabolism. Deterioration in glucose homeostasis and lipid metabolism that usually occurs in ad libitum fed rodents with aging is retarded by energy restriction in minipigs, at least in males. Miniature swine can be used as a model for the study of human endocrine related changes during aging.

6. REFERENCES

1. Hsu, C.K., 1982, Uses of pigs in biomedical research: strengths and limitations, in: *Pig Model for Biomedical Research*, (H.R. Roberts, and W.J. Dodds, eds.), Pig Research Institute of Taiwan, Miaoli, Taiwan, Republic of China, pp. 3-10.
2. Pond, W.G., 1991, Of pigs and people, in: *Swine Nutrition* (E.R. Miller, D.E. Ullrey, and A.J. Lewis, eds.), Butterworth-Heinemann, Boston, pp.3-23.
3. Tumbleson, M.E. (ed.), 1986, *Swine in Biomedical Research*, Volumes 1, 2 and 3. Plenum Press, New York.
4. Miller, E.R., D.E. Ullrey, and A.J. Lewis (eds.), 1991, *Swine Nutrition*, Butterworth-Heinemann, Boston.
5. Roberts, H.R., and W.J. Dodds (eds.), 1982, *Pig Model for Biomedical Research*, Pig Research Institute of Taiwan, Miaoli, Taiwan, Republic of China.
6. Bustad, L.K., and R. O. McClellan (eds.), 1966, *Swine in Biomedical Research*, Frayn, Seattle.
7. Yu, B.P., Suescun, E.A., and Yang, S.Y., 1992, Effect of age-related lipid peroxidation on membrane fluidity and phospholipase A2: Modulation by dietary restriction, *Mech. Age. Devel.* 65:17-33.
8. Laganiere, S., and Yu, B.P., 1993, Modulation of membrane phospholipid fatty acid composition by age and food restriction, *Gerontology* 39:7-18.
9. Masoro, E.J., 1990, Animal models in aging research, in: *Handbook of the Biology of Aging*, 3rd ed. (E.L.Schneider and J.W. Rowe, eds.), Academic Press, San Diego, pp. 72-94.
10. Fernandes, G., 1995, Effects of calorie restriction and omega-3 fatty acids on autoimmunity and aging, *Nutr. Rev.* 53:572-579.
11. Liepa, G.U., Masoro, E.J., Bertrand, H.A., and Yu, B.P., 1980, Food restriction as a modulator of age-related changes in serum lipids, *Am. J. Physiol.* 238:E253-E257.
12. Choi, J.H., and Yu, B.P., 1989, The effect of food restriction on kidney membrane structures of aging rats, *Age* 12:133-136.
13. Masoro, E.J., 1985, Nutrition and aging: a current assessment. *J. Nutr.* 115:842-848.
14. Yu, B.P., 1994, How diet influences the aging process of the rat, *Proc. Soc. Exp. Biol. Med.* 205:97-105.
15. Rao, G., Xia, E., Nadakavukaren, M.J., and Richardson, A., 1990, Effect of dietary restriction on the age-dependent changes in the expression of antioxidant enzymes in rat liver, *J. Nutr.* 120:602-609.
16. Masoro, E.J., 1992, Retardation of aging process by food restriction: an experimental tool, *Am. J. Clin. Nutr.* 52:1520S-1522S.
17. Ausman, L.M., and Russel, R.M., 1990, Nutrition and aging, in: *Handbook of the Biology of Aging*, 3rd ed. (E.L. Schneider and J.W. Rowe, eds.), Academie Press, Inc., San Diego, pp. 384-406.
18. Masoro, E.J., 1992, A dietary key to uncovering aging processes, *News Physiol. Sci.* 7:157-160.
19. Masoro, E.J., 1991, Animal models in aging research, in: *Handbook of the Biology of Aging*, 3rd ed. (E.L.Schneider and J.W. Rowe, eds.), Academic Press, San Diego, pp. 72-94.
20. Kagawa, Y., 1978, Impact of westernization on the nutrition of Japanese: Changes in physique, cancer, longivity, and centenarians, *Prev. Med.* 7:205-217.

21. Lopez, S.A., 1984, Metabolic and endocrine factors in aging, in: *Risk Factors for Senility* (H. Rothschild, and C.F. Chapman, eds.), Oxford University Press, New York, pp. 205-219.

22. Bhathena, S.J., Berlin, E., Judd, J., Nair, P.P., Kennedy, B.W., Jones, J., Smith, P.M., Jones, Y.,Taylor, P.R., and Campbell, W.S., 1989, Hormones regulating lipid and carbohydrate metabolism in premenopausal women: modulation by dietary lipids, *Am. J. Clin. Nutr.* 49:752-757.

23. Bhathena, S.J., Berlin, E., Judd, J.T., Kim, Y.C., Law, J.S., Bhagavan, H.N., Ballard-Barbash, R., and Nair, P.P., 1991, Effects of omega-3 fatty acids and vitamin E on hormones involved in carbohydrate and lipid metabolis in men, *Am. J. Clin. Nutr.* 54:684-686.

24. Owens, P.C., Conlon, M.A., Campbell, R.G., Johnson, F.J., King, R., and Ballard, F.J., 1991, Developmental changes in growth hormone, insulin-like growth factors (IGF-I and IGF-II) and IGF-binding proteins in plasma of young growing pigs, *J. Endocrinol.* 128:439-447.

25. Lee, C.Y., Bazer, F.W., Etherton, T.D., and Simmen, F.A., 1991, Ontogeny of insulin-like growth factors (IGF-I and IGF-II) and IGF-binding proteins in porcine serum during fetal and postnatal development, *Endocrinology* 128:2336-2344.

26. McCusker, R.H., Cohick, W.S., Busby, W.H., and Clemmons, D.R., 1991, Evaluation of the developmental and nutritional changes in porcine insulin-like growth factor-binding protein-1 and -2 serum levels by immunoassay, *Endocrinology* 129:2631-2638.

27. Klindt, J., and Stone, R.T., 1984, Porcine growth hormone and prolactin: Concentrations in the fetus and secretory patterns in the growing pig, *Growth* 48:1-5.

28. Klindt, J., Porcine growth hormone and prolactin secretion: the first month of postnatal life, *Growth* 50:516-525.

29. Hausman, G.J., Campion, D.R., and Buonomo, F.C., 1991, Concentrations of insulin-like growth factors (IGF-I and IGF-II) in tissues of developing lean and obese pig fetuses, *Growth Dev. Aging* 55:43-52.

30. Kosco, M.S., Bolt, D.J., Wheaton, J.E., Loseth, K.J., and Crabo, B.G., 1987, Endocrine responses in relation to compensatory testicular growth after neonatal hemicastration in boars, *Biol. Reprod.* 36:1177-1185.

31. Ford, J.J., and Schanbacher, B.D., 1977, Luteinizing hormone secretion and female lordosis behavior in male pigs, *Endocrinology* 100:1033-1038.

32. Ford, J.J., 1983, Serum estrogen concentrations during postnatal development in male pigs, *Proc. Soc. Exp. Biol. Med.* 174:160-164.

33. Buononmo, F.C., and Klindt, J., 1993, Ontogeny of growth hormone (GH), insulin-like growth factors (IGF-I and IGF-II) and IGF-binding protein-2 (IGFBP-2) in genetically lean and obese swine, *Domestic Anim. Endocrin.* 10:257-265.

34. Mersmann, H.J., 1991, Characteristics of obese and lean swine, in: *Swine Nutrition* (E.R. Miller, D.E. Ullrey, and A.J. Lewis, eds.), Butterworth-Heinemann, London, pp.75-89.

35. Althen, T.G., and Gerrits, R.J., 1976, Pituitary and serum growth hormone levels in Duroc and Yorkshire swine genetically selected for high and low backfat, *J. Anim. Sci.* 42:1490-1497.

36. Mersmann, H.J., Pond, W.G., and Yen, J.T., 1982, Plasma glucose, insulin and lipids during growth of genetically lean and obese swine, *Growth* 46:189-198.

37. Statistical Analysis System Institute, Inc., 1988, *SAS/STAT user's guide: vesion 6.03*, Cary, NC, SAS Institute, Inc.

38. Andres, R., 1971, Aging and diabetes, *Med. Clin. N. Am.* 55:835-846.

39. Harris, M., 1982, The prevalence of diabetes, undiagnosed diabetes and impaired glucose tolerance in the United States, in: *Genetic and Environmental Interactions in Diabetes Mellitus* (H.S. Mehish, J. Hanna, and S. Baba, eds.), Excerpta Medica, Amsterdam, pp. 70-76.

40. Davidson, M.B., 1979, The effect of aging on carbohydrate metabolism: a review of English literature and a practical approach to the diagnosis of diabetes mellitus in the elderly, *Metabolism* 28:688-705.

41. Hershcopf, R.J., Elahi, D., Andres, R., Baldwin, H.L., Raizes, G.S., Schoeken, D.D., and Tobin, J.D., 1982, Longitudinal changes in serum cholesterol in men: an epidemilogic search for an etiology, *J. Chronic Dis.* 35:101-114.

42. Goldberg, A.P, and Hagberg, J.M., Physical exercise in the elderly, in: *Handbook of the Biology of Aging, 3rd ed. (E.L. Schneider, and J.W. Rowe, eds.), Academic Press, San Diego, pp. 407-428.*

43. Harman, D., 1993, Free radicals and age related diseases, in: *Free Radicals in Aging* (Yu, B.P., ed.), CRC Press, Boca Raton, pp. 206-222.

44. Floyd, R.E., 1993, Basic free radical biochemistry, in: *Free Radicals in Aging* (Yu, B.P., ed.), CRC Press, Boca Raton, pp. 40-55.

45. Pryor, W.A., 1987, The free radical theory of aging revisited: a critique and a suggested disease specific theory, in: *Modern Biological Theories of Aging* (H.R. Warner, R.N. Butler, R.L. Sprott, and E.L. Schneider, eds.), Raven Press, New York, pp. 89-112.

46. Exton-Smith, A.N., 1972, Physiological aspects of aging: relationship to nutrition, *Am. J. Clin. Nutr.* 25:853-859.

47. Rowe, J.W., Minaker, K.L., Pallotta, J.A., and Flier, J.S., 1983, Characterization of the insulin resistance of aging, *J. Clin. Invest.* 71:1581-1587.

48. Fink, R.I., Kolterman, O.G., Griffin, J., and Olefoky, J.M., 1983, Mechanism of insulin resistance in aging, *J. Clin. Invest.* 71:15 23-1525.

49. Cher, M., Bergman, R.N., Pacini, J., and Porte, D., Jr., 1985, Pathogenesis of age related glucose intolerance in man: insulin resistance and decreased beta-cell function, *J. Clin. Endocrinol. Metab.* 60:13-20.

50. Harris, M.I., Hadden, W.C., Knowler, W.C., and Bennett, P.H., 1987, Prevalence of diabetes and impaired glucose tolerance and plasma glucose levels in US population aged 20-74 yr., *Diabetes* 36:523-534.

51. Werman, M.J., and Bhathena, S.J., 1993, Restricted food intake ameliorates the severity of copper deficiency in rats fed a copper deficient, high fructose diet, *Med. Sci. Res.* 21:309-310.

52. Saari, J.T., Johnson, W.T., Reeves, P.G., and Johnson, L.K., 1993, Amelioration of effects of severe dietary copper deficiency by food restriction in rats, *Am. J. Clin. Nutr.* 58:891-896.

53. Werman, M.J., and Bhathena, S.J., 1996. Effects of changes in dietary energy density and the amount of fructose on indices of copper status and metabolic parameters in male rats. *J. Clin. Biochem.* In Press.

54. Chipalkatti, S., De, A.K., and Aiyar, A.S., 1983, Effect of diet restricition on some biochemical parameters related to aging in mice, *J. Nutr.* 113:944-950.

55. Yu, B.P., Lee, D.W., Marler, C.G., and Choi, J.-H., 1990, Mechanism of food restriction: protection of cellular homeostasis, *Proc. Soc. Exp. Biol. Med.* 193:13-15.

56. Johnson, B.C., and Good, R.A., 1990, Chronic dietary restriction and longevity, *Proc. Soc. Exp. Biol. Med.* 193:4-5.

57. Fernandes, G., Venkatraman, J., Khare, A., Horbach, G.J.M.J., and Friedrichs, W., 1990, Modulation of gene expression in autoimmune disease and aging by food restriction and dietary lipids, *Proc. Soc. Exp. Biol. Med.* 193:16-22.

58. Eaton, R.P., and Schade, D.S., 1982, Hormonal antagonism of insulin, in: *Diabetes and Obesity* (R.N. Brodoff, and S.J. Bleicher, eds.), Williams and Wilkins, Baltimore, pp. 27-34.

59. Bhathena, S.J., 1992, Fatty acids and diabetes, in: *Fatty Acids in Foods and Their Health Implications* (C.K. Chow, ed.), Marcel Dekker, New York, pp. 823-855.

60. Berlin, E., Banks, M.A., Bhathena, S.J., Peters, R.C., and Johnson, W.A., 1996, Aging and miniature swine heart and liver plasma membranes, in: *Advances in Swine in Biomedical Research* (M.E. Tumblson, and L.B. Schook, eds.), Plenum Press, New York, 581-593.

61. Lee, A.T., and Cerami, A., 1991, Modifications of proteins and nucleic acids by reducing sugars:possible role in aging, in: *Handbook of the Biology of Aging*, 3rd ed. (E.L. Schneider, and J.W. Rowe, eds.), Academic Press, San Diego, pp. 116-130.

62. Migeon, C.J., Keller, A.R., Lawrence, B., and Shepard, T.H., 1957, Dehydroepiandrosterone and androsterone levels in human plasma. Effect of age and sex, day to day and diurnal variations, *J. Clin. Endocrinol. Metab.* 17:1051-1062.

63. Barrett-Conner, E., Kahn, K.T., and Yen S.S.C., 1986, A prospective study of dehydroepiandrosterone sulfate, mortality and cardiovascular disease, *N. Engl. J. Med.* 315:1519-1524.

64. Yen, T.T., Allan, J.A., Pearson, D.V., Acton, J.M., and Greenberg, M.M., 1977, Prevention of obesity in A^{vy}/a mice by dehydroepiandrosterone, *Lipids* 12:409-413.

65. Coleman, D.L., Leiter, E.H., and Appleweig, N., 1984, Therapeutic effects of dehydroepiandrosterone metabolites in diabetes mutant mice (C57BL/KSJ-db/db), *Endocrinology* 115:239-243.

66. Coleman, D.L., Leiter E.H., and Schwitzer, R.W., 1982, Therapeutic effects of dehydroepiandrosterone (DHEA) in diabetic mice, *Diabetes* 31:830-833.

67. Gansler, T.S., Muller, S., and Cleary, M.P., 1985, Chronic administration of dehydroepiandrosterone reduces pancreatic B-cell hyperplasia and hyperinsulinemia in genetically obese Zucker rats, *Proc Soc. Exp. Biol. Med.* 180:155-162.

68. Wichmann, U., Wichmann, G., and Krause, W., 1984, Serum levels of testosterone precursors, testosterone and estradiol in 10 animal species, *Exp. Clin. Endocrinol.* 83:283-290.

69. Stone, B.A., and Seamark, R.F., 1985, Steroid hormones in uterine washings and in plasma of gilts between days 9 and 15 after oestrus and and between days 9 and 15 after coitus, *J. Reprod. Fert.* 75:209-221.

70. Campion, D.R., McCusker, R.H., Buonomo, F.C., and Jones W.K., Jr., 1986, Effect of fasting neonatal piglets on blood hormone and metabolite profiles and on skeletal muscle metabolism, *J. Anim. Sci.* 63:1418-1427.

49

AGING AND MINIATURE SWINE HEART AND LIVER PLASMA MEMBRANES

Elliott Berlin, PhD,[1]* Melanie A. Banks, PhD,[1] Sam J. Bhathena, PhD,[1] Renee C. Peters, BS,[1] and Wesley A. Johnson, DVM[2]

[1] Metabolism and Nutrient Interactions Laboratory
Beltsville Human Nutrition Research Center
Agricultural Research Service
U.S. Department of Agriculture
Beltsville, Maryland 20705
[2] Center for Food Safety and Applied Nutrition
Division of Toxicological Studies
Beltsville Research Facility
Food and Drug Administration
U.S. Department of Health and Human Services
Laurel, Maryland 20708

1. ABSTRACT

Age related changes in heart and liver plasma membranes were assessed over a wide age range, prepubertal, < 0.5 yr; young, 0.5 to 2.5 yr; middle aged, 5.9 to 10 yr; and old, 11.5 to 13.9 yr, in male and female Hormel-Hanford miniature swine fed the same energy restricted, low fat, cholesterol free, stock diet continuously post weaning. Mid bilayer heart plasma membrane fluidity, by diphenylhexatriene (DPH) polarization, was increased in old pigs, but fluidity in polar membrane domains was not increased in the old pigs. Age had no effect on fluidity by DPH in liver cell membranes, but surface domains were influenced by age. Liver cell membrane fluidity, by polar probes, followed a U shaped pattern with age and reached a maximum in the middle aged minipig group. In heart, middle aged pigs were

* Reprint requests to Dr. Elliott Berlin, MNIL, BHNRC, ARS, USDA, Building 307, Rm. 323, BARC-East, Beltsville, MD 20705, (301) 504-8297.

Advances in Swine in Biomedical Research, edited by Tumbleson and Schook
Plenum Press, New York, 1996

reduced in total n-6 polyunsaturates, especially 18:2, and increased in saturates. Conversely, liver plasma membranes in middle aged pigs were elevated in unsaturated and reduced in saturated fatty acids. Heart membrane lipid peroxidation tended to be increased in middle aged and elderly pigs, perhaps due to increased membrane vitamin E in middle aged and old pigs. Thus, lipid peroxidation tended to increase over the lifespan of miniature swine, even when they were food restricted.

2. INTRODUCTION

The free radical theory of aging[1,2] has become the foundation for our understanding of the mechanism of development of many age associated chronic diseases, such as heart disease and cancer, as well as a mechanism for the aging process *per se*. Free radical reactions in biological systems increase with age[3] leading to peroxidized product accumulation[4] and structural damage to cells, accompanied by cellular dysfunction.[5] Age associated increased free radical production and lipid peroxidation in cellular membranes with resulting decreased fluidity have been demonstrated[6-8] in rodents and fish. Age associated alterations in cellular lipids and membrane fatty acids have been observed in rats.[9,10] It is likely that some of the physiological changes that occur with age are at least partially based on alterations within cellular membranes since cellular membrane fatty acid composition and fluidity are known to influence activities of membrane proteins, including enzymes and receptors.[11-13]

Chronic moderate restriction of food energy intake has been demonstrated to increase longevity in rats.[14] Energy restricted animals do not show the same membrane changes upon aging as *ad libitum* fed animals. Energy restricted rats have decreased long chain polyunsaturated fatty acid content,[10] attenuated membrane lipid peroxidation [8,10] and do not show a decline in membrane fluidity with age.[8]

We studied food restricted miniature swine, which have much longer life spans than rats, to determine how age affects heart and liver plasma membrane fatty acid composition and fluidity and each of several parameters associated with fatty acid oxidation in the heart. We assessed susceptibility to lipid peroxidation and vitamin E content in heart tissue homogenate and membrane isolates in consideration of the heart's usual high oxygen, hemoglobin/iron and long chain polyunsaturated fatty acid concentrations. Vitamin E concentration was measured since it and other antioxidants have been hypothesized as protective against degenerative diseases associated with aging and perhaps the aging process itself.[2,15,16]

The pig was selected for this work as it is considered an excellent animal model for investigation of human diseases because of its many anatomic, physiological, metabolic and nutritional similarities to humans. Pigs often have been used in cardiovascular studies and have been recommended as models for gerontological research.[17]

3. METHODS

3.1. Animals and Treatment

Miniature male and female Hormel-Hanford swine, age 0.4 to 13.9 yr were housed individually (in heated pens in winter) with indoor and outdoor runs with access to water *ad libitum*. The entire experimental protocol was approved by the FDA Institutional Animal Care and Use Committee. After weaning at 6 wk, animals were fed continuously the same low fat, cholesterol free stock diet (USDA 1160) for their entire lives except during

Table 1. Composition of diets (g/kg)

Ingredients	USDA 1160*	USDA 1180**
Corn meal	739	500
Alfalfa meal	50	---
Dicalcium phosphate	21	22
Limestone	3.5	3.5
Swine mineral mix	1	2
Selenium	0.5	0.5
Salt, iodized	5	5
Soybean meal (49%)	179	245
Swine vitamin mix	1	1.5
Whey (dried)	---	150
Sugar	---	50
Soybean Oil	---	20

*   Fed to adult animals at 680 to 900 g/day for maintenance
      and ad lib to prepubertal animals postweaning.
**   Fed to pregnant and lactating sows at 1150 to 1800 g/day
      and made available ad libitum to suckling piglets.

pregnancy and lactation when sows were fed a higher fat diet (USDA 1180) (Table 1). Suckling piglets also were allowed USDA 1180 *ad libitum*; weaned young animals, less than 6 mo old, were allowed USDA 1160 *ad libitum*. At 6 mo all animals were restricted to weight maintenance feeding, 680 to 900 g/da. Major nutrient contents of the USDA 1160 diet[18] were 16% protein, 75% carbohydrate and 2.7% fat (P/S = 3.8). The pigs were at different life stages: prepubertal (<6 mo), young (6 mo to 2.5 yr), middle aged (5.9 to 10 yr) and old (11.5 to 13.9 yr). Animals were sacrificed by electrocution; samples were excised from the heart and liver during necropsy, placed in cold, 4C, normal saline and kept on ice while being transported to the laboratory. Tissue samples always were taken from the same regions in these organs. Heart samples, 30 to 40 g, were chopped into small pieces with scissors, placed in ice cold 0.25 M sucrose and homogenized with a Polytron device. The homogenate was centrifuged at 16,000 x g for 30 min to collect a crude pellet containing subcellular organelles, intact cells and subcellular fractions trapped therein. Supernatant was removed as a crude cytosolic preparation. To isolate the subcellular organelles, the heart pellet was resuspended in 0.25 M sucrose and centrifuged through a discontinuous 0.25 to 2.0 M sucrose gradient at 100,000 x g for 1 hr in a Beckman L8-80 ultracentrifuge, using an SW-28 swinging bucket rotor. Mitochondrial, microsomal, cytosolic and plasma membrane fractions were frozen until analysis. Liver plasma membranes were prepared by the technique of Neville[19] as adapted.[20] Membrane fluidity was measured immediately upon isolation without freezing. Samples for MDA analysis were frozen and kept briefly at -20C, and samples for vitamin E and fatty acid analyses were stored at -85C.

3.2. Fluidity Measurement

Membrane fluidity was assessed by measuring fluorescence polarization of polar and nonpolar hydrocarbon probes in membrane samples. Steady state fluorescence polarization measurements were made using methods developed by Shinitzky and Barenholz.[21] Polarization data were taken as a function of temperature between 37 and 4C with diphenylhexatriene (DPH), its cationic derivative trimethylammoniumdiphenylhexatriene (TMA-DPH)[22] and its anionic derivative diphenylhexatrienepropionic acid (DPH-PA).[23] All measurements were made with the SLM 4800 fluorometer in the steady state mode and equipped with polarizers in the T optical format. Excitation and emission wavelengths were set at 366 and 430 nm, respectively. Appropriate sample dilutions were made to prevent light scattering errors; a Wratten 2A cutoff filter was used to prevent errors due to forward scattering of the excitation light.

3.3. Chemical Analyses

Membrane fatty acyl compositions were determined by capillary gas chromatography of corresponding methyl esters prepared by transesterification with methanolic HCl.[24] Chromatography was performed with a Hewlett-Packard Model 5890A gas chromatograph equipped with dual flame ionization detectors, a Model 7673A automatic sampler, a Model 3396A integrator and a Supelco SP-2560 fused silica 100 m, 0.25 mm ID capillary column with a 0.20 μm thick film.

Protein[25] and phosphorus[26] concentrations were determined colorimetrically. Vitamin E contents of heart homogenate and subcellular fractions were determined via HPLC according to the method of Desai[27] with modifications. A mixture consisting of sample, absolute ethanol and 25% ascorbic acid was prepared and held at 70C for 5 min before adding 10 N KOH. The mixture was incubated at 70C for 30 min and cooled at 4C, after which 2.5% ethanolic pyrogallol was added before extracting with hexane. Solvent was removed by evaporation under nitrogen; vitamin E containing residues were dissolved in absolute C_2H_5OH. α-Tocopherol contents were assayed by HPLC[28] with a Supelcosil LC-18 column (5 μm particle size, 4.6 mm x 25.0 cm) using a Spectra-Physics SP8810 isocratic pump, with methanol as solvent, and a Perkin-Elmer 650-10S spectrophotofluorometer equipped with a Hewlett Packard 3390A integrator for fluorescence detection and quantitative tocopherol assay. Excitation and emission wavelengths were 292 and 336 nm, respectively. Peak areas were quantitated upon calibration with authentic samples of tocopherols supplied by Hoffman-La Roche.

Lipid peroxidation in whole heart homogenates and plasma membrane isolates was assessed by determining the concentration of thiobarbituric acid reactive substances in the presence or absence of $FeCl_3$/ascorbate (5 mM each) according to the method of Tatum,[29] except that malondialdehyde was measured spectophotometrically at 535 nm and its concentration was calculated using the molar extinction coefficient for this compound.[30] Data are expressed in terms of malondialdehyde production per unit protein or phosphorus. Peroxidizability may be inferred from the inducible malondialdehyde production which was calculated as the difference between Fe^{3+}/ascorbate induced (total) and noninduced (baseline) lipid peroxidation.

3.4. Statistical Analyses

Data were subjected to analysis of variance and linear regression analysis for the various measurements using the SAS statistical programs.[31] The model included sources of variation due to age with data classified into 4 age groups: prepubertal, young, middle aged and old. The Student-Newman-Kuels test was used to determine differences in model classified composition and anisotropy data. Lipid peroxide data were subjected to the Kruskal-Wallis test and further analyzed by Duncan's multivariate ANOVA after logarithmic transformation.

4. RESULTS

Heart plasma membrane fatty acid compositions, as mole percentages, for the 4 major age groups are presented in Table 2. There were no differences in composition among young

Table 2. Miniature pig heart plasma membrane fatty acids ($\bar{x} \pm$ SEM)

AGES:	Prepub n = 11	Young n = 10	Middle n = 9	Old n = 6
Acid		mole percent		
12:0	0.13 ± 0.05[a]	0.04 ± 0.02[a]	0.04 ± 0.02[a]	0.02 ± 0.01[a]
14:0	2.24 ± 0.33[a]	1.82 ± 0.35[a]	2.32 ± 0.32[a]	1.99 ± 0.53[a]
15:0	1.44 ± 0.27[a]	1.39 ± 0.34[a]	1.31 ± 0.36[a]	1.72 ± 0.84[a]
16:0	33.41 ± 2.00[b]	36.44 ± 2.45[b]	46.11 ± 1.39[a]	35.63 ± 3.81[b]
16:1	6.08 ± 0.29[a]	6.30 ± 1.86[a]	4.38 ± 0.61[a]	4.70 ± 0.58[a]
18:0	19.19 ± 1.09[b]	20.77 ± 1.94[b]	24.95 ± 1.10[a]	18.55 ± 0.53[b]
18:1	24.97 ± 3.97[a]	22.07 ± 2.10[a]	17.03 ± 2.49[a]	26.15 ± 4.51[a]
18:2	9.71 ± 1.95[a]	7.95 ± 2.86[ab]	1.18 ± 0.36[b]	8.10 ± 2.25[ab]
18:3	0.08 ± 0.02[a]	0.09 ± 0.04[a]	0.03 ± 0.03[a]	0.13 ± 0.05[a]
20:0	0.63 ± 0.07[a]	0.68 ± 0.08[a]	1.06 ± 0.13[a]	0.82 ± 0.26[a]
20:3	0.14 ± 0.05[a]	0.09 ± 0.06[a]	0	0.04 ± 0.02[a]
20:4	1.13 ± 0.45[a]	1.15 ± 0.73[a]	0.29 ± 0.21[a]	0.79 ± 0.40[a]
20:5	0.09 ± 0.03[a]	0.17 ± 0.03[a]	0.20 ± 0.07[a]	0.15 ± 0.07[a]
22:5	0.09 ± 0.03[a]	0.15 ± 0.06[a]	0.04 ± 0.02[a]	0.10 ± 0.03[a]
22:6	0	0.08 ± 0.07[a]	0.02 ± 0.02[a]	0
24:0	0.65 ± 0.17[a]	0.75 ± 0.16[a]	1.00 ± 0.31[a]	1.08 ± 0.54[a]
24:1	0.03 ± 0.01[a]	0.07 ± 0.03[a]	0.03 ± 0.02[a]	0.03 ± 0.02[a]
Σn-3	0.26 ± 0.05[a]	0.49 ± 0.14[a]	0.30 ± 0.11[a]	0.37 ± 0.06[a]
Σn-6	10.98 ± 2.41[a]	9.18 ± 3.60[a]	1.47 ± 0.08[a]	8.92 ± 2.64[a]
ΣPUFA	11.23 ± 3.90[a]	9.67 ± 3.69[a]	1.78 ± 0.48[a]	9.29 ± 2.63[a]
ΣMUFA	31.09 ± 2.45[a]	28.43 ± 2.88[a]	21.44 ± 2.44[a]	30.89 ± 4.93[a]
UI	56.57 ± 4.14[a]	51.51 ± 8.29[a]	26.44 ± 2.73[b]	51.94 ± 8.21[a]

Values for the same fatty acid with different superscripts are significantly different (P < 0.05). Ages are defined as: prepub < 0.5 yr; 0.5 <young < 2.5 yr; 5.9 < middle < 10 yr; old > 11.5 yr. The unsaturation index, UI = $(X_{16:1} + X_{18:1} + X_{24:1}) + 2(X_{18:2}) + 3(X_{18:3} + X_{20:3}) + 4(X_{20:4}) + 5(X_{20:5} + X_{22:5}) + 6(X_{22:6})$

and old animals; however, middle aged pigs were elevated in saturated fatty acids at the expense of the polyunsaturates. Palmitate (16:0) and stearate (18:0) were elevated, and linoleate (18:2) was reduced in middle aged (5.9 to 10 yr old) pigs. Total n-6 polyunsaturated fatty acids tended to decrease in middle aged pigs, though not significantly. Laurate content alone was decreased in the oldest pigs. Results with these individual fatty acids are reflected in the unsaturation index, UI, which is calculated as a function of the sum of mole percentages of unsaturated fatty acids times the number of olefinic double bonds. UI was reduced in heart plasma membranes of the middle aged pigs, as compared with the other age groups. Liver plasma membrane fatty acid composition data are given in Table 3. In contrast with the heart data liver membranes from the young and middle aged pigs were elevated in

Table 3. Miniature pig liver plasma membrane fatty acids ($\bar{x} \pm$ SEM)

AGES:	Prepub n = 11	Young n = 10	Middle n = 9	Old n = 6
Acid	mole percent			
14:0	0.84 ± 0.09[a]	0.38 ± 0.23[a]	0.63 ± 0.21[a]	0.85 ± 0.25[a]
15:0	1.12 ± 0.21[a]	0.55 ± 0.27[a]	0.95 ± 0.35[a]	1.36 ± 0.41[a]
16:0	17.54 ± 0.89[a]	14.94 ± 1.07[ab]	10.69 ± 0.73[c]	13.86 ± 1.62[b]
16:1	0.36 ± 0.03[a]	0.23 ± 0.06[a]	0.22 ± 0.06[a]	0.14 ± 0.09[a]
18:0	49.16 ± 2.44[a]	37.80 ± 2.94[b]	37.61 ± 1.80[b]	45.43 ± 2.60[ab]
18:1	15.56 ± 1.15[a]	11.28 ± 0.81[b]	10.77 ± 0.58[b]	10.33 ± 1.05[b]
18:2	8.87 ± 1.37[b]	19.54 ± 3.21[a]	19.83 ± 1.74[a]	13.00 ± 1.82[ab]
18:3	0.61 ± 0.12[a]	0.50 ± 0.12[a]	0.72 ± 0.12[a]	0.67 ± 0.27[a]
20:0	0.55 ± 0.13[a]	0.40 ± 0.09[a]	0.53 ± 0.10[a]	0.83 ± 0.13[a]
20:3	0.59 ± 0.11[a]	1.15 ± 0.27[a]	0.99 ± 0.18[a]	1.12 ± 0.24[a]
20:4	2.63 ± 1.24[b]	10.36 ± 2.35[a]	12.59 ± 1.10[a]	7.59 ± 2.63[ab]
20:5	0.20 ± 0.07[a]	0.13 ± 0.09[a]	0.22 ± 0.10[a]	0.25 ± 0.18[a]
22:5	0.23 ± 0.21[a]	1.39 ± 0.52[a]	1.79 ± 0.41[a]	1.64 ± 0.72[a]
22:6	0.04 ± 0.04[a]	0.14 ± 0.06[a]	0.16 ± 0.06[a]	0.10 ± 0.05[a]
24:0	1.29 ± 0.21[bc]	0.88 ± 0.18[c]	1.76 ± 0.16[ab]	2.11 ± 0.36[a]
24:1	0.41 ± 0.12[a]	0.33 ± 0.08[a]	0.57 ± 0.12[a]	0.71 ± 0.16[a]
Σn-3	1.08 ± 0.27[b]	2.15 ± 0.52[ab]	2.88 ± 0.43[a]	2.65 ± 0.66[ab]
Σn-6	12.09 ± 2.51[c]	31.05 ± 3.91[a]	33.40 ± 1.73[a]	21.71 ± 4.47[b]
ΣPUFA	13.17 ± 2.73[c]	33.20 ± 4.01[ab]	36.29 ± 0.48[a]	24.37 ± 5.05[b]
ΣMUFA	16.33 ± 1.20[a]	11.84 ± 0.86[b]	11.56 ± 0.70[b]	11.19 ± 1.16[b]
UI	50.57 ± 8.21[b]	105.68 ± 12.78[a]	117.67 ± 5.86[a]	82.95 ± 16.41[a]

Values for the same fatty acid with different superscripts are significantly different (P < 0.05). Ages are defined as: prepub < 0.5 yr; 0.5 < young < 2.5 yr; 5.9 < middle < 10 yr; old > 11.5 yr. The unsaturation index, UI = $(X_{16:1} + X_{18:1} + X_{24:1}) + 2(X_{18:2}) + 3(X_{18:3} + X_{20:3}) + 4(X_{20:4}) + 5(X_{20:5} + X_{22:5}) + 6(X_{22:6})$

unsaturates and reduced in saturates. UI was lower for liver membranes from the prepubertal animals than from the others. Regression analyses yielded no relations between heart membrane unsaturation and age, but linear regression analyses of liver membrane fatty acid data in mole percentages vs age in yr yielded the equations:

$$MUFA = 11.65 + 0.34(AGE) \text{ with } P < 0.02 \text{ and } R = 0.418$$

$$PUFA = 33.26 - 1.75(AGE) \text{ with } P < 0.0006 \text{ and } R = 0.576$$

$$UI = 107.5 - 4.87(AGE) \text{ with } P < 0.002 \text{ and } R = 0.535$$

Heart plasma membrane fluidity, as can be inferred from the reciprocal of the anisotropy data for the neutral probe, DPH, (Table 4) was increased in old animals, despite the absence of an increased level of unsaturation in this group. Results for the polar probes, TMA-DPH and DPH-PA, indicated nonsignificant trends toward decreased fluidity in heart membranes from the middle aged group. Data in Table 4 are grouped as indicated for different life stages of the animals; however, regression analyses of all heart membrane anisotropy data for each probe vs age, in da, yielded the following equations:

$$r_{DPH-PA} = 0.237 + 2.87 \times 10^{-6}x(age) \text{ with } P < 0.06 \text{ and } R = 0.316 \text{ and}$$

$$r_{TMA-DPH} = 0.271 + 1.14 \times 10^{-5}x(age) \text{ with } P < 0.08 \text{ and } R = 0.442.$$

Though the data for the polar probes yielded these relations approaching statistical significance no significant relations existed between the DPH anisotropy and age. Analysis of the fluorescence data for 10 and 4C showed similar relationships. Linear regression analyses of the relation between probe fluorescence and fatty acid unsaturation yielded equations approaching statistical significance between r, the steady state anisotropy, for the polar probes and UI, the unsaturation index:

Table 4. Plasma membrane fluorescence anisotropy at 37C ($\bar{x} \pm$ S.E.M.)

Probe	Prepub n = 12	Young n = 9	Middle n = 10	Old n = 6
		Heart Data		
DPH	0.127 ± 0.008[a]	0.126 ± 0.008[a]	0.137 ± 0.008[a]	0.096 ± 0.005[b]
TMA-DPH	0.287 ± 0.005[a]	0.262 ± 0.011[a]	0.314 ± 0.002[a]	0.296[a]
DPH-PA	0.238 ± 0.014[a]	0.236 ± 0.004[a]	0.253 ± 0.006[a]	0.243 ± 0.004[a]
		Liver Data		
DPH	0.136 ± 0.023[a]	0.119 ± 0.014[a]	0.118 ± 0.013[a]	0.128 ± 0.009[a]
TMA-DPH	0.216 ± 0.030[a]	0.195 ± 0.014[ab]	0.187 ± 0.015[b]	0.207 ± 0.021[ab]
DPH-PA	0.023 ± 0.031[a]	0.222 ± 0.017[a]	0.205 ± 0.015[a]	0.230 ± 0.021[a]

Values for the same fluorescent probe with different superscripts are significantly different (P < 0.05). Only one TMA-DPH anisotropy value was obtained for heart membranes from old miniature pigs. Ages are defined as: prepub < 0.5 yr; 0.5 <young < 2.5 yr; 5.9 < middle < 10 yr; old > 11.5 yr.

Table 5. Heart tissue homogenate and plasma membrane lipid peroxidation

AGES	Baseline	Induced	Inducible
		Plasma Membrane Data	
Prepub	5.6 ± 2.0	16.3 ± 6.1	10.7 ± 4.2
Young	114.7 ± 66.8	186.1 ± 96.0	71.4 ± 31.4
Middle	93.0 ± 44.1	234.0 ± 101.1	141.0 ± 67.3
Old	370.1 ± 348.4	623.4 ± 570.6	253.3 ± 222.3
		Heart Homogenate Data	
Prepub	1717 ± 394	1953 ± 410	254 ± 48
Young	1568 ± 331	1929 ± 333	361 ± 125
Middle	1102 ± 62	1296 ± 75	193 ± 34
Old	2458 ± 828	2913 ± 957	455 ± 148

Data are $\bar{x}$ ± SEM of nmol malondialdehyde per mmole lipid phosphorus for age groups defined as: prepub < 0.5 yr; 0.5 < young < 2.5 yr; 5.9 < middle < 10 yr; old > 11.5 yr. Inducible lipid peroxidation is defined as the difference between the induced and baseline values. There were no differences among the age groups for all measurements.

$$r_{DPH-PA} = 0.252 - 2.33 \times 10^{-4}(UI); \text{ with } P < 0.06 \text{ and } R = 0.33; \text{ and}$$

$$r_{TMA-DPH} = 0.318 - 6.93 \times 10^{-4}(UI); \text{ with } P < 0.10 \text{ and } R = 0.42,$$

but no significant relation was observed for r_{DPH} and UI. Examination of liver membrane fluidity data indicated a general trend toward increased fluidity in the middle aged group of animals. The only significant effect was in the reduced $r_{TMA-DPH}$ in middle aged animals. There were no significant relations noted upon linear regression analysis of the liver membrane anisotropies vs age. Thus, there were distinctly different age related patterns in the heart and liver membranes.

Lipid peroxidation (Table 5) tended to increase with age, though the differences among age groups were not significant. This trend was apparent under both baseline (noninduced) and Fe^{3+}/ascorbate induced conditions, and in the calculated inducible difference. Variances in malondialdehyde data appear nonhomogeneous, with variance becoming particularly large in plasma membrane data for the oldest group of pigs; hence, these data are difficult to evaluate by analysis of variance. Logarithmic transformation of these data and ANOVA showed significant ($P < 0.05$) differences for log (malondialdehyde concentration) among age groups with prepubertal animals showing less peroxidation. The Kruskal-Wallis ranking test with Wilcoxon scores yielded $P > \chi2 = 0.048$ for plasma membrane baseline malondialdehyde data. Thus, at least this variable shows a tendency to increase with age.

Vitamin E, as ng α-tocopherol/mg protein, in various heart subcellular fractions are given in Table 6. The microsomal fraction was the highest and the plasma membrane was the lowest, other than the cytosol. Total cell homogenate vitamin E concentration, as well as plasma membrane fraction, tended to increase with age, becoming higher in middle aged and old pigs, (Table 7) as a function of molar phospholipid content. The nonsignificant reduction in cytosolic vitamin E concentration in old pigs (Table 6) may reflect increased mobilization of vitamin E to plasma membrane in older pig hearts (Table 7).

Table 6. Vitamin E contents of heart subcellular fractions, ng α-tocopherol/mg protein

FRACTION	Prepub n = 12	Young n = 10	Middle n = 10	Old n = 6
Homogenate	48.9 ± 10.8	52.6 ± 12.0	66.4 ± 13.5	62.7 ± 12.7
Membrane	10.8 ± 1.9	11.1 ± 4.0	23.1 ± 6.7	16.3 ± 6.3
Microsomes	71.2 ± 15.8	59.5 ± 10.2	50.0 ± 12.1	66.4 ± 26.2
Mitochondria	63.4 ± 12.3	42.2 ± 8.2	60.1 ± 12.8	47.0 ± 16.5
Cytosol	8.5 ± 1.9	9.0 ± 1.9	8.0 ± 1.2	4 ± 1.3

Values are x̄ ± SEM for miniature pigs in age groups defined as: prepub
< 0.5 yr; 0.5 <young < 5.8 yr; 5.9 < middle < 10 yr; old > 10 yr. There
were no differences among the age groups for all cell fractions.

5. DISCUSSION

Swine have been recommended[17] as excellent atherosclerosis and gerontology research models because of their many anatomical and physiological similarities to man, however few aging studies have been conducted in this species. In an earlier study, Berlin *et al.*[32] examined the effect of age on fluidity and composition in lipoproteins from male and female miniature swine. Most recent investigations[6,8,10,33-37] into metabolic alterations associated with aging, including nutritional factors, have been conducted in rodents because of their shorter life span which makes it easier to control studies and maintain procedures constant.[38] Some of these studies[6,37] in rats were focused on heart tissue. Many investigators consider rats old after 2 yr but the average life span of a laboratory rat is 3 yr;[39] hence a 2 yr old rat would correspond to the middle aged pigs in the present study.

We[15,39] demonstrated there was minimal incorporation of dietary n-3 fatty acids into heart plasma membranes in middle aged female swine from high fat diets though liver cell plasma membrane and brain synaptosomes in the same animals showed incorporation of dietary n-3 fatty acids. In the present study, where age alone was the experimental variable, substantial differences also were observed in heart and liver membrane fatty acid with the liver containing more polyunsaturated fatty acids of both the n-3 and n-6 types. All data reported here were from food restricted animals; no *ad libitum* fed control group was available for comparison. Effects observed with these food restricted aging pigs would be expected to parallel those reported for food restricted aging rats. Aging (2 yr old) *ad libitum* fed rats usually show conversion of membrane 18 carbon PUFA to longer chain PUFA, such as arachidonate and docosahexaenoate. Food restriction in the rat, up to 2 yr, resulted in neither a progresssive loss of the essential fatty acids, 18:2 and 18:3, nor an increase in the

Table 7. Age and heart molar vitamin E content (nmole α-tocopherol/μmole phospholipid)

Prepub n = 12	Young n = 10	Middle n = 10	Old n = 6
		Heart Homogenate Data	
0.371 ± 0.085[a]	0.370 ± 0.084[a]	0.422 ± 0.081[a]	0.489 ± 0.109[a]
		Plasma Membrane Data	
0.647 ± 0.302[b]	1.068 ± 0.275[ab]	1.907 ± 1.040[ab]	3.800 ± 1.535[a]

Values are x̄ ± SEM for minipigs in age groups defined as: prepub < 0.5 yr;
0.5 <young < 2.5 yr; 5.9 < middle < 10 yr; old > 11.5 yr.

longer chain polyunsaturates, 20:4 and 22:6, in plasma lipids,[40] heart,[8,10,37] kidney[41] and splenocytes and bone marrow cells.[9] We show different results in liver and heart membranes. Comparison of data for middle aged pigs with published data for 2 yr old rats shows substantial differences. We exhibited increased saturation in the heart and increased unsaturation in the liver in contrast with maintenance of fatty acyl composition in various rat tissues. Neverthless, in middle aged miniswine heart membranes, UI was reduced; in older animals, UI was not increased beyond the level of the young groups. With liver membranes, we showed no differences between groups of pigs (Table 3) but UI was related inversely to age in accordance with the equation:

$$UI = 107.5 - 4.87(AGE) \text{ with } P < 0.002 \text{ and } R = 0.535.$$

Laganiere and Yu[10] postulated that food restriction modulates the aging process in animals by reducing the potential for peroxide formation by inhibiting the conversion of the precursor fatty acids, 18:2 and 18:3, to the longer chain acids with multiple double bonds hence reducing the available substrate for peroxidation. A similar reduction in peroxidation in our pigs would be expected in terms of our UI data but we had no *ad libitum* fed control group. Increased peroxidation was indeed observed as occurring in the heart with age.

The free radical theory of aging[1] is based on an extensive literature describing the formation of the superoxide radical followed by lipid peroxidation, loss of membrane integrity and cellular degeneration as organisms age. Lipid peroxidation has been demonstrated to decrease membrane fluidity, presumably through the resultant cross linking of structural biomolecules.[42] Sawada *et al.*[7] observed increased superoxide formation and lipid peroxidation in brain and liver homogenates from aging, free living male salmon. They associated the accompanying loss in plasma membrane fluidity in both organs with compositional changes resulting from production of superoxide radicals. Wahnon *et al.*[43] observed a decrease in membrane fluidity upon aging, independent of compositional changes. Yu *et al.*[8] demonstrated with rat liver microsomal and mitochondrial membranes that food restriction mitigated the increased lipid peroxidation and phospholipase A_2 activity and decreased membrane fluidity that occur with aging. Fernandes[37] reported higher superoxide dismutase levels in the splenic cytosol fraction in calorie restricted rats than in *ad libitum* fed animals. In the present study with food limited aging pigs of both sexes, heart membrane peroxidation (Table 5) tended to increase with age but fluorescence anisotropy data (Table 4) for different age groups did not show a lipid peroxidation induced decrease in membrane fluidity.

In the present study of food restricted miniature swine, equations relating r_{DPH} to age were not significant for both liver and heart membranes and the only grouped data showing a significant difference was the lower DPH fluidity observed in heart membranes from the oldest group of pigs. Perhaps food restriction effectively minimized any age effects on membrane fluidity. Differences among results for several fluorescent probes probably are related to their location in the cell membrane. Nonpolar hydrocarbon DPH intercalates into any hydrophobic region of the membrane between the fatty acyl chains and/or much deeper in midbilayer regions oriented parallel to the surface of the bilayer. Polar probes are associated with phospholipid headgroups and extend partially into the bilayer between the fatty acyl chains. DPH-PA is associated with, or covalently bound to, phospholipids of the exofacial leaflet of the membrane bilayer; TMA-DPH is associated with phospholipids of the inner membrane leaflet.[44] Hence the plasma membrane domain wherein fluidity is subject to age associated decreases, probably is nearer to the surface in the segment of the fatty acyl chains that is close to the headgroups. Prior reports of age related loss in membrane fluidity are subject to similar interpretations as Yu *et al.*[8] used TMA-DPH and Sawada *et al.*[6,7] used *trans* parinaric acid, which is also a linear probe capable of binding covalently binding to, or associating with, phospholipid headgroups.

Membrane fluidity is associated with lipid composition with: fatty acid unsaturation usually increasing fluidity,[45] cholesterol usually decreasing fluidity,[46] lipid peroxidation decreasing fluidity[42] and vitamin E modulating fluidity in a variable manner as a function of concentration.[47] Increased heart plasma membrane fluidity in the old pigs, as determined with DPH (Table 4), cannot be attributed to fatty acids; this group did not exhibit more unsaturation (Table 2) than the younger pigs. The major compositional change which may impinge on fluidity was the increase with age in the membrane vitamin E per mole of phospholipid (Table 7). Membrane vitamin E increased with age over the entire period of the study according to the equation:

$$E = 0.471 + 6.8 \times 10^{-4} \text{ (age); with P} < 0.002 \text{ and R} = 0.531.$$

This was not true of the heart homogenate. Vitamin E has been reported as causing both increased and decreased order in membranes and model liposomes.[47] At physiological membrane tocopherol concentrations, vitamin E fluidizes the membrane but, at exceptionally higher concentrations, vitamin E decreases fluidity. In earlier studies with human subjects, Berlin et al.[48,49] reported decreased lipoprotein fluidity following vitamin E supplementation; however, no measurements were made of membrane fluidity in any cells from those subjects. It is most likely the increased DPH determined fluidity in the oldest animals was due to the elevation in vitamin E. It is unclear why the higher vitamin E in these membranes was not accompanied with higher levels of fatty acyl unsaturation but was accompanied with higher baseline levels and inducible levels of malondialdehyde, thus suggesting a prooxidant role for vitamin E. Bowry et al.[50] documented such activity under some experimental conditions. Vitamin E concentrations in the membrane compared favorably with values of one α-tocopherol molecule per several thousand fatty acid molecules.[47] Kornbrust and Mavis[51] reported molecular α-tocopherol:PUFA ratios of 1:200 in highly oxygenated tissues such as heart and lung. Using our data in Tables 2 and 7, we observe α-tocopherol:PUFA ratios of 1:347, 1:172, 1:19 and 1:49 in the 4 successively older groups of pigs. Hence there should have been adequate membrane vitamin E to prevent peroxidation. Perhaps the extent of baseline peroxidation would have been even greater were it not for the vitamin E present. The increasing inducible MDA production may serve as an indicator of peroxidation that did not occur physiologically.

In conclusion, we found that changes in fatty acid composition, susceptibility to lipid peroxidation, membrane fluidity and membrane α-tocopherol content occur in the plasma membrane fraction of heart tissue from aging pigs, even when they are food restricted. Age associated changes in liver plasma membranes also occurred, but the changes in liver and heart membrane fatty acids were in opposite directions. The mechanisms accounting for these changes need further clarification. Our work with aging miniature swine emphasizes the need for comparative studies in different organs and different species when using animal models for human aging research.

6. REFERENCES

1. Harman, D., 1987, The free radical theory of aging, in: *Clinical and Nutritional Aspects of Vitamin E*, (O. Hayaism and M. Mino, eds.) Elsevier Science-Publishers B.V., Amsterdam, pp. 285-393.
2. Ames, B.N., Shiggnaga, M.K., and Hagen, T.M., 1993, Oxidants, antioxidants, and the degenerative diseases of aging, *Proc. Natl. Acad. Sci. USA* 90:7915-7922.
3. Harman, D., 1981, The aging process, *Proc. Natl. Acad. Sci. USA* 78:7124-7128.

4. Katz, M.L., and Robinson, W.G., 1985, Nutritional influences on autooxidation, lipofuscin accumulation, and aging, in: *Free Radicals, Aging, and Degenerative Diseases,* (J.E. Johnson, ed.) Alan R. Liss Inc. New York.

5. Thomas, J.A., 1994, Oxidative stress, oxidant defense, and dietary constitutents, in: *Modern Nutrition in Health and Disease,* (M.E. Shils, J.A. Olsen, and M. Shike, eds.) Vol. 1., Lea and Febiger, Philadelphia, PA. Vol. 1, pp. 501-512.

6. Sawada, M., Sester, U., and Carlson, J.C., 1992, Superoxide radical formation and associated biochemical alterations in the plasma membrane of brain, heart, and liver during the lifetime of the rat, *J. Cellular Biochem.* 48:296-304.

7. Sawada, M., Sester, U., and Carlson, J.C., 1993, Changes in superoxide radical formation, lipid peroxidation, membrane fluidity and cathepsin B activity in aging and spawning male chinook salmon (*Onchoryhnus Tschawytscha*), *Mech. Age. Devel.* 69:137-147.

8. Yu, B.P., Suescun, E.A., and Yang, S.Y., 1992, Effect of age-related lipid peroxidation on membrane fluidity and phospholipase A_2: modulation by dietary restriction, *Mech. Age. Devel.* 65:17-33.

9. Laganiere, S., and Fernandes, G., 1991, Study on the lipid composition of aging Fischer-344 rat lymphoid cells: Effect of long term calorie restriction, *Lipids* 26:472-478.

10. Laganiere, S. and Yu, B.P., 1993, Modulation of membrane phospholipid fatty acid composition by age and food restriction. *Gerontology* 39:7-18.

11. Murphy, M.G., 1990, Dietary fatty acids and membrane protein function, *J. Nutr. Biochem.* 1:68-79.

12. Berlin, E., Bhathena, S.J., Judd, J.T., Nair, P.P., Peters, R.C., Bhagavan, H.N., Ballard-Barbash, R., and Taylor, P.R., 1992, Effects of omega-3 fatty acid and vitamin E supplementation on erythrocyte membrane fluidity, tocopherols, insulin binding, and lipid composition in adult men, *J. Nutr. Biochem.* 3:392-400.

13. Berlin, E., Bhathena, S.J., Judd, J.T., Clevidence, B.A., and Peters, R.C., 1994, Human erythrocyte membrane fluidity and insulin binding are independent of dietary *trans* fatty acids. *J. Nutr. Biochem.* 5:591-598.

14. Masoro, E.J., 1990, Animal models in aging research, in: *Handbook of the Biology of Aging, 3rd edition* (E.L.Schneider and J.W. Rowe, eds.) Academic Press, San Diego, CA pp. 72-94.

15. Tappel, A.L., 1968, Will antioxidant nutrients slow aging processes?, *Geriatrics* 23:97-105.

16. Packer, L. and Fuchs, J., (eds.), 1993, *Vitamin E in Health and Disease,* Marcel Dekker, Inc., New York.

17. Hsu, C.-K., 1982, Uses of pigs in biomedical research: Strengths and limitations, in: *Pig Model for Biomedical Research,* (H.R. Roberts, and W.J. Dodds, eds.), Pig Research Institute Taiwan, Miaoli, Taiwan, China pp. 3-10.

18. Khan, M.A., Earl, F.L., Farber, T.M., Miller, E., Husain, M.A., Nelson, E., Gertz, S.D., Forbes, M.S., Rennels, M.L., and Heald, F.P., 1977, Elevation of serum cholesterol and fatty streaking in egg yolk-lard fed castrated miniature pigs, *Exp. Mol. Pathol.* 26:63-74.

19. Neville D. M. Jr., 1968, Isolation of an organ specific antigen from cell surface membrane of rat liver. *Biochim. Biophys. Acta* 154:540-552.

20. Berlin, E., McClure, D., Banks, M.A., and Peters, R.C., 1994, Heart and liver fatty acid composition and vitamin E content in miniature swine fed diets containing corn and menhaden oils, *Comp. Biochem. Physiol.* 109A:53-61.

21. Shinitzky, M. and Barenholz, Y., 1978, Fluidity parameters of lipid regions determined by fluorescence, *Biochim. Biophys. Acta* 515:367-394.

22. Kuhry, J-G., Fonteneau, P., Duportail, G., Maechling, C., and Laustriat, G., 1983, TMA-DPH: A suitable fluorescence polarization probe for specific plasma membrane fluidity studies in intact living cells. *Cell Biophys.* 5:129-140.

23. Trotter, P.J. and Storch, J., 1989, 3-[*p*-(6-Phenyl)-1,3,5- hexatrienyl]-phenylpropionic acid (PA-DPH): Characterization as a fluorescent membrane probe and binding to fatty acid binding proteins, *Biochim. Biophys. Acta* 982:131-139.

24. Matusik, E.J., Reeves, V.B., and Flanagan, V.P., 1984, Determination of fatty acid methyl esters. Elimination of tissue-derived contamination and artefacts and preparation of a support-coated acid modified polyester liquid phase, *Anal. Chim. Acta* 166:179-188.

25. Lowry, O.H., Rosebrough, N.J., Farr, A.L., and Randall, R.J., 1951, Protein measurement with the Folin phenol reagent, *J. Biol. Chem.* 193:265-275.

26. Bartlett, G.R., 1959, Phosphorus assay in column chromatography. *J.Biol. Chem.* 234:466-468.

27. Desai, I.D., 1984, Vitamin E analysis methods for animal tissues, *Methods Enzymol.* 105:138-147.

28. Lehmann, J. and Martin, H.L., 1982, Improved direct determination of alpha and gamma tocopherols in plasma and platelets by liquid chromatography with fluorescence detection, *Clin. Chem.* 28:1784-1787.

29. Tatum, V.L., Changrit, C., and Chow, C.K., 1990, Measurement of malondialdehyde by high performance liquid chromatography with fluorescence detection, *Lipids* 25:226-229.

30. Rietjens, I.M.C.M., Pollen, M.C., Hempenin, R.A., Gijbels, M.J., and Alink, G.M., 1986, Toxicity of ozone and nitrogen dioxide to alveolar macrophage: Comparative study revealing differences in their mechanism of toxic action, *J. Toxicol. Environ. Health* 19:558-568.

31. Helwig, J.T. and Council, K.A., 1988, *SAS/STAT User's Guide*, Release 6.03, SAS Institute, Cary, NC.

32. Berlin, E., Khan, M.A., Henderson, G.R., and Kliman, P.G., 1985, Influence of age and sex on composition and lipid fluidity in miniature swine plasma lipoproteins, *Atherosclerosis* 54:187-203.

33. Dinh, T.K.L., Bourre, J.M., and Durand, G., 1993, Effect of age and α-linolenic acid deficiency on Δ6 desaturase and liver lipids in rats, *Lipids* 28:517-523.

34. Dinh, T.K.L., Bourre, J.M., Dumont, O., and Durand, G., 1995, Comparison of recovery of previously depressed hepatic Δ6 desaturase activity in adult and old rats, *Ann. Nutr. Metab.* 39:117-123.

35. Maniongui, C., Blond, J.P., Ulmann, L., Durano, G., Poisson, J.P.,and Bezard, J., 1993, Age-related changes in Δ6 and Δ5 desaturase activities in rat liver microsomes, *Lipids* 28:291-297.

36. Barzanti, V., Battino, M., Baracca, A., Cavazzoni, M., Cocchi, M., Noble, R., Maranesi, M., Turchetto, E., and Lenaz, G., 1994, The effect of dietary lipid changes on the fatty acid composition and function of liver, heart, and brain mitochondria in the rat at different ages, *British J. Nutr.* 71:193-202.

37. Fernandes, G., 1995, Effects of calorie restriction and omega-3 fatty acids on autoimmunity and aging, *Nutr. Rev.* 53:572-579.

38. Hollander, C.F., 1984, Aging in the laboratory animal, *Maturitas* 6:253-258.

39. Berlin, E., Kim, C.S., McClure, D, Banks, M.A.,and Peters, R.C., 1995, Dietary ω3 fatty acids are differentially incorporated into brain, liver, and heart membranes in miniature swine fed menhaden oil and corn oil. Second International Congress of International Society for the Study of Fatty Acids and Lipids. #156.

40. Liepa, G.U., Masoro, E.J., Bertrand, H.A., and Yu, B.P., 1980, Food restriction as a modulator of age-related changes in serum lipids, *Am. J. Physiol.* 238:E253-E257.

41. Choi, J.H. and Yu, B.P., 1989, The effect of food restriction on kidney membrane structures of aging rats, *Age* 12:133-136.

42. Curtis, M.T., Gilfor, D., and Farber, J.L., 1984 Lipid peroxidation increases the molecular order of microsomal membranes, *Arch. Biochem. Biophys.* 235:644-649.

43. Wahnon, R., Mokady, S., and Cogan, U., 1989, Age and membrane fluidity, *Mech. Age. Devel.* 50:249-255.

44. Kitagawa, S., Matsubayashi, M., Kotani, K., Usui, K., and Kametani, F., 1991, Asymmetry of membrane fluidity in the lipid bilayer of blood platelets: fluorescence study with diphenylhexatriene and analogs, *J. Membrane Biol.* 119, 221-227.

45. Stubbs, C.D. and Smith, A.D., 1984, The modification of mammalian membrane polyunsaturated fatty acid composition in relation to membrane fluidity and function, *Biochim. Biophys. Acta.* 779:89-137.

46. Yeagle, P.L., 1985, Cholesterol and the cell membrane, *Biochim. Biophys. Acta.* 822:267-287.

47. Zimmer, G., Thurich, T., and Scheer, B., 1993, Membrane Fluidity and Vitamin E, in: *Vitamin E in Health and Disease*, (L. Packer, and J. Fuchs, eds.) Marcel Dekker, Inc., New York, pp. 207-222.

48. Berlin, E., Bui, T.B., Sainz, E., Sundaram, S.G., Manimekalai, S., and Goldstein, P.J., 1986, Vitamin E modulation of lipoprotein fluidity, *Ann. NY Acad. Sci.* 463:80-82.

49. Berlin, E., Lehmann, J., and Judd, J.T., 1988, Vitamin E supplementation and lipoprotein fluidity in normal human subjects, *Adv. Diet Nutr.* 2:193-198.

50. Bowry, V.W., Ingold, K.U., and Stocker, R., 1992, Vitamin E in human low density lipoprotein: When and how this antioxidant becomes a pro-oxidant, *Biochem. J.* 288:341-344.

51. Kornbrust, D.J., and Mavis, R.D., 1980, Relative susceptibility of microsomes from lung, heart, liver, kidney, brain and testes to lipid peroxidation: correlation with vitamin C content, *Lipids* 15:315-322.

50

COMPARING PORCINE MODELS OF CORONARY RESTENOSIS

Robert N. Willette,[1*] Hong Zhang,[1] Calvert Louden[2] and Robert K. Jackson[3]

[1,3] Cardiovascular Pharmacology
[2] Toxicology and Laboratory Animal Sciences
SmithKline Beecham Pharmaceuticals
King of Prussia, Pennsylvania

1. INTRODUCTION

The use of swine to study the vascular response to injury has gained wide acceptance over the past 10 yr. Our approach to the porcine model has been focused on the evaluation of novel therapeutic approaches to the treatment of arterial restenosis. It is from this perspective that the following discussion was prepared. Thus, the purpose of this brief review is to present a rationale for the use of the pig, describe the general technique, compare and contrast some of the variations in the use of the model and highlight relevant similarities to the human problem of coronary restenosis.

2. RATIONALE AND PRACTICAL CONSIDERATIONS

Swine are the one readily available animal species in which cardiovascular anatomy and physiology resembles humans. There are many similarities between the species. In humans and pigs, size and distribution of arteries are similar,[1] with blood flow through the right and left anterior descending coronary arteries supplying nearly 80% of the myocardium in swine.[2] Similarities in end artery coronary anatomy also make the pig an excellent model for the study of regional myocardial ischemia.[3] In the 25 to 30 kg pig, the heart to body size ratio is identical to that of humans.[2] Blood pressure values,[4] heart rate[4] and lipoprotein patterns[5] also are similar. Swine have a platelet coagulation system that is more closely related to humans than most other mammalian species used in cardiovascular research.[6,7] However, one notable exception is that a number of cyclic peptide and nonpeptide glycoprotein IIb/IIIa (fibrinogen receptor) antagonists are powerful inhibitors of platelet aggre-

*Reprint requests to: Robert N. Willette, SmithKline Beecham Pharmaceuticals, P.O. Box 1539, King of Prussia, PA 19406, (610) 27000-6052.

Advances in Swine in Biomedical Research, edited by Tumbleson and Schook
Plenum Press, New York, 1996

gation in the dog, nonhuman primate and human, but not in the pig (A.J. Nichols, personal communication).

Pigs are one of the few animals that develop atherosclereosis naturally and have a high incidence of end organ lesions.[8] Like humans, the severity of atherosclerosis can be increased dramatically by feeding diets high in fat and cholesterol. Atherosclerotic plaques observed in domestic[9] and miniature[10] pigs bear a strikingly close resemblance to the human condition. The combination of an atherogenic diet and arterial injury in the pig creates a lesion more typical of human atherosclerosis than diet alone.[8,11]

There are also some very practical considerations which favor the use of pigs for restenosis studies. First, the size of the vasculature and heart of miniature swine and juvenile domestic breeds make pigs ideal for developing and testing intravascular device prototypes similar in size and configuration to the model that may be used in humans. Secondly, standard interventional cardiac techniques can be applied in the pig using the same catheters and equipment used for human procedures.

Although the pig is not the most widely used model of restenosis, it is has become the model of choice for studying the effects of luminal injury on the neointimal response in coronary arteries. The response to injury in elastic arteries, e.g., the carotid artery, can be studied effectively in rats, rabbits and dogs; however, coronary artery size in pigs allows the response to be investigated in a muscular artery. It has been reported that the local proliferative response elicited by interventional techniques, i.e., PTCA and stenting, is identical in appearance in both pig and human coronary vessels.[12] In particular, human and porcine proliferative coronary lesions have a similar matrix and cellular density. In the pig, neointimal thickness resulting from vascular injury is directly proportional to the depth of the arterial injury, similar to the relationship that is thought to exist in human coronary arteries.[12,13] This proportionality can be useful when quantifying the efficacy of drug effect on the neointimal response to injury.

As has been established, it's anatomy and physiology make the pig an excellent animal model for cardiovascular disease in man. However, the size of swine and the cost of maintaining them in a laboratory setting can present definite drawbacks. For example, domestic breeds have been bred for meat production; they can gain up to 1 kg/da and reach weights of more than 500 kg. Animals of this size are difficult to handle and house in a standard biomedical research facility.

Fortunately, most cardiovascular research can be accomplished in pigs of a much smaller size. Young domestic pigs weighing between 23 and 55 kg and miniature and micro breeds are used routinely. Because of their size, the miniature (25 to 40 kg at 6 mo of age) and micro (16 to 20 kg at 6 mo of age) breeds are becoming a standard for atherosclerosis research and other longer term studies. For short term studies, juvenile domestic pigs in the 25 to 30 kg body weight range are used.

In our laboratories, we conducted a number of studies using specific pathogen free (SPF) pigs; although the purchase price is higher than conventionally raised farm pigs for research ($275 vs $170), there are certain advantages. SPF swine were developed in the pork industry to produce a healthier animal that would achieve market weight faster than those in conventional production operations. The primary advantage to cardiovascular researchers is that the SPF pig is free of respiratory and pulmonary pathogens, such as *Mycoplasma hyopneumoniae*, which could confound research results. To maintain registry as an SPF herd, periodic testing and necropsies are performed on the herd. If SPF pigs are used, it must be considered that they cannot be housed in the same area with conventionally raised swine. SPF pigs may be highly susceptible to organisms carried asymptomatically by the conventional farm animal.

3. PORCINE MODEL(S) OF RESTENOSIS

3.1. Technique

The following is a discription of the technique employing the placement of oversized stents in normal pig coronary arteries. The procedure is essentially that described by Schwartz et al.[12,14] and Rodgers et al.[11] who demonstrated the utility of the model for the study of restenotic events. Relevant modifications of the general technique will be highlighted.

Juvenile domestic pigs (Yorkshire crossbred from a SPF stock; Barton Farms, NJ) weighing 25 to 30 kg are sedated with ketamine (33 mg/kg, IM) and atropine sulfate (60 ug/kg, IM). The central ear vein is cannulated for the induction of anesthesia with thiamylal sodium (8 mg/kg, IV). Each animal is intubated, placed in a dorsal recumbent position and ventilated with isoflurane in oxygen sufficient to maintain surgical anesthesia. Coronary angiography is accomplished by first isolating the right common carotid artery and introducing an 8-9F hemostatic sheath. Under flouroscopic guidance, common Judkin PTCA guides are reversed for engaging the LAD and the RCA, e.g., JR4 for the LAD and JL4 for the RCA. The guide catheter diameter is useful for estimating vessel diameter. Following identification by the injection of contrast media, a 0.014 in guidewire is passed into the target artery and a deflated 3.0 mm PTCA balloon carrying a compressed tantalum or stainless steel stent is delivered to a predetermined coronary site based on size. The stent is deployed by a single 10 atm inflation (30% > artery diameter) and the PTCA balloon is deflated and removed. Subsequent angiography is used to confirm successful stent deployment. The right common carotid artery is ligated, all other wounds are repaired and the pig is prepared for recovery and observation.

Various combinations of anticoagulant, antiplatelet, vasodilator and antiarrhythmic agents are used routinely in this procedure.[12,14,15] In our laboratory, 1 da prior to the stent/angioplasty procedure each animal received verapamil (60 mg) and aspirin (500 mg) orally and was fasted 16 to 18 hr. Fifteen min prior to coronary catheterization, heparin (200 IU/kg, IV), aspirin (5 mg/kg, IV) and bretylium (50 mg, IV) were administered. Nitroglycerin (100 ug) was administered into the coronary artery just prior to removing the guide catheter. Heparin was again administered (200 IU/kg, IV) at the end of the procedure and buprenorphine (0.05 mg/kg, IM) was administered as a postoperative analgesic to smooth recovery. Using this procedure, >90% of animals prepared survive ≥30 da. Most mortalities occur within the postoperative period (<48 hr).

Four wk after stenting, animals were sacrificed and hearts were perfusion fixed with neutral buffered formalin at systemic pressures for 1 hr. Hearts were immersed in fixative for 24 hr. Stented coronary arteries were removed carefully into 70% ethanol and prepared for histologic sectioning and staining. If vessels were to be prepared with stents in situ, special embedding, sectioning and staining techniques were required. One method, based on that described by Karas et al.,[15] was to harden tissue by embedding in glycol methacrylate (GMA). The infiltration and polymerization procedure can take up to 4 da to attain the adequate hardness. Tissue was cut into 4 equal pieces with a diamond wafering saw, reembedded and cut at 5 μm with a tungsten carbide knife. Sections prepared this way were stained routinely with toluidine blue (Fig. 1C).

Once histologic sections were prepared for microscopic examination, standard digital image anlysis techniques were used to quantify neointimal response (area or thickness), lumen area and degree of stent injury based on strut position in the vessel wall.[14] One injury scoring method is illustrated in Fig. 2. Scoring was as follows: 1 = minimal damage, when the stent strut was located within the lumen or compressing the internal elastic lamina (IEL),

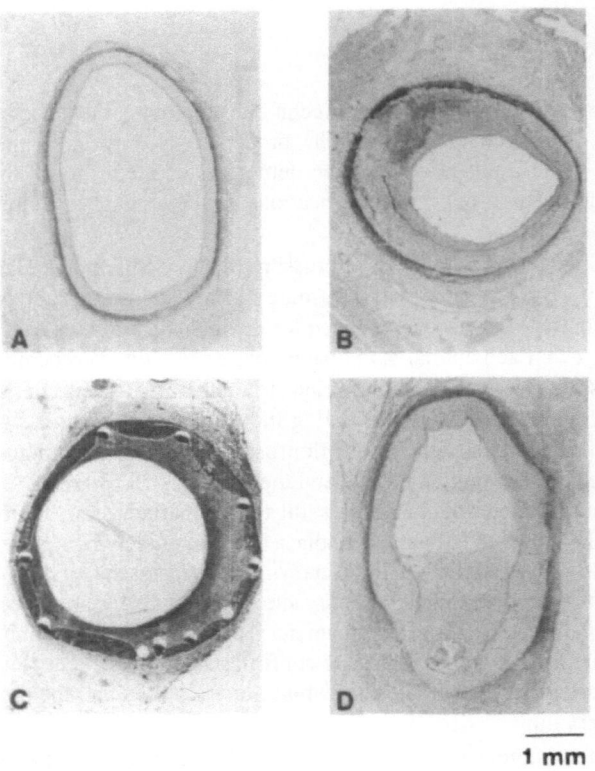

1 mm

Figure 1. Histologic sections obtained from representative coronary arteries in 4 separate studies were prepared 28 da after injury. Normal (sham; A), balloon overinflation (B) and platinum single strut stent (D) were stained using the Verhoeff-Van Giesen technique. Histologic sections from the traditional, multistrut stent design (platinum) were embedded in glycol methacrylate and stained with toluidine (C). This allowed for processing of the tissue with the stent embedded in the vessel wall.

2 = moderate damage, when the stent strut had ruptured the IEL and was located in the media and 3 = severe damage, when the stent strut had ruptured the external elastic lamina (EEL). When traditional multiple strut stents were used, the score from each strut was averaged or summed and averaged over the length of the stent segment.

To address issues of stent cost and availability, we used the identical procedure described above to deliver multistrut and single strut platinum research stents (0.01 in; NuMed, Hopkinton, NY) to produce coronary artery injury (Figs. 3 and 4, respectively). Lesions observed with these experimental platinum stents were similar qualitatively and quantitatively to those produced by tantalum and stainless steal stents (Fig. 1C and 1D). In addition, the single strut stent (SSRS) was removed easily from fixed vessels and therefore required no special histologic preparation (Fig. 1D). Because no metal remains in vessels following removal of the single strut, magnetic resonance imaging (MRI) microscopy also can be used for lesion analysis (Fig. 5). Lesion thickness determined by MRI analysis and histologic analysis correlates well in severely, moderately and mildly injured vessels (Fig. 6).

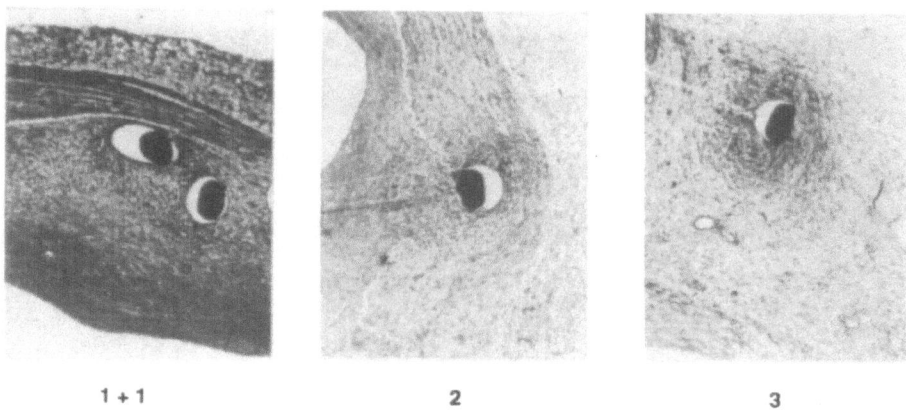

1 + 1   2   3

Figure 2. Vessel wall injury produced by stent expansion was quantified based on the position of each stent strut in the vessel wall.

Another variation in the technique has been the use of simple balloon overinflation as a method for producing coronary injury.[12,16,17] Typically, the vessel was overinflated 25 to 75% with a 3.5 mm PTCA balloon catheter for 20 sec at 10 atm and repeated 2 to 3 times. With this technique, neointimal formation occurred only when the internal elastic lamina was ruptured (Fig. 1B). Acute mortality increased when using a 4.0 mm PTCA balloon in 30 kg pigs. In our laboratory, the overall incidence of neointimal formation following overinflation (25 to 75%) was 45% (n = 33) vs 100% (n = 68) in stented coronary arteries.

Other variations in the model include coronary stenting following preinjury (endothelial abrasion) of the target coronary vessel and in the setting of prolonged hypercholesterolemia.[11,18] However, in a much needed study by the Baylor group,[19] it was found the degree and character of neointimal thickening observed 4 wk after stenting in normolipemic swine was similar to that observed in swine fed a high cholesterol diet with or without prior endothelial abrasion. In all cases, the neointima consisted primarily of smooth muscle cells and matrix (collagen). Intracoronary stenting (not cholesterol levels nor underlying vascular injury) was the major stimulus to induce neointimal thickening in porcine models of restenosis.

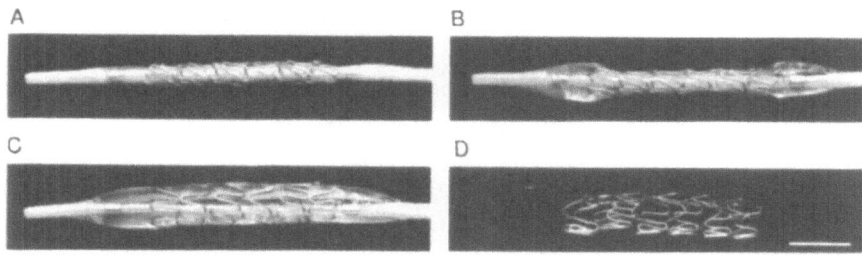

Figure 3. A flexible platinum stent with a design similar to that of a tantalum Medtronic-Wiktor stent. In A, the platinum stent is compressed onto a 3.0 mm angioplasty catheter which is inflated partially (B), expanded fully (C) and in the deployed configuration (D). The line represents 0.5 cm.

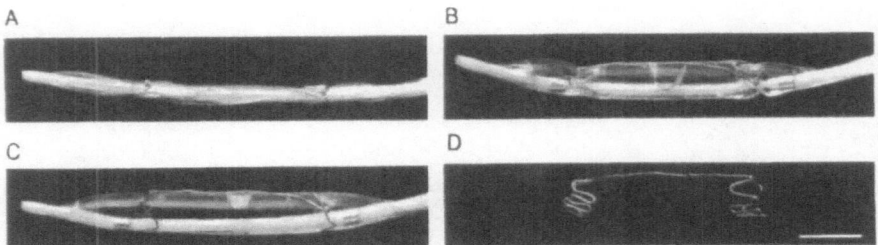

Figure 4. A platinum single strut research stent (SSRS) designed for producing a reproducible vascular lesion suitible for MRI and routine histology. The line represents 0.5 cm.

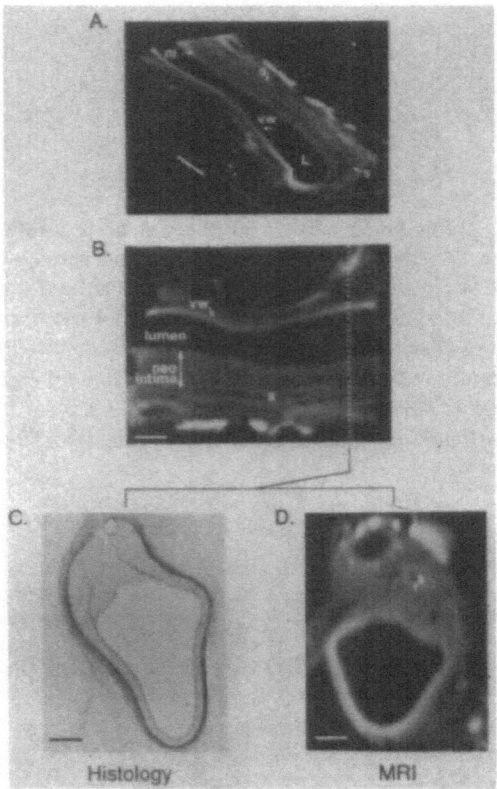

Figure 5. *Ex vivo* magnetic resonance image (MRI) obtained from a porcine left anterior descending coronary artery following removal of a platinum single strut research stent. Note the remaining longitudinal strut canal (S). Oblique (A) and longitudinal (B) renderings illustrate the uniform profile of the neointima (bar = 1 mm) formed after 28 da. The line in C represents 0.6 mm.

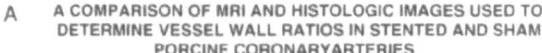

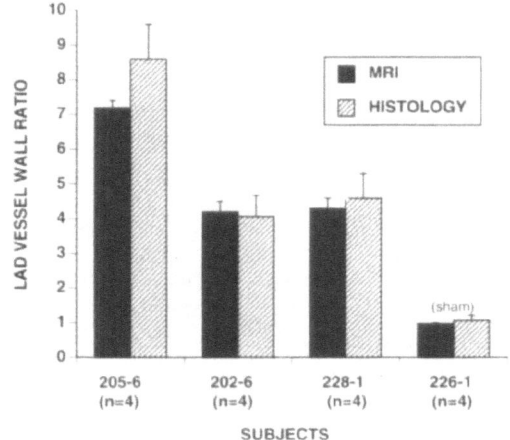

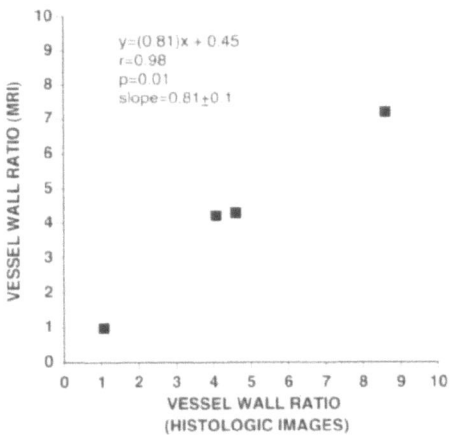

Figure 6. Vessel wall ratios (lesion thickness/normal wall thickness) derived from MRIs and histologic sections from the same animals were compared. Vessel wall ratios were similar in histologic sections and MRIs (A) and the slope of the relationship was 0.81 ± 0.1 (B).

The 3 porcine models of coronary restenosis described above are compared in Table 1. We found the lesion incidence to be low (50%) in the balloon overinflation model when compared to multiple strut stents and single strut stents (100%). We also found the lesion size to be more variable ($\pm 30\%$) with the overinflation and multiple strut stent methods when compared to the single strut stent ($\pm 11\%$). Finally, the multiple strut stent was not suitable for MRI and routine histologic analysis. Overall, the single strut stent possessed some advantages in the porcine model of coronary restenosis, especially when evaluating potential therapeutic agents.

Table 1. Comparing methods of coronary injury in the porcine model of restenosis

	Balloon overinflation (n=24)	"Traditional" Stenting (n=17)	SSRS(n=8)
Lesion incidence	+(55%)	+++(100%)	+++(100%)
Lesion variability(standard error)	+(30%)	+(30%)	+++(11%)
NMR Microscopy(suitability)	+++		+++
Histologic Preparation	+++		+++
Injury Response Relationship(r value)	++(0.75)	++(0.57)	+++(0.82)

Legend: + = least desirable, ++ = better, +++ = best

3.2. Lesion Characterization

In the porcine stent model of coronary restenosis, each stent strut represents a discrete site of vessel injury capable of initiating a neointimal response. This is illustrated by the uniform neointimal formation evoked by the introduction of a single stent strut (Fig. 5). At 4 wk following stenting, the neointimal response (thickening) is related to the stent induced

A. SSRS

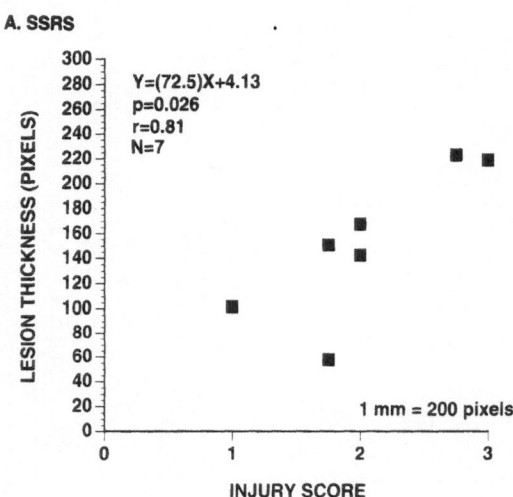

B. "Traditional" Stent Design

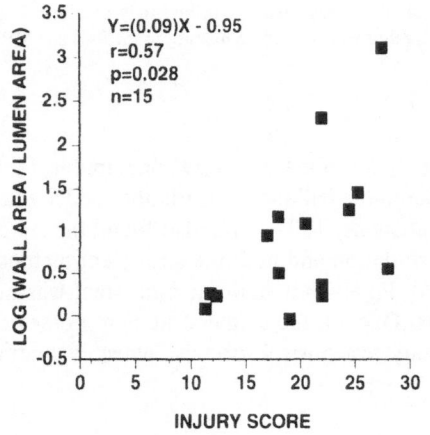

Figure 7. Injury response relationships were determined in porcine coronary arteries 28 da after deployment of platinum single strut stents (SSRS in A) or traditional multiple strut stents (B). The injury score was related linearly to the lesion thickness in the SSRS group (A) and the log (lesion wall area/lumen area) in the multiple strut stent group (B).

injury of the vessel wall. Specifically, the deeper the stent strut is driven into the vessel wall the greater is the neointimal response.[14] A simplified injury scoring scheme[14] has allowed the neointimal response to injury to be normalized, quantified and compared for the evaluation of potential antirestenotic agents. An injury response relationship has been observed with tantalum, stainless steel and platinum stents in the porcine stent model of coronary restenosis (Fig. 7). Consistent with these observations in the pig, histopathologic evidence in humans suggests that the depth of angioplasty injury increases the likelihood of restenosis.[13] The injury relationship does not exist in the canine stent model of coronary restenosis where the neointimal response to stenting is minimal.[20]

An injury response relationship also has been described at 4 wk in the porcine overinflation model of coronary restenosis.[17] In this model, the distance between the ruptured ends of the internal elastic lamina is related linearly to the resultant neointimal area. There was no difference in the injury response relationship in feeder pigs and Yucatan minipigs following balloon overinflation.

Neointima formed in the stent model of coronary restenosis often is referred to as a proliferative lesion. Various investigators[12,15,16,21-23] have attempted to define the genesis of this lesion; 4 stages of development generally are appreciated. The first stage has been referred to as thrombotic in which platelets accumulate and mural thrombi form within the first hr at injury sites.[16,21] These mural thrombi, containing platelets, fibrin and erythrocytes, become less prevalent after 24 to 48 hr. An occlusive thrombus rarely is observed, probably because of the effective antiplatelet and anticoagulant protocols. The second stage, cellular recruitment, commences 3 to 4 da after injury. At this time, endothelial regrowth along the luminal boarders is apparent and is complete by approximately da 7. Numerous monocytes and lymphocytes attach to endothelium and migrate deeply into the thrombus. Monocytes are thought to be responsible for subsequent resorption of the thrombus. The third stage, cellular proliferation, overlaps cellular recruitment and is thought to be initiated by the release of cytokines and growth factors from activated monocytes. By da 7, approximately 28% of the cells accumulating within the lesion are proliferating cell nuclear antigen (PCNA) positive smooth muscle cells.[23] The number of PCNA positive cells declines steadily over the next 4 wk. Thus, the maximum proliferative phase precedes the maximum lesion size by 2 to 3 wk. However, PCNA positive smooth muscle cells with a distinct secretory phenotype still are present in the lesion at 4 wk (Fig. 8). The precise origin of proliferating smooth muscle cells remains unclear, with possible sites including normal adjacent media, injured media and adventitia. The final phase accounting for the production of neointimal mass in this model may be termed migratory/synthetic and is thought to include luminal migration of synthetic smooth muscle cells which expand the lesion by producing extracellular matrix. At 4 wk, the neointima is composed primarily of smooth muscle cells and matrix (Fig. 3A). In this regard, the neointimal lesion in the porcine model and the human condition appear to have a similar proportion of cells to matrix.[12]

One potential component of the lesion unique to stent models of restenosis is chronic inflammation associated with a foreign body reaction. Materials used for stents in this model, i.e., stainless steel, tantalum and platinum, generally are thought to be biologically inert. Schwartz et al.[12] reported only a minimal amount of chronic inflammation, which was judged to be an insignificant factor in lesion development. However, in a later study by Karas et al.,[15] most of the tantalum stent struts were surrounded by inflammatory infiltrates composed of lymphocytes, eosinophils, histocytes (macrophages) and occaisional multinucleated giant cells. We made similar observations when using platinum stents. One common feature in these studies was the inflammatory reaction surrounding stent struts occurred only following deep vascular

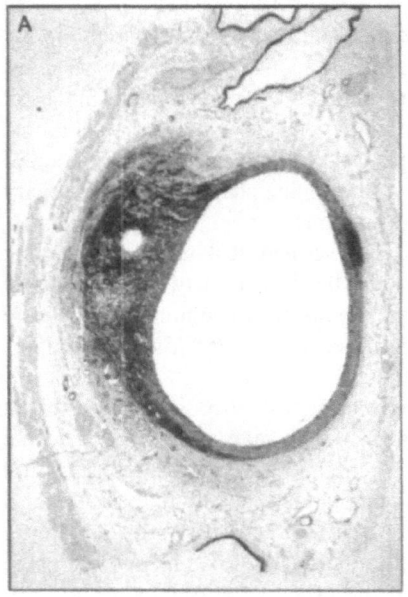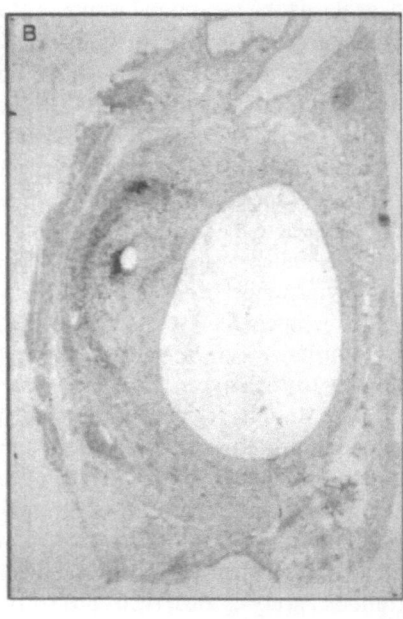

1 MM

Figure 8. Histologic sections were prepared for immunohistochemistry 28 da after deploying a platinum single strut stent in the porcine LAD. Smooth muscle alpha-actin like immunoreactivity was apparent throughout the media, neointima and adventitia (A). At 28 da, PCNA like immunoreactivity was localized sparsely throughout the neointima (B).

injury involving adventitia. This inflammatory reaction often was associated with thickening of the adventitia (Fig. 1C). Chronic inflammatory reactions have been observed following deep balloon overinflation injury in the pig and in some atherosclerotic postangioplasty human coronary artery specimens.[13] It appears that deep vascular injury, unlike intimal or medial damage, activates histocytes and undifferentiated mesenchymal cells in the adventitia.

4. CONCLUSION

Coronary artery restenosis in the pig represents an important model for understanding vascular response to injury.[24,25] Lesions produced in this model closely resemble human coronary restenotic vessels produced by balloon angioplasty or stent procedures. The major clinical problems associated with these procedures, thrombosis and intimal hyperplasia, are important processes for lesion development in the porcine model. In addition, the predictable relationship between coronary injury and neointimal response in the porcine model make it attractive for evaluating approaches to limit restenosis. The challenge is to identify and regulate important restenotic mechanisms in the pig model which will have a probability of direct application to the treatment of human coronary restenosis.

5. ACKNOWLEDGMENTS

The authors are grateful to R. Kapada for his expert help with MRI studies, C. Sauermelch and M. Mithchell for assistance with the angioplasty procedures, R. Coatney, for help with surgical procedures in the pig and W.J Crowell and L. Schultz for their help in preparing the manuscript. We are grateful to Mr. Alan J. Tower at NuMed, Inc. for the generous supply of platinum stents.

6. REFERENCES

1. Lee, K.T., 1986, Swine as animal models in cardiovascular research, in: *Swine in Biomedical Research* Volume 3 (M.E. Tumbleson, ed.), Plenum Press, New York, pp. 1481-1496.
2. Hughes, H.C., 1986, Swine in cardiovascular research, *Lab. Anim. Sci.* 36:348-350.
3. Horneffer, P.J., Gott, V.L., and Gardner, T.J., 1986, Swine as a cardiac surgical model, in: *Swine in Biomedical Research* Volume 1 (M.E. Tumbleson, ed.), Plenum Press, New York, pp. 321-325.
4. Smith, G.S., Smith, J.L., and Mameesh, M.S., 1964, Hypertension and cardiovascular abnormalities in starved-refed swine, *J. Nutr.* 82:172-182.
5. Mahley, R.W., and Weisgraber, K.H., 1974, An electrophoretic method for the quantitative isolation of human and swine plasma lipoproteins, *Biochemistry* 13:1964-1969.
6. Folts, J.D., and Rowe, G.G., 1983, Acute thrombus formation in stenosed pig coronary arteries, causing sudden death by ventricular fibrillation, *Circulation* 68(Suppl III):264.
7. Leach, C.M., and Thorburn, G.D., 1982, A comparative study of collagen-induced thromboxane release from platelets of different species: Implications for human atherosclerosis models, *Prostaglandins* 24:47-59.
8. Jokinen, M.P., Clarkson, T.B., and Prichard, P.W., 1985, Recent advances in molecular pathology: animals models in atherosclerosis research, *Exp. Molec. Pathol.* 42:1-28.
9. Getty, R., 1965, Gross and microscopic occurrence and distribution of spontaneous atherosclerosis in the arteries of swine, in: *Comparative Atherosclerosis* (A. Robert, and E. Straus, eds.), Harper and Row, New York, pp. 11-20.
10. Zugibe, F.T., 1965, Atherosclerosis in the miniature pig, in: *Comparative Atherosclerosis* (A. Robert, and E. Straus, eds.), Harper and Row, New York, pp. 37-42.
11. Rodgers, G.P., Minor, S.T., and Robinson, K., 1990, Adjuvant therapy for intracoronary stents: investigations in atherosclerotic swine, *Circulation* 82:560-569.
12. Schwartz, R.S., Murphy, J.G., and Edwards, W.D., 1990, Restenosis after balloon angioplasty: a practical proliferative model in porcine coronary arteries, *Circulation* 82:2190-2200.
13. Virmani, R., Farb, A., and Burke, A.P., 1994, Coronary angioplasty from the perspective of atherosclerotic plaque: morphologic predictors of immediate success and restenosis, *Am. Heart J.* 127:163-179.
14. Schwartz, R.S., Huber, K.C., Murphy, J.G., Edwards, W.D., Camrud, A.R., Vlietstra, R.E., and Holmes, D.R., 1992, Restenosis and the proportional neointimal response to coronary artery injury: results in a porcine model, *J. Am. Coll. Cardiol.* 19:267-274.
15. Karas, S.P., Gravanis, M.B., Santoian, E.C., Robinson, K.A., Anderberg, K.A., and King, S.B., 1992, Coronary intimal proliferation after balloon injury and stenting in swine: an animal model of restenosis, *J. Am. Coll. Cardiol.* 20:467-474.
16. Steele, P.M., Chesebro, J.H., and Stanson, A.W., 1985, Balloon angioplasty: natural history of the pathophysiological response to injury in a pig model, *Circ. Res.* 57:105-112.
17. Humphrey, W.R., Simmons, C.A., Toombs, C.F., and Shebuski, R.J., 1994, Induction of neointimal hyperlasia by coronary angioplasty balloon overinflation: comparison of feeder pigs to Yucatan minipigs, *Am. Heart J.* 127:20-31.
18. Buchwald, A.B., Unterberg, C., Nebendahl, K., Grone, H.J., and Wiegand, V., 1992, Low-molecular-weight heparin reduces neointimal proliferation after coronary stent implantation in hypercholesterolemic minipigs, *Circulation* 86:531-537.
19. Grinstead, W.C., Rodger, G.P., Mazure, W., French, B.A., Cromeens, D., Van Pelt, C., West, S.M., and Raizner, A.E., 1994, Comparison of three porcine restenosis models: the relative importance of hypercholesterolemia, endothelial abrasion, and stenting, *Coronary Artery Dis.* 5:425-434.

20. Schwartz, R.S., Edwards, W.D., Bailey, K.R., Camrud, A.R., Jorgenson, M.A., and Holmes, D.R., Jr., 1994, Differential neointimal response to coronary artery injury in pigs and dogs: implications for restenosis models, *Arterioscler. Thromb.* 14:395-400.
21. Schwartz, S.M., deBois, D., and Obrien, E.R.M., 1995, The intima, *Circ. Res.* 77:445-465.
22. White, C.J., Ramee, S.R., Banks, A.K., Mesa, J.E., Chokshi, S., and Isner, J.M., 1992, A new balloon-expandable tantlum coil stent: angiographic patency and histologic findings in an atherogenic swine model, *J. Am. Coll. Cardiol.* 19:870-876.
23. Carter, A.J., Laird, J.R., Farb, A., Kufs, W., Wortham, D.C., and Virmani, R., 1994, Morphologic characteristics of lesion formation and time course of smooth muscle cell proliferation in a porcine proliferative restenosis model, *J. Am. Coll. Cardiol.* 24:1398-1405.
24. Schwartz, R.S., and Holmes, D.R., 1994, Pigs, dogs, baboons and man: lessons for stenting from animal studies, *J. Interven. Cardiol.* 7:355-368.
25. Schwartz, R.S., Edwards, W.D., Huber, K.C., Antoniades, L.C., Bailey, K.R., Camrud, A.R., Jorgenson, M.A., and Holmes, D.R., 1993, Coronary restenosis: prospects for solution and new perspectives from a porcine model, *Mayo Clin. Proc.* 68:54-62.

51

SINCLAIR MINIATURE SWINE MELANOMA AS A MODEL FOR EVALUATING NOVEL LYMPHOGRAPHY CONTRAST AGENTS

David K. Johnson,[1*] Erik R. Wisner,[2] Stephen M. Griffey,[2]
Adele R. Vessey,[1] and Patrick J. Haley[1]

[1] Nycomed Inc.
Imaging and Laboratory Animal Medicine
466 Devon Park Drive
Wayne, Pennsylvania 19087-8630
[2] University of California Davis
Medical Center, Department of Radiology
Radiology Research FOLB-IIE
2421 45th Street, Sacramento, California 95817

1. ABSTRACT

Malignant melanoma is an heritable, nonsex linked trait in the strain of Sinclair miniature swine. Cutaneous melanoma lesions arise at or shortly after birth in up to 54% of the piglets. These tumors readily metastasize to lymph nodes, lung and liver, yet may regress spontaneously due to immune mechanisms. The genetic predisposition, apparent spontaneous transformation and similar histomorphology to human melanomas endorse this model for investigations of human melanomas. In addition, this model allows for assessment of melanoma metastasis to regional lymph nodes using novel iodinated contrast agents.

Using subcutaneously administered iodinated nanoparticles (indirect CT lymphography), studies were conducted to characterize the Computed Tomography (CT) appearance of normal lymph nodes compared to lymph nodes infiltrated with metastatic melanomas. An iodinated suspension was injected subcutaneously into distal extremities of normal Sinclair miniature swine. Contrast agent was injected subcutaneously in a ring pattern around each lesion in age matched miniature swine with cutaneous melanomas. Pre and 24 hr postinjection CT images were obtained through opacified regional nodes. The CT appearance of normal and cancerous nodes was characterized and correlated histologically. Since the cortex

---

* Reprint requests to: Dr. David K. Johnson, Nycomed Inc., 466 Devon Park Drive, Wayne, PA 19087-8630, (610) 225-4272

Advances in Swine in Biomedical Research, edited by Tumbleson and Schook
Plenum Press, New York, 1996

and medulla in swine lymph nodes are reversed, as compared to human nodes, this characteristic must be taken into account when interpreting imaging data.

Normal opacified lymph nodes exposed to contrast agent were larger than nonopacified, unexposed, normal contralateral nodes. There was a marked, uniform increase in attenuation of the lymph node medulla and good demarcation between medulla and less opacified cortex. Cancerous nodes with macrometastases generally were larger than opacified normal contralateral nodes. Typical architectural changes in cancerous nodes included incomplete opacification associated with small to massive filling defects, disruption and irregularity of the opacified medullary zone and irregular foci of opacification within the cortex. Altered lymph node architecture was appreciated in cancerous nodes on indirect CT lymphographic studies after injection of iodinated contrast agents. Sinclair swine bred for the trait of melanoma development provide a good model for studies of lymphographic contrast agents.

2. INTRODUCTION

Two radiographic techniques currently are used to detect cancer metastases in lymph nodes. Nonenhanced CT relies entirely on size criteria, and therefore cannot accurately diagnose an inflamed, uninvolved node or a normal sized involved node. Alternatively, radiographic images are obtained after direct injection of ethiodized oil (Ethiodol®) into the distal lymphatic ducts. Because of morbidity and difficulty of this procedure, it is performed infrequently and few physicians are skilled in the techniques. An improved strategy would be the use of indirect Computed Tomography (CT) lymphography, using a contrast agent for subcutaneous or submucosal administration. Its application would be used to evaluate metastasis to lymph nodes of primary tumors or for which biopsy/dissection of the lymph nodes is part of other diagnostic or treatment protocols. An iodinated contrast agent has been formulated to meet these criteria. Prior studies in normal animals has demonstrated indirect CT imaging of lymph nodes by subcutaneous and submucosal routes of administration.[1-7] The goal of this study was to evaluate the swine melanoma model for the characterization of this contrast agent's ability to distinguish normal vs abnormal (metastatic) lymph nodes in an animal model with a high incidence of cancer.

3. ANIMAL MODEL

Sinclair miniature swine were selected for study since they have an heritable trait of naturally occurring melanoma with metastasis to regional draining lymph nodes. The source of melanoma bearing swine for this study was the Sinclair Research Center, Columbia, MO. This strain originated at the Hormel Institute in Minnesota and was bred at the University of Missouri to achieve a miniature size. In 1967, a melanoma was reported and, since then, by selective breeding, a strain with a high incidence of melanomas has been established.[8-12] Melanomas are present at or shortly after birth and typically metastasize to regional lymph nodes that drain the primary lesion site. Lesions appear as atypical melanocytic hyperplasia or raised black tumors of an invasive malignant melanoma similar to human superficial spreading melanoma. The incidence in piglets sired from parents both with melanomas is 54% at the time of birth with an increasing incidence of 85% by 1 yr of age.[13] The swine melanomas demonstrate an immunologically associated spontaneous regression. Inheritance involves at least 2 loci: the Swine Leucocyte Antigen complex (SLA) and an independently segregating autosomal locus with homozygous alleles for tumor initiation and development. Swine and human melanomas possess a common cross reactive, tumor associated antigen.

Lymphocytes from melanoma bearing swine are cytotoxic for human melanoma cells.[14,15] Features in common with human melanomas are the spontaneous appearance, melanocytic lesions capable of malignant transformation, similar histopathology, pattern of metastasis and heritable nature. Differences are that swine tumor development is not related to ultraviolet radiation, complete tumor regression occurs in most swine and melanomas are observed only in darkly pigmented pigs. These swine serve as an excellent model for cancer metastasis to regional lymph nodes. However, the anatomy of the porcine lymph node is unique in that the cortical and medullar region of the node are reversed, with the cortex located centrally and the medulla located peripherally. Afferent lymphatics enter the node at the hilum and efferent lymph fluid flow through the node is centrifugal. Because of this anatomical feature, the interpretation of CT films must be taken into account when comparing the appearance of swine and human lymph nodes filled with contrast agent.

4. MATERIALS AND METHODS

4.1. Contrast Medium

The contrast medium formulated for this study was a sterile suspension of water insoluble, hydrolyzable ester of diatrizoic acid which is approximately 50% iodine by weight. Average particle size was less than 350 nm.

4.2. Animals

Six immature (approximately 5 kg) Sinclair swine born with cutaneous melanomas were selected. A description of their lesions is described in Table 1.

Animals were induced and maintained with 1 to 3% halothane, with oxygen inhalant anesthesia for all imaging procedures. Two ml of contrast material was injected subcutaneously in a ring pattern around each lesion (maximum of 0.5 ml per injection) to opacify regional lymph nodes receiving afferent lymphatics from the primary lesion. A precontrast injection CT study and a 24 hr, postinjection study of the appropriate anatomical regions of interest were obtained.

Table 1. Cutaneous sites of primary melanomas

Animal ID	Sites of lesions
1965	Anterior scapular areaPosterior scapular area
	Dorsal area of head
1964	Middorsal area of backRight flank
	Right mandible
1956	Left rear leg, lateral area
	Middorsal area of back
1961	Middorsal area of back
	Left rear leg, medial area
1949	Midcervical area
	Proximal ulnar area
1975	Right maxillary area

4.3. Computed Tomography

CT studies were performed using a General Electric 8800 computed tomography unit operating at 120 kVp and 250 or 160 mAs, depending on anatomical location. Pre and 24 hr postcontrast injection, 10 mm collimated, contiguous, transverse images were obtained in the anatomic regions of interest appropriate to the contrast injection site. Additional 3.0 mm collimated images were obtained through lymph nodes identified on the postcontrast injection survey study for quantitative data analysis. Serial dilutions of meglumine diatrizoate, at iodine concentrations within the range expected for opacified lymph nodes, were included to verify calibration stability. Images were archived on optical disk and magnetic tape for image processing and recorded on radiographic film using a multiformat camera.

4.4. Necropsy Examination

Each animal was euthanized by an overdose of pentobarbital (125 mg/kg IV) after the last postcontrast injection imaging study. Opacified lymph nodes from melanoma-positive swine were placed in 10% buffered formalin for histologic examination. Lymph nodes were prepared for embedding by making a single cut centrally along the long axis of the lymph node and sectioning parallel with the cut surface. Microscopic examination was performed on 5 micron thick Hematoxylin and Eosin stained tissue samples. Lymph nodes were first classified as either cancerous or not cancerous by a veterinary pathologist. Cancerous nodes were divided further into those with up to 25% tumor replacement and those with greater than 25% tumor replacement by estimating the percent of cross sectional area of each lymph node that was replaced by melanoma.

5. RESULTS

5.1. Imaging

Normal lymph nodes which contained contrast agent showed uniform enhancement or opacification of the medullary regions with slightly less opacification of the cortex. Because cortex tends to be located centrally, the enhanced CT image of contrast exposed normal nodes presented a variable ring pattern. Such ring patterns were most visible in large nodes; whereas, small nodes had more uniform opacification and lacked medullary cortical demarcation. Nodes bearing metastatic melanomas were enlarged and showed enhancement of normal regions which lacked tumor, along with regions devoid of contrast material induced enhancement, termed filling defects. Because contrast material does not penetrate tumor tissue, such filling defects are indicative of tumor metastases. Additional CT findings in tumor bearing nodes included expansion of the central cortical region, discontinuity of medulla, nonuniform opacification of both cortex and medulla, and poor delineation of the corticomedullary junction (Fig. 1).

5.2. Histopathology

Lymph nodes containing metastatic melanoma were characterized by irregular, densely cellular, coalescing sheets of neoplastic melanocytes. In some nodes, the cells formed well demarcated but irregular cuffs surrounding centrally located afferent lymphatic channels (Fig. 2).

In other nodes, the above pattern was accompanied by a poorly delineated, diffuse infiltrate of neoplastic cells which penetrated between cortical lymphocytes and obliterated the normal cortical and medullary architecture. Neoplastic cells were enlarged and pleomor-

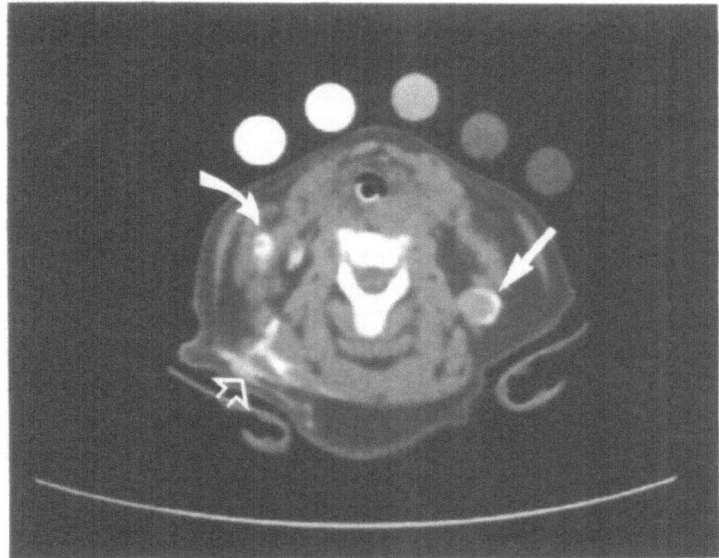

Figure 1. Normal (curved arrow) and cancerous (straight solid arrow) cervical lymph nodes on a 3 mm collimated 24 hr postcontrast CT image. Conspicuity of the nodes is increased due to contrast opacification. Internal architecture of the cancerous node has been disrupted and the corticomedullary junction is defined poorly. Interstitially deposited contrast media also is evident near an injection site.

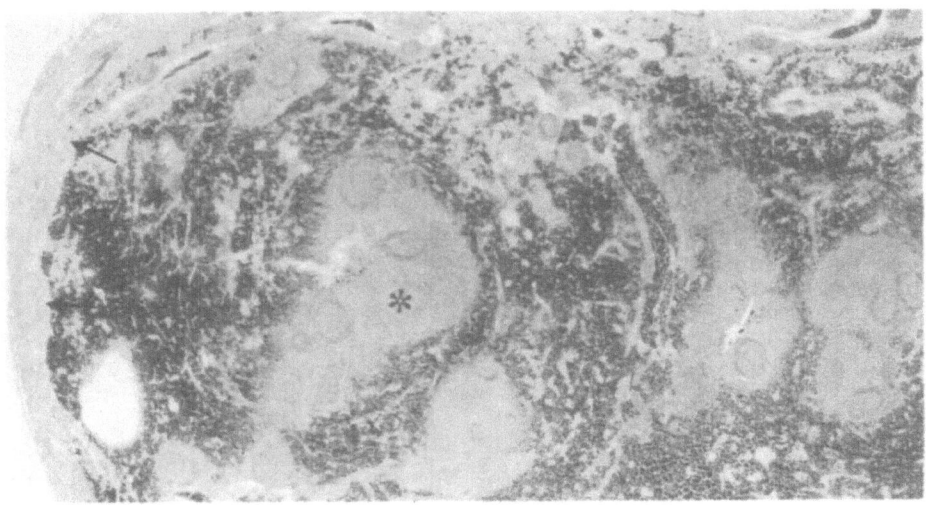

Figure 2. Photomicrograph of a cervical lymph node from a Sinclair Miniature Pig with metastatic malignant melanoma. Note the nodular accumulation of large, heavily pigmented neoplastic cells within the central cortex. Areas of normal appearing cortex and paracortex ( ) are isolated by the extensive sheets of surrounding neoplastic cells. Peripherally, a rim of normal macrophage filled medullary tissue can be seen (arrows).

phic with ample cytoplasm filled with abundant, large, darkly pigmented granules. Single
neoplastic cells also could be seen scattered within otherwise normal cortical tissue.

6. CONCLUSIONS

The Sinclair swine melanoma model was useful for defining the imaging characteristics of a novel lymphography contrast agent. Indirect CT lymphography using iodinated
contrast agents could improve significantly the accuracy of lymph node lesion detection, as
opposed to unenhanced procedures, by increasing lymph node conspicuity, depicting alterations in lymph node internal architecture and quantifying differences in contrast uptake in
normal and cancerous nodes. Melanoma bearing Sinclair swine prove a reliable, well
characterized, naturally occurring animal model of metastatic neoplasia which can be used
effectively to determine the efficacy of diagnostic imaging lymphographic agents.

7. REFERENCES

1. Wisner, E.R., Katzberg, R.W., Koblik,P.D., Shelton, D.K., Fisher, P.E., Griffey, S.M., Drake, C., Harnish,
   P.P., Vessey, A.R., Haley, P.J., Sarpotdar, P.P., and Rajagopalan, N., 1994, Iodinated nanoparticles for
   indirect computed tomography lmphography of the craniocervical and thoracic lymph nodes in normal
   dogs, *Acad. Radiol.* 1:377-384.
2. Wisner, E.R., Katzberg, R.W., Koblik, P.D., McGahan, J.P., Griffey, S.M., Drake, C.M., Harnish, P.P.,
   Vessey, A.R., and Haley, P.J., 1995, Indirect computed tomography lymphography of subdiaphragmatic
   lymph nodes using iodinated nanoparticles in a normal canine model, *Acad. Radiol.* 2:405-412.
3. Wolf, G.L., Gazelle, G.S., McIntire, G.L., Bacon, E.R., Toner, J.L., Cooper, E.R., and Haley, P.J., 1994,
   Percutaneous computed tomography lymphography in the rabbit by subcutaneously injected nanoparticles, *Acad. Radiol.* 1:352-357.
4. Wolf, G.L., Na, G.C., Gazelle, G.S., McIntire, G.L., Cannillo, J., Bacon, E.R., and Halpem, E., 1994,
   Time-lapse quantitative computed tomography lymphography: Assessing lymphatic function in vivo,
   Acad. Radiol. 1:358-363.
5. Wisner, E.R., Katzberg, R.W., Griffey, S.M., Drake, C.M., Haley, P.J., and Vessey, A.R., 1996, Indirect
   computed tomography lymphography using iodinated nanoparticles: Time and dose response in normal
   canine lymph nodes, *Acad. Radiol.* In Press.
6. Wisner, E.R., Katzberg, R.W., Griffey, S.M., Haley, P.J., Johnson, D.K., and Vessey, A.R., 1996,
   Characterization of normal and cancerous lymph nodes on indirect CT lymphographic studies after
   interstitial injection of iodinated nanoparticles, *Acad. Radiol.* In Press.
7. Wolf, G.L., Rogowska, J., Hanna, G.K., and Halpern, E.F., 1994, Percutaneous CT lymphography with
   perflubron: Imaging efficacy in rabbits and monkeys, *Acad. Radiol.* 1:501-505.
8. Hook, R.R., Jr., Berkelhammer, J., and Oxenhandler, R.W., 1982, Animal model of human disease:
   Sinclair swine melanoma, *Am. J. Pathol.* 108:130-133.
9. Millikan, L.E., Boylon, J.L., Hook, R.R., Jr., and Manning, P.J., 1974, Melanoma in Sinclair swine: A
   new animal model, *J. Invest. Dermatol.* 62:20-30.
10. Flatt, R.E., Middleton, C.C., Tumbleson, M.E., and Perez-Mesa, C., 1968, Pathogenesis of benign
    cutaneous melanomas in miniature swine, *J. Am. Med. Assoc.* 153:936-941.
11. Goldschimdt, M.H., 1985, Benign and malignant melanocytic neoplasms of domesitc animals, *Am. J.
    Dermatol.* 7:203-212.
12. Beattie, C.W., Tissot, R., and Amoss, M.S., Jr., 1988, Experimental models in human melanoma research:
    A logical perspective, *Seminars in Oncology* 15:500-511.
13. Hook, R.R., Jr., Aultman, M.D., Adelstein, E.H., Oxenhandler, R.W., Millikan, L.E., and Middleton, C.C.,
    1979, Influence of selective breeding on the incidence of melanomas in Sinclair miniature swine, *Int. J.
    Cancer* 24:668-672.
14. Misfeldt, M.L., and Grimm, D.R., 1994, Sinclair miniature swine: An animal model of human melanoma,
    Vet. Immunol. Immunopathol. 43:167-175.
15. Tissot, R.G., Beattie, C.W., and Amoss, M.S., Jr., 1987, Inheritance of Sinclair swine cutaneous malignant
    melanoma, *Cancer Res.* 47:5542-5545.

52

THE YUCATAN MINIATURE PIG MODEL OF VENTRICULAR SEPTAL DEFECT

M. Michael Swindle,* Robert P. Thompson, Alison C. Smith,
George B. Keech, Blase A. Carabello, Wolfgang Radtke, Derek Fyfe, and
Paul C. Gillette

Departments of Comparative Medicine
Pediatric Cardiology, Cardiology, Anatomy
    and Cellular Biology
Medical University of South Carolina
171 Ashley Ave.
Charleston, South Carolina 29425-2211

1. ABSTRACT

An heritable model of ventricular septal defect (VSD) was developed in Yucatan miniature swine. Retrospective reviews of the parent herd and prospective test matings in an experimental herd indicated that the defect is most likely polygenic in transmission. Clinically the defect may be detected by a characteristic mid or holosystolic murmur auscultated over the lower third to fifth intercostal spaces on either side of the thorax. Subpopulations of the animals develop a failure to thrive syndrome as neonates and juveniles. Adults may develop pulmonary hypertension if the shunt is significant. The defects are described morphologically as perimembranous, with smaller numbers of muscular outlet and doubly committed subarterial defects. Associated defects include atrial septal defect (ASD), aortic arch anomalies and patent ductus arteriosus (PDA). The defect has been characterized by echocardiography in fetal through adult stages and by cardiac catheterization procedures in adults and neonates. Digitalized computer imagery has been used in embryologic studies of the heart. Studies are ongoing in the use of transvenous and transarterial methods of closure using interventional catheter methodology. The model is similar to the most common form of VSD in humans and useful for anatomic, diagnostic and therapeutic studies.

---

* Reprint requests to: Dr. Michael Swindle, Comparative Medicine, Medical University of South Carolina, 171 Ashley Avenue, Charleston, SC 29425, (803) 792-3625.

2. INTRODUCTION

Approximately 0.1 to 0.2% of live human births are affected with isolated ventricular septal defect (VSD), making it one of the most common congenital anomalies. The condition is variable in its severity and may result in mortality or chronic clinical disease associated with the left to right shunt. Severe shunts (>1.5:1.0, Qp:Qs) may require surgical correction. It can be associated with other cardiac anomalies such as tetralogy of Fallot. The various syndromes involving a VSD have been associated with both genetic and environmental factors.[1]

VSD has been described in domestic swine, but not as a reproducible model.[1,2] Case reports have been made in many species but only the dog,[3] chicken[1] and Yucatan miniature pig[4,5] have been studied in any detail as models of the human condition. This manuscript is a review of the information gathered about the condition in the Yucatan miniature pig over the last 10 yr and its development as a genetically reproducible model, which is now available commercially for experimental studies.

3. CLINICAL HISTORY AND CLINICAL SIGNS

The VSD was first discovered as an incidental finding in Yucatan miniature swine (*Sus scrofa domestica*) in a commercial breeding herd (Charles River Laboratories, Andover, MA) after their purchase of the stock herd from Colorado State University. A retrospective analysis of the founding stock pedigree traced the origin of the defect to a single founding boar.[4]

VSD may be detected by auscultation of a midsystolic or holosystolic murmur over the lower third to fifth intercostal space on either side of the thorax. Most of the animals do not express clinical signs; however, some animals may experience sudden death, failure to thrive, pulmonary hypertension and associated heart failure. This may be due to the sedentary lifestyle of swine and signs might be acerbated by exercise testing. When heart failure occurs in neonates and juveniles, it is difficult to detect until acute respiratory failure secondary to pulmonary edema is noted. Cyanosis secondary to respiratory distress is difficult to ascertain clinically because of the normal dark pigmentation of the species. When respiratory distress is noted, the administration of parenteral furosemide 10 mg/kg bid and supportive care is effective if instituted promptly. Small defects tend to close spontaneously with maturity.

4. GENETIC STUDIES

A retrospective analysis of the animals in the founding herd revealed there was not a sex difference in incidence and that an increased incidence of affected offspring occurred with an increased degree of relatedness and inbreeding. The mean inbreeding coefficient for the affected litters was 0.042 compared to 0.025 for normal litters. The degree of relatedness of the sire and dam for affected litters was 0.0757 compared to 0.037 for normal litters.[4]

After reviewing the pedigree of the parent herd, a prospective series of matings was performed in an experimental herd established at the Medical University of South Carolina in 1987 (Fig. 1). Test matings included unrelated individuals, brother x sister and parent x offspring pairings. As in the parent herd, no sex related differences in incidence were noted. Autosomal dominant was eliminated because of the ratios of affected offspring from affected matings, evidence that the defect does not skip generations and that it can be transmitted from nonaffected matings. However, this does not preclude completely the possibility of incomplete penetrance. Autosomal recessive was eliminated because the ratios of affected offspring in test matings were too high and the higher incidence of the defect in parents, siblings, offspring and closely related relatives. Polygenic or multifactorial inheritance is

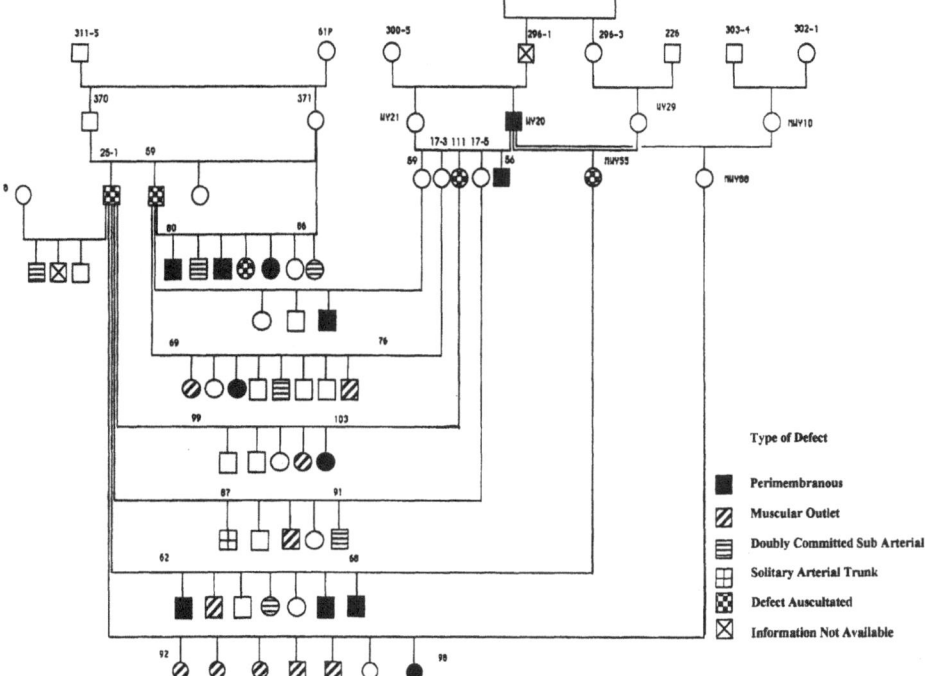

Figure 1. Pedigree chart of the experimental herd of pigs with VSD.

exclusionary in its determination using Mendelian genetic techniques and is thought to be the most probable mode of transmission with this defect.[4]

An attempt was made to produce an inbred line of these animals to further define the genetic pattern; however, infertility developed in the F-5 generation and the effort was discontinued in favor of obtaining pregnant sows and affected offspring from selected breedings in the parent herd. This process of obtaining affected animals has been highly successful.

5. HEMODYNAMIC AND ANGIOGRAPHIC STUDIES

Neonatal, juvenile and adult animals have been characterized using cardiac catheterization and angiographic imaging technologies. Left to right shunts were detected in all animals with a functional VSD using oximetric techniques. Shunts of 1.5:1.0 (Qp:Qs) or greater magnitude were considered to be hemodynamically significant in experimental studies. Angiographic images were obtained by selective placement of pigtailed catheters in the ventricle and administration of radiopaque dyes via an automated injector while recording at 60 frames/sec (Fig. 2).[4,6]

Neonatal pigs with VSD and failure to thrive syndrome were determined to have physiologic abnormalities similar to human infants and were placed in a paired study to determine the pathogenesis of the condition. In this experiment, complex hemodynamic studies, with load manipulation and beta blockade in the porcine model, were compared to clinical hemodynamic studies in human infants. The failure to thrive syndrome was associated with a decreased systemic output and forward stroke volume that were secondary to failure of the end diastolic volume to increase adequately. This was in contrast to conventional thinking that the pathogenesis was related to impaired contractile function, excess afterload or reduced preload.[6]

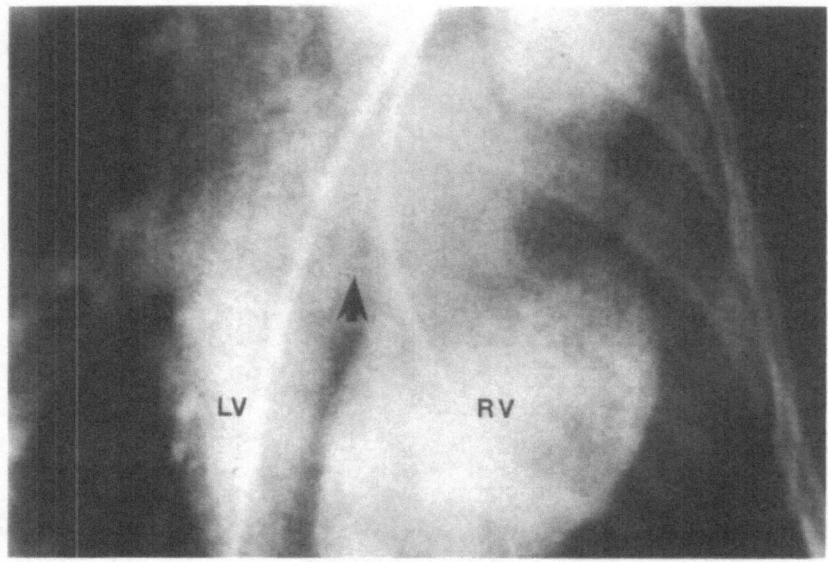

Figure 2. Angiocardiogram of a perimembranous VSD (arrow) from the right dorsal ventral oblique position.

Animals without clinical signs did not differ from controls in other experimental studies at any age, except for adults greater than 12 mo of age with significant left to right shunts. They were detected to have pulmonary hypertension hemodynamically with characteristic pulmonary changes in small and midsized arteries which were detected histologically.[4]

6. MORPHOLOGIC AND NECROPSY STUDIES

A retrospective pathology review was conducted of 17 hearts with VSD from neonates thru adults from the commercial breeding herd. Of these, 11 had an isolated perimembranous VSD (Fig. 3). The remaining 6 animals had the following concomitant defects: patent foramen ovale (PFO)-4, PFO and patent ductus arteriosus (PDA)-1, and coarctation of the aorta-1. The defects were typically elliptical or oval in shape. Histologically the borders of the defects were associated with branches of the cardiac conduction system and were noninflammatory lesions (Fig. 4).[4,7]

In prospective studies, neonatal and fetal pigs were studied using 2 dimensional (2-D) and color flow Doppler echocardiography (Fig. 5). Studies VSD morphology were correlated with necropsy studies. In the neonatal study 4 to 18 da old piglets were studied for the presence of VSD and associated defects. In 29 affected neonates, perimembranous, muscular outlet and doubly committed subarterial defects ranging in size from 1 to 6 mm were diagnosed and characterized. The defects were confirmed in all 29 at necropsy and comparisons were made of the sizes for shrinkage after formalin fixation. Use of echocardiography with a 5 Mhz transducer tended to overestimate the size of the defect diameter by 21% (0.6 mm).[8]

In utero fetal echocardiography was performed on 68 third trimester fetuses using the same equipment as described above except it was performed intraoperatively on the gravid uterus following laparotomy. Diagnosis of the defects as small as 1 mm were

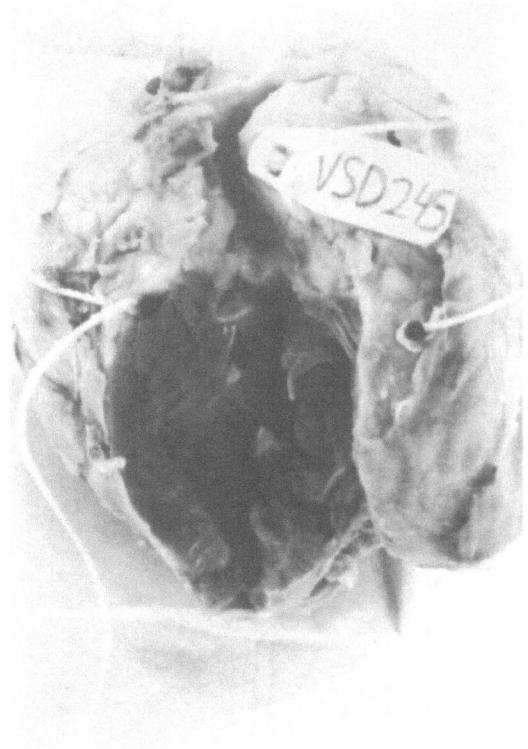

Figure 3. Perimembranous VSD (arrow) viewed from the left ventricle of a formalin fixed heart.

performed correctly in the 15 affected fetuses in the study and results were correlated at necropsy in all of the fetuses. This study has made it possible to image and characterize VSDs at an early stage for ongoing therapeutic studies. Histologic analysis of first trimester fetuses indicated that the probable time for normal closure of the fetal VSD is 25 to 27 days of gestation (Fig. 6).[9,10]

Seventy three neonates from 15 litters in the experimental colony were studied at necropsy. Fifty two of the hearts had a VSD. The defects were perimembranous in 34, muscular outlet in 12 and doubly committed juxtaarterial in 6. Associated defects included PFO in 12, solitary arterial trunk in 1 and hypoplasia of the aorta and PDA in 1. This represented a shift in the location and morphology of the defect in the experimental herd as compared to the parent herd. This shift may be due to the effect of concentrating a polygenic trait in a line of pigs undergoing inbreeding. Long term studies would be required to identify whether different genes are involved in the different morphologies of VSD.[7]

7. SUMMARY

VSD remains one of the most common congenital cardiac anomalies in humans. The most common anatomic locations in humans and swine when viewed from the left side passing into the right side of the heart are perimembranous, muscular and doubly committed

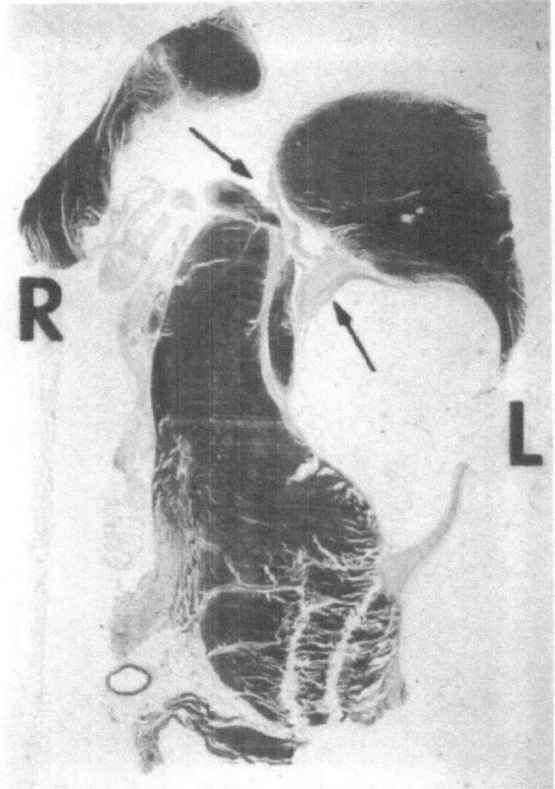

Figure 4. Histologic section of a perimembranous VSD (arrow). Note the continuity of the edges of the defect with the cardiac conduction tissue.

and juxtaarterial. Perimembranous defects had the central fibrous body as part of their border, muscular defects had septal musculature surrounding the defect and doubly committed and juxtaarterial defects had fibrous continuity between the leaflets of the aortic and pulmonary valves. All of the defects in the Yucatan model were located high in the aortic outflow tract of the left ventricle.

Hemodynamically, porcine VSD can be characterized the same as human VSD except that severe signs were not noted unless the animals were stressed, usually by shipment or procedures requiring anesthesia. Death under general anesthesia was noted most commonly in adults and acute respiratory distress was the most common clinical sign in neonates and juveniles. Neonatal swine with failure to thrive syndrome developed growth retardation greater than 2 SD from the mean for their age group. These animals, more than any other group, were demonstrated to have significant hemodynamic compromise. They also shared a common pathogenesis for the occurrence of the condition with human patients. Pulmonary hypertension is one of the earliest noted hemodynamic abnormalities with a significant shunt from a VSD. Pulmonary hypertension was noted only in the failure to thrive group and in adults greater than 12 mo of age. Histologic examination of the lungs following necropsy revealed changes of pulmonary hypertension only in the adults. Not all animals studied, including some with sudden death, were

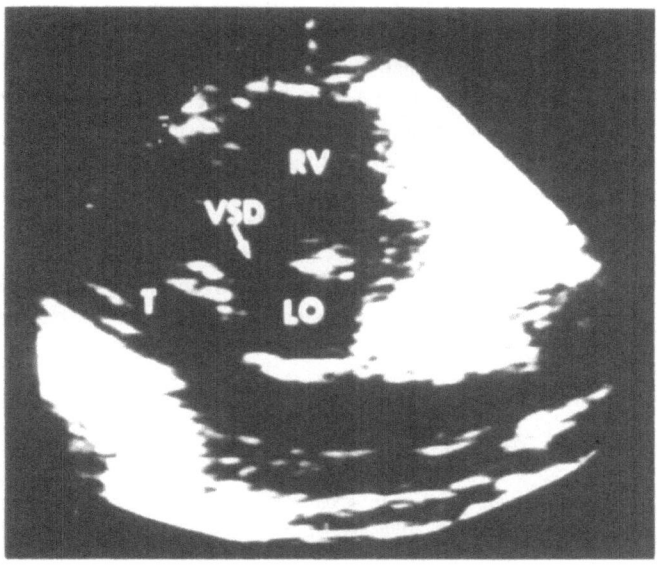

Figure 5. Long axis two dimensional echocardiogram of a perimembranous VSD (arrow). RV = Right Ventricle T = Tricuspid valve LO = Outflow tract.

studied in the catheterization lab and the incidence of pulmonary hypertension in the juvenile animals may be higher than we detected.

We demonstrated the feasibility of correctly performing the diagnosis and correct morphologic description of the VSD in neonatal and fetal swine with defects as small as 1

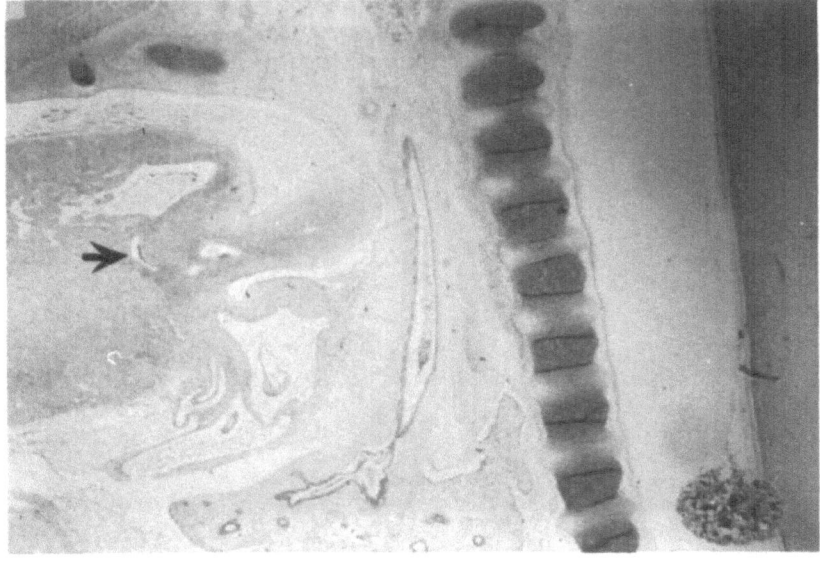

Figure 6. Cross section of the heart of a 32 day fetus showing an open VSD. (arrow)

mm using noninvasive 2-D and color flow Doppler echocardiography. These techniques are useful for studying the course of therapeutic strategies such as interventional catheter techniques.

Transvenous and transarterial catheter techniques involving the use of implantable coils delivered into the VSD are being conducted. This type of technique offers promise for closure of the defect without having to resort to open heart surgery. The anatomic location, morphology and physiologic characteristics of the defect make it valuable in the study of these therapeutic interventions.

This model of VSD shares important similarities to one of the most common congenital cardiac anomalies in humans. This model has been useful in the study of both basic and applied scientific protocols to study the pathogenesis and potential therapeutic interventions. The location of the VSD is predictable and the model can be reproduced reliably by selective breeding techniques.

8. REFERENCES

1. Pyle, R.L., 1979, Ventricular septal defect, in: *Spontaneous Animal Models of Human Disease*, Volume 1, (E.J. Andrews, B.C. Ward, and N.H. Altman, eds.), Academic Press, New York, pp. 69-70.
2. Hsu, F.S., and Du, S.J., 1982, Congenital heart diseases in swine, *Vet. Pathol.* 19:676-686.
3. Patterson, D.F., 1980, Genetic aspects of cardiovascular development in dogs, in: *Etiology and Morphogenesis of Congenital Heart Disease*, (R. Van Praagh, and A. Takao, eds.), Futura Publishing, Mt. Kisco, NY, pp. 1-19.
4. Swindle, M.M., Thompson, R.P., Carabello, B.A., Smith, A.C., Hepburn, B.J.S., Bodison, D.R., Corin, W., Fazel, A., Biederman, W.W.R., Spinale, F.G., and Gillette, P.C., 1990, Heritable ventricular septal defect in Yucatan miniature swine, *Lab. Anim. Sci.* 40: 155-161.
5. Swindle, M.M., Thompson, R.P., Carabello, B.A., Smith, A.C., Green, C., and Gillette, P.C., 1992, Congenital cardiovascular disease, in: *Swine as Models in Biomedical Research*, (M.M. Swindle, ed.), Iowa State University Press, Ames, IA, pp. 176-184.
6. Corin, W.J., Swindle, M.M., Spann, J.F., Jr., Frankis, M., Biederman, W.W.R., Smith, A.C., Taylor, A., and Carabello, B.A., 1988, The mechanism of decreased stroke volume in children and swine with ventricular septal defect and failure to thrive, *J. Clin. Invest.* 82: 544-551.
7. Ho, S.Y., Thomspon, R.P., Gibbes, S., Swindle, M.M., and Anderson, R.H., 1991, Ventricular septal defects in a family of Yucatan miniature pigs, *Int. J. Cardiol.* 33: 419-426.
8. Johnson, T.B., Fyfe, D.A., Thompson, R.P., Kline, C.H., Swindle, M.M., and Anderson, R.H., 1993, Echocardiographic and anatomic correlation of ventricular septal defect morphology in newborn miniature pigs, *Am. Heart J.* 125:1067-1072.
9. Swindle, M.M., Wiest, D.B., Smith, A.C., Garner, S.S., Case, C.C., Thompson R.P., Fyfe, D.A., and Gillette, P.C., 1996, Fetal surgical protocols in Yucatan miniature swine, *Lab. Anim. Sci.* 46: 90-95.
10. Dungan, L.J., Wiest, D.B., Fyfe, D.A., Smith, A.C., and Swindle, M.M., 1995, Hematology, serology and serum protein electrophoresis in fetal miniature Yucatan swine: normal data, *Lab. Anim. Sci.* 45: 285-289.

53

AN EXTERNAL THORACIC DUCT VENOUS SHUNT TO ALLOW FOR LONG TERM COLLECTION OF LYMPH AND BLOOD IN THE CONSCIOUS PIG

H. Vince Mendenhall,[*] Christopher Horvath, Marek Piechowiak, Lisa Johnson, and Kim Bayer

Mason Laboratories
57 Union Street
Worcester, Massachusetts 01608

1. ABSTRACT

Studies involving the pharmacokinetics of quinolones, cytokines or various growth factors require an animal model which allows for continuous lymph and/or blood sampling. Commonly used procedures to achieve this end result in continuous drainage of the lymph. These procedures are prone to infection, catheter occlusion and developing physiologic abnormalities. To address these problems, a method of creating an external thoracic duct venous shunt has been developed to allow for convenient collection of thoracic duct lymph and/or blood in the conscious pig for extended periods of time, while maintaining normal lymph flow. Under strict aseptic procedures, a silastic catheter is placed in the right external jugular vein and exteriorized in the interscapular region. A coated polyurethane catheter is placed in the thoracic duct through a right thoracotomy and exteriorized in the midline posterior thoracic region. The two catheters are connected to a 3 way stopcock. Flow of lymph from the thoracic duct back to the external jugular vein is thus reestablished. A specially prepared aluminum jacket is placed over the pig to prevent disruption of the catheters and stopcock. The catheters are flushed twice daily with heparin. Lymph and/or blood samples can be collected from the animal while conscious at any convenient or study required time points. This shunt system has remained open and functional for more than 14 da in more than 35 pigs.

[*] Reprint requests to: Dr. H. V. Mendenhall, Mason Laboratories, 57 Union Street, Worcester, MA, 01608, *(508) 791-0931.*

2. INTRODUCTION

Studies involving the transport of connective tissue metabolites from tissues to the circulation, or the pharmacokinetics of quinolones, cytokines or various growth factors, require a conscious animal model which allows for continuous lymph and/or blood sampling by methods that do not affect lymph flow.[1-4] After subcutaneous administration, plasma concentrations of macromolecules are dependent on capillary and lymphatic absorption processes[5] which can be affected by anesthesia.[3] Chronic catheter placement allows for sample collection for a sufficient time to enable characterization of plasma and lymph concentrations of cytokines in conjunction with the use of long half life derivatives, slow release formulations on, multiple dosing regimens. Commonly used procedures to achieve this end result in continuous drainage of the lymph.[6-16] These procedures are prone to infection, catheter occlusion, and developing physiologic abnormalities. To address these problems, methods for creating an external thoracic duct venous shunt have been described that potentially would allow for convenient collection of thoracic duct lymph and/or blood in dogs and pigs.[17-19] Unfortunately, even these preparations have been reported to remain patent for less than 7 da after creation. Since the pig has many anatomical and physiological similarities to man,[20] and may be more suitable than the dog for studies involving such instrumentation, we modified published techniques for use in Yorkshire and Yucatan swine in an attempt to prolong the patency of such preparations. We implanted a hydrogel coated polyurethane catheter in the thoracic duct, and administered heparin systemically and locally, both intra and postoperatively. In addition, a wearable aluminum jacket was placed on the animals to protect the externalized instrumentation. To accomplish our objective, a series of pilot studies were performed to evaluate and modify the strain and gender of pigs, anesthetic regimen, surgical procedures, catheter type and maintenance techniques and sample collection times.

3. MATERIALS AND METHODS

3.1. Animals

Thirty nine female Yorkshire pigs, weighing 15.8 to 24.3 kg, and 7 male Yucatan pigs, weighing 20.1 to 21.2 kg, were used in this study. Animals were examined for signs of disease or injury upon arrival, and were held for at least 7 da prior to release from quarantine. Animals were housed individually in stainless steel pens bedded with rubber mats and wood shavings. Animal rooms were maintained at temperatures of 66 to 75F and relative humidities of 31 to 56%. A 12 hr light/dark cycle was employed and the room underwent a minimum of 10 fresh air changes/hr. Pigs were fed Purina Porcine Chow Grower 5084™ twice daily, supplemented with washed, fresh produce and provided fresh filtered water ad libitum. All other husbandry conditions were maintained as described in the Guide for the Care and Use of Laboratory Animals.[21] During the holding period, a physical examination was performed on each pig and rectal swabs for culture and fecal samples for occult blood and ova and parasite determinations were collected. Pigs were handled daily to acclimate them to close human contact and were acclimated to wearing a <2.0 kg protective jacket made from aluminum.

3.2. Surgery

3.2.1. Preoperative Procedures.

3.2.1.1. Fasting. All food, including the diet supplement, was withheld overnight prior to surgery. Water was available ad libitum.

3.2.1.2. Preanesthesia. To better visualize the thoracic duct at the time of surgery, the animals were given ≈200 ml vegetable oil and ≈200 ml milk with ≈30 g butter saturated with Sudan III by stomach tube approximately 1 hr prior to induction of anesthesia. To facilitate this gavage procedure, animals were tranquilized with acepromazine (0.15 mg/kg, IM) approximately 30 min prior to gavage.

3.2.1.3. Anesthesia and Antibiotic Administration. Animals were anesthetized initially with a mixture of ketamine HCl (25 mg/kg), atropine (0.04 mg/kg), butorphanol (0.55 mg/kg) and xylazine (2 mg/kg) administered IM. They were intubated and maintained in anesthesia with isoflurane inhalant anesthetic, delivered through a volume regulated respirator. A percutaneous catheter was placed in an ear vein for intraoperative heparin administration. Cefotaxime™ (50 mg/kg, IV) was administered.

3.2.1.4. Surgical Preparation. The neck, right thoracic area and back were clipped of all hair; the animal was positioned in left lateral recumbency with a towel placed under the animal to elevate its chest. The operative area was cleaned with 3 alternating scrubs of povidone iodine scrub solution and 70% isopropyl alcohol, with a final application of povidone iodine solution that was allowed to dry. The right neck, chest and back were draped appropriately for aseptic surgery.

3.2.2. Surgical Procedures.
An incision was made over the right jugular fossa and the right external jugular vein was mobilized. It was ligated cranially, and a previously prepared silastic catheter was inserted and advanced caudally approximately 10 cm so that the tip would be positioned in the superior vena cava, close to the right atrium. The jugular vein, with the indwelling catheter, was ligated to help maintain its position. A blood sample was taken to determine the baseline activated clotting time (ACT) using the Hemochron® Model 801 Whole Blood Coagulation System. The catheter was tunnelled subcutaneously to an area midway between the scapulae. The catheter was immobilized with multiple sutures to the subcutaneous fascia. An infusion of lactated Ringer's solution (10 ml/kg/hr) was started through this catheter. Heparin (300 U/kg) was administered IV and an ACT taken approximately 5 min later. Additional heparin was given, if necessary, to raise the ACT to approximately 4 times baseline, in an attempt to reduce postoperative clotting within the thoracic duct catheter. This administration of heparin during surgery was discontinued later, due to deaths associated with postoperative hemorrhage within the thoracic cavity and because it was found to be unnecessary. The incision in the neck was closed with 2/0 Vicryl™ in a continuous pattern to appose the muscular layers and subcutaneous tissues. Three/0 Vicryl™ in a subcuticular pattern was used to close the skin.

A right lateral thoracotomy was performed through the 5th to 7th intercostal space and a rib was removed. The thoracic duct was identified subpleurally, between the aorta and the vertebral bodies, and carefully dissected from the surrounding tissue for approximately 3 cm. The thoracic duct was ligated cranial to the insertion point. An appropriately sized silastic catheter (pigs #1 to 20) or a hydrogel coated polyurethane catheter (Hydracath™*, Access Technologies, Skokie, IL) (pigs #21 to 46) was inserted into the duct and advanced

1 to 2 cm caudally, so the tip was positioned at approximately the level of the 10th thoracic vertebral body. (Hydracath™ catheters were substituted for the silastic ones, because of their reportedly decreased thrombogenicity, in an effort to prevent clotting of the thoracic lymph fluid and to diminish the use of systemic heparin.) A ligature was tied around the duct and catheter. The catheter was fixed to the thoracic wall with 4/0 braided nylon and filled with heparin (5,000 IU/ml) to prevent coagulation. The catheter was passed through the right 7th intercostal space to a site 15 to 20 cm caudal to the site of exit of the jugular catheter. The thoracic duct catheter was attached to the jugular vein catheter by a single 3 way stopcock. The thoracotomy was closed in standard fashion with 0 Vicryl™ suture around the ribs in an interrupted pattern, 2/0 Vicryl™ suture in the underlying muscles and subcutaneous tissues in layers, and 3/0 Vicryl™ suture in the skin in a continuous subcuticular pattern. The stopcock was secured to the back of the animal with several sutures of #2 monofilament nylon. The exit site was closed tightly with 2/0 monofilament nylon and irrigated with povidone iodine solution. In early animals, another ACT was taken, and an appropriate dose of heparin was given to adjust the postoperative ACT to approximately 4 times baseline. This procedure was discontinued after the 4th animal in this series, once the appropriate dose of heparin was established. The aluminum protective jacket was replaced on the animal and it was allowed to recover from anesthesia, being extubated once the swallowing reflex had returned. The jackets were worn continuously after surgery, checked twice daily and replaced or adjusted as necessary.

3.2.3. Postoperative Procedures.

3.2.3.1. Postoperative Analgesia. Animals were given either buprenorphine (0.1 mg/kg, IV) prior to extubation and then twice daily (8 to 18 hr apart) or butorphanol (0.5 mg/kg, IV) 4 times daily, for at least 3 consecutive da. If required, as evidenced by inappetence, additional analgesic was administered.

3.2.3.2. Catheter Flushes. For the first 19 animals, jugular and thoracic duct catheters were flushed by retrograde infusion of heparin (1 ml at 5000 U/ml) into the lymph catheter and antegrade infusion of heparin (2.5 ml at 1000 U/ml) into the venous catheter twice daily. When the Hydracath™ catheter was used, heparin flushes were decreased for the lymph catheter (0.5 ml at 5000 U/ml) and jugular catheter (1.25 ml at 1000 U/ml). This also was done to decrease the incidence of postoperative hemorrhage.

3.2.3.3. Antibiotic Administration. Cefotaxime™ (50 mg/kg, IV) was administered twice daily for 3 da after surgery.

3.3. Clinical Observations

Clinical observations were performed daily beginning the day after surgery, and continued for the duration of the study. Evidence of any clinical effects were noted, with observations including, but not limited to, changes in skin and hair, eyes and mucous membranes, respiratory system, circulatory system, central nervous system, somatomotor activity, behavior pattern, occurrence of tremors, convulsions, salivation, diarrhea or lethargy. Additionally, written comments related to the clinical appearance or behavior of the animals were entered in the study file. Surgical incision and catheter exit sites were cleaned and disinfected at least daily. Any signs of infection, inflammation and general integrity were noted and treated appropriately.

3.4. Blood Samples

Blood samples were collected from the jugular vein via the stopcock at the times indicated in Table 1. The samples for hematology and clinical chemistry parameters were collected at approximately the same time each day (morning). Blood samples for clinical chemistry were collected prior to feeding.

3.5. Lymph Samples

Lymph was collected from the thoracic duct via the stopcock at the times indicated in Table 1. After day 15, catheter patency was determined, but no samples were taken.

3.6. Euthanasia

All surviving pigs were euthanized at the completion of the studies, or when lymph fluid could no longer be aspirated for 3 consecutive days, whichever came first. Euthanasia was performed in accordance with accepted American Veterinary Medical Association guidelines,[22] by first administering a mixture of ketamine HCl (25 mg/kg) and xylazine (5 mg/kg), via IM injection, followed by sodium pentobarbital (390 mg/kg, IV).

4. RESULTS AND DISCUSSION

4.1. Animal Disposition

4.1.1. Yucatans. All these animals were implanted with silastic thoracic duct catheters. Three died of internal bleeding attributed to excessive heparin administration and 2 were euthanized after the lymph catheter clotted in less than 14 da.

4.1.2. Yorkshires. Of the first 4 pigs to undergo surgery, only 1 died of internal bleeding. Intraoperative heparinization and ACT determinations were discontinued beginning with the 5th pig in this series. Even so, a second pig was found dead as a result of intrathoracic bleeding. In response, the amount of heparin used to flush the catheters postoperatively was reduced by half. No additional heparin related deaths occurred. Four

Table 1. Blood and Lymph Collection Timepoints

Group 1	Blood			Lymph
		Clinical pathology		Clinical pathology
Time	Hematology	Clinical chemistry	Pharmacokinetics	Pharmacokinetics
Day 1	X	X	X	X
Da 1; min 5, 15, 30 and 90; hr 3, 5, 8, 11, 15 and 18	-	-	X	X
Da 2; hr 24	X	X	X	X
Da 4 to 15	X	X	-	X
Volume Anticoagulant	0.5 ml EDTA	1.8 ml None	2 ml EDTA	2 ml EDTA

animals died variable periods of time after surgery. Some of these pigs had respiratory distress or arrest or had evidence of laryngeal edema when examined in a limited necropsy. Their deaths were attributed to complications of surgery and/or anesthesia.

4.2. Surgical Instrumentation

There was considerable anatomic variation among animals in the location, diameter and arborization of the thoracic ducts. Therefore, the size and length of catheter implanted in each animal was variable. As a result, the flow of lymph through the catheters also was variable, and may have affected the tendency for the lymph fluid to clot. Lymph fluid was collected most readily after the pigs had recently eaten, presumably due to increased gastrointestinal lymph flow.

4.3. Duration of Lymphatic Catheter Patency

Thirty six animals lived until day 15. The lymph catheters in several animals were clotted for variable periods, then became patent again. Overall, the silastic thoracic duct catheter remained patent after day 3 in 26 pigs, but in only 8 animals until day 15. These animals were allowed to survive until day 35, at which time all catheters were still patent. Six of the 10 Hydracath™ catheters remained patent until day 35. The other 4 were of smaller size (3.5 Fr) and remained patent an average of 8.5 da.

4.4. Clinical Observations

Swelling, edema, discharge and signs of catheter tract infections were rare.

5. DISCUSSION

Originally, it was thought the use of the protective cage, and intraoperative and frequent postoperative flushing of the catheters with heparin would limit the incidence of catheter occlusion. This in fact did prove do be the case in the majority of animals; however, this practice also was found to yield a significant incidence of fatal postoperative hemorrhage. When the silastic catheters were replaced with the hydrogel coated polyurethane catheter, the patency rate was found to improve markedly with minimal use of heparin and the mortality rate decreased to zero. This catheter did prove to be less thrombogenic and thus less likely to result in clotting of the thoracic duct lymph. Nevertheless, 4 pigs receiving the smaller diameter of these catheters developed clots shortly after implantation.

6. CONCLUSIONS

The creation of an external thoracic duct external jugular vein shunt for the collection of lymph and blood samples in the conscious pig for periods of up to 14 da is possible. To achieve this end, we feel that the use of a 5 Fr hydrogel coated polyurethane catheter in the thoracic duct, combined with minimal flushing of the catheters with heparin, will increase the likelihood of achieving long term (>7 da) patency. Additionally, the use of an aluminum protective cage covering the back of the pig will help to prevent disruption of the catheters. This cage obviates the need for protective bandaging, a practice which can in itself increase the likelihood of disruption and local infection.

7. REFERENCES

1. Bocci, V., Pessina, G.P., Paulesu, L., and Nicoletti, C., 1988, The lymphatic route. VI. Distribution of recombinant interferon-alpha$_2$ in rabbit and pig plasma and lymph, *J. Bio. Resp. Mod.* 7:390-400.
2. Bocci, V., Pessina, G.P., Nicoletti, C., and Paulesu, L., 1990, The lymphatic route. VII. Distribution of recombinant human interleukin-2 in rabbit plasma and lymph, *J. Bio. Reg. Homeo. Agents* 4:25-29.
3. Muranishi, S., 1991, Drug targeting towards the lymphatics, *Adv. Drug Res.* 21:1-38.
4. Bocci, V., Carraro, F., Zeuli, M., Naldini, A., and Calabresi, F., 1993, The lymphatic route. VIII. Distribution and plasma clearance of recombinant human interleukin-2 after SC administration with albumin in patients, *Biotherapy* 6:73-77.
5. Supersaxo, A., Hein, W.R., and Steffen, H., 1990, Effect of molecular weight on the lymphatic absorption of water-soluble compounds following subcutaneous administration, *Pharm. Res.* 7:167-169.
6. Bollman, J.L., Cain, J.C., and Grindlay, J.H., 1949, Techniques for the collection of lymph from the liver, small intestine, or thoracic duct of the rat, *J Lab. Clin. Med.* 33:1349-1352.
7. Martin, D.P., and Leiseca, S.A., 1977, Thoracic duct cannulation of the rhesus monkey (Macaca mulatta) for lymphocyte collection: thoracic approach, *Lab. Anim. Sci.* 27:1017-1023.
8. Butterfield, A.B., Lumb, W.V., and Litwak, P., 1976, Surgical preparation of miniature swine for atherosclerosis research, *Am. J. Vet. Res.* 37:1519-1523.
9. Jakab, F., Sugar, I., and Szabo, G., 1980, Cannulation of the cervical thoracic duct in the rat, *Lymphology* 13:1844-1850.
10. Manolas, K.J., Farmer, H.M., Cussen, M., and Welbourn, R.B., 1983, An experimental model for simultaneous chronic sampling of portal and systemic blood and gastrointestinal lymph via cannulae in conscious swine, *Cornell Vet.* 73:333-339.
11. Frank, W.L., Stuhldreher, D., Muchnik, S., Ray, P., and Guinan, P., 1992, A new technique for the cannulation of the rat thoracic duct, *Lab. Anim. Sci.* 42:526-527.
12. Uhley, H., Leeds, S.E., and Sampson, J.J., 1963, A technique for collection of right duct lymph flow in unanesthetized dogs, *Proc. Soc. Exp. Biol. Med.* 112:9844-9850.
13. Nelson, A.W., and Swan, H., 1969, Long-term catheterization of the thoracic duct in the dog, *Arch. Surg.* 98:883-886.
14. Doemling, D.B., and Steggerda, F.R., 1960, Chronic thoracic venous shunt preparation in dogs, *J. Appl. Physiol.* 15:745-746.
15. Giarardet, R.E., and Benninghoff, D.L., 1973, Surgical techniques for long-term study of thoracic duct lymph circulation in dogs, *J. Surg. Res.* 15:168-174.
16. Lascelles, A.K., and Morris, B., 1961, Surgical techniques for the collection of lymph from unanesthetized sheep, *Quart. J. Exp. Physiol.* 52:200-205.
17. Jensen, L.T., Olesen, H.P., Risteli, J., and Lorenzen, I., 1990, External thoracic duct-venous shunt in conscious pigs for long term studies of connective tissue metabolites in lymph, *Lab. Anim. Sci.* 40:620-624.
18. Brown, C.S., and Hardenbergh, E., 1951, A Technique for sampling lymph in unanesthetized dogs by means of an exteriorized thoracic duct-venous shunt, *Surgery* 29:502-507.
19. Manolas, K.J., Farmer, H.M., and Cussen, M., 1983, An experimental model for simultaneous chronic sampling of portal and systemic blood and gastrointestinal lymph via cannulae in conscious swine, *Cornell Vet.* 74:333-339.
20. Swindle, M.M., 1984, Swine as replacement for dogs in the surgical teaching and research laboratory, *Lab. Anim. Sci.* 34:383-385.
21. *Guide for the Care and Use of Laboratory Animals*, 1985, NIH Publ. #86-23. 22. Report of the American Veterinary Medical Association (AVMA) Panel on Euthanasia, 1993, *J. Am. Vet. Med. Assoc.* 202:229-249.

54

PREGNANT YUCATAN MINIATURE SWINE AS A MODEL FOR INVESTIGATING FETAL DRUG THERAPY

Donald B. Wiest,[1*] M. Michael Swindle,[2] Sandra S. Garner,[1]
Alison C. Smith,[2] Paul C. Gillette[3]

[1] Departments of Pharmaceutical Sciences
[2] Comparative Medicine and
[3] Pediatric Cardiology
Medical University of South Carolina
171 Ashley Ave.
Charleston, South Carolina

1. INTRODUCTION

Initially it was felt that during maternal drug therapy the placenta served as a protective barrier preventing fetal exposure. This concept was invalidated as a result of the thalidomide tragedy in the early 1960s. Maternal thalidomide use in the first trimester of pregnancy resulted in approximately 7,000 limb anomalies, with or without gastrointestinal or cardiac malformations.[1] As a result of this disaster, research was focused on toxicologic and teratogenic effects of drugs on the fetus. Despite 3 decades of research, approximately 20 drugs and chemicals, e.g., isotretinoin, valproic acid, phenytoin, polychlorinated biphenyls, have been confirmed to be teratogenic. Most drugs, when used appropriately, do not pose a threat to the fetus.[2] In the 1970s, therapeutic benefits started to be recognized in the pharmacologic management of fetal disorders. The discovery inspiring research in this area was the observation that maternal beclomethasone administration enhanced fetal lung maturity decreasing the incidence of neonatal respiratory distress syndrome.[3] Considerable data have accumulated to indicate that nearly all maternally administered drugs enter the fetal circulation and may exert pharmacologic effects on the fetus. These observations have led to the increasing use of the maternal route of drug administration to manage the compromised fetus. Pharmacologic management has extended into the treatment of fetal arrhythmias[4], toxoplasmosis[5], congenital adrenal hyperplasia[6] and preventing vertical transmission of HIV during pregnancy.[7] We will discuss factors that influence the kinetics of drug

---

[*] Reprint requests to: Dr. Donald Wiest, Department of Pharmaceutical Sciences, QF220, Medical University of South Carolina, 171 Ashley Ave., Charleston, SC 29425, (803) 792-3118.

Advances in Swine in Biomedical Research, edited by Tumbleson and Schook
Plenum Press, New York, 1996

transfer to the fetus, pharmacologic management of fetal tachyarrhythmias and the application of pregnant Yucatan miniature swine as a model for investigating fetal antiarrhythmic
drug therapy.

2. FACTORS INFLUENCING PLACENTAL DRUG TRANSFER

From a physicochemical view, essentially all drugs equilibrate to some extent in the
fetal circulation following maternal administration. The level of fetal exposure is based on
the pharmacokinetic behavior of the drug within the maternal placental fetal unit (MPFU).
Factors influencing placental drug transfer include the physicochemical properties of the
drug as well as the physiologic characteristics within the MPFU.

2.1. Mechanisms of Placental Drug Transport

As with most biological membranes movement of drugs across the placenta occurs
primarily by passive diffusion. Facilitated diffusion, i.e., carrier mediated, and active
transport, i.e., carrier mediated and energy dependent, are presumed to have a minimal role
in placental drug transfer. Passive diffusion is described by Fick's law which states that the
rate of diffusion across a membrane is proportional to the difference in drug concentration
on each side of the membrane, that is:

$$\text{rate of diffusion } (R/t) = K \, A(C_m - C_f)/D$$

where K = diffusion coefficient of the drug; C_m and C_f represent the maternal and fetal blood
concentrations, respectively; D = thickness of the membrane and A = surface area. Based on
Fick's law, passage of the same drug and the rate of passage will change throughout gestation
due to alterations in placental thickness and surface area. The concentration or amount of
drug reaching the placenta will be affected by the maternal dose, method of drug administration, e.g., IV bolus vs. oral administration, and maternal, placental, fetal pharmacokinetic
parameters including protein binding.[8,9] Fick's equation does not consider uterine blood flow
which increases with increasing gestational age. Undoubtedly, uterine blood flow would be
a rate limiting step in placental drug transfer.

2.2. Physicochemical Properties

The rate of diffusion also is influenced by the physicochemical characteristics of the
drug. Generally drugs with a molecular weight (MW) greater than 500 have incomplete or
slow placental transfer. Agents such as heparin (MW > 1000) are relatively impermeable.
However, most drugs have a molecular weight less than 500; therefore, minimizing the
importance of this physical characteristic. Lipid solubility is considered one of the most
important factors for drug diffusion. Ionized particles pass slower than nonionized. For
example antipyrine which is minimally ionized at physiologic pH, crosses the placenta
immediately.[10] Generally, drugs that are highly ionized at physiologic pH, i.e., bases with
high pKa's and acids with low pKa's, will cross the placenta poorly. However, this is not
always the case. Ampicillin and methicillin are strong acids and have complete placental
transfer.[11,12] With the fetal pH being 0.10 to 0.15 units lower than maternal pH, ion trapping
or fetal accumulation is possible. Weak bases will be more ionized (more dissociated) in the
fetal circulation leading to potentially higher drug concentrations than observed on the
maternal side. Fetal acidosis would pronounce this effect.[13] Protein binding is another
important determinant in placental drug transfer. This area is difficult to address due to the

number of influencing factors on drug protein binding which include temperature, pH, association constant (Ka), dissociation rate constants (K_1) and changing levels of both fetal and maternal plasma proteins as a function of gestational age. In addition, the protein binding capacity of fetal serum is decreased when compared with maternal serum. At this time, attempting to predict the influence of protein binding for a given drug on the degree of placental transfer cannot be determined.[14]

2.3. Maternal and Fetal Physiologic Factors

The amount of drug available for placental transfer is not only a function of the physicochemical properties discussed but also of maternal hemodynamics, pharmacokinetics, stage of pregnancy and uterine blood flow. As uterine blood flow increases during pregnancy, the amount of drug being transferred may increase, especially with highly lipid soluble compounds such as narcotics and psychotropic agents. Drug distribution may vary widely during pregnancy due to 30% expansion of maternal blood volume.[15] This will produce lower peak maternal drug concentrations following the same dose, which in turn decreases the driving force for transplacental passage. Similarly, gastric emptying time is delayed. This extends the time for complete drug absorption resulting in lower peak drug concentrations. Maternal renal blood flow and glomerular filtration rate increases 30 to 50% during pregnancy.[16] This would enhance the elimination rate of renally excreted drugs.

Several changes occur in the developing fetus that alter drug disposition. Examples of these include decreasing percentages of total body water and increases in body fat as gestation progresses. This would effect the distribution of lipophilic and hydrophilic drugs. Another consideration is the ability of the fetal liver to metabolize drugs. This has been established for a number of xenobiotic compounds beginning as early as the 16th wk of gestation.[17] Although fetal metabolic capacity is low and variable when compared to the adult liver, it should be taken into consideration.[18]

3. PHARMACOLOGIC MANAGEMENT OF FETAL ARRHYTHMIAS

Fetal arrhythmias represent one of the most frequent indications for transplacental pharmacologic management. Fetal tachyarrhythmias occur in 0.4 to 0.6% of all pregnancies.[19] Ten percent of these arrhythmias are considered potential sources of morbidity. Irregular rhythms, i.e., extrasystoles, account for 85% of the arrhythmias reported. Most are atrial in origin and most likely represent atrial premature beats. These rhythms are benign in 98 to 99% of the time. Approximately 10% of fetal arrhythmias are sustained rapid heart rates (>180 bpm). Supraventricular tachycardia (SVT) is the most common and most successfully treated form of fetal tachycardia.[20,21] When SVT is sustained and associated with nonimmune hydrops fetalis, the prognosis is extremely poor with perinatal mortality reaching 50 to 98%. Premature birth is encountered in approximately 80% of cases.[22,23] Hydrops fetalis associated with SVT necessitates a prompt diagnosis and aggressive management. Three approaches to the management of fetal arrhythmias presently are available: 1) deliver by caesarean section or induction of labor with the outcome of treating the neonate directly, 2) transplacental drug therapy or 3) direct injection of an antiarrhythmic into the fetus, e.g., IM, IV, or intraamniotic. Clearly, if the fetus is not at a viable gestational age, inutero therapy would be the only option.

The transplacental pharmacologic management of SVT in the human fetus primarily has involved the use of digoxin.[24,25] Digoxin achieves a variable fetal to maternal ratio (F/M) of 0.1 to 0.9. Interpretation of digoxin concentration data is complicated by the presence of digoxin like immunoreactive substance (DLIS) which is found in the fetal and maternal circulation. This substance cross reacts with digoxin immunoassays.[26] In cases where digoxin has not been effective, various other agents have been added or substituted with varying degrees of success. These include verapamil, flecainide, amiodarone, propranolol, procainamide and sotalol.[27-30]

Limited information is available concerning the transplacental passage of antiarrhythmic agents. Most of the available data were generated from case reports and generally are limited to single time point determinations of maternal and cord blood antiarrhythmic levels at the time of delivery. These data have confounding variables such as the influence of anesthesia, surgery in cesarean sections and changes in uteroplacental blood flow during labor, all of which may influence placental drug transfer. These therapeutic trials generally have preceded controlled laboratory investigations in appropriate animal models. Hence, prior to therapeutic investigations in the human fetus and neonate, it is of considerable importance to determine influences of biologic maturation on placental drug transfer, maternal and fetal drug disposition and pharmacodynamic responses.

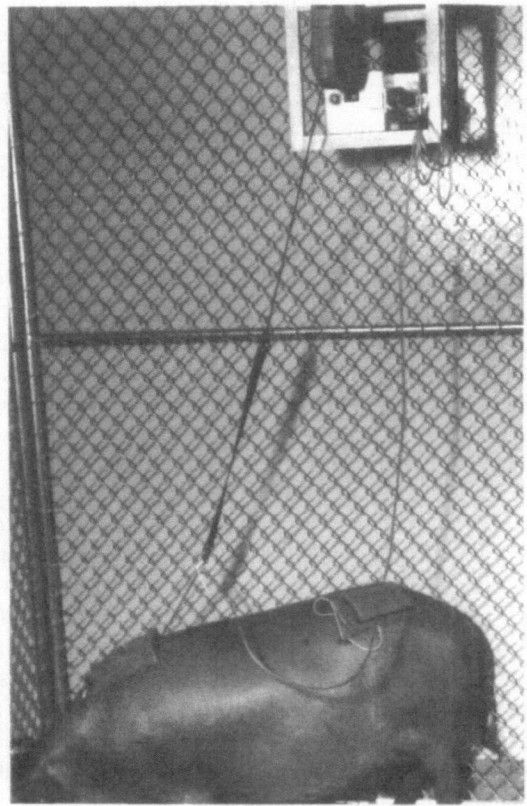

Figure 1. Intravenous moricizine infusion system displaying the syringe pump, tethering system to retain IV tubing above the sow and canvas pouch where vascular access ports are located. The infusion system is attached to the top of the cage.

4. PREGNANT YUCATAN MINIATURE SWINE FOR INVESTIGATING FETAL ANTIARRHYTHMIC THERAPY

An animal model using pregnant Yucatan miniature swine was developed to assess the feasibility of using moricizine for the treatment of resistant fetal tachycardia. Moricizine is a phenothiazine derivative with Class 1-C antiarrhythmic activity. This agent was selected because of its favorable adverse effect profile and lipid solubility which would enhance transplacental passage.

Using the surgical procedure previously described,[31] sows at 95 to 100 da gestation had 6 French silastic catheters implanted into an external jugular vein and internal carotid artery. Vascular access ports were placed on the dorsum in a canvas pouch sutured to the skin. Following 5 to 7 postoperative days, fetal surgery was performed for the placement of 4 French silastic catheters into the external jugular vein and internal carotid artery. Five days following the fetal surgery, unsedated sows received moricizine, 1 mg/kg/hr, infused IV for 5 hr using a syringe pump and weight/pulley device attached to the pen.

Serial maternal and fetal arterial samples were obtained during and post infusion. A terminal cesarean section was performed 4 hr into the infusion. Concurrent terminal moricizine samples were obtained when fetal catheter patency was lost. Immediately prior to euthanasia, fetal and maternal blood samples were obtained to characterize normal hematology, serology and serum protein.[32] Of the 17 sows and fetuses, 16 and 8 completed the protocol, respectively. Maternal sow pharmacokinetic data revealed: clearance (Cl) = 1.42 $\pm$ 0.29 l/kg/hr, volume of distribution (Vd) = 1.48 $\pm$ 0.6 l/kg, half life ($t_{1/2}$) = 0.98 $\pm$ 0.3 hr. Steady state maternal moricizine levels were 0.68 $\pm$ 0.13 mcg/ml. Fetal to maternal ratios (F/M) were 2.17 $\pm$ 0.97 (p$\leq$ 0.05). Two ewes and their fetuses [gestational age: 133 (#1) and 118 (#2) da; term 140 da] completed the protocol providing the following maternal data for #1 and #2, respectively: Cl = 1.44, 0.76 l/kg/hr; Vd = 1.44, 0.27 l/kg; $t_{1/2}$ = 0.24, 0.69, hr; steady state moricizine concentration = 0.6, 1.2 mcg/ml and F/M ratios of 0.32 $\pm$ 0.2 and 0.71 $\pm$ 0.2. Fetal moricizine concentrations were detectable within 30 min in both the swine

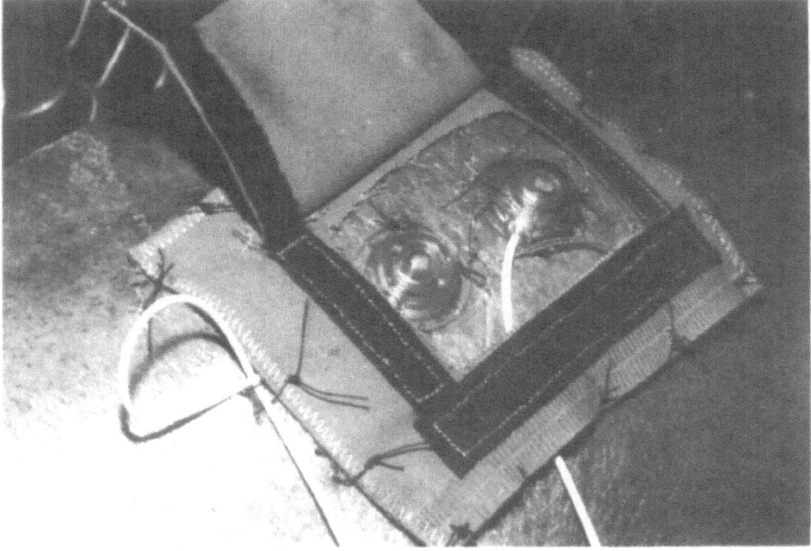

Figure 2. Close up of fetal vascular access ports attached to the sows flank.

and ewe models, demonstrating rapid transplacental passage. Upon termination of the infusion, fetal samples declined rapidly and were undetectable 2 hr post maternal infusion. Preliminary data in the ewe indicates F/M ratios were approximately 50% less than observed in the swine model. Based on the fetal uptake of moricizine after maternal administration in the swine model, further investigation into the antiarrhythmic efficacy is warranted. We also demonstrated the feasibility of using chronically catheterized Yucatan miniature swine for the investigation of other pharmacologic agents and their effects on the fetus.

5. SUMMARY

An appropriate animal model for investigating placental drug transfer is necessary due to ethical considerations. An advantage of the Yucatan miniature swine model is that the pig is a polytocous species in which individual fetuses are large enough to permit chronic catheterization and repeated blood sampling.[33] A second advantage is that the piglet and human neonate have been reported to be similar with respect to cardiovascular, respiratory, renal, hematologic[34] and hepatic mixed function oxidase systems.[35] Because the diffuse epitheliochorial type of placentation in the swine is more similar to humans then the cotyledonary placentation of the sheep, this fetal model may be more appropriate to study pharmacologic agents and their effects on the fetus.

Data generated from animal studies cannot be directly extrapolated to the pregnant human due to the differences in placentation. However, insight can be provided by these investigations into the pharmacokinetics of drugs in the MPFU. Due to the physiologic similarities between the human and swine, pharmacodynamic data gathered in both the fetus and mother is of value.

REFERENCES

1. Lenz, W., 1966, Malformations caused by drugs in pregnancy, *Am. J. Dis. Child.* 112:99-106.
2. Koren, G., 1990, Teratogenic drugs and chemicals in humans, in: Maternal-Fetal Toxicology (G.Koren, ed.), Marcel Dekker, New York, pp. 15-27.
3. Liggins, G.C., and Howie, R.N., 1972, A controlled trial of antepartum glucocorticoid treatment for the prevention of the respiratory distress syndrome in premature infants, *Pediatrics* 50:515-525.
4. Azancot-Benity, A., Jacqz-Aigrain, E., Guirgis, N.M., Decrepy, A., Oury, J.F., and Blot, P., 1992, Clinical and pharmacologic study of fetal supraventricular tachyarrhythmias, *J. Pediatr.* 121:608-613.
5. Hohlfeld, P., Daffos, F., Thulliez, P., Aufrant, C., Couvreur, J., MacAleese, J., Descombey, D., and Forestier, F., 1989, Fetal toxoplasmosis: Outcome of pregnancy and infant follow-up after in utero treatment, J. Pediatr. 115:765-769.
6. Pang, S., Pollack, M.S., Marshall, R.N., and Immken, L., 1990, Prenatal treatment of congenital adrenal hyperplasia due to 21-hydroxylase defciiency, *New Engl. J. Med.* 322:111-115.
7. Connor, E.M., Sperling, R.S., Gelber, R., Kiselev, P., Scott, G., O'Sullivan, M.J., VanDyke, R., Bey, M., Shearer, W., Jacobson, R.L., Jimenez, E., O'Neill, E., Brazin, B., Delfraissy, J.F., Culnane, M., Coombs, R., Elkins, M., Moye, J., Stratton, P., and Balsley, J., 1994, Reduction of maternal-infant transmission of human immunodeficiency virus type 1 with zidovudine treatment, *New Engl. J. Med.* 331:1173-1180.
8. Levy, G., 1981, Pharmacokinetics of fetal and neonatal exposure to drugs, *Obstet. Gynecol.* 58:9s-16s.
9. Waddell, W.J., and Marlowe, C., 1981, Transfer of drugs across the placenta, *Pharmac. Ther.* 14:375-390.
10. Challier J.C., Guerro-Millo M.,Nandakumaran, M., and Gerbaut, L., d'Athis P.H. 1985, Clearance of compouds of different moleculr size in the human placenta in vitro, *Biol. Neonate.* 48:143-148.
11. MacAulay, M.A., Abou-Sabe, N., and Charles, D., 1966, Placental transfer of ampicillin, *Am J. Obstet. Gynecol.* 96:943-953.
12. Depp R., Kind A.C., Kirby, W.M.M., and Johnson, W.L., 1970, Transplacental passage of methicillin and dicloxacillin in the foetus and amniotic fluid, *Am. J. Obstet. Gynecol.* 107:1054-1057.
13. Ward R.M., 1992, Maternal drug therapy for fetal disorders, *Semin. Perinatol.* 16:12-20.

14. Green T.P., O'Dea R.F., and Mirkin, B.L., 1979, Determinants of drug disposition and effect in the fetus, *Ann. Rev. Pharmacol. Toxicol.* 19:285-322.

15. Berlin, N.I., Goetsch, C., Hyde, G.M., and Parson, R.J., 1953, The blood volume in pregnancy as determined by P^{32} labeled red blood cells, *Surg. Gynecol. Obstet.* 97:173-176.

16. Lindheimer, M.D., and Katz, A.I., 1973, Pregnancy and the kidney, *J. Reprod. Med.* 11:14-18.

17. Juchau, M.R., 1976, Drug biotransformation reactions in the placenta, in:*Perinatal Pharmacology and Therapeutics*, (B.L. Mirkin, ed), Academic Press, New York, pp. 77-118.

18. Pelkonen, O., Kaltiala, E.H., Larmi, T.K., and Karki, N.T., 1973, Comparison of activities of drug-metabolizing enzymes in human fetal and adult livers, *Clin. Pharmacol. Ther.* 14:840-846.

19. Southhall, D.P., Richards, J., Hardwick, R., Shinebourne, E.A., Gibbens G.L., Thelwall-Jones, H., Swiet, M.D., and Johnston, P.G., 1980, Prospective study of fetal heart rate and rhythm patterns, *Arh. Dis. Child.* 55:506-511.

20. Kleinman, C.S., Copel, J.A., Weinstein, E.M., Santulli, T.V., and Hobbins, J.C., 1985, In utero diagnosis and treatment of fetal supraventricular tachycardia, *Semin. Perinatol.* 9:113-129.

21. Reed, K.L., 1989, Fetal arrhythmias: Etiology, diagnosis, pathophysiology, and treatment, *Semin. Perinatol.* 13:294-304.

22. Etches, P.C., and Lemons, J.A., 1979, Nonimmune hydrops fetalis: Report of 22 cases, including three siblings, *Pediatrics.* 64:326-332.

23. Kleinman, C.S., Donnersten, R.L., DeVore, G., Jaffe, C.C., Lynch, D.C., Berkowitz, R.L., Talner, N.S., and Hobbins, J.C., 1982, Fetal echocardiography for evaluation of in utero congestive heart failure: A technique for study of nonimmune fetal hydrops, *New Engl. J. Med.* 306:568-575

24. Nagashima, M., Asai, T., Suzuki, C., Matsushima, M., and Ogawa, A., 1986, Intrauterine supraventricular tachyarrhythmias and transplacental digitilisation, *Arch. Dis. Child.* 61:996-1000.

25. Azancot-Benisty, A., Jacqz-Aigrain, E., Guirgis, N.M., Decrepy, A., Oury, J.F., and Blot, P., 1992, Clinical and pharmacologic study of fetal supraventricular tachyarrhythmias, *J. Pediatr.* 121:608-613.

26. Weiner, C.P., Landas, S., and Persoon, T.J., 1987, Digoxin-like immunoreative substance in fetuses with and without cardiac pathology, *Am. J. Obstet. Gynecol.* 157:368-371.

27. Younis, J.S., and Menachem, G., 1987, Insufficient transplacental digoxin transfer in severe hydrops fetalis, *Am. J. Obstet. Gynecol.* 157:1268-1269.

28. Arnoux, P., Seyral, P., and Llurens, M., 1987, Amiodarone and digoxin for refractory fetal tachycardia, *Am. J. Cardiol.* 59:166-167.

29. Weiner, C.P., and Thompson, M.B., 1988, Direct treatment of fetal supraventricular tachycardia after failed transplacental therapy, *Am. J. Obstet. Gynecol.* 158:570-573.

30. Wren, C., and Hunter, S., 1988, Maternal administration of flecainide to terminate and suppress fetal tachycardia, *Br. Med. J.* 269:249.

31. Swindle, M.M., Wiest, D.B., Smith, A.C., Garner, S.S., Case, C.C., Thompson, R.P., Fyfe, D.A., and Gillette, P.C., Fetal surgical protocols in Yucatan miniature swine, *Lab. Anim. Sci.* in press.

32. Dungan, L.J., Wiest, D.B., Fyfe, D.A., Smith, A.C., and Swindle, M.M., 1995, Hematology, serology and serum protien electrophoresis in fetal miniature Yucatan swine: normal data, *Lab. Anim. Sci.* 45:285-289.

33. Randall, G.C.B., 1986, Chronic implantation of catheters and other surgical techniques in fetal pigs, in: *Swine in Biomedical Research* (M. E. Tumbleson, ed), Plenum Press, New York, pp. 1179-1185

34. Glauser E., 1966, Advantages of piglets as experimental animals in pediatric research, *Exp. Med. Surg.* 24:181-90.

35. Dvorchik, B.H., 1981, Nonhuman primates as animal models for the study of fetal hepatic drug metabolism, in: *Drug Metabolism in the Immature Human* (L.F. Soyka, and G.P. Redmond, eds.), Raven Press, New York, pp. 145-162.

55

A SWINE MODEL FOR DETERMINING THE BIOAVAILABILITY OF LEAD FROM CONTAMINATED MEDIA

Stan W. Casteel,[1][*] Ross P. Cowart,[1] Christopher P. Weis,[2]
Gerry M. Henningsen,[2] Eva Hoffman,[2] William J. Brattin,[3]
Matthew F. Starost,[1] John T. Payne,[1] Steven L. Stockham,[1]
Stephen V. Becker[4] and James R. Turk[1]

[1] College of Veterinary Medicine
University of Missouri
Columbia, Missouri 65211
[2] U.S. EPA Region VIII (8HWM-SM)
999 18th Street
Denver, Colorado 80202-2466
[3] Roy F. Weston, Inc.
215 Union Blvd
Lakewood, Colorado 80228
[4] Environmental Studies
University of Illinois
Springfield, Illinois 62794-9243

1. INTRODUCTION

1.1. Bioavailability Concepts (Absolute vs Relative Bioavailability)

Bioavailability is the portion of a chemical dose that enters the systemic circulation from an administered dosage form. Enteric absorption depends on the physical and chemical properties and associated matrix, e.g., soil, slag, food, water, of the chemical. For our purposes, two separate connotations of the term bioavailability will be clarified. Absolute bioavailability is synonymous with the oral absorption fraction (AF_o) for a specific chemical from its associated matrix. For example, if lead (Pb) from lead acetate (PbAc) is 50% absorbed from drinking water and lead from lead sulfide (PbS) is 25% absorbed from soil, the absolute bioavailabilities or AF_os of lead would be 50 and 25%, respectively. Relative

[*] Reprint requests to: Dr. Stan W. Casteel, VMDL, College of Veterinary Medicine, 1600 E. Rollins, *Columbia, MO 65211, (314) 882-6811.*

bioavailability (RBA) refers to the absorption of one chemical form compared to some other reference form. For example, if lead acetate in drinking water is the reference form of lead, then the RBA of lead from lead sulfide would be 25/50, or 50% compared to lead from lead acetate.

1.1.1. Application of Bioavailability Concepts to Risk Calculations. Use of the relative bioavailability approach is convenient for making adjustments to risk calculations at Superfund sites because risks are calculated from reference doses (RfDs) or slope factors (SFs), both of which are based on administered (not absorbed) doses of some particular chemical in some particular matrix. If environmental contaminants differ either in chemical form and/or their associated matrix, then it may be necessary to adjust toxicity values (RfD, SF) appropriately. To further illustrate these concepts, if the RfD for lead were based on exposure to lead acetate in water (50% absorption) and exposure to lead sulfide in soil (25% absorption), the adjustment would be:

$$RfD(PbS) = RfD(Pb[Ac]_2) \times AF_o(PbS)/AF_o(Pb[Ac]_2)$$
$$= RfD(Pb[Ac]_2) \times 0.5$$

This issue is particularly germane to the bioavailabilities of other toxic metals such as arsenic, mercury and cadmium and their associated matrices.

1.2. Lead Poisoning

Lead poisoning in children is a serious and preventable childhood environmental threat.[1] Increasing concern for excessive exposure of young children and pregnant women to Pb results from epidemiologic studies indicating that low levels of Pb affect fetal and childhood development. Davis and Svendsgaard[2] concluded that exposure to Pb levels sufficient to produce blood Pb concentrations of 10 to 15 µg/dl, and possibly lower, were linked to undesirable developmental outcomes in human fetuses and children. A 1988 report to Congress stated that 3 million children in the United States have blood Pb concentrations exceeding 15 µg/dl and that 4 million fetuses are estimated to be at risk for Pb toxicity during the next 10 yr.[3]

1.2.1. Adverse Effects of Lead on the CNS of Children. The detrimental impact of low level Pb exposure on children's cognition and behavior has played a key role in assessing the public health importance of this toxicant. Low levels of Pb exposure may induce delayed cognitive development, reduced intelligence quotient, impaired hearing and neurobehavioral deficiencies.[2] In response to epidemiologic and clinical studies demonstrating adverse health effects of Pb at blood concentrations as low as 10 µg/dl,[4-9] the Centers for Disease Control (CDC) recommended in 1991 that children with ≥ 15 µg/dl receive educational intervention and frequent monitoring of blood Pb levels, and that when a large proportion of children in a community have levels in the range of 10 to 14 µg/dl, Pb poisoning prevention activities be initiated.[10] Sources previously deemed safe are now being reconsidered and bioavailability of low level sources is of considerable importance.

1.2.2. Exposure of the Fetus to Lead. In utero exposure to Pb is an emerging area of concern for public health officials. Fetal exposure to Pb is inferred from maternal measurements of blood lead (PbB) during pregnancy. In pregnant women, there is a decrease in mean blood Pb from wk 12 to wk 20 of gestation and an increase from wk 20 to parturition.[11] Mobilization of bone Pb, increased gut absorption and increased resorption by the kidney were postulated to explain the upward trend in blood Pb during the second half of pregnancy. This trend in PbB was intensified in mothers with low dietary calcium intake during pregnancy. Women with higher lifetime exposures to Pb, and a consequent increase in bone Pb and calcium-deficient women may have higher PbB levels during the second half of pregnancy thereby increasing exposure of the fetus. Decreasing the risk of fetal exposure may require a lifetime reduction of Pb exposure in child bearing women.

1.3. Lead Contaminated Soils

Centuries of mining and processing have resulted in the redistribution of Pb in the environment, making it an ubiquitous multimedia contaminant. Successful reductions of Pb in environmental media such as air are the result of regulated decline in use of leaded gasoline. However, lead contaminated soil in and around urban areas remains a persistent problem.[12,13] Of specific concern is the contamination of residential soils.[14-16]

1.3.1. Soil Ingestion by Children. Soil ingestion by children is an important issue in assessing public health risks associated with exposure to lead contaminated soils. Soil ingestion is incidental to hand to mouth activity and represents the principal direct pathway for exposure to nondietary sources of lead in contaminated areas. Several studies have addressed this concern by providing estimates of the amount of soil ingested by children.[17-21] Determining the enteric bioavailability of Pb in a diverse range of soils and other media will allow scientifically derived data to dictate appropriate remedies at these sites to reduce the risk of childhood Pb exposure to acceptable levels.

1.4. Effects of Particle Size, Matrix, and Chemical Species on Enteric Lead Bioavailability

In general, the systemic availability of equivalent oral doses of Pb is matrix and chemical species dependent.[14,16,22] Specifically, the gastrointestinal absorption of multimedia sources of Pb is influenced not only by physiological factors within the digestive tract, but also by chemical and physical variables such as chemical species, particle size and matrix association.

2. ADVANTAGES OF THE SWINE MODEL FOR BIOAVAILABILITY DETERMINATIONS

2.1. Metal Bioavailability and Its Relationship to the Swine Model

The issue of metal bioavailability is presumably important at many contaminated sites, but especially so in and around mining operations and their affiliated industries. The inorganic compounds of potential concern commonly found at mining and smelting sites, e.g. lead, arsenic and cadmium compounds, often exist, in part, as poorly soluble sulfides, particles of inert mineral (tailings, waste rock) or in a vitreous matrix (slag), all of which tend to reduce the enteric bioavailability of the metal. Similar concerns exist in many inner cities and other high traffic areas where decades of leaded gasoline use have resulted in soil contamination from automobile emissions.

Most metal sulfides are much less soluble than their oxygenated analogues. Although these solubility differences contribute to differences in bioavailability, there are valid reasons why simple *in vitro* solubility measurements are an inappropriate basis for estimating the magnitude of differences in gastrointestinal (GI) absorption. First, *in vitro* systems fail to replicate adequately the dynamic conditions in the GI tract. Second, dissolved metals are removed continuously from the GI tract by active and passive absorption mechanisms, so equilibrium conditions in the GI tract are never achieved.

A somewhat more useful approach is to evaluate the dissolution kinetics of metal species from their associated matrix. Differences in the dissolution rates of metals could be important if two conditions are met: (1) dissolution (not intestinal transport) is the rate limiting step in absorption and (2) transit time through the GI tract is rapid enough that dissolution does not have time for completion. However, this method is not ideal; even if these assumptions do apply for some chemicals or media, the difficulty in replicating GI conditions *in vitro* precludes using the ratio of dissolution rates as a basis for quantitative estimation of relative bioavailability.

The most straightforward and reliable method to determine either absolute or relative bioavailability of a chemical in a medium is to measure it in either humans or animal models. Two basic experimental approaches are available. The first approach is to measure urinary and/or fecal excretion following a known dose, then calculate absorption either as 100 percent minus the percentage in feces or as the percentage in urine. This approach underestimates true absorption if there is tissue retention and/or biliary excretion, e.g. lead, but often is useful for chemicals that are both well absorbed and efficiently excreted in urine, e.g. arsenic. The second approach is to measure the area under the blood concentration vs time curve (AUC) and compare this to the AUC for the same dose given by the IV route. Alternatively, net accumulation of metals in storage tissues (bone for lead and kidney for cadmium) can be used to estimate absorption fraction. Comparison of blood or tissue response is used most often for chemicals that are absorbed less extensively and/or retained in the body for extended periods, e.g. lead, cadmium, mercury.

Use of an appropriate animal model for investigating the enteric bioavailability of Pb in young children necessitated selection based on similar age and anatomical and physiological characteristics. Pigs as animal models are remarkably similar to humans with respect to their digestive tract, nutritional requirements, bone development and mineral metabolism.[23] Immature pigs have been used successfully as a model for gastrointestinal function of children.[24] The size and tractable nature of young pigs facilitates repeated blood sampling of ample volume (5 to 7 ml) for analysis and archiving, and various surgical procedures including IV catheter placement with attached vascular access ports.

2.2. Feeding Behavior Considerations

Feeding behavior and its connection with the presence or absence of ingesta in the stomach has a large impact on the enteric bioavailability of lead. The presence of ingesta in the stomach clearly reduces the absorption of lead by our swine model as it does in humans.[12] Because rabbits are continuous feeders, the stomach of the healthy rabbit is never empty. This behavior sustains the presence of gastric flora to digest cellulose and release essential nutrients and vitamins from plant material. It follows that continuous feeding and the presence of gastric flora will buffer gastric pH thereby hindering the dissolution of lead contaminated materials. The continuous presence of ingesta and associated flora in the rabbit's stomach assures the presence of ligands for ionic lead and other metals will not be interrupted. By contrast, pigs, like humans, tend to ingest food intermittently allowing the stomach to evacuate periodically. This physiology is consistent with the way in which children most likely ingest lead contaminated materials—between meals when the gastric pH is lowest. Modeling the maximal exposure that might reasonably be expected by assessing bioavailability of lead laden soil or other material on an empty stomach is possible only in species with periodic feeding behavior.[25]

Coprophagy and the anatomical and physiological differences associated with this behavior in rats and rabbits prevents accurate bioavailability determinations because of the reingestion of lead contaminated feces. Denying this behavior leads to nutrient deficiencies in these species. Further complications are connected with the relatively high biliary excretion of Pb by rats.

2.2.1. Gastrointestinal Physiology. Calcium is absorbed in the upper small intestine by a variety of energy dependent and independent mechanisms. Active transport mechanisms for calcium in the GI tract parallel increased calcium requirements for bone growth and maturation of other tissue. Calcium binding proteins involved in these mechanisms have a similar, if not greater, affinity for lead.[26] Rapid postnatal developmental changes in the active transport mechanism for calcium across the intestinal barrier as seen in juvenile rats makes this species less desirable for the conduct of Pb bioavailability studies.[27] Sexual maturity in the rat occurs at about 7 wk of age. This developmental change in rats occurs concurrently with the cessation of active calcium transport. Since the childhood population is clearly the segment of concern for lead exposure, assessment of Pb bioavailability in an animal model during or following changes in the active absorption mechanism is inappropriate for credible understanding. The average age of puberty in swine ranges from 4 to 9 mo, a distinct advantage in subchronic exposure studies.

2.3. Other Advantages of the Swine Model

The similarity of immature swine to the childhood population in physiologic age and body weight also is an advantage as is the ease of serial blood sampling without risk of anemia.

3. MATERIALS AND METHODS

3.1. Test Animals

Fifty-two Line-26 intact males (7.5 to 8.5 kg), were purchased from a Pig Improvement Corporation (Franklin, KY) facility in Missouri. Dosing and care of the pigs were performed in compliance with the animal care and use protocol approved by the University of Missouri Animal Care and Use Committee in accordance with provisions of the "Guide for the Care and Use of Laboratory Animals", NIH publication No. 86-23, 1985. Pigs were ear tagged and placed in individual stainless steel pens with wire floors and nipple waterers and fed a commercial pelleted swine diet (Super Pig, MFA Agri Services, Columbia, MO) for the first da of acclimation. Acclimation continued for 6 additional days during which the diet was changed gradually to 100% of a specially formulated low lead (< 0.2 ppm Pb) diet (Zeigler Bros, Gardners, PA) by da -3. This nutritionally complete diet was fed in equally divided portions (half in the morning and half in the afternoon) at a daily rate of 5% of the mean body weight of all pigs in each group. This method of feeding was used to simplify the procedure and maintain a uniform body weight between pigs within a group. Feed quantities and Pb doses were weight adjusted every 3 da. In addition to daily visual inspection of the pigs, samples of blood were collected on da 0, 7 and 16 as a part of the swine health monitoring protocol for complete blood counts.

3.2. Generalized Experimental Design

As an example, an experiment was designed to determine the relative bioavailability of Pb in two soils, 1 containing 1,600 and the other 6,300 ppm Pb (analytical), by comparing them to an orally administered soluble Pb salt (lead acetate=PbAc) that was fully available for absorption. We measured the biological responses (area under the blood Pb concentration time curve, and terminal liver, kidney, and bone lead concentrations) produced by several doses of Pb from PbAc and from a lead contaminated soil. The amount of Pb in the soil that was as bioavailable as Pb from oral PbAc was calculated using the average ratio of doses of Pb from PbAc and from the test soil that produced equivalent biological responses. This method is not dependent on knowing either the absolute amount of Pb that was absorbed or how the biological response depends on the amount of Pb absorbed into the blood. It is a relative comparison of how much of the reference Pb from PbAc vs test soil Pb it required to produce identical biological responses.

On da -4, pigs were assigned randomly to one of 11 treatment groups: (Grp 1) negative control (n=2), (Grp 2) 25 μg oral Pb from lead acetate/kg body weight/da (μg PbAc/kg/da) (n=5), (Grp 3) 75 μg PbAc/kg/da (n=5), (Grp 4) 225 μg PbAc/kg/da (n=5), (Grp 5) 75 μg Pb from soil 1 matrix/kg/da (n=5), (Grp 6) 225 μg Pb from soil 1 matrix/kg/da (n=5), (Grp 7) 450 μg Pb from soil 1 matrix/kg/da (n=5), (Grp 8) 75 μg Pb from soil 2 matrix/kg/da (n=5), (Grp 9) 225 μg Pb from soil 2 matrix/kg/da (n=5), (Grp 10) 675 μg Pb from soil 2 matrix/kg/da (n=5), and (Grp11) 100 μg of Pb from intravenous PbAc/kg/da (n=5). Dosing was initiated on the morning of da 0 and terminated on da 15.

3.3. Quality Control and Validation Procedures

All studies were conducted according to Good Laboratory Practice guidelines of the EPA. Quality control samples, including blanks, duplicates, spikes and check samples from the CDC, were prepared and assigned encoded numbers in the same fashion as test samples, to mask their identity from analytical personnel. If spike recovery was outside the 80 to 120% acceptance range, a second set was prepared and reanalyzed.

Leachate from containers, reagents, solutions and equipment items were analyzed to ensure minimal Pb contamination. It is imperative that Pb concentration in these materials be minimized to prevent contamination of dosing materials or blood and tissue samples collected for analysis. Maximum acceptable concentration of Pb in animal drinking water is 10 μg/l. Leachate from containers can be prepared for Pb analysis by placing 2 ml of 2% (v/v in double distilled water) ultrapure nitric acid (Ultrex[R], JT Baker, Phillipsburg, NJ) in the container, agitating or inverting it several times so the inner surface is rinsed, then allowing the solution to stand overnight. Maximum acceptable levels of Pb in the collected 2 ml leachates over acid blanks were: 5 μg/l for blood tubes (Vacutainer[R], Becton Dickinson, Rutherford, NJ), 5 μg/l for 15 ml blue top polypropylene graduated tubes (Falcon[R] tubes, Fisher Scientific, Pittsburgh, PA), 20 μg/l for 10 ml polyethylene vials (Fisherbrand[R], Fisher Scientific, Pittsburgh, PA), 20 μg/l for teflon digestion vessels (Savillex[R], Minneapolis, MN) and 20 μg/l for plastic bags (Whirl-Pak[R], Nasco). Reagents and solutions checked for Pb include: ultra pure water (Fisher Scientific) Triton X-100 (Sigma Chem Co, St. Louis, MO), lanthanum oxide (Fisher Scientific), and ammonium phosphate dibasic (J T Baker, Phillipsburg, NJ). A 5 ml aliquot of each reagent or aqueous solution was placed in a Falcon[R] tube for Pb analysis. Maximum acceptable Pb concentration for reagents was set at 5 μg/l.

3.4. Collection and Characterization of Lead Contaminated Soils

Approximately 1 Kg of each test soil or material was collected and air dried with minimal agitation at temperatures not exceeding 60c. Soils were sieved with nylon, lead free sieves. A 10 mesh sieve was placed on top followed by a 60 mesh sieve (< 250 μm) and the collection tray. Samples were tapped and swirled, not forced or ground, until smaller particles fell through the top sieve; the 10 mesh sieve containing the coarse soil fraction was removed and discarded. Tapping and swirling of the remaining 60 mesh sieve continued until the smallest particle fraction (< 250 μm) was separated into the collection tray.

The micromineralogy of the lead bearing test soils was ascertained by electron microprobe analysis while lead concentrations were determined by x-ray fluorescence.[28]

3.5. Dosing, Feeding, and Sampling

Doses were divided and delivered daily starting at 9 am and 3 pm, 2 hr before morning (11 am start time) and afternoon (5 pm start time) feeding. To facilitate precise dosing, feeding and blood sampling, daily time details were followed with a 3 min interval allocated per pig for each procedure. For example, pig number 206 was bled between 8:00 and 8:03 am, dosed with the 1st half of the calculated daily dose between 9:00 and 9:03 am, fed with the 1st half of the calculated daily feeding (2.5% of group mean body weight) between 11:00 and 11:03 am, dosed with the 2nd half of the calculated daily dose between 3:00 and 3:03 pm and fed the 2nd half of the calculated daily feeding between 5:00 and 5:03 pm. This routine was followed sequentially for each pig on a daily basis for the specified number of dosing days, e.g., 15 da. Calculation of doses and feed was based on the group mean body weight, adjusted every 3 da. Oral delivery of PbAc was achieved by placing 20 to 100 μl of an appropriate concentration of stock solution (5, 20 or 100 μg/μl) into a depression in a 4 to 6 g mass of feed moistened with enough double distilled water to form a dough like consistency. After the PbAc solution was absorbed, the doughball was squeezed in on itself and subsequently hand fed. Pigs in each dose group received the same volume of solution based on their respective group mean body weight. Similarly, dosing with lead contaminated soil was performed by placing the soil mass (±5% weighing precision) into the 4 to 6 g mass of moistened feed as described previously. Prior to removal of soil dose aliquots, the approximately 1 l bottles containing the bulk soil samples were mixed on a roller (U.S. Stoneware, East Palestine, OH) at 8 rpm for 30 min to ensure collection of homogenous doses.

The group dosed IV with PbAc solution (Grp 11 in our example) received 100 μg Pb/kg/da in 2 equal doses (50 μg/kg/dose) by injecting 0.1 ml/kg of an aqueous solution containing 500 μg of Pb from PbAc/ml through an indwelling venous catheter with an attached subcutaneously implanted vascular access port (Access Technologies, Skokie, IL). Following dose injection, catheters were flushed with 4 to 5 ml of heparinized saline.

Blood samples for Pb analysis were collected on da -3, 0, 1, 2, 3, 5, 7, 9, 12 and 15. Five to 8 ml samples were collected from the anterior vena cava with needle and syringe and dispensed into collection tubes (Vacutainer[R], Becton Dickinson, Rutherford, NJ) containing potassium EDTA. To ensure adequate mixing of blood and anticoagulant, each tube was inverted 6 times.

Pigs were euthanized and necropsied on da 16. Gross examination included evaluation of the brain, lungs, heart and the GI and urinary tracts. Fifty to 100 g of the right medial liver lobe were removed from each pig and placed in a plastic bag (Whirl-Pak[R], Nasco), labeled and stored in the freezer at -70C until preparation for Pb analysis. The right kidney was handled in like manner while the right femur was removed and flensed before storage.

3.6. Preparation and Analysis of Biological Samples

One ml (±0.05 ml) of blood was pipetted into a polyethylene tube followed by 9 ml of matrix modifier consisting of 0.2% v/v nitric acid, 0.5% v/v Triton X-100 and 0.2% w/v ammonium phosphate in double distilled water. One g (±0.05 g) samples of kidney cortex or liver were placed in Teflon digestion containers with 2 ml of 70% nitric acid. Screw caps were applied and containers placed in a 90C oven for 12 to 18 hr. Following digestion, container contents and rinsate were placed in a 10 ml volumetric flask and brought to volume with double distilled water, then transferred to polyethylene tubes with screw caps.

Whole right femurs were placed in Coors crucibles (Fisher Scientific, Pittsburgh, PA) and dried in an oven (Precision Scientific, Chicago, IL) at 100C overnight. The dried bones were ashed in a muffle furnace (Barnstead/Thermolyne, Dubuque, IA) set at 450C for 48 hr then ground into a fine powder with mortar and pestle. Two hundred mg aliquots were removed and dissolved in 10 ml of 1:1 (v/v) ultrapure nitric acid in double distilled water. One ml of the dissolved bone solution was diluted to 10 ml in a polyethylene tube by adding 9 ml of bone matrix modifier. The bone matrix modifier consisted of an 853 ppm lanthanum solution derived from a stock solution. The lanthanum stock solution (1,706 ppm lanthanum) was prepared by dissolving 2.0 g of lanthanum oxide in about 250 ml of double distilled water, adding 160 ml of ultrapure nitric acid brought to 1.0 L volume with double distilled water. The stock solution was used to prepare the 853 ppm solution as needed by mixing one volume with an equal volume of double-distilled water.

Samples for Pb analysis were submitted to the analytical lab in blinded fashion, except for identification of the sample matrix, by assigning each prepared sample an unique encoded number. Pb analysis was by a Perkin Elmer[R] (Norwalk, CT) model 5100 spectrometer with graphite furnace. Quantitation of Pb in blood followed a modified method developed at the CDC.[29]

4. RESULTS

4.1. Speciation of Lead in Lead Contaminated Soils

Lead species present in low lead soil (1,600 ppm total Pb) consisted primarily of iron lead sulfate (35.9%), lead phosphate (21.7%), manganese lead oxide (18.1%), iron lead oxide (17.9%), and iron lead silicate (5.5%). High lead soil (6,300 ppm total Pb) contained iron lead sulfate (58%), lead phosphate (13%), iron lead oxide (11%), lead sulfate (6.3%), lead organics (5.4%), manganese lead oxide (2.7%), iron lead silicate (1.6%) and lead sulfide (1.3%).

4.2. Biological Response vs Dose

These biological responses (AUC, liver, kidney, and bone lead loading concentrations) revealed that low lead soil (1600 ppm Pb) had an absolute bioavailability ranging from 19% based on the area under the PbB concentration time curve vs dose, to 10% based on calculations from bone Pb loading vs dose. Similarly, the high lead soil (6300 ppm Pb) had an absolute bioavailability ranging from 17% based on the area under the PbB concentration time curve vs dose, to 11% based on calculations from bone Pb loading vs dose.

5. DISCUSSION

5.1. Model Application to Risk Assessment

Data from these types of studies support a departure from EPA's default assumptions regarding Pb bioavailability.[30] The impetus for this departure is to provide additional scientific evidence in support of EPA's integrated exposure uptake biokinetic (IEUBK) model and site specific data generated from Superfund site test soils. This is consistent with EPA's efforts to strengthen the scientific credibility of the agency's risk assessment procedures thereby facilitating the selection of reasonable and appropriate remedies at Superfund sites to reduce risks of childhood Pb exposure to acceptable levels.

5.1.1. Model Application Example. We were able to refine the Pb bioavailability from the EPA default value of 30 to 19% for the low lead soil and 17% for the high lead soil based on an area under the PbB time curve (AUC) vs dose. The lower bioavailabilities determined for these sites enables the consideration of higher levels of Pb in residential soils while maintaining safe exposure levels for public health protection. As a consequence, public health protection can be achieved with considerably less economic impact on industry and government.

Increasing importance of quantitative risk assessment, and the associated regulations permitting some level of risk, emphasizes the necessity for greater accuracy in these determinations and the need for accurate measurements of effective dose. The utility of this model lies in its ability to reduce the uncertainties associated with risk assessment of contaminated media in general by determining the actual bioavailability of the toxicant in a species closely related to humans.

6. REFERENCES

1. U.S. EPA, Office of Pollution Prevention and Toxics, 1993, EPA acting to prevent childhood lead poisoning, *Chems. Prog. Bull.* 14:1-17.
2. Davis, J. M., and Svendsgaard, D. J., 1987, Low-level lead exposure and child development, *Nature* 329:297-300.
3. U.S. HHS, Agency for Toxic Substances and Disease Registry, Public Health Service, 1988, The nature and extent of lead poisoning in children in the United States: A report to Congress, Atlanta, GA.
4. Schwartz, J., and Otto, D., 1987, Blood lead, hearing thresholds, and neurobehavioral development in children and youth, *Arch. Environ. Health* 42:153-160.
5. Bellinger, D., Leviton, A., Waternaux, C., Needleman, H., and Rabinowitz, M., 1987, Longitudinal analyses of prenatal and postnatal lead exposure and early cognitive development, *New Engl. J. Med.* 316:1037-1043.
6. Dietrich, K. N., Krafft, K. M., Bornschein, R. L., Hammond, P. B., Berger, O., Succop, P. A., and Bier, M., 1987, Low-level fetal lead exposure effect on neurobehavioral development in early infancy, *Pediatrics* 80:721-730.
7. McMichael, A.J., Baghurst, P. A., Wigg, N. R., Vimpani, G. V., Robertson, E. F., and Roberts, R. J., 1988, Port Pirie cohort study: Environmental exposure to lead and children's abilities at the age of four years, *New Engl. J. Med.* 319:468-475.
8. Needleman, H. L., and Gatsonis, C. A., 1990, Low-level lead exposure and the IQ of children: A meta-analysis of modern studies, *J. Am. Med. Assoc.* 263:673-678.
9. Needleman, H. L., Schell, A., Bellinger, D., Leviton, A., and Allred, E. N., 1990, The long-term effects of exposure to low doses of lead in childhood: An 11-year follow-up report, *New Engl. J. Med.* 322:83-88.
10. Centers for Disease Control, 1991, Preventing lead poisoning in young children. A statement by the Centers for Disease Control, 4th rev., CDC, Atlanta, GA.
11. Rothenberg, S. J., Karchmer, S., Schnaas, L., Perroni, E., Zea, F., and Alba, J. F., 1994, Changes in serial blood lead levels during pregnancy, *Environ. Health Perspect.* 102:876-880.

12. Rabinowitz, M. B., and Bellinger, D. C., 1988, Soil-lead blood-lead relationship among Boston children, *Bull. Environ. Contam. Toxicol.* 41:791-797.

13. Aschengrau, A., Beiser, A., Bellinger, D., Copenhafer, D., and Weitzman, M., 1994, The impact of soil lead abatement on urban children's blood lead levels: Phase II results from the Boston lead-in-soil demonstration project, *Environ. Res.* 67:125-148.

14. LaVelle, J. M., Poppenga, R. H., Thacker, B. J., Giesy, J. P., Weis, C., Othoudt, R., and Vandervoort, C., 1991, Bioavailability of lead in mining wastes: An oral intubation study of young swine. *Chem. Spec. Bioavail.* 3:105-111.

15. Cook, M., Chappell, W. R., Hoffman, R. E., and Mangione, E. J., 1993, Assessment of blood lead levels in children living in a historic mining and smelting community, *Am. J. Epidemiol.* 137:447-455.

16. Gulson, B. L., Davis, J. J., Mizon, K. J., Korsch, M. J., Law, A. J., and Howarth, D., 1994. Lead bioavailability in the environment of children: Blood lead levels in children can be elevated in a mining community, *Arch. Environ. Health.* 49:326-331.

17. Binder, S., Sokal, D., and Maughan, D., 1986, Estimating the amount of soil ingested by young children through tracer elements, *Arch. Environ. Health* 41:341-345.

18. Clausing, P., Brunekreff, B., and van Wijnen, J. H., 1987, A method for estimating soil ingestion in children, *Int. Arch. Occup. Environ. Med.* 59:73-82.

19. Calabrese, E., Barnes, R., Stanek, E.J., 1989, How much soil do young children ingest: An epidemiologic study, *Reg. Toxicol. Pharmacol.* 10:123-137.

20. Davis, S., Waller, P., Buschom, R., Bailou, J., and White, P., 1990, Quantitative estimates of soil ingestion in normal children between the ages of 2 and 7 years: Population-based estimates using aluminum, silicon, and titanium as soil tracer elements, *Arch. Environ. Health* 45:112-122.

21. van Wijnen, J. H., Clausing, P., and Brunekreef, B., 1990, Estimated soil ingestion by children. *Environ. Res.* 51:147-162.

22. Davis, A., Ruby, M. V., and Bergstrom, P. D., 1992. Bioavailability of arsenic and lead in soils from the Butte, Montana, mining district, *Environ. Sci. Technol.* 26:461-468.

23. Dodds, J. W., 1982, The pig model for biomedical research, *Fed. Proc.* 41:247-256.

24. Miller, E. R., and Ullrey, D. E. 1987, The pig as a model for human nutrition, *Ann. Rev. Nutr.* 7:361-382.

25. Weis, C. P., and LaVelle, J. M., 1991, Characteristics to consider when choosing an animal model for the study of lead bioavailability, *Chem. Speciation Bioavail.* 3:113-119.

26. Fullmer, C.S., Edelstein, S. and Wasserman, R.H., 1985, Lead-binding properties of intestinal calcium binding proteins, *J. Biol. Chem.* 260:6816-6819.

27. Mooradian, A.D., and Song, M.K., 1989, Age-related alterations in duodenal calcium transport in rats, *Mech. Ageing Dev.* 47:221-227.

28. Davis, A., Drexler, J. W., Ruby, M. V., and Nicholson, A., 1993, Micromineralogy of mine wastes in relation to lead bioavailability, Butte, Montana, *Environ. Sci. Technol.* 27:1415-1425.

29. Miller, D. T., Paschal, D. C., Gunter, E. W., Stroud, P. E., and D'Angelo, J., 1987, Determination of lead in blood using electrothermal atomization atomic absorption spectrometry with a L'vov platform and matrix modifier, *Analyst* 112:1701-1704.

30. National Research Council, 1994, Default Options, In Science and Judgement in Risk Assessment, Ch. 6, pg. 105.

56

EVALUATION OF SINCLAIR MINIATURE SWINE AS AN OSTEOPENIA MODEL

Guy F. Bouchard,[1*] Rogely W. Boyce,[2] Carol L. Paddock,[2]
Edward Durham[1] and Chada S. Reddy[3]

[1] Sinclair Research Center, Inc
Columbia, Missouri 65203
[2] Rhône-Poulenc Rorer Pharmaceuticals
Collegeville, Pennsylvania
[3] Veterinary Biomedical Sciences
University of Missouri
Columbia, Missouri 65211

1. INTRODUCTION

Postmenopausal osteoporosis is a chronic, disabling disease. The high prevalence of osteoporosis in elderly people, particularly women, is becoming increasingly important as baby boomers are reaching retirement age and life expectancy in industrialized countries is increasing.[1,2] In addition to the reduced quality of life suffered by affected people, osteoporosis is a substantial contributor to the burden of health care costs frequently due to prolonged treatment and hospitalization from osteoporotic fractures.[1,2] Although risk factors contributing to osteoporotic fractures, such as gonadal failure following menopause, lower peak bone mass, lack of physical activity, genetic and nutrition, are well identified, the deficit in understanding of osteoporosis pathogenesis makes the search for therapeutic agents difficult.[3,4]

Understanding the pathogenesis of osteoporosis has been hampered by lack of predictable animal models. Traditionally, the rat has been the species most commonly used in the study of osteoporosis due to practical advantages such as reduced cost, short lifespan and the availability of genetically well defined stocks or strains.[5,6] Several considerations, including a lack of cortical remodeling, delayed epiphyseal closure in the aging animal, absence of age related osteopenia, small body size, low blood volume and high metabolic rate, are serious limitations to the rat model.[5,6] In addition, the pharmacokinetics of drugs is often different between rats and humans. In that respect, the rat generally is reserved for screening or pilot projects. Miniature swine have been evaluated as a possible osteopenia

* Reprint requests to: Dr. Guy Bouchard, Sinclair Research Center, Inc., 5701 South Sinclair Rd, Columbia, MO 65203, (314) 446-6464.

model. They are polyestrous, omnivorous, small in body size and have lamellar bone and trabecular and cortical remodeling similar to humans. In addition, the anatomy and physiology of several organ systems such as skin, cardiovascular, gastrointestinal and urogenital are similar to humans.[7,8] Different routes of administration (oral, parenteral or transdermal) of tested compounds are performed easily.

The combination of ovariectomy and mild dietary calcium restriction resulted in 7 to 10% reduction in miniature gilt vertebral bone mass which was associated with alteration in cancellous bone microstructure.[9] The reduction in trabecular bone and the alteration in microstructure appeared primarily due to trabecular perforation. The perforation of trabecular elements occurred in concert with exaggerated resorptive cell function at the level of the remodeling unit. A similar pathogenesis for microstructural changes occurring in women around menopause has been proposed. Calcium restricted and ovariectomized Sinclair miniature swine appear to be a useful model of osteopenia and trabecular plate perforation and a promising model for the study of the influence of microstructural changes on bone biomechanics. In an attempt to evaluate the suitability of miniature swine as an animal model to study osteopenia, we have conducted a series of experiments to further define miniature swine and to standardize their use in osteoporosis research.

2. PEAK BONE MASS

The first experiment consisted of determining the peak bone mass of Sinclair miniature swine and to evaluate the variation in body composition in different age groups of Sinclair(S-1) miniature swine. Peak bone mass of a species is an index of skeletal maturity and it is central into the development of an osteopenia animal model. To achieve this objective, we used 48 female Sinclair miniature swine. Sow age ranged from 297 to 2815 da. Animals were divided into 8 groups based on age. Age ranges in da for each age group were as follows: 297 to 372, 509 to 583, 734 to 826, 903 to 994, 1114 to 1117, 1223 to 1331, 1442 to 1490 and 2100 to 2815, which corresponded approximately to 1, 1.5, 2, 2.5, 3, 3.5, 4 and >6 yr, respectively.

Bone mineral content (BMC), bone mineral density (BMD) and body composition consisting of lean mass (LM) and body fat (BF) were measured using a dual energy x-ray absorptiometer (DEXA; Hologic QDR-2000 Plus). Sedated animals were placed supine with the front and hind limbs maintained in cranial and caudal positions using adhesive tape. The lumbar spine was scanned at the level of L1 through L4; a whole body scan also was performed. The proximal to midshaft portion of the left femur was scanned by pulling the limb in a 45° angle craniolaterally while maintaining the animal supine. All results are presented as mean ± SD. Age related differences in BMC, LM, BF, BMC+LM, body weights, percent of fat for the whole body and BMD for lumbar spine were determined using analysis of variance. Means were separated using the protected LSD rule with $\alpha = 0.05$.

Overall, BMC of the combined first 4 lumbar vertebrae (L1 to L4) increased until it peaked at 3 and 4 yr of age (P<0.05) and decreased at 6 yr (P<0.05). BMD of L1 to L4 increased until 2.5 yr of age (P<0.05) and leveled off thereafter (P>0.05). When vertebrae were considered individually without regard for age difference, the BMD of L1 and L2 was unchanged with age (P>0.05); whereas, that for L3 and L4 decreased progressively (P<0.05). The BMD of L3 and L4 also were different from each other for all ages (P<0.05).

Whole body BMC, LM and BMC added to LM (BMC+LM) increased until age 3 (P<0.05) and leveled off thereafter (P>0.05). The whole body fat and body weight increased up to 3 yr of age (P<0.05) and leveled off thereafter (P>0.05) with the exception of 3.5 yr old miniature sows, which had a lower body fat and body weight than the other age groups older than 3 yr. DEXA estimated body weight correlated well with actual body weight (r^2=0.99519, P=0.0001).

Based on lumbar spine BMD and whole body BMC and LM, female Sinclair(S-1) miniature swine reach adulthood between 2.5 and 3 yr of age. Lumbar BMC and femoral BMC and BMD peak between 2 and 4 yr of age and decrease after 4 yr of age presenting a trend similar to the bone change profile seen in women. Increase in body weight after 3 yr of age, although not significant, was due to body fat accumulation which could be corrected with diet adjustment. BMD of lumbar vertebrae decreased progressively from L1 through L4, although L1 and L2 had similar bone mineral density.

3. SHORT TERM BONE LOSS IN YOUNG ADULT MINIATURE SWINE

The purpose of this study was to evaluate changes in vertebral cancellous bone microstructure, histomorphometry and calcitropic hormones following treatment with 2 therapeutic bone agents in a model of osteopenia in miniature swine. We examined dose and regimen dependent effects of salmon calcitonin and a conjugated estrogen on vertebral structure and cancellous bone remodeling in calcium restricted ovariectomized Sinclair miniature swine. Previously, we demonstrated that mild dietary calcium restriction in combination with estrogen deficiency results in a 7 to 10% decrease in spinal BMD which is accompanied by an increase in erosion depth and alterations in vertebral trabecular connectivity.[9] Because this animal mimics aspects of microstructural changes in peri-menopausal bone loss, it was utilized further to characterize changes in the 3 dimensional distribution of connectivity in response to calcium restriction and estrogen deficiency and to assess the modulating effects of these antiresorptive therapies. Seventy two Sinclair minipigs, age 4 mo, were fed diets containing either 0.9 or 0.75% calcium for 6 mo. At 10 mo of age, pigs were either sham operated (Sham) or ovariectomized (OVX). During the 6 mo postOVX period, pigs were fed their respective diets and assigned to a treatment group as listed in table 1.

Following termination at 16 mo of age, BMD was measured on excised specimens of vertebrae (T12 to L2) with a Lunar DPX-L. Appropriately sampled histological sections of lumbar vertebrae were prepared and connectivity was estimated using the ConnEulor principle.[10] Histomorphometric analyses were performed according to the methods of Eriksen.[11] To attempt to correlate changes in cancellous bone turnover with secondary changes in calcitropic hormones in response to selected treatments, a separate study was conducted to examine the acute effects of salmon calcitonin on serum calcium, parathyroid hormone (PTH) and calcitriol. Age matched minipigs were administered either placebo, 1.67 or 0.83 IU/kg salmon calcitonin and sequential serum samples were collected for analyses.

Vertebral BMD was reduced by approximately 6% in the OVX group compared with Sham 0.9%. BMD was increased in 3 of the 4 salmon calcitonin treatment groups and in the conjugated estrogen group. Erosion depth was increased in the OVX group and treatment with salmon calcitonin or conjugated estrogen reduced erosion depth. Bone formation rate, tissue referent and activation frequency estrogen, estimates of cancellous bone turnover were increased in OVX pigs administered 0.83 IU/kg salmon calcitonin 5 da/wk. These values also tended to be increased in the other salmon calcitonin treatment groups. These changes likely were related to the acute decrease in serum calcium and increases in serum PTH and calcitriol which were demonstrated to occur following calcitonin administration. Vertebral connectivity density was reduced in the OVX group compared with the 2 sham control groups

Table 1. Experimental design for the short term bone loss in young adult
Sinclair miniature swine experiment

Group	Surgery	Dietary Calcium	Treatment[§]	Regimen
1	Sham operated	0.9%	Placebo	5 da/wk
2	Sham operated	0.75%	Placebo	5 da/wk
3	Ovariectomized	0.75%	Placebo	5 da/wk
4	Ovariectomized	0.75%	Cltn: 1.67 IU/kg (100 IU/60 kg)	5 da/wk
5	Ovariectomized	0.75%	Cltn: 1.67 IU/kg (100 IU/60 kg)	3 da/wk
6	Ovariectomized	0.75%	Cltn: 0.83 IU/kg (50 IU/60 kg)	5 da/wk
7	Ovariectomized	0.75%	Cltn: 0.83 IU/kg (50 IU/60 kg)	3 da/wk
8	Ovariectomized	0.75%	CE: 0.625 mg/60 kg	5 da/wk

[§]Cltn = salmon calcitonin; CE = Conjugated Estrogen.

with loss of trabecular elements being largely concentrated under the vertebral endplates. Treatment salmon calcitonin or conjugated estrogen generally blunted this reduction.

4. ACCELERATED BONE LOSS IN MINIATURE SWINE

The objective of this preliminary study was to develop an accelerated bone loss model in miniature swine. Bone loss occurs over a prolonged period of time in existing large animal models. A large animal model with rapid bone loss would be useful for the evaluation of drugs aiming at restoring bone loss in advanced cases of osteoporosis or to mimic bone loss suffered due to paralysis or other mobility deficits in the human. To achieve this objective, we used 2 injured Sinclair miniature sows with limited use of their posterior limbs, causing immobilization. Miniature sow BI had clinical signs of possible luxation or fracture of the lumbar vertebrae and miniature sow RGT had a ruptured gastrocnemius tendon. Miniature sow BI was mostly in lateral recumbency and had to be supported daily to eat, while miniature sow RGT could sit and move around without discomfort using her front legs. Bone density and body composition of the animals was assessed regularly during the following 7 mo with a dual energy X-ray absorptiometer (Hologic 2000+). During the 7 mo period, miniature sow BI lost approximately 12.4 kg of body weight of which 11.1 kg was fat, 1.0 kg was lean mass and 0.3 kg was bone mass representing a loss of 21.4% of her body weight, 55.2% of her body fat, 2.9% of her lean body mass and 18.2% of her bone mass. During the same period, miniature sow RGT gained 0.4 kg of body weight. She actually lost 1.3 kg of body fat and 0.05 kg of bone mass and gained 1.7 kg of lean body mass representing a gain of 0.5% in body weight and 4.4% in lean mass and a loss of 3.4% of her body fat and 3.3% of her bone mass. Both animals (BI and RGT) lost BMD in the lumbar spine (L1 to L4), to the extent of 12.9 and 6.5%, respectively. Greater loss of BMD, however, occurred in the femur (global analysis), 18.7 and 20.3% for miniature sows BI and RGT, respectively. The majority of bone loss was seen within 2 mo after study initiation or 3 mo after the injuries occurred. The immobilized miniature sow may mimic the bone changes observed in hemiplegic or paraplegic human patients and could be used as an animal model to study bone loss following these conditions as well as bed rest, limited exercise or disuse. Gastrocnemectomy would appear to be the better animal model since 1) no wastage was seen in the animal with tendon rupture, 2) the animal was comfortable in spite of the injury, and 3) bone mass loss was similar to the recumbent miniature sow suffering from a back injury.

5. CONCLUSION

Sinclair miniature swine reach full bone maturity between 2.5 to 3 yr of age. The miniature swine osteopenia model presented in this manuscript uses a young adult. Although growth artefacts may be seen in the histomorphometry analysis, this miniature swine osteopenia model provides bone loss in only 6 mo with changes in connectivity similar to what is seen in women early during menopause. Also, they respond to 2 therapeutic agents used in the treatment of osteoporosis. We suggest that miniature swine are suitable as a potential model of accelerated bone loss to evaluate therapies using anabolic bone agents.

6. REFERENCES

1. Norris, R.J., 1992, Medical costs of osteoporosis, *Bone* 13:S11-S16.
2. Kanis, J.A., and Pitt, F.A., 1992, Epidemiology of osteoporosis, *Bone* 13:S7-S15.
3. Chestnut, C.H., 1991, Theoretical overview: bone development, peak bone mass, bone loss, and fracture risk, *Am. J. Med.* 91(Suppl. 5B):2S-4S.
4. Heaney, R.P., 1991, Effect of calcium on skeletal development, bone loss, and risk fractures, *Am. J. Med.* 91(Suppl. 5B):23S-28S.
5. Kalu, D.N., 1991, The ovariectomized rat model of postmenopausal bone loss, *Bone Mineral* 15:175-192.
6. Rodgers, J.B., Monier-Faugere, M.-C., and Malluche, H., 1993, Animal models for the study of bone loss after cessation of ovarian function, *Bone* 14:369-377.
7. Hannon, J.P., Bossone, C.A., and Wade, C.E., 1990, Normal physiological values for conscious pigs used in biomedical research, *Lab. Anim. Sci.* 40:293-298.
8. Swindle, M.M., 1984, Swine as a replacements for dogs in the surgical teaching and research laboratory, *Lab. Anim. Sci.* 34:383-385.
9. Mosekilde, L., Weisbrode, S.E., Safron, J.A., Stills, H.F., Jankowsky, M.L., Ebert, D.S., Danielsen, C.C., Soggaard, C.H., Franks, A.F., Stevens, M.L., Paddock, C.L., and Boyce, R.W., 1993, Calcium-restricted ovariectomized Sinclair(S-1) minipigs: an animal model of osteopenia and trabecular plate perforation, *Bone* 4:379-382.
10. Gundersen, H.J.G., Boyce, R.W., Nyengaard, J.R., and Odgaard, A., 1993, The ConnEulor: unbiased estimation of connectivity using physical dissectors under projection, *Bone* 14:217-222.
11. Eriksen, E.F., 1986, Normal and pathological remodeling of human trabecular bone: three dimensional reconstruction of the remodeling sequence in normals and metabolic bone disease, *Endocrine Rev.* 7:379-408.

57

NEOVASCULARIZATION OF THE ISCHEMIC MYOCARDIUM BY CARDIOMYOPLASTY

Its Study Using the Casting Method and Selective Acute Myocardial Infarction in Swine

Jose M. Borrego, MD, PhD,[*] Antonio Ordonez, MD, PhD,
Ana Hernandez, MD, and Jose Perez, MD, PhD

Division of Cardiovascular Surgery and Center for Cardiovascular
  Research
Hospital Universitario "Virgen del Rocio"
U. Los Minaretes 3-4-D, 41020 Seville, Spain

1. INTRODUCTION

In spite of medical and surgical advances, ischemic cardiopathy is a disease with a dramatic evolution and increasing incidence in western industrialized countries. Due to a life expectancy increase, new effective antianginous drugs and advances in intervention cardiology, myocardial ischemic pathology is experiencing major transformations. Incidence, clinical types of presentation and therapeutic attitudes are changing. Ischemic cardiopathy presentation frequently is diagnosed in elderly patients. With improving diagnostic evaluation, numerous patients with multiple vascular affection, who do not allow any kind of direct intervention, are detected. Therefore, ischemic cardiopathy and decreased ventricular function are associated more and more frequently.

Elderly patients with multiple affected vessels, many coronary atheromatic lesions and severe ventricular dysfunction, with no possibility of conventional surgical revascularization, are numerous. For such patients with advanced ischemic cardiopathy, refractory to pharmacological treatment and without possibility of revascularization by using conventional techniques, heart transplant would be their final therapeutic option. Nevertheless, the advanced age of these potential recipients, together with donor shortage, are decisive limitations. Consequently, heart transplant is becoming more and more selective for a determined patient group. Cardiomyoplasty has been introduced as a surgical alternative to heart transplant for these patients. Latissimus dorsi

[*] Reprint requests to: Dr. Jose M. Borrego, Division of Cardiovascular Surgery and Center for Cardiovascular Research, Hospital Universitario "Virgen del Rocio", U. Los Minaretes 3-4-D, 41020 Seville, Spain.

muscle which is transformed and transferred into thorax, enveloping the heart and beating in a synchronized way, has been shown to be effective for biological ventricular assistance.[1-3]

We expound a new cardiomyoplasty application for patients with end stage ischemic cardiopathy, not revascularizable and with contraindicated heart transplant. Therefore, it is necessary to achieve a reliable, reproducible acute infarction model in swine to be able to study numerous new techniques in advanced cardiology and cardiac surgery.

Traditionally, the model used has been artery coronary occlusion, both chronic and acute, requiring advanced surgical equipment and highly trained staff.[4,5] Furthermore, any subsequent surgical procedure is difficult due to adhesion formation in the thoracic cavity. Another disadvantage is that the infarcted zone size is unpredictable. Because of these reasons, postoperative morbidity/mortality is high when this method is used.[6-8] In some experiments,[9] an angioplasty balloon equipped catheter was used. Once inside the coronary artery, the balloon was inflated to induce a reversible myocardial ischemia. Although this method has definite advantages, there is a risk of causing greater damage than expected to vessel wall during occlusion with the balloon.

There is another experimental method which does not require open chest surgery; however, it also involves catheterization. In this latter method, coronary lesions are produced by the injection of microspheres into the coronary arterial system.[10] These lesions are irreversible, they may be distal, depending on size of microspheres used, and are multiple. These features mean this model is more similar to human diffuse coronary disease than that produced by occlusion of a major coronary artery. The study of diffuse coronary disease is of great importance given the difficulties experienced in surgical revascularization of patients with this condition. Results achieved using these methods to create an experimental model of acute myocardial infarction are satisfactory. The major drawback is the high mortality secondary to serious arrhythmias, especially as a result of postinfarction ventricular fibrillation. This mortality reaches almost 100% when pigs are used as experimental subjects.[10]

We used swine as the biological model to create an acute myocardial infarction experimental model by injecting microspheres into the coronary tree. To increase survival rate and improve results achieved to date with these techniques, we modified the generally used model; the pharmacological treatment has been intensified to avoid high mortality rate resulting from arrhythmias. Also, we evaluated myocardial revascularization through neo-formed vessels from latissimus dorsi muscle enveloping the heart[11-13] and determined if the neoformed vessels were functional, effective and sufficient to revascularize and take oxygen and nutrients to ischemic myocardium.

2. MATERIALS AND METHODS

To conduct this research study, we used an experimental model of myocardial infarction on which we performed a cardiomyoplasty and morphological and functional studies after 4 mo. In 10 Large White 26 kg swine, anesthesia was induced with IV nembutal doses of 50 mg/kg body weight and maintained with halothane and air through orotracheal intubation connected to a respirator. Selective acute myocardial infarction was accomplished by means of microsphere impact. By using this experimental model, we tried to approach ischemic cardiopathy with no possibility of revascularization and multiple occlusion of small caliber vessels.[10] A microsphere impact on the distal coronary vascular region instantaneously led to acute ischemia and acute myocardial infarction. Myocardial necrosis was determined by electrical and enzymatical assessment. To prevent ventricular fibrillation, IV lidocaine perfusion was used prophylactically in doses of 2 mg/kg body weight 10 min prior to infarction.

The procedure was begun by introducing a Judkins 5F catheter through the right femoral artery and into the left coronary artery ostium. The Seldinger technique was used and catheterization was monitored radioscopically. Catheter position was checked by selective left coronariography after injecting the contrast medium. To maintain patency, a pressure bag containing 500 ml physiological serum with 2500 U Na heparin was connected to the catheter throughout catheterization.

A Tracker 3F catheter was introduced through the washing system valve and guided radioscopically. Once inside the anterior descending artery, the catheter was guided until the middle/distal third of the artery was reached. This catheter was flushed continuously with a solution of 500 ml physiological serum containing 500 U Na heparin. We used 250 to 366 micron contour emboli microspheres. Prior to intracoronary injection, they were diluted in radiological contrast medium so their introduction into the selected could be confirmed. When the catheter position was confirmed, 5 ml of the microsphere suspension were introduced into the artery via the catheter. Each animal received a prophylactic dose of 5,000,00 IU penicillin.

Once a biological model group with no revascularizable ischemic cardiopathy was achieved, cardiomyoplasty, using left latissimus dorsi muscle, was performed on each pig. To prove our hypothesis that cardiomyoplasty can be a source of neoformed vessels nourishing the ischemic myocardium, morphological and functional studies were conducted.

2.1. Morphological Study

Several techniques were used. We did selective angiographies of the grafted muscle artery, checking radiologically neovessels and performing simultaneously a coronariography to show the relation between both vascular trees. Angioarchitecture of vascular system plastic cast (casting method), both coronary systems and latissimus dorsi muscle vessels is effective for macroscopic study[14] showing the infarcted area and interconnections between the cardiomyoplasty vascular system and ischemic myocardium in different colors in the plastic cast (Fig. 1). Using scanning electron microscopy (SEM), the methacrylate cast obtained after corrosion or casting was studied.[15-19] New vessel morphology and characteristics could be observed (Fig. 2), showing their interconnections and guessing the origin of their formation (vascular buds previously existing from collateral coronaries or new vessels). Using optical microscopy, neovessel size, infarcted area location, neoformed vessel density and quantity per surface area, and their penetration from epicardium to endocardium, were studied. By injecting latissimus dorsi muscle with a coloring (India ink), we confirmed the existence of flow through the cardiomyoplasty (Fig. 3). By studying the India ink marked piece, neoformed vessel diameter and structure were evaluated, to assess the probability of maintaining adequate flow. To perform morphological techniques, the other 5 animals were injected with India ink as a vascular marker. An angiographic assessment was performed on each animal.

2.2. Functional Study

Revascularization effectiveness was determined by studying regional venous flow in 10 animals under 2 conditions. For total basal flow, our biological model was an ischemic heart with cardiomyoplasty performed 4 mo earlier. Therefore, total basal flow was the sum of coronary artery flow plus subscapular artery flow of embedded muscle. Also, we evaluated postocclusion or temporary postligation of the artery which hypothetically was to provide blood flow to the cardiomyoplasty. According to Unger et al.[20] flow provided to a region is equal to differential flow, meaning the existing

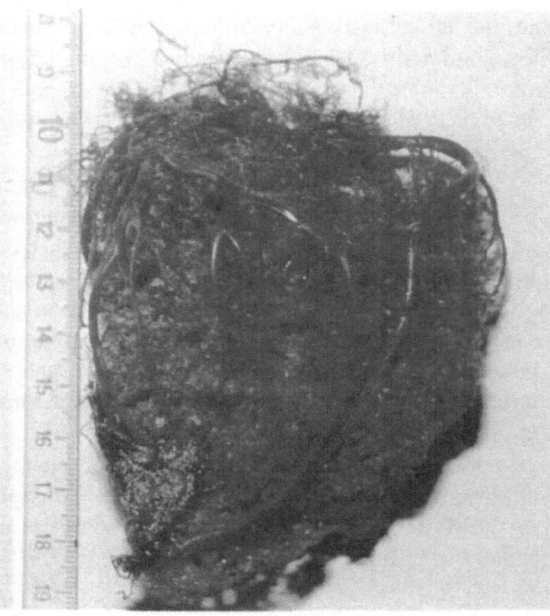

Figure 1. Plastic vascular cast of cardiomyoplasty vascular tree (red) and latissimus dorsi (blue).

difference between basal and subscapular artery postocclusion flow. To carry out this functional study, a #11 (Nihon Kodhen) electromagnetic flow meter was placed into the coronary sinus. Average blood flow circulating through this coronary venous system was measured continuously. The device was inserted a few cm before the union of

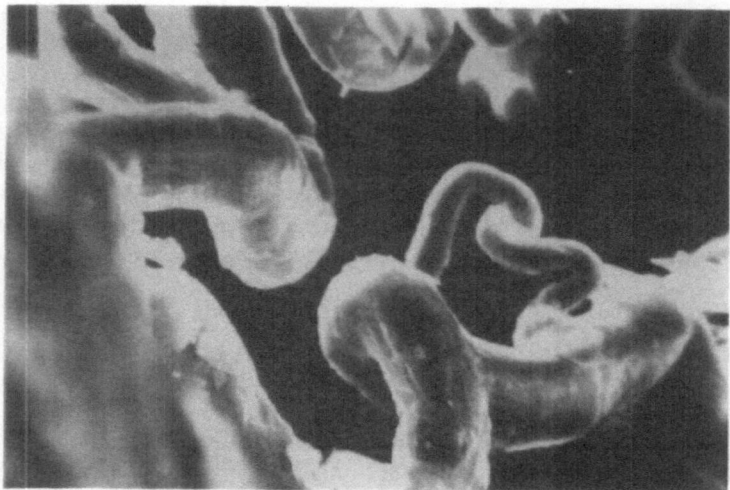

Figure 2. Visualization of neoformed vessels through scanning electron microscopy, noting it sinusoidal path and its penetration toward myocardial thickness.

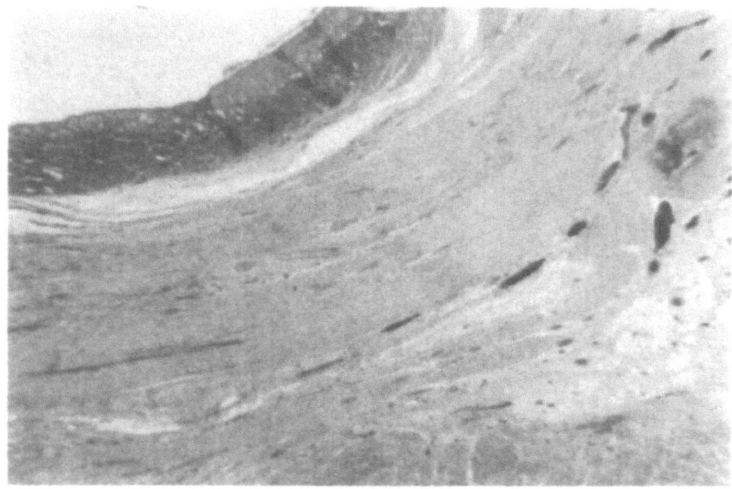

Figure 3. By injecting latissimus dorsi muscle with a coloring (India ink), with the help of optical microscopy, we detect the coloring into ischemic myocardium to confirm the existence of flow through cardiomyoplasty.

right auricle and cava to take measurements with no possibility of contamination by flows from other vessels.

3. RESULTS

During injection of the microspheres, all animals developed sinusoidal tachycardia (170 ± 20 beats/min) and a drop in arterial pressure (60 ± 15 mm Hg). In the electrocardiogram, this was reflected as an increase in the ST segment of the heart anterior aspect. All animals, treated prophylactically with lidocaine 10 min before performing the procedure, survived acute myocardial infarction. The episodes of extrasystole were less numerous and none of them suffered tachycardia nor ventricular fibrillation; all of the animals survived the acute phase. Four wk after performing the procedure, these animals continued to survive and a new ECG was performed. Signs of infarction, Q waves and inversion of the T waves in the anterior aspect of the heart, were observed. As a result of development of this experimental model, we proved there was ischemic myocardium revascularization, starting from cardiomyoplasty, in 100% of the cases. Average waiting time between cardiomyoplasty and neovessel confirmation was 120 ± 30 da.

From a morphological point of view, neovessels were detected macroscopically with angiographies and the casting method. Also, they were shown through optical, electron and scanning microscopies. Revascularized area corresponded to infarcted area. From a functional viewpoint, flow increase to ischemic myocardium was confirmed in 100% of the cases. Average flow provided by neovessels was 51 ± 7 ml/min/100 g myocardial tissue. Basal flow, the sum of coronary and subscapular artery flows, and postocclusion flow, which is

Table 1. Basal flow achieved from 10 experiences, as well as postligation and differential flow.

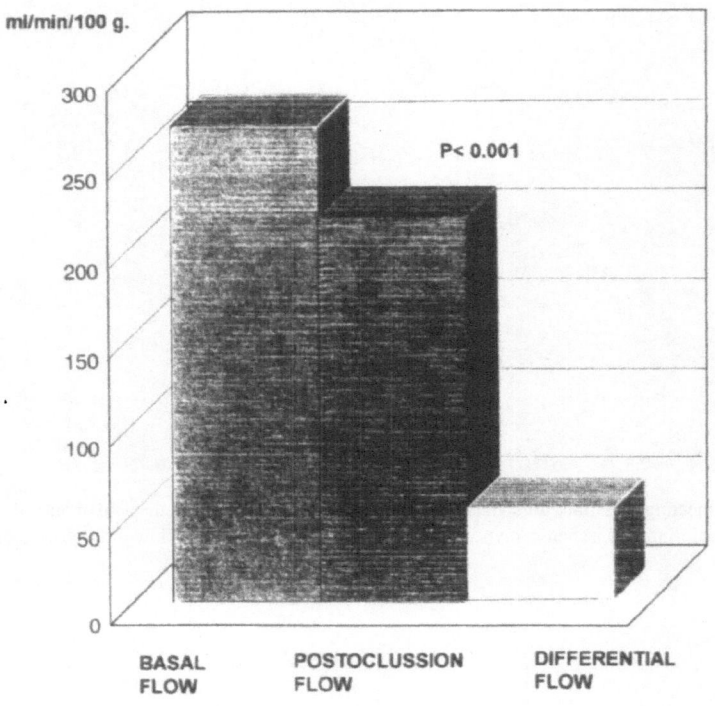

equal to flow responsible only for coronaries once cardiomyoplasty flow is removed, are listed in Table 1. There was a difference (p<0.001) between basal flow (266 ml/min/100 g) and postocclusion flow (215 ml/min/100 g).

4. DISCUSSION

Our aim was to develop a selective acute myocardial infarction swine model without open chest surgery. By injecting microspheres into coronaries postsurgery, morbidity and mortality rates (respiratory insufficiency, infection, mediastinitis, hemorrhage, sternal dehiscence, etc) are avoided. These occur with myocardial infarction models using coronary artery ligation. The present method has been developed to achieve a myocardial infarction biological model that could be used for surgical procedures without necessity of a second operation.

After injecting microspheres, there was a multifocal necrosis area representing an irreversible lesion. This did not occur when an angioplasty balloon was used. With the latter technique, a larger caliber vessel was occluded temporarily, an ischemia being induced which was reversible once the balloon was withdrawn.[9] For this reason, this technique usually is used for acute phase studies of myocardial ischemia.

Using this highly selective technique of catheterization, we reduced the infarction size, the occluded vessel diameter depending only on the size of microspheres used and on the volume of suspension injected. Furthermore, multiple lesions can be produced in the microcirculation, this providing closer approximation to diffuse coronary disease in humans and enabling important studies to be conducted. We modified the technique described by other authors[10] who used injected microspheres to induce myocardial infarction. The highly selective catheterization of the coronary artery with a Tracker 3F catheter allowed us to reach the middle or distal third of the vessel. This allowed a more precise selection of the infarcted zone and improved postoperative morbidity/mortality.

As indicated by survival rates, combining this technique with the prophylactic use of lidocaine 10 min before performing the procedure was found to give the best results. As a result of its electrophysiological properties,[21,22] this drug was especially useful in ischemic and ventricular fibers. It reduced ventricular excitability and increased the threshold of ventricular fibrillation under conditions of infarction. These effects also have been studied in other drugs, such as Amyodarone.[23]

The technique of highly selective catheterization, although somewhat more complicated than the conventional technique, provided a more limited zone of infarction to be produced. This, together with prophylactic treatment of the animals with lidocaine, allowed the creation of a high survival rate biological model of acute selective myocardial infarction without open chest surgery. There have been many attempts to revascularize ischemic myocardium throughout cardiovascular surgery evolution. Some years before the aorta-coronary graft technique was started, Vineverg managed to revascularize an ischemic heart by using pericardiopexia with talcum. Myocardial blood flow was increased by means of scarifications in cardiac muscle, provoking inflammatory reactions with vascularization increase, and even directly implanting an inner mammary artery in ischemic muscle.

Cardiomyoplasty is indicated for patients with end stage cardiopathy and severe impairment of ventricular function without any other surgical alternative, including heart transplant.[24,25] Patients with more actual indication will be those suffering dilated myocardiopathy and not revascularizable ischemic cardiopathy, with serious ventricular dysfunction, elderly and/or for whom heart transplant is contraindicated. Cardiomyoplasty effectiveness to improve ventricular function in helping insufficient myocardium by means of the contraction of another muscle previously trained in it, has been shown on numerous occasions, both experimentally and clinically.[26-28] To date, its usefulness for successful myocardial revascularization has not been described. In our experience, myocardial infarction was induced; afterwards, cardiomyoplasty, using latissimus dorsi muscle, was performed on 10 pigs. Myocardial revascularization, only in the ischemic area, was shown in 100% of the cases, both morphologically and functionally.

Vessel neoformation, starting from cardiomyoplasty, can be explained through 2 mechanisms. First, because of the myocardium's need for oxygen, its own myocardial ischemia is the best stimulus for blood vessel formation. Adenosine presence in ischemic muscle favors formation and development of vascular buds coming from nonischemic nearby areas. Second, any inflammatory process involves vascularization increase through a cicatrization phenomenon. Cicatrix is one of the areas containing a greater number of vessels throughout the human organism since new tissue formation implies appearance of new blood vessels. Implanting a healthy, well vascularized muscle in an ischemic area, like a myocardial infarction region, has advantages for the appearance of new vessels, oxygen supply and the cicatrization process.

In conclusion, cardiomyoplasty not only improves ventricular function but also favors ischemic muscle revascularization. We can add a new cardiomyoplasty application for patients without a possibility of conventional surgical revascularization.

5. REFERENCES

1. Lorusso, R., La Canna, G., Metra, M., Zogno, G., Giubini, R., Bonandi, L., Picchioni, A., Marzollo, P., Visioli, O., and Alfieri, O., 1994, Dynamic cardiomyoplasty as a biomechanical support in end-stage chronic heart failure: a debated surgical option, *J. Cardiothorac. Vasc. Anesth.* 8(Suppl. 2):94.

2. Kratz, J.M., Johnson, W.S., Mukherjee, R., Hu, J., Crawford, F.A., and Spinale, F.G., 1994, The relation between latissimus dorsi skeletal muscle structure and contractile function after cardiomyoplasty, *J. Thorac. Cardiovasc. Surg.* 107:868-878.

3. Lorusso, R., Zogno, M., La Canna, G., Metra, M., Sandrelli, L., Borhette, V., Maisano, F., and Alfieri, O., 1993, Dynamic cardiomyoplasty as an effective therapy for dilated cardiomyopathy, *J. Cardiac. Surg.* 8(Suppl.):177-183.

4. Vanoli, E., Hull, S.S., Jr., Foreman, R.D., Ferrari, A., and Schwartz, P.J., 1994, Alpha-1-adrenergic blockade and sudden cardiac death, *J. Cardiovasc. Electrophysiol.* 5:76-89.

5. Zimmer, H.G., 1992, Development and modulation of experimental right ventricular hypertrophy in rats, *J. Cardiovasc. Pharmacol.* 20(Suppl. 1):S1-S6.

6. Lindpaintner, K., Lu, W., Niedermajer, N., Schieffer, B., Just, H., Ganten, D., Drexler, H., 1993, Selective activation of cardiac angiotensinogen gene expresion in postinfarction ventricular remodeling in the rat, *J. Molec. Cell. Cardiol.* 25:133-143.

7. Hartford, C.G., Rogers, G.G., Marcos, E.F., Rosendorff, C., 1993, Influence of Ketanserin on regional myocardial blood flow after myocardial infarction in baboons, *J. Cardiovasc. Pharmacol.* 21:144-148.

8. Van-Woerkens, L.J., Van-der-giessen, W.J., and Verdouw, P.D., 1992, The selective bradycardic effects of zatebradine (UL-FS 49) do not adversely affect left ventricular function in conscious pigs with chronic coronary artery occlusion, *Cardiovasc. Drug Ther.* 6:59-65.

9. Cohen, M.V., Yang, X.M., Liu, Y., Snell, K.S., and Downey, J.M., 1994, A new model of controlled coronary artery occlusion in conscious rabbits, *Cardiovasc. Res.* 28:61-65.

10. Beyer, M., Eggeling, T., Hoffer, H., Mierdl, S.E., Beyer, U., and Hannekum, A., 1992, Experimental selective myocardial infarction in the dog without open-chest-surgery, *Res. Exp. Med.* 192:169-175.

11. Mannion, J.D., Buckman, P.D., Magno, M.G., and Dimeo, F., 1992, Collateral blood flow from skeletal muscle to normal myocardium, *J. Surg. Res.* 53:578-587.

12. Beyer, M., Hoffer, H., Eggeling, T., Goertz, A., Mierdl, S., and Hannekum, A., 1992, Cardiomyoplasty to improve myocardial collateral blood supply as an alternative to transplantation in intractable angina, *J. Heart Lung Transplant.* 11:S189-S191.

13. Mannion, J.D., Magno, M.G., Buckman, P.D., DiMeo, F., Greene, R., Bowers, M., McHugh, M., and Menduke, H., 1993, Acute electrical stimulation increases extramyocardial collateral blood flow after a cardiomyoplasty, *Ann. Thorac. Surg.* 56:1351-1358.

14. Ordonez, A., Rafel, E., Perez-Bernal, J.B., Borrego, J.M., Hernandez, A., and Gutierrez, E., 1996, Angioarchitecture of the coronary tree: contributions to its study with the casting method, *Res. Surg.* In Press.

15. Aharinejad, S.H., and Lametschander, A., 1992, Microvascular corrosion casting, in: Scanning Electron Microscopy. Techniques and Applications, *Edi. Springer Neurscheinumg* pp. 300.

16. Lametschander, A., Lametschander, U., and Weiger, T., 1990, Scanning electron microscopy of vascular corrosion cast technique and applications: updated revies, *Scanning Microsc.* 4:889-940.

17. Aharinejad, S., Bock, Lametschander, A., Franz, P., and Firbas, W., 1991, Sphincters in the rat pulmonary veins. Comparison of scanning electron and transmission electron microscopic studies, *Scanning Microsc.* 5:1091-1096.

18. De Andres, A.V., Munoz-Chapuli, R., Sans-Coma, V., and Garcia-Garrido, L., 1992, Anatomical studies of the coronary system in elasmobranchs: II. coronary arteries in hexanchoid, squaloid, and carcharhinoid sharks, *Anat. Rec.* 233:429-439.

19. Konerding, M.A., 1991, Scanning electron microscopy of corrosion casting in medicine, *Scanning Microsc.* 5:851-865.

20. Under, E.F., Sheffield, C.D., and Epstein, S.E., 1990, Creation of anastomoses between an extracardiac artery and the coronary circulation. Proof that myocardial angiogenesis occurs and can provide nutritional blood flow to the myocardium, *Circulation* 82:1449-1466.

21. Bellemin-Baurreau, J., Poizot, A., and Armstrong, J.M., 1994, An in vitro method for the evaluation of antiarrhythmic and antiischemic agents by using programmed electrical stimulation of rabbit heart, *J. Pharmacol. Toxicol. Methods* 31:31-40.

22. Bertini, G., Giglioli, C., Rostagno, C., Conti, A., Russo, L., Taddei, T., and Paladini, B., 1993, Early out-of-hospital lidocaine administration decreases the incidence of primary ventricular fibrillation in acute myocardial infarction, *J. Emerg. Med.* 11:667-672.

23. Anastasiou-Nana, M.I., Nanas, J.M., Rapti, A., Poyadjis, A., Stathaki, S., and Mouleopulos, S.D., 1994, Effects of amiodarone on refractory ventricular fibrillation in acute myocardial infarction: experimental study, *J. Am. Coll. Cardiol.* 23:253-258.

24. Taguchi, S., Yozu, R., Iseki, H., Shimizu, H., Takahashi, R., and Kawada, S., 1994, A new method of biventricular assistance using a single skeletal muscle powered ventricle (preliminary study), *Pn. J. Artif. Organs* 23:10-13.

25. Chiu, R.C.-J., Odim, J.N.K., Burgess, J.H., Blundell, P.E., Steward, J.A., Rabinovitch, M.A., Williams, H.B., Robinson, R.J.S., Templeton, A., Fetterley, L., and Lu, A.J., 1993, Responses to dynamic cardiomyoplasty for idiopathic dilated cardiomyopathy, *Am. J. Cardiol.* 72:475-479.

26. Capouya, E.R., Gerber, R.S., Drinkwater, D.C., Pearl, J.M., Sack, J.B., Aharon, A.S., Barthel, S.W., Kaczer, E.M., Chang, P.A., and Laks, H., 1993, Girdling effect of nonstimulated cardiomyoplasty on left ventricular function, *Ann. Thorac. Surg.* 56:867-871.

27. Chachques, J.C., Acar, C., Tapia, M., Guibourt, P., Fiemeyer, A., Bensasson, D., Berrebi, A., Grare, P., Bechara, M., Baron, J.F., and Carpentier, A., 1994, Intermediate results of cardiomyoplasty, *Arch. Mal. Coeur. Vaiss.* 87:49-56.

28. Bellotti, G., Moraes, A., Bocchi, E., Arie, S., Medeiros, C., Moreira, L.F., Jatene, A., and Pileggi, R., 1993, Late effects of cardiomyoplasty on left ventricular mechanics and diastolic filling, *Circulation* 88:304-308.

58

VACCINATION STUDY WITH THE SINCLAIR MINIATURE SWINE - EFFECT OF VACCINE DOSE AND LITTER

Guy Bouchard,[1]* Edward Durham,[1] Boh Chang Lin,[2] Susan Turnquist,[3] and Chada Reddy[3]

[1] Sinclair Research Center, Inc.
Columbia, Missouri 65203
[2] MPV Laboratories, Inc.
Ralston, Nebraska 68127
[3] University of Missouri
Columbia, Missouri 65211

1. ABSTRACT

Prophylactic protection of miniature swine, critical to the outcome of research using these animals, received little attention in the past. Lack of approved commercially available vaccines contributed to our lack of knowledge of the efficacy of vaccination and the observed anaphylactoid reactions following the use of conventional swine bacterin. To evaluate the serologic response and clinical effect of various doses of a commercial vaccine (*Actinobacillus pleuropneumoniae-Erysipelothrix rhusiopathiae*), 20 titer negative male Sinclair miniature swine (between 41 and 43 da old) from 5 litters were divided randomly in 4 groups based on their litter of origin and were administered 2 IM doses of the vaccine (21 da apart) at 2 and 2, 1 and 2, 1 and 1, and 0.5 or 1 ml. Behavior changes and rectal temperatures were monitored following vaccination. Two different muscle vaccination sites were used to detect any immediate or delayed local reactions. Body weights and blood samples were collected at 0, 21 and 42 da. Serology titers were determined using an ELISA assay specific to *Actinobacillus pleuropneumoniae* serotypes 1, 5 and 7, which are present in the commercial vaccine. The highest dose regimen yielded a higher serologic titer than the other regimens only 21 da after the second vaccination. The litter of origin had a strong effect on post vaccination serologic titers with some litters having very low titers mainly after the first vaccination. No clinical effects were noted following vaccination, but nodules were palpated at the vaccination site which gradually regressed within 2 mo. Body weights and rectal

* Reprint requests to: Dr. Guy Bouchard, Sinclair Research Center, Inc., 5701 South Sinclair Rd, Columbia, MO 65203 (314) 446-6464.

temperature elevation did not affect the titers. The litter of origin heavily influences the outcome of the vaccination and the highest dose regimen is necessary to provide the best protection.

2. INTRODUCTION

The preventive medicine programs designed for miniature swine are scarce and generally are designed from their larger cousins, domestic swine.[1] No commercial vaccines currently are approved for miniature swine.[1] The efficacy of the domestic swine vaccine in miniature swine remains undetermined and anaphylactoid reactions have been observed after prophylactic administration of commercially available bacterin.[2]

Anaphylactic and anaphylactoid reactions are not a rare problem in swine following vaccination,[3] but the severity of the reaction encountered in Sinclair miniature swine following routine vaccination suggests that miniature swine might be more sensitive to the large dose of bacterin designed for larger domestic swine.[2] Nineteen 3 wk old Sinclair miniature piglets from 3 different litters and weighing 2.5 to 4 Kg, were vaccinated with 2 commercial bacterin (*Bordetella bronchiseptica*, *Erysipelotrix rhusiopathiae* and *Pasteurella multocida* bacterin* and *Haemophilus parasuis*, *Haemophilus pleuropneumonia* and *Pasteurella multocida* bacterin[†]) according to label instructions. Less than 1 hr following vaccination, the miniature pigs were vomiting and dyspneic and exhibited ataxia followed by lateral recumbency. Eleven of the piglets died despite intensive therapy with epinephrine, antihistamine, corticosteroid and oxygen.[2]

The objectives of this study were to evaluate the effects of different doses of a commercial bacterin and to identify the effect of other variables such as litter and body weights on the humoral response and clinical signs of Sinclair miniature swine following vaccination.

3. MATERIALS AND METHODS

Twenty male Sinclair miniature swine were divided randomly in 4 groups based on their litter of origin and body weight; they originated from 5 different litters. All swine were between 41 and 43 da old at the initiation of the study.

The treatment consisted of administering 2 IM doses, at 21 da apart, of the commercially available combined *Actinobacillus pleuropneumoniae-Erysipelothrix rhusiopathiae* vaccine.[‡] The volumes of the vaccine for the first and second dose for groups 1, 2, 3 and 4 were of 2 and 2, 1 and 2, 1 and 1, or 0.5 and 1 ml, respectively. Animals were observed for behavior changes following vaccination and rectal temperature was monitored both before and 45 min after vaccination. The first and second vaccination sites, in the right and left hind leg muscle, respectively, were examined for any immediate or delayed local reactions. Body weights and blood samples were collected at 0 (before first vaccination), 21 (before booster vaccination) and 42 da. Serology titers were determined using an ELISA assay specific to *Actinobacillus pleuropneumoniae* serotypes 1, 5 and 7, present in the commercial vaccine. Serology titers were not determined for *Erysipelothrix rhusiopathiae*.

---

* Titan 3, Pitman-Moore, Mundelein, Illinois

† Parapleuroshield-P, Grand Laboratories, Inc. Larchwood, Iowa

‡ Pneumopac-ER, Shering-Plough Animal Health, Kenilworth, NJ.

3.1. Preparation of Microtiter Plates

A whole culture of *Actinobacillus pleuropneumoniae* (App), serotypes 1, 5 and 7 was prepared by inoculating fresh medium (250 ml of TSB supplemented with fresh yeast extract, bovine serum and NAD) with frozen stock of App (2 ml) and incubating at 37C overnight, then using a 6 ml inoculum from the overnight culture to reinoculate a fresh medium (TSB as above) for 3 to 40 hr at 37C. The concentration of cells was adjusted to 1×10^9 cells per ml and cells were killed with formalin. The whole killed culture was stored at 4C overnight before use.

One hundred µl of the whole App culture was added to each well of a 96 well microtiter plate. Plates were incubated overnight at 4C and then washed 3 times with PBS (150 µl per well). Plates were blocked with 100 µl per well of a blocking buffer containing PBS with 1% bovine serum albumin. Plates were incubated for 30 min at 37C and washed 3 times (150 µl per well) with PBS containing 0.05% Tween 20 (PBS + T). They were then ready for use in the ELISA.

3.2. Performing the ELISA

Seventy µl of a dilution buffer (0.05% Tween 20, 0.5% bovine serum albumin in PBS) was added to all wells in a 96 well microtiter plate that had not been coated. Seventy µl of a diluted serum (test serum as well as control serum was diluted in advance at 1:50 in a sterile 5 ml glass tube with a dilution buffer) was added to the first 2 well row (8 samples per plate) and 2 fold dilutions were made using a Titertek multichannel pipette to make 1:100 to 1:3200 dilutions. The PBS + T buffer was shaken gently out of the freshly prepared coated plates and 50 µl of serum sample of all dilutions was transferred from the dilution plate to the coated plate in duplicate. Plates were incubated at 37C for 90 min in a moist pan covered with foil. Plates were washed 3 times with PBS + T. Fifty µl of conjugate antibody (alkaline phosphatase conjugate goat antiswine IgG obtained from KPL) at 1:1000 (10 µg/10 ml diluent buffer) dilution was transferred in each well of the coated plate. Plates were incubated at 37C for another 90 min in a humid environment. Plates were washed 3 times with PBS + T. One hundred µl/well of substrate solution (2 tablets of Sigma 104-105 phosphatase substrate in 10 ml of diethanolamine buffer which was prepared by dissolving 97 ml of diethanolamine, 0.2 g of sodium azide and 0.1 g of magnesium chloride in 800 ml of distilled water, pH 9.8) was added and the plates were incubated at room temperature. The color reaction was stopped by adding 30 µl per well of 5N NaOH as OD_{410} of the positive control wells (diluted 1:100) reached approximately 1.30.

The assay was not considered valid if the OD_{410} of the negative control wells (diluted 1:100) reached 0.30. Each plate contained the following control wells: (1) wells with a known negative pig serum sample, (2) wells with a known positive pig serum sample and (3) wells in which pig serum was replaced by diluent buffer. Any serum sample at any dilution was determined to have positive reactivity with App antigens as long as its OD_{410} reading at that particular dilution was above the cut off point of 0.40. The antibody titer of the test serum was defined as the reciprocal of the highest dilution of the test serum showing positive reactivity.

3.3. Statistics

The effect of litter and treatment (different vaccine doses) were tested using analysis of variance. Means were compared using the least significant difference (LSD) rule, with $\alpha = 0.05$. Variables included were the rectal temperature differential (related to both vaccination), serologic titers at 21 and 42 da, serologic titer differential (between 42 and 21 da), and body weights (0, 21 and 42 da). The statistical analysis for the serologic titer differential was not different compared to the serologic titer at day 42. Thus, the results of the serologic titer differential were not reported. The rectal temperature differential was calculated by substracting the rectal temperature measured 45 min after vaccination from the one measured right before vaccination. Regression analysis between serology titers and rectal temperature differential was calculated. Pearson correlation coefficients also were calculated for all variables listed. Because there were only 2 piglets in litter number 4, we calculated the statistics for the effect of litter with and without litter number four. Both sets of statistics had similar findings, thus, we only reported the set of statistics that included all 5 litters.

4. RESULTS

All serologic titers were negative before the first vaccine inoculation for the 3 serotypes of *Actinobacillus pleuropneumoniae*. On da 21, each treatment group had elevated serologic titers, but not different (P>0.05). However, on da 42, serologic titers continued to rise and treatment group 1 had higher serologic titers than the other treatment groups (P<0.05). No differences in the serologic titers were seen between treatment groups 2, 3 and 4 (P>0.05). The treatment did not have any effect on body weight or rectal temperature differential Table 1.

The litter of origin had a significant effect on the serologic titers on da 21 and 42 (P<0.05). Some litters had very poor responses to vaccine inoculation while others responded well to the challenge. Furthermore body weights on da 0 and 21 were different between litters (P<0.05), but rectal temperature differentials were not (Table 2).

Serologic titers correlated negatively with rectal temperature differential on da 21 based on regression analysis (P<0.05). As rectal temperature differential increased, serologic titer decreased. The same correlation was not seen on da 42 (P>0.05). Based on the Pearson correlation coefficients, body weights did not correlate with serologic titers (P>0.05) for any time period. No behavior changes were noted in piglets following vaccination. None of the piglets manifested signs of anaphylactic or anaphylactoid reaction or appeared depressed after vaccination. Palpable nodules in muscles at the vaccination site were present 2 wk following vaccination in 19 of 20 piglets. Nodules measured 0.5 to 1.5 inches in diameter and did not seem to hinder the piglets in their physical activity. The nodules completely regressed in all piglets within 2 mo without complications.

5. DISCUSSION

Only 41 to 43 da old male piglets were used in this study because of availability. The commercial vaccine that we used is recommended for 28 da old piglets with a booster 21 da later. However, it is important to realize that domestic piglets are about 15 to 20 kg at 28 da of age while the Sinclair miniature swine are only 3 to 5 kg. In this study, the miniature swine used weighed between 2.7 and 6.7 kg.

The vaccine dose had an effect at 42 da. Group 1, which received the highest doses of vaccine at 0 and 21 da (2 ml), had the highest serologic titers at 42 da. Group 2 which received 1 and 2 ml at 0 and 21 da, respectively, had lower serologic titers at 42 da compared to group 1 and similar serologic titers to the other groups which received lower vaccine doses. The higher vaccine dose at day 0 of group 1 must have carried a priming effect to yield a higher serologic titer at day 42 in this group.[5]

A strong litter effect was noticed on serum titers at both time periods evaluated while the treatment effect was significant only at 42 da. Body weights did not correlate with

Table 1. Effect treatment on serology, body weight and rectal temperature differential (Mean ± SD).

Trtmt Grps[†]	N	Serologic Titers Da 0	Serologic Titers Da 21[‡]	Serologic Titers Da 42[‡]	Body Weight Da 0	Body Weight Da 21	Body Weight Da 42	Rectal T° Diff.[§] Da 21	Rectal T° Diff.[§] Da 42
1	5	0	360 ± 296	4160 ± 2147^a	5.58 ± 0.58	7.98 ± 1.13	8.46 ± 1.43	1.26 ± 1.18	0.44 ± 0.23
2	5	0	240 ± 329	1920 ± 716^b	5.00 ± 0.96	7.22 ± 1.51	7.72 ± 1.22	1.02 ± 0.36	0.48 ± 0.71
3	5	0	200 ± 187	1520 ± 1073^b	4.86 ± 1.47	7.06 ± 2.41	7.58 ± 2.57	0.94 ± 0.77	0.76 ± 0.55
4	5	0	120 ± 164	1760 ± 876^b	4.82 ± 1.31	7.14 ± 1.67	7.48 ± 1.95	1.92 ± 1.09	0.64 ± 0.31

[†] Treatment groups.
[‡] Means with different superscripts within the same column are different.
[§] Rectal temperature differential.

Table 2. Effect of litter of origin on serology, body weight and rectal temperature differential (Mean ± SD).

Trtmt Grps[†]	N	Serologic Titers Da 0	Serologic Titers Da 21[‡]	Serologic Titers Da 42[‡]	Body Weight Da 0	Body Weight Da 21[‡]	Body Weight Da 42	Rectal T° Diff.[§] Da 21	Rectal T° Diff.[§] Da 42
1	4	0	425 ± 287^a	4000 ± 1600^a	5.00 ± 1.04^{a,b}	6.98 ± 1.91^{b,c}	8.08 ± 2.06	1.35 ± 1.14	0.53 ± 0.59
2	5	0	440 ± 219^a	2880 ± 2086^b	5.30 ± 0.33^{a,b}	7.24 ± 1.00^b	8.44 ± 1.35	0.84 ± 0.82	0.64 ± 0.33
3	5	0	100 ± 100^b	1360 ± 537^b	6.02 ± 0.73^a	8.88 ± 1.02^a	8.22 ± 1.21	1.48 ± 0.91	0.48 ± 0.58
4	2	0	50 ± 71^{a,b}	800 ± 0^b	3.15 ± 0.64^c	4.65 ± 1.34^c	4.30 ± 0.99	0.55 ± 0.64	1.20 ± 0.42
5	4	0	25 ± 50^b	2000 ± 800^b	4.60 ± 1.00^b	7.30 ± 1.06^{a,b}	8.00 ± 1.14	1.90 ± 0.84	0.38 ± 0.22

[†] Treatment groups.
[‡] Means with different superscripts within the same column are different.
[§] Rectal temperature differential

serologic titers. Thus, the litter effect was not due to the difference in body weights observed at 0 and 21 da. Litter effect may be an underlining cause to vaccine failure. Vaccine failures generally are attributed to unsatisfactory vaccine, improper administration, disease incubation, immunosuppression, passive immunization or biological variation.[4]

No clinical signs of anaphylactoid or anaphylactic reactions were observed following vaccination. Anaphylactic reactions are of type 1 or immediate hypersensitivity. Anaphylactoid reactions mimic anaphylactic reactions. These reactions are not mediated by immune mechanisms and are caused by the direct action of toxic substances such as endotoxin.[5] No prior sensitization is necessary in anaphylactoid reactions as opposed to anaphylaxis. Clinical signs of anaphylaxis induced experimentally in pigs include dyspnea, coughing, yawning, incoordination, patchy erythema, retching, vomiting, tenesmus and edema of the face and eyelids.[6] Gross lesions in these pigs included gastric fundic hemorrhage, pulmonary edema, mesocolic edema, gastric submucosal edema, effusion into serous cavities and intestinal hyperemia.

The loss of Sinclair miniature piglets reported previously was likely to be due to anaphylactoid reactions since they were never vaccinated before.[2] These animals received 2 vaccines; the vaccine may have caused the anaphylactoid reaction. We conducted a small study using 6 unrelated male Sinclair miniature piglets that received either the bacterin containing *B. bronchiseptica, E. rhusiopathiae* and *P. multocida* (group A, n=2), or the bacterin containing *H. parasuis, A. pleuropneumoniae and P. multocida* (group B, n=2), or, lastly, both bacterin in combination (group C, n=2). Within 20 min, all of the piglets in groups A and C were vomiting and exhibiting ataxia, thus implicating the bacterin containing *B. bronchiseptica, E. rhusiopathiae* and *P. multocida*. None of the piglets which received the other bacterin manifested any clinical signs of anaphylaxis. The anaphylactoid reaction in the miniature piglets was not due to the extra dose of bacterin but the presence of specific antigen present in one of the bacterin.

Nodules were present in the muscles at the vaccination site. The nodules regressed slowly without apparent clinical disturbance during the 2 mo following the vaccine administration. Attention should be paid to the small size of miniature swine during vaccination and different vaccination sites should be sought to avoid additional inflammation at the vaccination site. Oil based adjuvants are known to cause more inflammation and granuloma or abscess formation at the injection site than insoluble salt adjuvants.[7] An oil base adjuvant is used in the *Actinobacillus pleuropneumoniae* commercial vaccines because it yields better serologic titers than insoluble salt adjuvants.[7]

The highest vaccine doses yielded better serologic titer for *Actinobacillus pleuropneumoniae*; the litter of origin contributed to the difference in serologic titers observed and the oil based adjuvant produced nodules which regressed without clinical disturbances within 2 mo after the vaccine administration.

6. REFERENCES

1. Braun, W., 1993, Helping your clients raise healthy potbellied pigs, *Vet. Med.* 88:414-428.
2. Turnquist, S.E., Bouchard, G., and Fisher, J.R., 1993, Naturally occurring systemic anaphylactic and anaphylactoid reactions in four groups of pigs injected with commercially available bacterin, *J. Vet. Diagn. Invest.* 5:103-105.
3. Blood, D.C., and Radostits, O.M., 1989, Allergy and anaphylaxis. in: *Veterinary medicine: A textbook of the diseases of cattle, sheep, pigs, goats, and horses,* 7th Edition (D.C., Blood, and O.M., Radostits, eds.), University Press, Oxford, England, pp. 89-91.
4. Tizard, I., 1987, *Veterinary Immunology: An Introduction*, 3rd Edition, W.B. Sounders, Philadelphia, PA, pp. 185-199.

5. Eyre, P., 1972, Pharmacological aspects of hypersensitivity I domestic animals: a review, *Vet. Res. Commun.* 4:83-98.
6. Nielsen, N.O., 1986, Edema disease, in: Diseases of swine, 6th Edition, (A.D. Leman, B. Straw, W. Mengeling, S. D'Allaire, D. Taylor), Iowa State University Press, Ames, IA, pp. 528-540.
7. Straw, B.E., Shin, S., Callihan, D., and Petersen M., 1990, Antibody production and tissue irritation in swine vaccinated with *Actinobacillus* bacterin containing various adjuvants, JAVMA 196(4):600-604.

59

EARLY CHARACTERIZATION OF PANEPINTO MICRO/MINIATURE SWINE FOR USE AS TRANSGENIC ANIMAL MODELS

Victoria Hampshire,[*] John Bacher, Melvin Dennis, Axel Wolff, and Melissa Yarko

National Center for Research Resources
Veterinary Resources Program
National Institutes of Health
Bethesda, Maryland 20892

1. ABSTRACT

The science of producing transgenic animals is advancing rapidly with the resulting manipulated offspring being utilized for biomedical modeling, possible xenotransplantation to humans and agricultural livestock production. The pig is one species under investigation which could have application in any of the aforementioned categories. We will highlight general facts regarding transgenic livestock, aspects of microinjection and our current work performed on developing the Panepinto micro size pig as an animal model for research requiring genetic manipulation. Specifically, a pilot study was undertaken with a hybrid cross of Yucatan micropigs and Vietnamese potbellied pigs as models for potential transgenic work using superovulatory techniques, follicular harvest and reimplantation. The advantage of using a small, easily managed animal needed to be balanced against the low reproductive efficiency of these pigs as compared to domestic breeds. The pilot study and subsequent followup work demonstrate our early experiences with the model and the direction in which this investigation should proceed.

2. INTRODUCTION

Animal models of human disease are valuable tools for studying pathophysiology, pathogenesis and treatments of a variety of conditions, with potential benefits being applicable, often both to man and animals. In the case of morbidity due to an underlying genetic

[*] Reprint requests to: Victoria Hampshire, VMD, Chief, Carnivore Ungulate Unit, Veterinary Resources Program, NIH, Building 28, Room 104, 9000 Rockville Pike, Bethesda, MD 20892, (301) 496-9201.

Advances in Swine in Biomedical Research, edited by Tumbleson and Schook
Plenum Press, New York, 1996

defect, the models have been dependent on spontaneous mutations and successive breeding of affected offspring.[1] This process is dependent on chance and must be conducted over an extended period, especially if livestock with long generation times are involved. Transgenics, in which genetic material is either inserted or altered, will result in the immediate offspring theoretically expressing this change genotypically and possibly phenotypically. The phenotypic expression is paramount in obtaining a useful clinical model. Results of changes to specific loci can be visualized rapidly and therapeutic intervention can be examined.

Livestock species in general are anticipated to be utilized more frequently for both biomedical research and agricultural production through the use of genetic engineering methodologies.[2] The relatively recent development of mice as founder animal stock for models of human disease has been limiting as far as the ability to provide the investigator with a useful clinical model for chronic disease studies. Long term intravascular access, clinical intervention and modern treatment strategies are difficult in rodent models. In some cases, genotypic expression of disease may not correlate with clinical disease or phenotypic expression. Canine, feline and primate models often are able to rid themselves of chronic instrumentation and inherent problems associated with repetetive anesthetic episodes and human risk during the use of primate models often makes aggressive clinical treatment during stages of morbidity difficult to perform. Attributes in biomedical research models which are desirable for the clinician are docile nature, ease of handling in patient care areas such as MRI, PET and radiation biology suites as well as the ability to meet PHS guidelines for housing criteria in laboratory animal facilities.

The domestic pig has been a successful study animal for years due to its comparable body size and digestive system similar to human beings. Investigations with porcine animal models include organ transplantation, cardiovascular studies including atherosclerosis, gastric ulcers, toxicology, diabetes, dermatology, alcoholism, nutrition and nephrology.[5] Large size and difficulty of manipulating and housing domestic swine, which can weigh upwards of 250 kg at maturity, prompted a search for a downsized or miniature pig as early as the 1940's.[6] The initial goal was to produce animals which, as adults, were equal to human adults in size and to be physiologically and anatomically normal. Various breeds and crosses of breeds have resulted in an array of miniature pigs which weigh 70 to 120 kg when mature. This size allows for ease in restraint, blood collection, lifting onto tables and housing in research facilities which usually are limited in animal housing space and generally geared toward canine subjects.

Development of miniature pigs has been underway for approximately 45 yr and has yielded some common strains such as Yucatans, Sinclair, Hormel, Hanford and the NIH SLA miniature pig.[7] European and Asian strains such as the Göttingen and Meishan pig are attractive for biomedical research; however, import and quarantine restrictions complicate ease of purchase for novel approaches where competition in the scientific arena may necessitate expeditious acquisition of animals. In the United States, the Yucatan is popular due to its small size and gentle nature. It has a sparse haircoat, applicable for topical absorption studies, which results in less odor, easier blood vessel location and greater applicability for skin studies.[8] Crossing the Panepinto micro strain with the Vietnamese Potbellied pig, resulted in F1 adults which weigh up to 70 kg, have a moderate haircoat and gentle disposition.[7]

Considering the number of years involved in selectively breeding animals to achieve the desired smaller strains, and in taking into consideration the researcher's needs at large metropolitan institutions, it would be advantageous to accelerate this process. Not only selection for size but specific disease states such as diabetes or hypertension which were developed previously developed by crossing offspring could benefit the researcher from direct genetic manipulation and establishment of founder lines for these disease conditions. Creating transgenic animals, however, is complex and requires a high skill level especially the precision involved in transferring genes to zygotes via microinjection. An alternative method under study is that of transferring genes by embryonic stem (ES) cell lines which allows precise manipulation of specific genes in the genome. The pig model has been used successfully for reintroducing an ES cell genome into a swine germ line.[4]

To facilitate the process of collection of large numbers of embryos suitable for microinjection procedures, exogenous gonadotropins have been used. These techniques have been utilized for 30 yr both in agriculture for estrus synchronization of large herds and, more recently, in biomedical research. The use of these hormones in NIH SLA miniature swine was studied in detail by Diehl *et al.*,[10] in the mid 1980s. They used pregnant mare's serum gonadotropin (PMSG) and human chorionic gonadotropin (HCG) to control the rate and time of ovulation, and the synthetic progestin, altrenogest, to synchronize the estrus cycle. The exogenous hormonal regimen was shown to be as effective in miniature pigs as in standard domestic breeds but the condition of cystic endometrial hyperplasia (CEH) and oviductal abnormalities which have been acquired through two decades of inbreeding in the NIH pig reduces this strain's fertility and makes it a less suitable subject for reproductive research.[11]

Successful transgenic studies utilizing domestic pigs have produced noteworthy findings starting in the early 1990's. Wall *et al.*[9] in an examination of milk proteins, successfully inserted a mouse gene into a swine genome and found the protein expressed in the milk of the offspring and the mouse gene incorporated in the mammary gland tissue. Biopsies of the piglets' tails were taken and polymerase chain reaction with primers specific to the gene under study were employed to identify successful transfer to the F1 animals. The positive results of such a study in the agricultural world point to the application of a transgenic pig for producing foreign proteins in its milk. In the biomedical arena, O'Donnell *et al.*[2] recently demonstrated the ability to produce founder pigs of domestic strains for the production of human hemoglobin, also suggesting the relative advantages of a large animal model whereby large quantities of human products can be produced.

Another area of intense investigation has been to explore the feasibility of xenotransplantation , utilizing pigs as organ donors for humans.[3, 12, 13] Presently, the prospect is clouded by hyperacute rejection of the animal organ by the human recipient. Specifically, high titer antibodies activate serum complement on the lumen of the vascular endothelium which causes stenosis of the vessel and lack of viability of the grafted organ.[3] Investigations are underway to produce transgenic pigs that express human terminal complement inhibitor, especially on the surface of blood vessels.[3] Another approach utilizes pigs which incorporated genes expressing human complement regulatory protein decay accelerating factor, which decreases the human complement activation.[4] Of the transgenic offspring, 65% transcribed the message and none showed any problems or abnormalities due to possession of the transgene.[13] This type of manipulation should make a xenostransplant less prone to rejection and allow the production of pigs specifically developed as clinical organ donors.[3, 4]

Rather than entering the body cavity by laparotomy or laparoscopy, another novel approach for incorporating foreign genetic material in the swine genome is that of directly injecting naked plasmid DNA into the epidermis.[5] Hengge *et al.*[14] investigated the skin as a site for genetic manipulation. Due to its accessibility, the potential ease of detecting the desired gene's expression, great morphologic and histologic similarity to humans and large surface area, this organ system has served as an ideal model for local gene therapy. This group successfully inoculated pig epidermis with plasmid DNA and have produced transient expression of biologically active factors in epidermal keratinocytes using NIH SLA miniature swine. Therapeutic benefits of such a procedure could include establishment of

α-interferon production in the skin to combat tumors and viral lesions. A vaccination system could be devised whereby genes of pathogens are expressed in the skin with subsequent immune response and protection. Again, the use of the pig for this type of study has proven to be advantageous due to its integument's similarity with human skin and the resulting ability of the animal's skin to take up and express the DNA after direct injection.[14]

Microswine with characteristics such as we have seen in the Panepinto hybrid strain afford the investigator a supple, soft skin surface which is similar to humans and may be manipulated easily by injection of DNA or the application of DNA containing chemical vehicles. Drawbacks of the superovulation and ova collection for transgenic animal production in microswine include the difficulty of collecting large numbers of fertilized ova, having readily available synchronized embryo recipients, the need for repeated surgical interventions, the high skill level required for microinjection of ova, the variability of expression possible in any animal model and economic constraints. Some of these concerns have been discussed, such as the use of exogenous hormones to enhance yields via superovulation and the synchronization of recipients. These methods are not foolproof and many animals must be used. The additional problems associated with minipigs, including low ovulation rates, low yields of fertilized ova, oviductal adhesions and CEH make their use more challenging. Still, the overwhelming advantages of the small adult swine model makes the further pursuit of synchronization and superovulation studies in hybrid strains imperative. Biomedical research programs demand suitable mammalian models for the study of human disease. Swine are particularly desireable because a wide body of literature exists regarding cardiovascular, digestive, dermatologic and behavioral systems. In addition, the pig has become the choice chronic animal study model due to the ease of maintaining long term intravascular access, maintenance of portable instrumentation devices, ease of housing and docile temperament. This demonstrable similarity in organ systems between humans and swine, coupled with the superior chronic experimental model in the miniature pig as compared to canine, ovine or rodent species fueled the initial search at NIH for the appropriate strain of micro sized pig for ova harvest and genetic manipulation.

The purpose of this study was to investigate the feasibility of utilizing an outbred strain of microswine for transgenic studies. Since transgenic efficiencies in swine are low (reportedly 6% at best) compared to mice, and overall reproductive efficiency is notably lower in mini and micro swine strains, a hybrid strain was attractive. The Panepinto micro size swine was selected due to the reported high reproductive efficiency of this small outbred micropig. Early characterization of this model included a pilot study of dose response to conventional superovulatory drugs and description of embryo characteristics in a subpopulation of embryos which were obtained. Examination of the ability to use donor gilts for second harvest procedures also was performed.

3. METHODS

3.1. Donor Synchronization

Estrous synchronization, superovulation and ova recovery was conducted using previously described techniques [9.] Gilts were synchronized for ova harvest using altrenogest (Regumate, Hoechst Roussel, Somerville, NJ) 7 ml orally for 21 da in the feed. Superovulation was accomplished in study groups using either high dose, 2000; medium dose, 1000; or low dose, 400 IU PMSG (Sigma Chemical, St. Louis, MO) subcutaneously 24 hr after the last dose of the altrenogest. Seventy nine hr later, (103 hr after the last dose of Regumate) each gilt was given 500 IU HCG (Steris Laboratories, Phoenix, AZ) IM. Each gilt was heat checked daily by introduction to a boar and bred in standing estrous between 24 and 48 hr following HCG administration. The boar to gilt ratio was 1/5 for this study. A 2 yr old 70 kg boar of the F1 hybrid generation was used.

3.2. Surgical Procedures

The micro sized pigs were fasted 24 hr before surgery but permitted free access to water. General anesthesia was induced between 16 and 20 hours following breeding with a mixture of ketamine (Vetalar, Fort Dodge Laboratories, Fort Dodge, IA), 20 mg/kg IM; xylazine (Rompum, Mobay Animal Health Division, Shawnee, KS), 2 mg/kg IM; butorphanol tartrate (Torbugesic, Fort Dodge Laboratories, Fort Dodge, IA), 0.2 mg/kg IM; and atropine (Elkins-Sinn, Cherry Hill, NJ) 0.05 mg/kg IM. Gilts were intubated with a 9 or 10 mm endotracheal tube (Endotracheal tube, Mallinckrodt, Glens Falls, NY). A 20 g, 1 1/4" angiocath (Deseret Pharmaceutical, Sandy, UT), was inserted into an ear vein for administration of supportive agents. Each animal received lactated ringers solution dripped at a rate of 30 ml/min during the surgical procedure. Anesthesia was maintained with a 1 to 2% isoflurane (Aerrane, Anaquest, Madison, WI) and oxygen at a flow of 4 l/min. A sterile lubricant ophthalmic ointment (Tearfair, The Butler Company, Columbus, OH), was placed in the eyes to prevent corneal drying and irritation. The abdomen was clipped and gilts were placed in dorsal recumbency on the operating table. The abdomen was scrubbed with hexachlorophene foam (Septisol, Vestal Laboratories, St. Louis, MO) and draped for aseptic surgery.

An 8 to 12 cm midline laparotomy was made between the umbilicus and pubis that permitted exposure and manipulation of both ovaries, oviducts and uterine horns. One ovary was brought out of the abdominal cavity and the follicles were counted. Ruptured sites were counted and compared to total follicular number. The oviduct was cannulated from the infundibular end with silastic tubing (Dow Corning Corporation, Midland, MI), 0.040 in internal diameter x 0.085 in outer diameter. The tubing had a bead of medical adhesive silicone type A (Silastic, Dow Corning Corporation, Midland, MI) on the end which was introduced into the oviduct. The bead was readily palpable and helped to determine the depth of insertion of silastic tubing into the oviduct. It also helped to prevent the catheter from slipping out of the oviduct during flushing, thus preventing the need to place a ligature around the oviduct. A 23 ga blunt needle was attached to a 20 ml syringe containing warmed Beltsville Embryo Culture Medium (made by the NIH media unit, Bethesda, MD). The needle was inserted into the tip of the uterine horn and threaded into the distal oviduct. The silastic catheter was held in place gently by squeezing it between the thumb and index finger while the oviduct was flushed with 10 to 15 ml of medium. Medium was collected in small petri dishes (60 x 15 mm) by means of the silastic catheter. The procedure was repeated for the other oviduct.

The number of embryos harvested was counted using a stereoscope (30 magnification). A fabricated glass pipet, heated and pulled to make a fine tip, was attached to the silastic tube with a mouth piece for collection of embryos. Embryos were placed in microcentrifugation tubes to avoid the possibility of losing them.

3.3. Embryo Evaluation

Embryos were transported to a nearby laboratory and were checked for developmental stages following centrifugation at 15,000 g for 15 min. The embryos were pipetted and transferred to slides with depression wells containing a drop of media and examined at 300x magnification. Estimated loss of embryos during centrifugation procedures was assessed at this time. All procedures were conducted with the use of an upright DIC microscope (Carl Zeiss, Westbury, NY) magnification.

4. RESULTS AND DISCUSSION

4.1. Superovulation and Embryo Harvest

The results of the superovulation and harvest trials are shown in Tables 1 and 2 and Fig. 1. All superovulated gilts demonstrated consistent superovulatory patterns and estrus behavior. Generally, standing heat was detected most reliably between 24 and 48 hr post HCG administration, although gilts in group A, high dose PMSG, were observed in standing heat closer to 24 hr as compared to those in mid, low, and control groups at 48 hr. A higher number of cystic follicles was also seen in the high dose gilts which may have accounted for variability in estrous behavior.

Average time from HCG administration to laparotomy procedures ranged from 65 to 72 hr. An ANOVA analysis of the follicular numbers, ovulatory structures and total embryos captured in the oviductal washings was performed for groups in the high, mid and low dose compared to unstimulated gilts. Statistical significance ($p<.05$) was achieved in high and medium doses of PMSG. There was no benefit to the administration of the high dose over the medium dose range. The unstimulated group was small in this pilot study due to a concommittent desire to perform technical manipulations on the embryos and the attempted use of the first group of gilts as repeat donors in an unstimulated model. The number of ova

Table 1. Comparison of follicular numbers, ovulatory structures and embryo capture in high (group A), medium (group B) and low dose (group C)

Pig I.D.	Age/m	PMSG/HCG (IU)	HCG-SEa(h)	HCG-sur(h)	#follicles	#ovulatory structures	total embryos captured
A)							
1007	11	2000/50	NEb	65	41	36	23
1011	11	2000/50	25	65	33	19	16
1006	11	2000/50	26	66	24	21	16
1009	11	2000/50	25	73	30	23	14
				Ave:	32	24	17
B)							
1004	14	1000/50	48	64	32	28	21
1052	8	1000/50	48	61	27	17	16
1057	8	1000/50	48	61	28	22	6
1056	8	1000/50	48	61	48	42	N/Aa
1061	9	1000/50	50	70	32	31	9
1058	9	1000/50	50	70	25	15	15
1008	9	1000/50	50	70	46	41	13
1053	9	1000/50	50	72	27	27	N/A
1059	9	1000/50	50	72	26	24	N/A
1060	9	1000/50	50	72	51	51	N/A
				Ave:	34	29	13
C)							
1091	10	400/500	50	66	9	9	0
1088	10	400/500	48	66	18	16	8
1092	10	400/500	48	66	21	17	16
1093	10	400/500	48	66	13	3	3
				Ave:	23	11	6

[a]Not staged
[b]Not in estrous and not bred; staging not applicable

Table 2. Follicular numbers, ovulatory structures and embryo capture values using synchronized unstimulated gilts

Pig ID	Age (m)	IU HCG	h HCG to Surgery	#follicular structures	# ovulatory structures	Total captured
1005	14	500	60	12	8	8
1012	13	500	64	18	9	3
1013	500	500	64	20	20	8
				Ave:	17	7

or embryos staged at 300x varied between groups and was dependent on technique and inherent losses in the laboratory before, during and following centrifugation of the embryos. In general, when staged in the operating room at 30x, total numbers of embryos captured was averaged and found to be 53% of total ovulatory structures in the high dose group, 43% for the mid dose range and 26% for low dose and unstimulated groups. The relative numbers of embryos harvested between gilts within each group varied between 27 and 94% of the ovulatory structures counted. This finding did not correlate with variability in time between HCG administration and embryo harvest as expected; however, the window of time between 65 and 70 hr post HCG administration appeared to be optimal for obtaining higher absolute numbers of embryos. Some losses were encountered between examination of embryos in the operating room at 30x and examination following centrifugation of the conical tube in the laboratory under 300x magnification. In general, of embryos staged in the high dose group, 23 embryos were examined. Eleven were fertilized at the one cell stage of development. Twelve were found at two cell or greater stages of development. In the mid dose group, 49 embryos were examined under high magnification at 300x and 46 were found to be at the

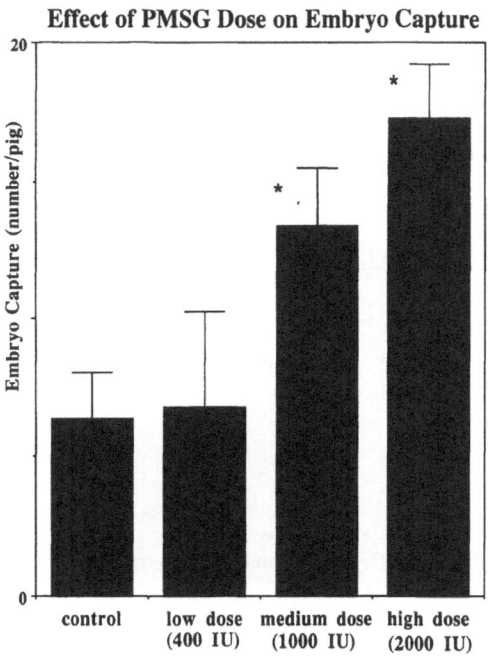

Figure 1. Statistical representation of dose response to high, medium and low dose PMSG in Panepinto micro swine as compared to unstimulated gilts.

single cell stage, 2 at the 2 cell stage and 1 at advanced stages of division. Evidence of fertilization included high numbers of sperm located inside the zona and by the evidence of ongoing division of embryos left to divide in the laboratory. Since this mid dose range was high compared to the standard dose used for domestic strains of swine (usually between 1250 and 2000 IU for the average 100 to 200 kg gilt), one gilt (1055 not shown in table) was stimulated using this medium dose range and bred with no surgical intervention, subsequently farrowing eleven piglets. Thus, optimal dose for stimulation in the hybrid Panepinto micro swine gilt (40 to 60 kg) appears to be 1000 IU PMSG. The dose on a unit per kg basis of this strain as compared to domestic stock is 2 to 3 times that of the average dose needed to stimulate domestic gilts in the 100 to 200 kg range. This would imply that this micro swine strain is somewhat more resistent to stimulation of follicular development than is its domestic counterpart. The low dose range of PMSG, which was comparable dose per weight to domestic gilts, offered no advantage in follicular development or ova numbers over control populations; however, the control group should be expanded to test the validity of this observation. Timing of embryo capture in all groups appeared to be optimal with the majority of embryos captured at single cell stages whereby vision of the male pronucleus was possible. This finding should verify that facilitation of the microinjection of foreign DNA at an opportune time is possible in this hybrid strain. Further data needs to be collected in the form of large numbers of microinjected ova to test the ability to transfer foreign DNA successfully into this strain of pig.

4.2. Second Surgical Attempt

The ability to perform a second harvest procedure on these small strains of gilts is important in reducing long term cost, especially when purchase of gilts is preferred to maintenance of a herd of animals. Four unstimulated gilts from group A were synchronized a second time 3 mo following the first surgical procedure with the intent of enlarging the control group. The findings upon laporatomy included fibrosis of the ovarian structures, adhesions between the serosal surfaces of the uterine horns, and oviductal abnormalities including adhesions of the fimbrial fascia to the follicular surfaces and prevention of full maturation of follicles. This finding was reported in similar studies by Pinkert[1] using inbred Yucatan gilts and is a disadvantage when compared to domestic gilts.

5. ACKNOWLEDGMENTS

The authors wish to thank Drs. Robert Wall and Vernon Pursel for preliminary visits to their laboratory and wise advise, Mr. Robert Joppy for help in oversight of breeding and Drs. Alan Schecter, Steven Shapiro and Ziu-Yao Lieu for early observations of embryos.

6. REFERENCES

1. Pinkert, C.A., and Murray, K., 1993, Superovulation and egg transfer in Yucatan miniature swine, *Anim. Repro. Sci.* 31:155-163.
2. Petters, R.M., 1994, Transgenic livestock as genetic models of human disease, *Repro. Fertil. Dev.* 6:643-645.
3. Fodor, W.L., Williams, B., Matis, L., Madri, J., Rollins, S., Knight, J., Velander, W., and Squinto, S., 1994, Expression of a functional human complement inhibitor in a transgenic pig as a model for the prevention of xenogeneic hyperacute organ rejection, *Proc. Nat. Acad. Sci.* 91:11153-11157.

4. Rosengard, A.M., Cary, N., Langford., Tucker, A., Wallwork, J., and White, D., 1995, Tissue expression factor in transgenic pigs-a potential approach for preventing xenograft rejection, *Transplantation* 59:1325-1333.

5. Scamark, R.F., 1994, Progress and emerging problems in livestock transgenesis - a summary perspective, *Repro. Fertil. Dev.* 6:653-657.

6. Panepinto, L.M., 1995, History, genetic origins, and care of Yucatan miniature and micropigs, *Lab. Anim.* 24:31-34.

7. England, D.C., and Panepinto, L., 1986, Conceptual and operational history of the development of miniature swine, in: *Swine in Biomedical Research,* Volume 1, (M.E. Tumbleson, ed.), Plenum, New York, pp. 17-22.

8. Panepinto, L.M., 1986, The Yucatan miniature pig: characterization and utilization in biomedical research, *Lab. Anim. Sci.* 36:344-347.

9. Wall, R.J., Pursel, V., Shamay, A., McKnight, R., Pittius, C., and Hennighausen, L., 1991, High level synthesis of a heterologous milk protein in the mammary glands of transgenic swine,*Proc. Natl. Acad. Sci.* 88:1696-1700.

10. Diehl, J.R., Lipetz, K., Stuart, L., and Wildt, D., 1986, Use of exogenous gonadotropin and embryo transfer to study reproductive efficiency in altrenogest synchronized miniature swine, in: *Swine in Biomedical Research,* Volume 1 (M.E. Tumbleson, ed), Vol. 1, Plenum, New York, pp. 135-142.

11. O'Donnell, K.J., Martin, M.J., Logan, J.S., and Kumar, R., 1993, Production of human hemoglobin in transgenic swine: an approach to a blood substitute, *Cancer Det. Prev.* 17: 307-312

12. Simon, G.A., 1993, The pig as an experimental animal in biomedical research, *Isreal J. Vet. Med.* 48:161-167.

13. White, D.J., Cozzi, E., Langford, G., Oglesby, T., Wang, M., Wright, L., and Wallwork, J., 1995, The control of hyperacute rejection by genetic engineering of the donor species, *Eye* 9:185-189.

14. Hengge,U.R., Chan, E., Foster, R., Walker, P., and Vogel, J., 1995, Cytokine gene expression in epidermis with biological effects following injection of naked DNA, *Nat. Gen.* 10:161-167.

60

MINIATURE SWINE BREEDS USED WORLDWIDE IN RESEARCH

Linda M. Panepinto[*]

P and S Associates
Masonville, Colorado 80541

1. ABSTRACT

Size, space, economic and other practical considerations limit the use of domestic farm swine in biomedical research. As a result, there has been a significant increase in the use of miniature swine worldwide for a broad range of biomedical research applications. However, there remains a relative dearth of information comparing the numerous strains and breeds in use today and differences among them. Genetic background, growth, rate, size at sexual and full maturity, behavior, husbandry requirements and other practical factors such as cost effectiveness all should be considered. All of these can vary significantly from breed to breed and may have a profound impact on the research being conducted. An understanding of these differences is important for investigators choosing an animal model as well as for facility managers, veterinarians and support staff.

2. INTRODUCTION

The last 10 years have seen an explosive increase in the research use of miniature swine worldwide. Numerous factors have contributed to this trend, including size, space, economic and other practical considerations. In addition, contemporary, user friendly methodologies and husbandry techniques have emphasized the humane handling of minipigs as laboratory animals rather than feedlot swine destined for the marketplace. The well documented similarities between swine and man have further enhanced the scientific validity of many studies conducted and their relevance to various disorders in the human population.[1-3]

Throughout the world, there are many breeds and strains of miniature swine. Some are indigenous to the areas in which they are found and some have been imported from other locations. Others have been created through selective breeding of existing breeds, the outcross mating of 2 or more strains as well as the concentrated inbreeding of certain strains

---

[*] Reprint requests to: Linda Panepinto, P&S Associates, Box 192, Redstone Canyon, Masonville, CO 80541, (970) 226-4865.

Advances in Swine in Biomedical Research, edited by Tumbleson and Schook
Plenum Press, New York, 1996

to achieve certain traits. In this report is a comprehensive review of those breeds and strains of miniature swine currently in use throughout the world. Some of the practical applications of methodology, husbandry, and management which have enhanced the user friendly aspects of applied minipig research will be presented.

3. BREEDS DEVELOPED IN THE USA

3.1. Yucatan Miniature and Micro Pigs

The Yucatan pig, which is native to the Yucatan Peninsula of Mexico, was first described for laboratory use in 1978.[4] This slate gray, hairless, docile natured pig was known as "Labco" in the early 1960's.[5] Most of the early genetic selection, development and characterization of Yucatans for laboratory use was conducted at the Colorado State University Swine Laboratory from 1972 to 1988. A comprehensive history of the genetic origins, history and care of Yucatan pigs recently was published[6] and included a detailed account of the breed's development since its importation in 1960. Also included is the development of the Yucatan micropig, initiated in Colorado in 1977 and described by Panepinto et al.[7] These pigs have been used widely in the U.S., Canada and Europe for a variety of research applications, including diabetes, cardiovascular related studies, nutrition, exercise, renal and metabolism.[8] Recently, they have been characterized as a model for ophthalmic studies by Saint-Macary[9] and hemodynamically compared to farm swine and Göttingen miniature pigs by Behnarkate.[10]

3.2. Sinclair Miniature Swine (Hormel Miniature)

The Sinclair Miniature Pig was developed originally at the Hormel Institute of the University of Minnesota in the 1950's as described by England in 1954.[11] These animals were known originally as Minnesota miniature pigs and Hormel miniature pigs and still may be referenced under those names. They were derived through the selective breeding of 4 feral strains: Piney Woods, Ras-n-Lansa, Catalina and Guam.[12,13] A Yorkshire domestic boar later was used to introduce white color into the strain.[13] Following their early genetic development at the University of Minnesota, significant characterization and development was conducted by Tumbleson's group at the Sinclair Comparative Medicine Research Farm at the University of Missouri.[14,15] There they were used for a broad spectrum of studies, with particular emphasis on melanoma and alcohol research. Selected animals were made available to other research facilities. Miniature pigs produced by the Sinclair Farm have been known as Sinclair(S-1) miniature swine. Miniature pigs from the same or similar genetic background, also may be referenced by the original strain names: Minnesota miniature and/or Hormel miniature. A comprehensive review of production characteristics of the Sinclair miniature pig was published recently by Bouchard[16] of the University of Missouri Sinclair Research Center. Minnesota miniature pigs also were used in the establishment of other breeds and strains, including the Göttingen, the FDA Hormel-Hanford strain, the NIH Minipig and the Minipig of the Czech Republic.

3.3. Hanford Miniature Swine

Hanford miniature swine were developed at the Hanford Laboratories (later known as Battelle Northwest Research Laboratories) in Richland, Washington.[5] Original foundation stock included a Pitman-Moore miniature boar and 2 Palouse females. A feral swamp hog and a Yucatan miniature boar later were incorporated into the breeding program which eventually developed a mature white pig weighing approximately 70 kg. Hanford pigs are especially valuable for studies requiring a non pigmented animal and have been extensively used for radiation and skin studies. Cardiac function and morphology was compared with the Yucatan miniature and micro strains by Smith et al.[17] Hanford miniature pigs have been used in the development of other strains, including the Munich minipig (Troll) and the FDA Hormel-Hanford strain.

3.4. Hormel-Hanford Strain

A crossbred strain of the Hormel and Hanford breeds was maintained and studied for many years at the FDA Research Facility in Beltsville, Maryland. These animals were used extensively for a wide variety of nutrition and drug metabolism studies at FDA throughout the 1970's and 1980's.[18,19] Unfortunately these animals were primarily used in house and generally were not available to outside research facilities.

3.5. NIH Minipig

An inbred strain of miniature swine known as the NIH SLA inbred miniature swine was developed by Sachs' group at NIH and described in detail by Lunney et al.[20] The strain was initiated with one Hormel pig and one Vita Vet miniature pig in the early 1970's, with one of the principle long term goals being the development of an organ transplantation animal model. Other areas of study for this inbred miniature swine strain have included transplantation rejection, disease resistance, transgenics, gene therapy and related studies.

3.6. Pitman-Moore

Pitman-Moore miniswine were named after the pharmaceutical company that originally developed them from feral pigs captured in the swamps of Florida.[5] They also were studied at the Hanford Laboratories as well as Vita Vet Labs (they have been referred to as the Vita Vet minipig) and were used in the development of other strains including the Hanford and the NIH minipig.

3.7. Panepinto Micro/Miniature Pig

A genetic selection and breeding program to outcross the Yucatan micro pig was initiated in Colorado by Panepinto and associates in 1990. Selected F-7 Yucatan pigs from the original Yucatan micro line[7] were used as foundation animals. These breeding animals had been maintained and bred for 13 yr by Panepinto and colleagues in Colorado under a carefully controlled genetic selection program. They were derived directly from stock originally obtained from the Thompson Research Farm in 1972, which were offspring of the original Yucatans imported from Mexico. This Colorado herd is unique, as no outside lines or strains were introduced genetically into this colony of Yucatan pigs between 1972 and 1995.

Selected F-7 Micro Line Yucatans were bred to Vietnamese miniature pigs and the progeny of numerous matings were evaluated to determine the optimal crosses. Subsequent matings have produced F-2 and F-3 generations of 3/4 and 7/8 Yucatan micropigs, known as Panepinto micro/miniature pigs. Birth weights are similar to purebred Yucatans, ranging from 450 to 1000 grams. However, reproductive performance is superior to either strain. Males and females can be bred successfully at the age of 4 mo and the litter average size of F-1 first litter gilts bred at 4 mo of age is 4.75. Subsequent F-2 litter size from those same F-1 gilts averaged 8.6 liveborn piglets with an average of 7.9 piglets weaned (19 litters).

This is higher than the 5 to 6 liveborn reported by Panepinto in 1981 for the F-2 progeny of Yucatan micropigs. Average litter size for sows from the original Yucatan miniature pig was reported at 6.12±1.78, while litter size for first litter Yucatan gilts averaged 5.96±1.03.[4]

The weight of the Panepinto micro/miniature pig is less than that reported for other miniature breeds, including the Yucatan micropig. Six month weights range from 10 to 14 kg; at 1 yr they range from 15 to 20 kg.

4. MINIATURE BREEDS DEVELOPED IN EUROPE

4.1. Göttingen Miniature Swine

Early genetic selection for development of the Göttingen miniature pig was conducted from 1960 to 1964 by Haring at the Friedland Experiment Station of the University of Göttingen in West Germany. Six Minnesota Minipigs were obtained from the Hormel Foundation in Austin, MN and 11 Vietnamese minipigs were acquired from German zoos as foundation stock for this program.[21] German Landrace pigs were utilized in the breeding program and 12 to 24 mo weights for the white line averaged 35 and 45 kg, respectively.[22] Multiplier units for Göttingen Miniature pigs have been set up in 9 European countries, Japan and Israel.[22] Göttingen pigs have been utilized for a broad range of research programs and in the development of other strains.

4.2. Munich Mini Swine (Troll)

The Munich mini swine, also known as "Troll" were developed by the crossing of 2 primary strains for 16 generations, Hanford miniature swine and native miniature swine from Columbia.[23] The Columbian swine are indigenous to Columbia and are used primarily for food in that area. Similar swine have been described in a National Academy of Sciences publication.[24] Selection criteria included smaller body mass, balanced muscle fat relationship, white skin and several other characteristics. The Troll minipig has been used for a variety of studies including acute radiation work[25] and cardiovascular studies.[26] They are produced in Germany.

4.3. Berlin Miniature Pigs

Berlin Miniature Pigs were developed and have been used primarily in Germany where they also have been crossed with Belgian Landrace domestic swine. Average litter size was reported at 8.31 and 7.02 for 2 groups studied in 1992.[27] Other reproductive traits of the Berlin Minipig have been described in a comprehensive 1993 publication by Seifert.[28]

4.4. Mini-Lewe

Developed in Czechoslovakia more than 20 yr ago, the Mini-Lewe miniature pig has been well characterized and used extensively in Eastern Europe as a model for mandibular and dental related studies.[29] Koppe et al.,[30] describe the Mini-Lewe as having a dentition sequence broadly identical to that of domestic breeds and the Pitman-Moore miniature pig. They confirmed that the Mini-Lewe skull is longer than that of the Vietnamese miniature pig with significant postnatal shifts in skull proportions. In 1976, an SPF colony of Mini-Lewe miniature pigs was started at Bad Langensalza, Czechoslovakia, using piglets derived by hysterectomy. By 1978 the colony consisted of 80 productive sows, managed under established SPF rearing conditions.[31]

4.5. Miniature Pig of the Czech Republic

In the early 1980's, the Veterinary Research Institute at Brno, Czechoslovakia, established a breeding program to develop a new miniature pig. Göttingen, Minnesota miniatures and Landrace domestic stock were used to establish the strain which is reported to mature at an adult weight less than 30 kg.[32]

4.6. Vietnamese Potbellied Miniature Pig

While the Vietnamese miniature pig is native to southeast Asia, this breed has been used as a laboratory animal primarily in Europe.[33] They have been utilized in the development of other strains of miniature swine, including the Göttingen, the Minisib and the Panepinto micro/miniature pig.[6] In the U.S. in the 1980's and 1990's they have become popular as companion animals and zoo exhibits.[34,35] This breed has a natural tendency for obesity with very short ears and small ear veins. These characteristics can be disadvantages to the researcher, as excessive weight in older animals is difficult for swine to lose and can lead to feet and leg problems. Smaller than average ears with limited vascular access can create difficulties for routine procedures, such as blood sampling and infusions. They are docile if handled when young, but it has been our experience that many require continued handling as adults to remain socialized.

4.7. Mini-Sib or Miniature Siberian Pig

This unique breed of miniature swine was developed by crossing Swedish Landrace, Vietnamese, Central European wild pigs and Central Asian wild pigs for 8 generations. A total of 5000 pigs were utilized for the developmental genetic selection program at the Academy of Sciences in Novosibirsk, Siberia.[36,37] The 12 month body weight averaged 29.8 ± 3.0 kg (males) and 35.4 ± 3.0 kg (females) for 8 males and 30 females. These animals have been characterized genetically and phenotypically as a laboratory animal and utilized in Russia for many years.

5. BREEDS DEVELOPED IN AUSTRALIA

5.1. Kangaroo Island Miniature Pigs

This is a feral strain of pig from Kangaroo Island off the coast of South Australia. They are believed to have been released onto the Island in 1801 at Port Jackson by a French navigator and explorer, Captain Nicholas Baudin.[38] The absence of natural predators on the Island most likely ensured their survival and successful reproduction in the wild. Parent stock for the development of a research colony were obtained from Kangaroo Island in 1976 by the CSIRO Division of Human Nutrition in Adelaide, Australia. A black feral sow from New South Wales was introduced in 1977 and a few of her offspring retained as part of the Kangaroo Island (K1) breeding pool. They are primarily black and white, with a mature weight of 98 ± 13 kg.[38]

6. MINIATURE BREEDS DEVELOPED IN ASIA

6.1. Ohmini

In 1945, a breeding program to develop a small pig was initiated by Hiroshi Ohmi at the Nippon research laboratory for domestic animals in Tochigi, Japan. Three different strains of Manchurian Chinese pigs were crossed systematically with Minnesota miniature swine, Hampshire and Duroc domestic swine.[39,40] One strain was crossed with a Landrace to produce white color for studies requiring nonpigmented animals. These pigs, known as Ohmini, weigh from 35 to 45 kg at 12 mo of age and have been utilized for both biomedical research and for food at the Ohmi restaurant in Japan.[41] The pigs are primarily black in color, have lean, angular bodies and a coarse hair coat. They have long, drooping ears, a long straight nose and wrinkled skin and are docile and quiet natured.

6.2. Clawn Minipig

This miniature strain was developed in Japan by crossing Göttingen and Ohmini breeds. They are mostly white, although some have black spots.[42] Females at 6 mo averaged 26 kg and at 12 mo averaged 49 kg. Males at 6 mo averaged 27 kg and at 12 mo their average weight was 50 kg.[43] An inbred strain was established between 1978 and 1989 after 7 generations of selective breeding.[44]

6.3. Lee Sung Miniature Pigs

Development of this breed was initiated in 1974 by Professor T. Y. Lee at the Department of Animal Husbandry, University of Taiwan. He selected native Taiwan small ear pigs and established an inbred population known as Lee Sung.[45] Twelve mo weights average approximately 30 kg in the 60% inbred population.[46]

6.4. Chinese and Taiwanese Miniature Pigs

There are numerous strains of miniature pigs native to China, including the China miniature pig, Beijing black pig, Manchurian minipigs and Meishan. Several scientists reported on their genetic and other characteristics as well as their suitability for biomedical research applications.[47,48] Native Taiwanese breeds include Taoyaun, Meinung, Short ear and Lan Yu.[49,50] Some of these strains have been utilized for the development of other breeds, such as the Ohmini and the Lee Sung. The Lan Yu Miniature Pig has been developed as an SPF animal.[51]

In addition to the numerous strains and breeds of miniature swine which have been developed and characterized for biomedical research applications, there are many other minipigs in various parts of the world. Some are native while some are feral and some have been crossed with domestic breeds to increase size for enhanced meat production. In some areas, certain strains have been domesticated for the production of food.[24] As the interest and use of miniature swine for the study of disease and other medical disorders in man continues to increase, it is likely that other strains will be developed.

7. CONTEMPORARY METHODOLOGY, HUSBANDRY AND MANAGEMENT

Of critical importance in the expanding use and acceptance of miniature swine as laboratory animals has been the refinement of humane, contemporary handling, husbandry and management methods. The shift away from a feedlot approach to the management of these animals has been an important factor in their growing popularity among researchers. A comprehensive swine laboratory program should address such factors as specialized nutritional and environmental requirements, housing, handling and all other aspects of maintaining these unique animals in a research setting. Handling, housing and methodology differences between acute and chronic study needs should be evaluated prior to the onset of all studies.

There have been numerous techniques and procedures developed which focus on humane handling during the conduct of research in miniature swine. Many of these, including the Panepinto sling, were described previously.[52] Also described were many other aspects of nutrition, management and husbandry relating specifically to swine in a laboratory environment. The use of the Panepinto Sling has expanded since then[53] and was described as a practical humane restraint method by Swindle *et al.*[54] A refinement has been developed in our facility and is pictured in Fig. 1. Named the *"Portable Panepinto Sling"*, it comfortably restrains swine weighing up to 50 kg in a lightweight, portable, collapsible sling which can be transported easily for field use or hung up for storage.

Smith *et al.*[55] described a chronic catheterization technique of the inferior vena cava in Yucatan miniature swine and Paschen[56] developed a method for blood sampling from the portal and hepatic veins in unrestrained, fully conscious miniature pigs. There are many other humane techniques, including the use of socialization as part of chronic handling procedures. A socialized pig which is amenable to human contact is easier to manage and handle for routine activities. In addition, the reduction of stress related to handling may allow for a better approximation of baseline physiological values. Also it is of benefit as a stress reducer to the scientific and technical staff handling the animals.

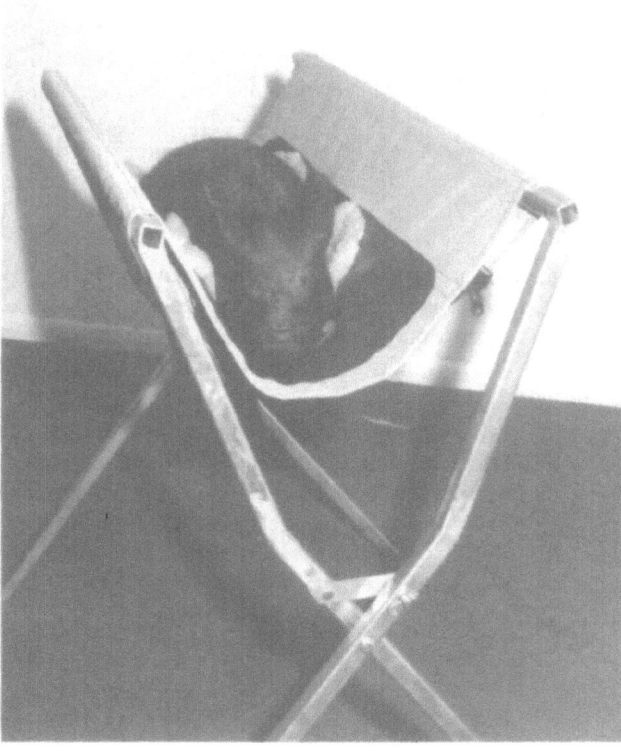

Figure 1. Portable panepinto sling.

The incorporation of environmental enrichment into certain laboratory routines and/or housing units may be of importance depending on several factors. If pigs are housed individually without any bedding material, the provision of other enrichments is critical for long term study animals. Pigs housed in indoor outdoor facilities or observed in a natural environment will spend time rooting and interacting with each other. The removal of all social and olfactory stimulation for extended periods in swine can be a source of stress. Typical behavioral indicators of such stress include hyperactivity, repeated banging of the caging or pen, chronic activation of waterer unit, aggressive behavior and pacing. Weight loss may result as well as other consequences of stress and hyperactivity and study results may be affected adversely.

Pigs should be housed in social groups whenever possible and care should be taken whenever animals are regrouped. This is important, as territorial dominance can cause conflict and the risk of injury in animals of all ages. It can cause problems in feeding among the less dominant animals. Regrouping should be conducted in neutral pens to minimize the negative effects of aggressive dominance. Metal feed pans and other protruding objects should be removed from the pen prior to regrouping and not returned until all animals have resolved their differences.

8. CONCLUSIONS

It is important to determine the type of miniature pig best suited for any proposed study. Certain breeds, such as the Ohmini, Göttingen, Yucatan and Panepinto pig are docile and ideally suited for frequent, close contact which may be required for long term studies. However, for studies requiring nonpigmented animals or animals with variable pigmentation, hair and skin color may be more important than behavior. The Hanford, some Hormels, selected Göttingens and Troll minipigs have white hair and skin. The Mini-Lewe has been characterized as a mandibular and dental swine model. There are, unfortunately, extensive restrictions and quarantine requirements for the import of swine into the U.S., which may limit or exclude the use of European and/or Asian breeds. Also, some of the research done in Eastern European and Chinese breeds is still difficult to access through traditional database resources.

As a result, detailed information regarding genetic development and research applications of breeds such as the Czech minipig is still somewhat limited. As access to electronic information resources expands, it may be easier to obtain such data as well as exchange information with some of these swine research groups in the Former Soviet Union, China and other areas of the world.

There are genetic, behavioral and other differences among the many miniature swine breeds and strains worldwide. Depending on geographic source, there may be differences among species. For example, most European strains of domestic and miniature pigs are *Sus scrofa*, while those of primarily Asian origins, such as the Ohmini and the Chinese breeds are defined as *Sus vittatus*. With this in mind, the specific breed or strain should be considered in advance depending on the type of research. Such factors as behavior, suitability for the proposed research, background information available on the pigs and herd health status should be considered.

9. REFERENCES

1. Bustad, L.K., 1966, Pigs in the laboratory, *Sci. Am.* 214:94-100.

2. Simon, G.A., 1993, The pig as an experimental animal in biomedical research, *Israel J. Vet Med.* 48:161-167.

3. Panepinto, L.M., and Swindle, M.M., 1986, Swine issue, *Lab. Anim. Sci.* 36(4):343-433.

4. Panepinto, L.M., Phillips, R.W., and Will, D.H., 1978, The Yucatan miniature pig as a laboratory animal, *Lab Anim Sci.* 28:308-313.

5. Bustad, L.K., Horstman, V.G., and England, D.C., 1966, Development of Hanford miniature swine, in: *Swine in Biomedical Research,* (L.K.Bustad and R.O. McClellan, eds.), Frayn, McClellan, Seattle, pp.769-744.

6. Panepinto, L.M., 1995, History, genetic origins and care of Yucatan miniature and micro pigs, *Lab. Anim.* 24(6):31-34.

7. Panepinto, L.M., and Phillips, R.W., 1981, Genetic selection for small body size in Yucatan miniatue swine, *Lab Anim Sci.* 31:403-404.

8. Panepinto, L,M, and Phillips, R.W., 1986, The Yucatan miniature pig: characterization and utilization in biomedical research, *Lab. Anim. Sci.* 36:344-347.

9. Saint-Macary, G., and Berthoux, C., 1994, Opthalmologic observations in the young Yucatan micropig, *Lab Anim Sci.* 44:334-337.

10. Behnarkate, M., Zanini, V., Blanc, R., Boucheix, O., Coyez, G., Genevois, J.P., and Pairet, M., 1993, Hemodynamic parameters of anesthetized pigs: a comparative study of farm piglets and Goettingen and Yucatan miniature swine, *Lab Anim Sci.* 43:68-72.

11. England, D.C., Winters, L.M., and Carpenter, L.E., 1954, The development of a breed of miniature swine-a preliminary report, *Growth* 17:207-214.

12. Dettmers, A.E., and Rempel, W.E., 1968, Minnestoa's miniature pigs, *Lab. Anim. Care* 18:104-109.

13. England, D.C., 1968, Genetic basis of and procedures for development of miniature swine, *Lab Anim Care.* 18:99-103.

14. Tumbleson, M.E., Middleton, C.C., Tinsley, O.W., and Hutcheson, D.P., 1969, Body weights and measurements of Hormel miniature swine from birth to nine months of age, *Lab. Anim. Care* 19:506-512.

15. Tumbleson, M.E., Dexter, MD., and Lowman, J.D., 1986, Fetal alcohol syndrome in miniature swine, in: *Swine in Biomedical Research,* Volume 1 (M.E. Tumbleson, ed.), Plenum, New York, London,pp. 597-609.

16. Bouchard, G., McLaughlin, R.M., Ellersieck, M.R., Krause, G.F., Franklin, C., and Reddy C.S., 1995, Retrospective evaluation of production characteristics in Sinclair miniature swine - 44 years later, *Lab. Anim. Sci.* 45:408-414.

17. Smith, A.C., Spinale, F.G., and Swindle, M.M., 1990, Cardiac function and morphology of Hanford miniature swine and Yucatan miniature and micro swine, 1990, *Lab. Anim. Sci.* 40:47-50.

18. Peggins, J.O., Shipley, L.A. and Weiner, M., 1984, Characterization of age-related changes in hepatic drug metabolism in miniature swine, *Drug Metab. Disp.* 12:379-381.

19. Friedman, L., Gaines, D.W., Newell, R.F., Sager, A.O., Matthews, R.N., and Braunberg, R.C., 1993, Body and organ growth of the developing Hormel-Hanford strain of male miniature swine, *Lab Anim.* 28:376-379.

20. Lunney, J.K., Pescovitz, M.D., and Sachs, D.H., 1986, The swine major histocompatibility complex: its structure and function, in: *Swine in Biomedical Research,* Volume 3 (M.E. Tumbleson, ed.), Plenum, New York, pp. 1821-1836.

21. Glodek, P., 1986, Breeding program and population standards of the Goettingen miniature swine, in: *Swine in Biomedical Research,* Volume 1 (M.E. Tumbleson, ed.), Plenum, New York, pp. 23-28.

22. Glodek, P., Bruns, E., Oldigs, B., and Holtz, W., 1977, The Goettingen miniature pig - a laboratory animal of world-wide importance. Part 1. Breeding programme and performance in the base population, *Zuchtungskunde.* 49:21-32.

23. Braeuer, H.M., Breeding and keeping according to their species of mini-pigs (Munich Minipigs Troll), *Tierlaboratorium* (Germany) Dialog File 203, pp. 1709-1722.

24. National Research Council, 1991, Micropigs, in: *Microlivestock,* (N. Vietmeyer, ed.), National Academy, Washington, D.C., pp. 63-72.

25. Siegl, R., 1986, Acute radiation in Troll miniature swine, *Thesis from Tierarztliche Faculty,* Ludwig Maximilians University, Munich, pp. 1-73.

26. Sandner, N., Dahme, E., Liebich, H.G., and Giesecke, D., 1990, Dietary oxycholesterols and arteriosclerosis: a study in pigs, *Adv. Anim. Phys. Anim. Nutr.* 20:60-73.

27. Seifert, H., Zobel, F., and Filler, J. 1992, Reproductive traits in Berlin miniature pigs, *Agrarwissenschaften* 41:9-13.

28. Seifert, H., and Reissmann, M., 1993, Development of model populations of Berlin miniature pigs and African dwarf goats as a basis for research on environmental stability and reproductive biology, *Archiv. Tierzucht.* 36:551-562.

29. Schumacher, K.U., Koppe, T., and Schumacher, B. 1988, Functional morphology of the maxillo-mandibular apparatus of the Mini-Lewe miniature pig; Intramascular nerve ramifications in the masticatory muscles of young animals, *Anatomischer Anzeiger.* 165: 371-377.

30. Koppe, T., Schumacher, B., and Schumacher, K.U., 1989, Functional morphology of the maxillo-mandibular apparatus of the Mini-Lewe miniature pig; Metric and dentition studies, *Anatomischer Anzeiger.* 169:35-39.

31. Gregor, G., Bohrisch, H.G., and Knorck, H., 1980, Experience with specific pathogen free (SPF) "Mini-Lewe" miniature pigs, *Monatshefte Veterinarmedizin.* 35:190-193.

32. Mraz, J., and Lojda, L., 1984, A new type of miniature pig, *Veterinarstvi* 34:73-74.

33. Otto, B., Jacob, K., and Schumacher, G.H., 1974, The Vietnamese miniature pig as a laboratory animal in stomatological research, *Zeitschrift Versuchstierkunde* 16:17-22

34. Braun, W.F., and Casteel, S.W., 1993, Potbellied pigs: miniature porcine pets, *Vet. Clin. North Am.* 23:1149-1177

35. Braun, W., 1993, Helping your clients raise healthy potbellied pigs, Vet Med. 43:414-428.

36. Tikhonov, V.N., and Panarina, L.M., 1981, Some genetic characters of "Mini-sibs" - the first Russian miniature pigs, breeding the Minisib and its morphological characters, *Genetika* (USSR)17:883-895.

37. Tikhonov, V.N., Gorelov, I.G., and Panarina, L.M., 1981, Some genetic characters of "Minisibs" - the first Russian miniature pigs; Immunogenetic characters of laboratory Minisibs, *Genetika* (USSR 17:1677-1689.

38. McIntosh, G.H., and Pointon, A., 1981, The Kangaroo Island strain of pig in biomedical research, *Australian Vet. J.* 57:182-185.

39. Ohmi, H., Ohmi, T., and Tomita, T., 1978, Breeding of a miniature pig strain Ohmini in Japan, *Collect. Breeding* 28:204-208.

40. Ohshima, S., Satoh, H., Nagase, S., Hirohashi, K., and Ohmi, H., 1973, A new strain of small-sized pig originating from a Chinese breed, *Exper. Anim.* 22(suppl):253-259.

41. Ohmi, 1988, *Personal communication* and on site visit in Japan by author.

42. Oishi, T., Nakamishi, Y., and Tanaka, K., 1991, Genetic analysis of "Clawn" miniature pigs by blood groups and biochemical polymorphisms, *Jap. J. Swine Sci.* 28:126-132.

43. Nakanishi, Y., Tojo, H., Hirohama, K., Ogawa, K., and Yamanouchi, C., 1981, Some aspects of growth, reproduction, lactational performance and semen characteristics in miniature pigs, *Bull. Faculty Ag. (Kogoshima Univ.)* 31:53-58.

44. Nakanishi, Y., Ogawa, K., Yanagita, K., and Yamauchi, C., 1991, Body measurements and some characters of inbred Clawn miniature pigs, *Jap. J. Swine Sci.* 28:211-218.

45. Sou, E., 1984, The miniature swine, Lee Sung - newly developed in Taiwan, *Anim. Exper. Lab. Anim.* 1:3-8.

46. Lee, T.Y., Sung, T.T., and Huang, T.M., 1983, A new miniature pig - Lee-Sung strain in Taiwan, Proc: *World Conf. Anim. Prod.*, pp. 67-68.

47. Wang, G.S., Guo, Y.P., and Chen, D.W., 1992, Studies on microflora in the guts of Chinese Miniature Pigs, *Acta Vet. Zootech. Sinica* 23:146-151.

48. Tanaka, K., Oishi, T., Kurosawa, Y., and Suzuki, S., 1983, Genetic relationship among several pig populations in East Asia analysed by blood groups and serum protein polymorphisms, *Anim. Blood Groups Biochem. Genetics* 14:191-200.

49. Oishi, T., Tanaka, K., Otani, T., and Tamada, S., 1988, Genetic variations of blood groups and biochemical polymorphisms in Meishan pigs, *Bulletin of the Nat'l Inst. of Anim. Industry* (Japan). No.48(1).

50. Watanabe, T. Hayashi, Y., Kimura, J., Yasuda, Y., Saitou, N., Tomita, T., and Ogasawara, N., 1986, Pig mitochondrial DNA: polymorphism, restriction map orientation and sequence data, *Biochem. Genetics* 24:385-396.

51. Wu, F.M., Shyu, J.J., and Lin H.K., 1984, Research and development of the Lan-Yu miniature pig. Establishment of a specific pathogen free laboratory, *Taiwan J. Vet Med Anim. Husb.* 44:1-6.

52. Panepinto, L.M., 1992, The minimum-stress physical restraint of swine and sheep in the laboratory, *ILAR Proc., Agric. Anim. in Res.* pp. 85-87.

53. Panepinto, L.M., 1986, Laboratory methodology and management of swine in biomedical research. In: *Swine in Biomedical Research,* Volume 1 (M.E. Tumbleson, ed.), Plenum, New York, pp. 97-109.

54. Swindle, M.M., Smith, A.C., Laber-Laird, K., and Dungan, L., 1994, Swine in biomedical research: management and models, *ILAR News.* 36:1-5.

55. Smith, D.M., Lieberman, R.P., Stribley, J.A., and Sharp, J.G., 1992, Chronic catheterization of the inferior vena cava in Yucatan miniature swine, *Lab. Anim. Sci.* 42:602-606.

56. Paschen, U., and Muller, M.J., 1986, Serial blood sampling from the portal and hepatic vein in conscious unrestrained miniature pigs, *Res. Exp. Med.* 186:87-92.

57. Panepinto, L.M., Phillips, R.W., Norden, S., Pryor, P.C., and Cox, R., 1983, A comfortable, minimum stress method of restraint for Yucatan miniature swine, *Lab. Anim. Sci.* 33:95-97.

58. Terris, J.M., and Simmonds, R.C., 1982, Description of a swine metabolsim unit for long-term studies, Lab. Anim. Sci. 32:302-303.

59. Terris, J.M., and Simmonds, R.C., 1982, A portable confinement unit for swine, *Lab. Anim. Sci.* 32:410-411.

60. Stephens, D.B., and Rader, R.D., 1983, Effects of vibration, noise and restraint on heart rate, blood pressure and renal blood flow in the pig, *J. Royal Soc. Med.* 76:841-847.

61. Cimini, C.M., and Zambraski, E.J., 1985, Non-invasive blood pressure measurement in Yucatan miniature swine using tail cuff sphygmomanometry, *Lab. Anim. Sci.* 35(4):412-416.

61

THE SIBERIAN MINIATURE PIG, ITS DEVELOPMENT, GENETICS, AND USE IN BIOMEDICAL RESEARCH

Vilen Tikhonov, DSci*

Institute of Cytology and Genetics
Russian Academy of Sciences
Russia 630090, Novosibirsk-90

1. ABSTRACT

The first form of laboratory minipigs, Minisibs (miniature Siberian swine), in the former USSR is the result of complicated hybridization with a system of crosses between domestic and wild pigs of European and Asiatic origin (Swedish Landrace, Black Vietnamese, *Sus scrofa* ferus, *Sus scrofa* nigripes) for 10 generations. "Minisibs" weigh 517 ± 6.5 g at birth, 4.7 ± 0.1 kg at 2 mo, 19.0 ± 0.9 kg at 6 mo (sexual maturity) and 29.8 ± 3.0 kg for males and 35.4 ± 0.1 kg for females at 12 mo (adulthood). More than 90% of Minisibs have white coat color, calm behavior, are tolerant to a wide range of environmental conditions and are omnivorous. Minisibs have a set of immunogenetic and cytogenetic features in which they differ from other known miniature pigs. One of these is the chromosome polymorphism; Minisibs have a karyotype of 2n = 38, 37 or 36 chromosomes. Therefore, we can use them for investigation of the role of definite chromosomes and to start genetic mapping.

Minisibs are similar to man in morphologic and physiologic parameters of many organs and in the course of various pathologic processes. For these reasons, Minisibs can be used as a model in a large scope of medical and biological research programs (Fig. 1). Minisibs are characterized by a wide immunogenetic polymorphism for all the studied blood group and sera systems thereby differing in this respect from wild and domestic pigs. When Minisibs were compared with American, West German and Japanese miniature pigs in relation to the blood group systems E, F, G and L, a distinction was demonstrated.[1,2] These immunogenetic peculiarities reflect adequately the complicated phylogenesis of Minisibs. They could be of use in studies of selection and genetics for modelling microevolutionary processes as well as for obtaining monospecific antisera reagents which detect a wide spectrum of erythrocyte and sera antigens.

---

*Reprint requests to: Dr. Vilen Tikhonov, Institute of Cytology and Genetics, Russian Academy of Sciences, Russia 630090, Novosibirsk-90, (7) 383-23564-57.

Figure 1. Adult Minisib boar and Swedish Landrace pig, one of the progenitors of this miniature breed and principal genetic source of white color coat, good fertility and calm behavior.

2. INTRODUCTION

In the world to date, there are more than 20 breeds of miniature pigs, used for biomedical and biological purposes.[1,3] Their genetic pool is represented by equal numbers of small aborigine breeds and specially developed breeds of dwarf pigs (Table 1).

In general, miniature pigs should be: a) small in size, which would be suitable for manipulations in laboratory conditions, economical with respect to food consumption and for housing needs, b) calm in behavior and c) have a white coat color.[4,5] The paradox is that selection for small size is more complicated than for large size. The process of this selection for minipigs is limited by the optimal size of the respective model and its parts; for example,

Table 1. Breeds of pigs for biomedical and biological purposes

Small (natural) aborigine pigs	Minipigs, developed by special selection
Yucatan pigs	Hormel
Corsican swine	Sinclair pigs
American-Esseks	Hanford pigs
Lee-sung	Yucatan swine
Taiwan	Beltsville White
Ossabaw pigs	Nebraska Pigs
Nepal Black pigs	Pitman-Moore
Hurrah pigs	Ohmini
Kakhetian	Gettingen
Svanetian	Mini-Lewe
	Minisibs
	Svetlogorskaya

Table 2. Standards for live weights of minipigs (kg)

Class-Category	newborn	Age, mo			
		2	6	12	36
Normal	0.70(0.6 to 0.8)	5.5(5.0 to 6.0)	22.5(20 to 25)	40(35 to 45)	70(60 to 80)
Small	0.45(0.3 to 0.6)	4.5(4.0 to 4.9)	17.5(15 to 19)	30(25 to 34)	50(40 to 59)
Very small	0.35(0.2 to 0.5)	3.5(3.0 to 4.0)	15(12 to 17)	20(15 to 25)	35(30 to 35)

size of organs for heterotransplantation and obtaining suitable amounts of serum. In our experience, we propose 3 body weights of animals (Table 2).

At present, there are reasons to expect development of specialized breeds of minipigs for different purposes.

3. MATERIALS AND METHODS

As a principal source of dwarf size genes, we chose and used Black Vietnamese Mask breed "I".[6] Adult animals of this breed have a small mass of approximately 100 kg (Fig. 2). In our opinion, more suitable for this goal was the Black-Spot Vietnamese breed but it was not available. Vietnamese breeds have a weak constitution and very intensive growth, especially of subcutaneous fat, at a young age. For correction of this, we used hybridization of the Vietnamese breed with *Sus scrofa* ferus (Fig. 3) and *Sus scrofa* nigripes (from the Belorussia-Poland border and from Middle Asia, respectively). However, the principal goal of this hybridization was artificial induction of chromosomal polymorphism in a pig population as a necessary condition for studies on the genetic function of definite chromosomes in pigs (Fig. 4). Unfortunately, all domestic pigs, irrespective of the origin of breeds, and more than 80% of wild boars, have a constant karyotype, 2n = 38. We had the luck to find some wild boars with karyotype polymorphism, which is explained by the presence of aberrant chromosomes with Robertsonian translocations suitable for the substitution of normal ones.[7,8]

Figure 2. Vietnamese Black mask breed "I", the principal source of the dwarf genes for Minisibs; as adults they weigh more than 100 kg.

Figure 3. *Sus scrofa* ferus, wild European boar, one of the progenitors of Minisibs, the principal source of karyotype polymorphism (in domestic pigs, there is a constant karyotype, 2n = 38). This wild boar is a source of the karyotype polymorphism in the population of miniature pigs.

We performed extensive cytogenetic studies on wild boars from various regions of the former USSR to find and identify donors with marker chromosomes. We intended to introduce these marker chromosomes into the genome of domestic pigs and, by substituting normal chromosomes for Robertsonian ones, we induced chromosomal intrapopulation polymorphism (2n = 38, 37 and 36). Karyotype analyses were performed in more than 100 wild boars of 4 species: *Sus scrofa* scrofa, Lin, 1758; *Sus scrofa* nigripes, Blanford, 1875; *Sus scrofa* attila, Thos, 1912; *Sus scrofa* ussuricus, Heude, 1888. There was karyotype polymorphism in 2 of the species, *Sus scrofa* scrofa and *Sus scrofa* nigripes. Their polymorphic karyotypes were 2n = 38, 2n = 37 and 2n = 36 (Figs. 5 and 6).

The results of Giemsa and C banding indicated this chromosomal polymorphism was due to centric fusion of chromosomes 16 and 17 in *Sus scrofa* nigripes and centric fusion of chromosomes 15 and 17 in *Sus scrofa* ferus.[9] Based on these cytogenetic observations, a program was implemented aimed at sequential substitution of domestic pig chromosomes 15, 16 and 17 by their genetic Robertsonian equivalents derived from wild boars.

Chromosome substitution was achieved by means of hybridization and back crosses. Donors of the aberrant chromosomes studied were Central European and Middle Asian wild boars, the Swedish Landrace and Vietnamese Black breed serving as recipients (Fig. 4). As a result of our efforts, the well known Swedish Landrace has been bred in Siberia for more than 20 yr, as well as the Vietnamese Black breed, an ancient breed of Asiatic origin. All of these breeds had advantages and disadvantages for our experimental purposes. Thus we produced dwarf laboratory pigs, which are the first pigs in the world remarkable for chromosomal polymorphism (2n = 38, 37, 36) with the involvement of Vietnamese and Swedish Landrace breeds and wild boars (Fig. 7). These pigs were named Minisibs (abbreviation for Miniature Siberian Swine). These miniature pigs demonstrated normal fertility (8 to 10 piglets/litter) with high viability of progeny.[6,10]

It is necessary to note differences in frequencies of gene alleles of the blood group systems E, G, L and especially F and others, including allotypes of blood sera, which were demonstrated by Minisibs in comparative tests with American and Japanese minipigs.[1,10] The most important genetic studies on Minisibs were performed by means of introducing

A GENEALOGICAL SCHEME OF THE CREATION OF LABORATORY PIGS
"MINISIBS" BY CROSSING DOMESTIC PIGS AND WILD BOARS

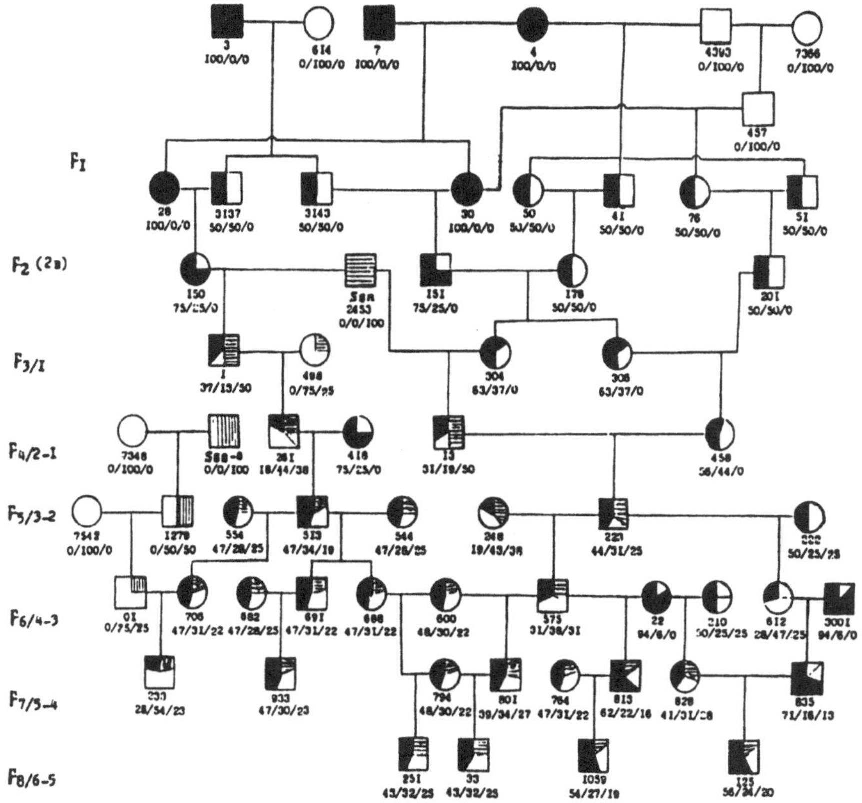

DESIGNATION. BLACK SQUARES AND CIRCLES – MALES AND
FEMALES VIETNAMEESE BLACK BREED, WHITE SQUARES AN CIRCLES –
LANDRACE BREED, THE HORIZONTAL HATCHING – SUS SCROFA NIGRIPES,
VERTICAL HATCHING – SUS SCROFA FERUS. THE ORIGIN OF ANIMAL
INDICATED UNDER THE NUMBER OA ANIMAL BY 3 FIGURES: THE SHARES
OF GENOMES VIETNAMEESE BLACK BREED, LANDRACE BREED AND WILD
BOARS.

Figure 4. A genealogical scheme of the creation of laboratory pigs Minisibs by crossing domestic pigs and wild boars, *Sus scrofa* ferus and *Sus scrofa* nigripes.

into their genome of some chromosomes, marked by Robertsonian translocations (Fig. 5). It allowed us to make the first mapping of identified autosomes (mapping of the locus of the blood group system G in chromosome 15 in the pigs).[11] Also, genetic distances of important loci of systems F and I from one to another and from centromeres were determined.[12] The most important medical investigations involved apolipoprotein metabolism, heterotransplantation of cardiac valves, genetic aspects of specific peptide functions responsible for alcohol addiction, medicines, e.g., Fluvoxsamin for correction of dipsomaniac behavior, and drug abuse.

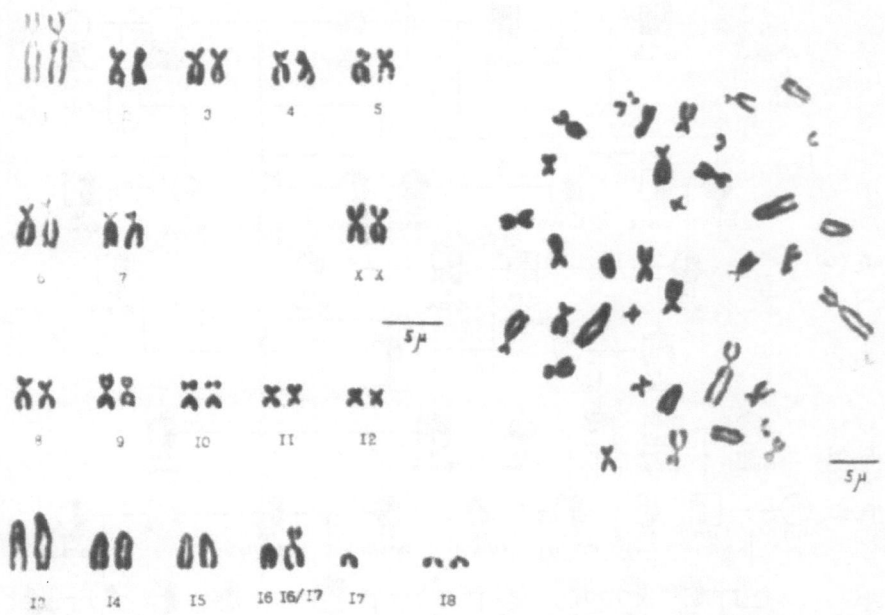

Figure 5. G banded karyotype of a female Minisibs with karyotype 2n = 37 Rb Tr 16/17.

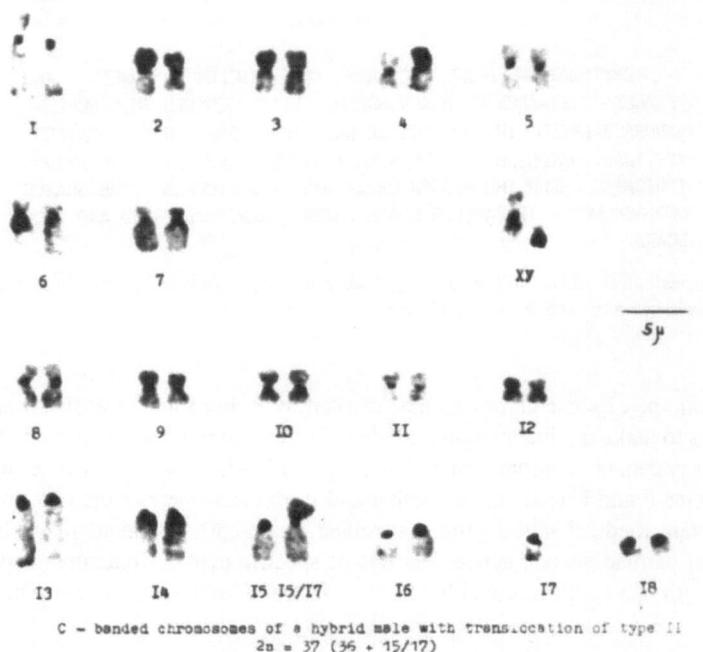

C - banded chromosomes of a hybrid male with translocation of type II
2n = 37 (36 + 15/17)

Figure 6. C banded chromosomes of a male Minisibs with karyotype 2n = 37 Rb Tr 15/17.

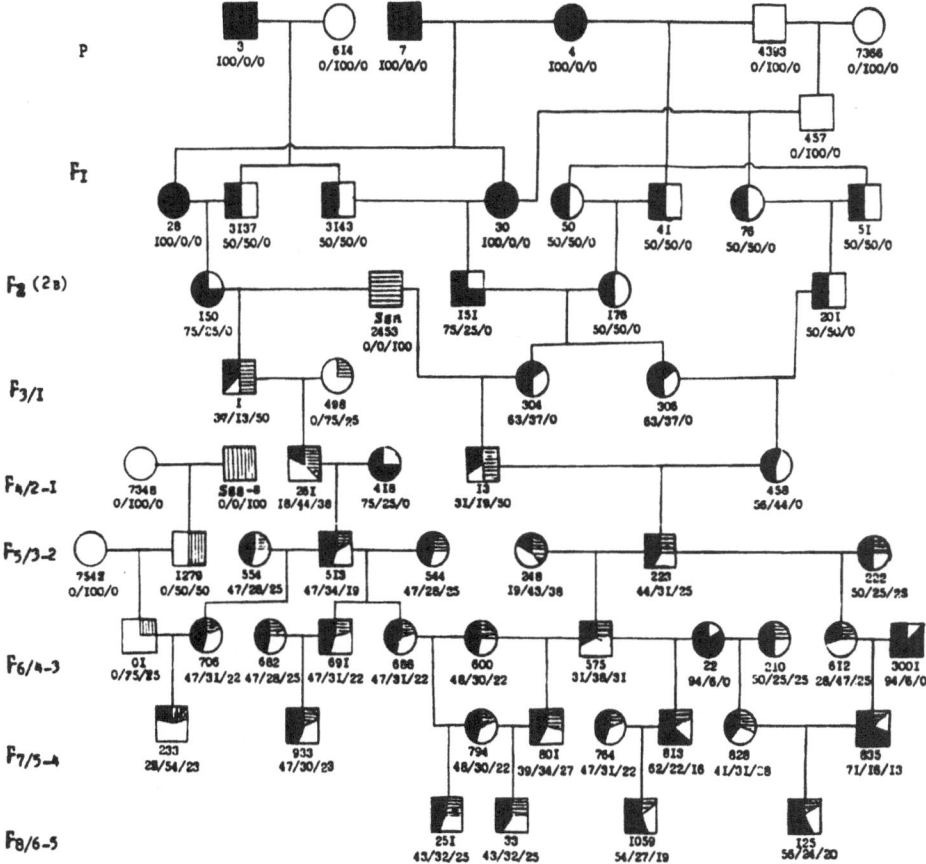

A GENEALOGICAL SCHEME OF THE CREATION OF
LABORATORY PIGS "MINISIBS" WITH KARYOTYPE POLYMORPHISM

DESIGNATION. SQUARES AND CIRCLES: THE WHITE ONES - BOARS AND SOWS WITH
NORMAL KARYOTYPES 2n = 38; WITH GORIZONTAL HATCHING - KARYOTYPES 2n = 36 (THE
HOMOZYGOTES FOR TRANSLOCATION RB 16/17); VERTICAL HATCHING - KARYOTYPES 2n =
36 (THE HOMOZYGOTE FOR TRANSLOCATION RB 15/17; HALF HATCHED SQUARES OR
CIRCLES) - HETEROZYGOTES FOR ONE OF THE TRANSLOCATION: 16/17 OR 15/17,(2n = 37)
AND FOR SYNTETIC KARYOTYPES (2n = 36), WHICH HAVE SIMULTANEOUSLY RB 16/17 AND
RB 15/17. UNDER THE NUMBER OF PIGS INDICATED THEIR ORIGIN, EXPRESSED AS THE
SHARES OF THE BREED GENOMES AND OF WILD BOAR, INVOLVED IN DEVELOPMENT OF
"MINISIBS" ("VIETNAMEESE / LANDRACE/WILD BOAR)

Figure 7. A fragment of the genealogical scheme for creation Minisibs with karyotype polymorphism.

4. RESULTS

The genetic linkage of the blood group system F locus with the epistatic white coat color locus "I" was discovered in the process of breeding white Minisibs in 1983.[12,13] A family analysis of association between the inheritance of alleles of F blood group system and alleles of white coat color "I" confirmed a close statistically reliable linkage between them. Special populations of black homozygous and white heterozygous animals with respective genotypes iiE E FbFb and IiE E FaFb, which are suitable for analytic crosses, were created. In this case, splitting in the locus E does not happen. That is why in linkage

Table 3. Study of genetic linkage of loci, determining the blood group F system and the system of color coat "I" (Analytic back crosses: Male N 807 I/i Fa/Fb x Females i/i Fb/Fb)

	Phenotypes and Genotypes of Progeny			
Traits	White		Nonwhite	
Color and blood group genotypes	I/i Fb/b	I/i Fa/b	i/i Fb/b	i/i Fb/a
Number of offspring observed	1	11	14	2
Expected in free segregation	7	7	7	7

$X^2 = 18.0$; $p<0.001$; $df = 3$

tables, the locus E is absent. In Table 3, data of analysis of 3 families (28 pigs) from boar N 807 demonstrate a close association of inheritance of genes "I" and Fa in the progeny. It is evidence of a genetic linkage of epistatic white coat color loci with the blood group system F.

Among 28 offspring, there are only 3 recombinated piglets (10.7% of crossing over); therefore, the hypothesis about the independent segregation is rejected ($p<0.001$). In addition, analytic crosses of 22 nonwhite iiFbFb females with 7 double heterozygous white boars with genotypes IiFaFb (cis position of the allele i with respect to allele Fb) were made. In 27 families, there were 234 piglets. Only 39 piglets were cross overs and it provides a possibility to make a convincing conclusion, that there is a genetic linkage of loci F and "I" ($p<0.0001$), but the distance between them is equal to 16.7 cM. The large distance and complicated interference of alleles determined a lot of coat color loci may account for the great diversity of results obtained on different breed populations.

The next step in intrachromosomal mapping of blood group F and white coat color loci on chromosome 17 of minipigs was made by us in 1985.[13] Thanks to the fact that chromosomes 15, 16 and 17 were substituted by the marker translocations Rb 15/17 and 16/17, chromosome 17 was identified reliably. A preliminary population analysis of 556 animals demonstrated that the frequency of the genotypes of system F and translocations did not differ from the theoretically expected one (Table 3).

For analytic back crosses, 5 boars were selected, which had centromeres of chromosome 17 and locus of the system F in cisposition and a linkage between them was discovered. Parents of the boars had the locus F on the chromosome which determined the antigen Fa and the centromere of chromosome 17 only in cisposition (genotype = Rb+Fa//Rb-Fb). All of these 5 diheterozygous boars were mated with dihomozygous females without translocations, which had genotype Fb/b. Among 118 newborn piglets, one part had normal chromosomes without translocations, and another part had chromosomes with Robertsonian translocations (Rb+)[7,14] or Rb.[7,10] Thanks to this situation, it was possible to determine exactly the chromosome which carried the locus F and estimate the recombination distance of this locus from the chromosome centromere (Table 4). Crosses of dihomozygous females with boars which had the identified translocation Rb in chromosome 17[7,10] demonstrated ($p<0.05$) the F locus is situated in Rb[7,10] or on chromosome 17 or 15 (Table 4). Crosses of dihomozygous females with boars, which had the identified translocation Rb in chromosome 17 (17/16), gave proof the F-locus is situated in Rb (17/15). The simultaneous association of the antigene Fa with Rb (17/15) and Rb (17/16) allows us to conclude the locus F is situated in chromosome 17 but not in chromosomes 15 or 16.

The conclusion that the locus F of blood group pigs of is situated just in chromosome 17 was confirmed by back crosses with minipigs which had both Rb 15/17 and Rb 16/17 (Table 4) ($p<0.001$). Data about recombined animals in back crosses (only 39 of 118) allows us to make an important conclusion about a long recombination distance between the locus

Table 4. Study of genetic linkage of loci, determined of blood group F system and the system of white coat color "I" by a family analysis (Analitic back crosses: males - double heterozygous on white color and on F-system Ii Fb/a, all females are double homozygotes for loci I and F: ii Fb/b)

Traits	Phenotypes and Genotypes of Progeny			
	White		Nonwhite	
Color and blood group genotypes	I/i Fb/b	I/i Fa/b	i/i Fb/b	i/i Fb/a
Number of offspring observed	88	17	22	107
Expected in free segregation	58.5	58.5	58.5	58.5

$X^2 = 107.3$; $p<0.0001$; df = 3

F and the centromere of the short acrocentric chromosome 17, namely 33.1 ± 4.3 cM (Table 5).

The mapping of the system F locus has allowed to make a new step and to map the locus "I", which determines the white coat color of pigs. According to data of Tikhonov et al.,[12] the recombination distance between loci F and "I" is 16.7 cM. Theoretically, locus I can be situated between the centromere of chromosome 17 and the locus F (16.4 cM from the centromere) or more distally from it (49.7 cM). However, since a combined inheritance of loci I and Rb[7,14] and Rb[7,10] in practice is not observed, it is possible to make a conclusion that the locus I of white coat color of pig is situated in chromosome 17 in the site, whose distance from the centromere of the chromosome is equal to 49.7 cM. This model investigation is the first case of simultaneous mapping of a blood group locus and of a locus, which determines the breed peculiarities in a definite chromosome at a known distance from the centromere in a large species of domestic animals.

There is another example of complicated genetic investigation, which allowed us to map another 2 blood group loci on identified chromosome 15 and to find between them a locus determining the embryonic lethality (and viability) (Table 5). Our Minisibs remarkable for chromosome polymorphism demonstrated the possibility of retaining a normal fertility (8 to 10 piglets/litter) and a high viability of domestic pigs after the substitution of their autosomes 15, 16 and 17 by genetic aberrant equivalents from *Sus scrofa* ferus and *Sus scrofa* nigripes. Artificial substitution does not affect the normal course of meiosis. The dwarf size of the minipigs with substituted chromosomes indicated that the studied chromosomes did not carry the gene determining growth and early maturity and controlling corresponding hormonal systems responsible for growth.

The use of Minisibs as recipients of aberrant translocated chromosomes allowed us for the first time in the history of animal breeding to produce an artificial synthetic form of laboratory pigs with chromosome polymorphism. This chromosome engineering is suitable for gene mapping.[13] Locus G was mapped by us at a distance of 15 to 20 cM from the

Table 5. Segregation in back crosses of diheterozygous males with dihomozygous females, type of crosses: Tr Rb+ Fa/Tr Rb- Fb x Tr Rb- Fb/Tr Rb- Fb

Genotype of system F	Karyotypes of Rb Tr (17,16 and 17,15)		Number of animals
	+/-	-/-	
F a/b	38	20	58
F b/b	19	41	60
	57	61	118

$X^2 = 12.21$; $p<0.001$; df = 1; number of recombinations = 39

centromere of chromosome 15. After our long term investigation, some data about localization of G locus of blood groups were confirmed in the literature.[15,16]

After creating a population of minipigs with special lines carrying Robertsonian translocations identified by alleles of 10 blood groups systems, 2 lines were derived from Ssf and Ssn (European and Asiatic wild boars). The mapping was made by us using special lines. For this we made back crosses in generations of 8 to 12 which had simultaneous heterozygosity for translocations only at loci of blood group system with cis and transpositions of marker alleles to respectively homozygous animals.

After only 7 yr of investigations we could make a conclusion about the linkage of loci E with loci G in chromosome 15.[14] Total data on all populations did not give a proof sufficient to make a conclusion about the statistical deviation from the theoretically expected data. Only the results of back crosses in one line, which has Robertsonian translocation in chromosomes 15 and 17 demonstrated a significant deviation from the normal situation, and made it possible to confirm the hypothesis about a linked inheritance of alleles of loci G and E (Table 6).

The linkage of loci E and G has been confirmed by data obtained additionally in crosses of heterozygous animals and those with cis and transposition of marker alleles Ga, Eedg and Ebdg, but only when the crossed animals had the translocation in chromosome 15, which may be explained by the suppression of crossing over. The recombination between loci E and G is 37 to 44% (Table 7).

The hypothesis about mapping of locus E in chromosome 17 was rejected. The independent segregation of alleles loci E and F, which have been mapped by us in chromosome 17, refutes this supposition. The linkage of loci E and G has been confirmed in another minipig population without translocation Rb15/17 (Yermolaev and Gorelov).[17] If locus E is situated in chromosome 15, the distance between loci E and G is equal to 40 cM. That is proof that the distance of locus E from the centromere of chromosome 15 being 50 to 60 cM.

Another important genetic discovery on Minisibs is the intrachromosomal localization of the gene of prenatal viability. Pig genome has a size of 2000 to 2300 cM, of which 90 to 95%, including 16 of 18 chromosomes, already are marked by 500 to 620 genes and linkages. These data were obtained during the last 10 yr as a result of collaboration with 20 laboratories in Europe and the USA. The strategy which is used in all programs of drawing up genetic maps is based on the study of linkage between loci by means of back crosses of animals belonging to as remote races as possible. In the USA and Europe, to localize the gene of multiple pregnancy, crosses of a fertile Chinese Meshansk breed of pigs (the average number of fetuses is 12 to 14 pigs at farrowing), with pigs of Large White and other European and American breeds, which have an average fertility 2 to 3 piglets less.

Table 6. Segregation in back crosses by using heterozygous animals with cis position of marker alleles of systems E and G (a Line Mini-Belfer, n = 150)

Genotypes of system G	Genotypes of progeny according to markers of system E		
	E+	E-	Total
Ga-positive	40	23	63
Ga-negative	38	49	87
	78	72	150

$X^2 = 5.74$; $p<0.05$; df = 1

Table 7. Segregation in analytic crosses by using heterozygous animals for systems E and G with transposition of marker alleles (Line Mini-Belfer, Type crosses: Gb/a E+/- x Ga/a E-/-)

Alleles of systems		Genotypes of progeny				Number of recombinations	Total of pigs	Crossing over, %
G	E	Gb/a E-/-	Gb/a E+/-	Ga/a E-/-	Ga/a E+/-			
G b	E edg	10	10	1	4	11	25	44.0 ± 9.3
G b	E bdg	9	5	5	8	10	27	37.0 ± 9.3
Total of animals		19	15	6	12	21	52	40.4 ± 6.8

During 20 yr of creating the laboratory minipigs (Table 8), we carried out many hybridizations of pigs of Swedish Landrace and Vietnamese breeds (fertility is 11 and 9 live pigs/litter, respectively) with wild boars of the European, Central Asian and Far East subspecies (*Sus scrofa* scrofa, *Sus scrofa* nigripes and *Sus scrofa* continentalis). For wild boars, low fertility and a small number of live piglets (on the average, less than 8) are typical. In contrast to domestic pigs, wild boars have a limited polymorphism for alleles of the blood group E system; simultaneously the wild boars, which were used in hybridization for 12 and more generations, possessed translocations in a few chromosomes.

The genetic parental forms and their descendants were studied for 12 and more generations by immunogenetic and cytogenetic methods with respect to markers and fertility. It allowed us to obtain informative hybrid litters for carrying out a family analysis by the markers of 10 blood groups genes simultaneously on the inheritance of the Robertson's translocations in several chromosomes. As a result, it became possible to study the linkage and draw up the first genetic maps of identified autosomes of the pig (Figs. 8 and 9).

The low fertility of wild boars is caused by the smaller number of follicles ovulating during the estrous cycle. Domestic pigs have a higher, than wild boars, so called potential fertility and simultaneously a considerable prenatal mortality rate of the embryos and fetuses; as a rule, more than 30%. On the contrary, in spite of the rather low potential fertility, typical for the ontogenesis of wild boars, just as for aboriginal pigs, is high postnatal viability. It is the combination of the latter with the high potential fertility of the Landrace pigs that accounts for the high heterosis of fertility, discovered by us, when creating Minisibs by means of crossing wild boars with Landrace, which agrees well with the theory of origin of the compensating complexes of genes.[18]

The high fertility in crosses of wild boars with Landrace and Vietnamese pigs is accompanied by the preservation in the genotype of the allele Eedg, transmitted from

Table 8. Effect of the long term hybridization of wild boar with Landrace on fertility and viability traits during the creation of Minisibs

Generation	Number of litters	Number of newborn pigs		Average fertility	Number of stillborn in a litter		Average number of piglets in a litter
		total	alive		abs.	%	
F 1	12	151	140	12.6	0.92	7.3	11.7
F 2b	15	188	173	12.5	1.00	8.0	11.5
F 3b	17	194	190	11.4	0.23	2.1	11.2
F 4b	26	320	297	12.3	0.88	7.2	11.4
F 5b	35	415	397	11.9	0.51	4.3	11.3

Figure 8. Pregnant adult Minisibs, average fertility of 8 to 9 live piglets/litter.

generation to generation from wild boar ancestors with a higher frequency than it is possible to expect according to theoretical calculations. Such a selective advantage, as well as the exceptionally wide spread of the allele Eedg of wild boars, may be the cause of the linkage of the blood group E system with the fertility appearing in the analysis.

In a recent paper by a group of American scientists about the discovery of the major gene for litter size,[19] as the gene of the fertility, it was shown this phenomenon is associated with polymorphism of the estrogen receptor gene. However neither in this nor in other publications are there data about the localization of this gene in any identified chromosome.

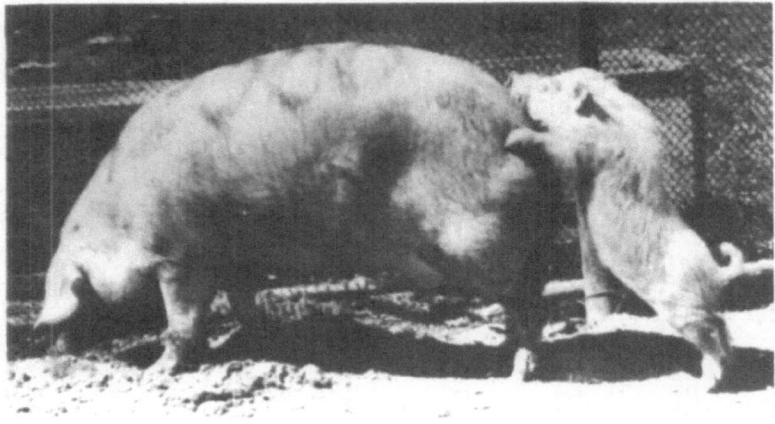

Figure 9. Adult Minisib boar near Landrace.

The fertility is determined not only by the number of ovulating follicles in one estrous cycle, but also by the number of fertilized ova and by the viability of embryos and fetuses during all of the period of intrauterine development (Table 8).

By analyzing back crosses of several years, 793 hybrid descendants were obtained. Segregation of the alleles of the system E at the highest level of significance showed the number of pigs with the identified allele E edg of the wild boar which is the ancestor in a number of the successive generations more than twice exceeds the number of E edg negative pigs. The supposition about the equal segregation of the alleles and free recombination, and about the pleiotropic effect of the allele E edg on the fertility is rejected by the data in Table 9 at the highest level of significance ($p<0.01$, $X^2>10$). Therefore, the locus E is linked closely with an unknown locus which controls the viability and is localized at a short distance from it.[20]

The higher viability of progeny inheriting the allele E edg of the Wild boar was proven in 2 more experiments by identification of 1020 animals, obtained in 91 crossings of experimental pigs at the Academy of Science and 664 animals on an industrial pig farm (Kemerovo region). Thus, the number of Eedg positive newborn pigs was by 1.54 times more than that of negative ones (Table 9). The close linkage between the loci of the system E blood group and prenatal viability also was proven in analysis of inheritance of other alleles of system E. In particular, E edf and E aeg Landrace and Vietnamese breeds, which have been used to develop Minisibs.

Thus, the data testify for the presence at a short distance from the locus E of a gene not described earlier, which controls the prenatal viability, and which we called VE (viability, linked with the E system). The gene VE is represented by two alleles: VEA and VEa which form 3 genotypes AA, Aa and aa. The aa genotype, which does not include the dominant allele A, does not provide for the normal prenatal development. That leads to reduction of the actual number of embryos in pregnancy, because these do not survive the pregnancy period.

Analyzing different types of crossings shows the number of cross over pigs with respect to E and VE systems makes up about 7%. The distance between the locus of the E blood group system and the locus of viability VE, 7 cM, can explain the real frequency, observed in all types of crossing.

The chromosome location of the gene of the E blood group system is interesting not only in theoretical aspects, but also is of economic importance, because there are data on the linkage of the corresponding gene with the gene VE (prenatal viability of pigs). Our discovery and location of the locus of prenatal viability VE in chromosome 15 probably is not the only one which controls fertility. Fertility depends upon many genetic and paragenetic factors, which determine the number and synchronism of ovulatory follicles, circumstances

Table 9. Effect of the allele E edg of wild boar on viability of hybrids progeny in back crosses (males E edg/bdg x females E bdg/bdg)

Genotypes of system E	edg/bdg	bdg/bdg
Number of alive newborn piglets	544	249
Expected number of piglets		
1) in free recombination	396	396
2) in close linkage of allele E edg with any factor of viability	595	198

of fertilization and many specific stages of embryonal and fetal periods of ontogenesis. Thanks to the close linkage of VE and E loci, and to the exceptional polymorphism of the latter, it can be used effectively for selection of fertile minipigs and domestic normal pigs by marking alleles of the E blood group system. As a result of the present study, the location of the gene of parental viability VE, which determines the fertility of pigs, is in chromosome 15 at a distance of 43 to 53 cM from its centromere with due regard for the earlier made genetic map.[11,12,14]

5. DISCUSSION

The examples of genetic investigation, development of Minisibs and mapping of chromosomes demonstrate a wide spectrum of opportunities for miniature pigs by creating a number of different specific lines, which will be suitable for many genetic and biomedical purposes. There are different minipigs with very different predispositions to alcohol, genetic susceptibility and tolerance, medical possibilities of treatment of some forms of hereditary alcoholism in the nursery.

Minipigs are suitable subjects for the study of different aspects of atherosclerosis and for the search for respective loci, which determine the conservation of small body size throughout life, including sexual maturity. These specific lines are needed for the study of different drug addictions as well as their prognosis and treatment. Methods of heteroimmunodiffusion and especially heterotransplantation of cardiac valves are developed in Siberia. Methods of skin heterotransplantation (with and without immunodepressants) are important because we have a high frequency of incidents with scalds and burns; medical aspects of using minipigs are so great this problem needs special consideration. We propose an international collaboration to use new technology for the development of specialized genetic lines of minipigs for different biomedical purposes.

6. REFERENCES

1. Tikhonov, V.N., 1990, Genetics of mini-pigs, *Novosibirsk* 209.
2. Tikhonov, V.N., 1991, Immunogenetics and biochemistry polymorphism of domestic and wild pigs, *Novosibirsk* 303.
3. Panepinto, L.M., and Philips, R.W., 1981, Genetic selection for small body size in Yucatan miniature pigs, *Lab. Anim. Sci.* 31: 403-404.
4. Bustad, L.K., and Horstman, V.G., 1986, Pigs: from B.C. to 2000 A.D.: from outhouse to penthouse, in: *Swine in Biomedical Research*, Volume 1 (M.E. Tumbleson, ed.), Plenum Press, New York, pp. 3-15.
5. Panepinto, L.M., and Philips, R.W., 1986, The Yucatan miniature pig: characteriation and utilization in biomedical research, *Lab. Anim. Sci.* 36: 344-347.
6. Tikhonov, V.N., and Panarina, L.M., 1981, Some genetic peculiarities of the USSR mini-pigs "Minisibs". 1. Laboratory miniature pig "Minisibs", its establishment and morpho-physiological features, *Genetics (Moscow)* 17: 883-895.
7. Tikhonov, V.N., and Troshina A.I., 1975, Chromosome translocations in karyotypes of wild boars *Sus scrofa* L. of the European and Asian regions of the USSR, *Theoret. Appl. Genet.* 45: 304-308.
8. Tikhonov, V.N., and Troshina, A.I., 1980, Marker chromosomal translocations in development of commercial Landrace X Wild Boars hybrids and Siberian mini-pigs, *Proc. 4th Eur. Colloq. Cytogenet. Domestic Anim.* pp. 242-248.
9. Tikhonov, V.N., and Troshina, A.I, 1978, Introduction of two chromosome translocations of *Sus scrofa* nigripes on *Sus scrofa* scrofa into the genome of *Sus scrofa* domestica, *Theoret. Appl. Genet.* 53: 261-264.
10. Tikhonov, V.N., Gorelov, I.G., and Panarina, L.M., 1981, Immunogenetic peculiarities of laboratory mini-pigs "Minisibs", *Genetics (Moscow)* 17: 1677-1689.
11. Tikhonov, V.N., and Nikitin, S.V., 1983, Mapping of locus of the blood group system G in the identified chromosome of domestic pigs, *Doklady USSR Acad. Sci.* 272: 214-216.

12. Tikhonov, V.N., Nikitin, S.V., and Astakhova, N.M., 1985, Mapping of loci of the blood group system F and white color coat I on the pig chromosome 17, *Doklady USSR Acad. Sci.* 280: 1258-1261.

13. Tikhonov, V.N., 1983, The genetic linkage of the blood group system F locus with the coat color I in swines, *Doklady USSR Acad. Sci.* 272: 719-722.

14. Tikhonov, V.N., Solodukha, K.I., and Bobovich, V.E., 1994, Intrachromosome mapping of the locus E of the blood group system of wild and domestic pigs, *Doklady USSR Acad. Sci.* 337: 697-699.

15. Fries, R., Rasmusen, B.A., Jarrell, V.L., and Maurer, R.R., 1984, Mapping of the gene for G blood group antigenes to chromosome 15 in swine, *Anim. Blood Gp. Biochem. Genet.* 15: 251-258.

16. Fries, R., Stranzinger, G., and Vogeli, P., 1983, Provisional assignment of the G blood group locus to chromosome 15 in swine, *J. Hered.* 74: 426-430.

17. Yermolaev, V.I., and Gorelov, I.G., 1993, Personal communication.

18. Strunnikov, V.A., 1974, The origin of the compensator complex of genes-one of principal resons of heterosis, *J. Gen. Biol. (Moscow)* 35: 666-677.

19. Rothschild, M.F., Jacobson, C., Vaske, D.A., Tuggle, C.K., Short, T.H., Sasaki, S., Eckardt, G.R., and McLaren, D.G., 1994, A major gene for litter size in pigs, *Proc. 5th World Congr. Genet. Appl. Livestock Prod.* pp. 225-228.

20. Bailey, N.T.J., 1961, *Introduction to the Mathematical Theory of Genetic Linkage* Clarendon Press.

62

PIGS AS MODELS FOR NUTRIENT FUNCTIONAL INTERACTION

Peter Reeds[1] and Jack Odle[2]

[1] Pediatrics
Baylor College of Medicine and USDA/ARS/CNRC
1100 Bates Street
Houston, Texas 77030
[2] Animal Sciences
North Carolina State University
Raleigh, North Carolina 27695-7621

The use of pigs in studies, not only of growth and development of mammalian neonates but also of degenerative problems that accompany aging, has a long history. In this section, scientists address the use of pigs in both respects. Partly because the pig will consume a wide variety of diets, there also has been a traditional emphasis on investigation of the impact of nutrition on these aspects of health. The continuing strength of this area of investigation is born out amply by the papers in this section and they are examples of how the research continues to expand in scope. In so doing, interest is turning increasingly towards studies that focus on mechanistic, including genetic, investigations. Specific foci of the nutrition papers are: a) metabolic development and the use of the pig as a model for studying human neonate organic nutrient metabolism, b) intestinal development with an emphasis on functional development, c) the role of the pig as a model for the impact of intestinal disease, particularly diarrheal illness, on intestinal function and d) the use of pigs in studies of genetic manipulation of lipid metabolism and relationships of lipid metabolism to circulatory diseases.

During the overall meeting, there was considerable discussion about advantages and disadvantages of the piglet as a model for human nutrition, growth, development and degenerative disease. In the most general sense, there seems little doubt that studies in the pig are of more direct relevance to the human than are studies in the rat and other rodents. As far as developmental studies are concerned, it is critical to remember the rodent is an altricial mammal and, as such, is born with many parts of its physiological function in a substantially immature state. This is particularly true of the intestinal system, and papers in this section (Buddington and Sangild) are focused on digestive function development, studied at the enzyme level, and intestinal growth, both *in utero* and in the perinatal period. At the functional level, there are potentially important details in which newborn pigs and humans differ, e.g., the level of expression of sucrase and maltase, but many aspects of neonatal pig small intestine development are similar to those of the human, and markedly different from rodents.

A major area of intestinal physiology in which the human and pig differ occurs during the first two days of life, a period which, in the pig, is marked by intense endocytosis of colostral proteins. This has been well established as being of critical importance to the acquisition of passive immunity and, as pointed out by Gomez, as well as by Krakowka, it is a critical factor in providing defenses against a variety of potential intestinal pathogens, such as rotavirus, a major factor in the morbidity of infants and children throughout the world. This aspect of porcine development also receives specific attention in the paper by Donovan, a paper concerned with the functional and growth promoting roles of colostral borne growth factors in general and of the insulin like growth factors in particular. This is an area of neonatal growth and development that has clinical relevance, particularly with regard to potential therapeutic interventions in infants, including those born prematurely (Sangild). Appropriate investigations demand invasive studies, hence a suitable animal model; this is precisely the area of mammalian growth and development in which studies in species where intestinal development is substantially immature at birth are inappropriate.

Other aspects of neonatal development in the pig also are different form those of humans. As pointed out by Pond, body triglyceride stores of the newborn pig are lower than those of the newborn human. On one hand, this observation reflects the fact that porcine white adipocyte undergoes its final differentiation immediately after birth; therefore, is renders the newborn pig as a good model for the study of the later stages of adipocyte maturation. In addition, as emphasized by Duee, the low fat stores of the neonatal pig make it particularly useful in the study of lipid and glucose metabolism interactions in the neonate. On the other hand, the very feature of porcine metabolic development that makes it an excellent model for *in vivo* study of general aspects of gluconeogenic control, might limit the degree to which the results can be extrapolated directly to the full term human neonate. Even so, in this important area of research, lower body fat stores of the newborn piglet render it a much more appropriate model for evaluating metabolic problems of the low birth weight human infant. This is a portion of the human population that continues to increase and presents a number of pressing clinical nutritional problems.

The piglet has two further advantages in the study of neonatal nutrition, i.e., size and growth rate. The former is a particular advantage if the focus is on metabolism in general because, as illustrated by Ball and Duee, the ability to maintain permanently catheterized and growing animals allows a broad spectrum of metabolic investigations to be carried out. Both also, illustrate the power of combining the piglet model with appropriate isotopic techniques. Although in some senses the high potential growth rate of the pig could be seen as a disadvantage when it is used as a model for the much more slowly growing human neonate, it confers substantial advantage to experimental design because it allows a wide range of nutritional and metabolic circumstances to be investigated. These advantages are put to use by Ball, who reported extensive studies concerned with the use of the pig to define nutrient requirements of infants receiving parenteral nutrition. While he focused on amino acid nutrition, it is important to emphasize that Ball and Duee both illustrated the facility with which the piglet can be used for *in vivo* studies of organic nutrient metabolism in healthy neonates.

The piglet has advantages over many other animal models in the study of disease, especially that of the intestinal tract. As we emphasized above, many aspects of intestinal structure and function are similar in the pig and the human and these similarities extend to their susceptibility to analogous organisms, including *E. coli* and rotavirus. Three papers (Krakowka, Gomez and Chandra) are focused on the world wide problem of diarrheal illness and its effects, not only on mortality, but also on intestinal growth and function. Chandra also commented on the impact of disease on intestinal resident microflora, another aspect of intestinal function in which there are similarities between the pig and the human. Finally, similarities in susceptibility of humans and pigs to the same organisms is used in the study of Krakowka, who reported the use of gnotobiotic pigs to study the pathogenesis of invasion by *Helicobacter pylori*, a pathogen that plays a critical role in human gastric disease. The key feature of all three papers is their emphasis on the human relevance of the porcine model and its use in closely focused and mechanistic studies.

Throughout these volumes, the similarities of the pig and the human at a variety of levels, from anatomic to genetic, receive attention. Further similarities between the two species is the wide variety of diets they consume, their propensity for a sedentary lifestyle and their susceptibility to degenerative diseases of the circulatory system. As pointed out by Pond, the use of the pig as a model for human cholesterol and lipoprotein metabolism, and its subsequent investigation as a model for blood vessel disease, has as long an history as its use in the study of growth and development. Sarwar also reported on interactions between dietary protein and fat sources and the circulating level of homocysteine, a methionine metabolite that has been identified from epidemiological studies as an independent risk factor for coronary artery disease.

In addition, the pig is a prolific species with a relatively short developmental period. Thus, it lends itself to major alterations in phenotype brought about by conventional breeding. Pond reported results of such interventions and their application in investigations of interactions between metabolic phenotype (as its relates to obesity, cholesterol and lipoprotein metabolism), nutritional environment, growth, central nervous system development and degenerative disease. By demonstrating the deleterious effects of derangements of cholesterol and lipid metabolism on both the newborn and the mature, and by relating these to measurable genotypes, he indicated, in a particularly apposite manner, the way in which research, using the pig as a model of human development and disease likely will develop. The size, nutritional and metabolic peculiarities of the pig allow increasingly subtle nutritional and metabolic experiments, while our increasing understanding of the genome of the species, particularly when coupled with suitable application of transgenic technologies will allow these investigations to move to the molecular level for the ultimate benefit of our understanding of both the pig and the human.

63

A PIGLET MODEL FOR NEONATAL AMINO ACID METABOLISM DURING TOTAL PARENTERAL NUTRITION

Ronald O. Ball,[1,2,3*] James D. House,[1,5] Linda J. Wykes,[6] and Paul B. Pencharz[1,3,4,5]

[1] Animal and Poultry Science
[2] Human Biology and Nutritional Sciences
University of Guelph
Guelph, Canada N1G 2W1
[3] Nutritional Sciences
[4] Paediatrics
University of Toronto
[5] Hospital for Sick Children
Toronto, Ontario, M5S 1A8
[6] School of Dietetics and Human Nutrition
McGill University, MacDonald Campus
Ste-Anne-de-Bellevue, Quebec, Canada, H9X 3V9

1. INTRODUCTION

Premature birth is a leading cause of infant morbidity and mortality. An important factor in the care of these neonates is the establishment of appropriate nutritional strategies. Low birth weight (LBW) infants often have an inability to tolerate oral feedings, due to a variety of factors, including short bowel syndrome, gastrointestinal surgery, chronic severe diarrhea, immature bowel function[1] and respiratory diseases.[2] In these instances, the provision of nutrition by parenteral means is necessary. Total parenteral nutrition (TPN) involves the infusion of amino acids, glucose, lipid, vitamins and minerals directly into the venous circulation. As with all neonatal nutrition regimens, the aim in providing TPN to LBW infants is to promote the growth of the infant, without imposing undue stresses on metabolic pathways or creating adverse long term outcomes.

A great many ethical and practical constraints exist regarding experiments in fragile, LBW infants. Most investigations designed to determine empirically nutrient requirements

* Reprint requests to Dr. Ronald O. Ball, Animal and Poultry Sciences, University of Guelph, Guelph, Ontario, Canada, N1G 2W1, (519) 824-4120.

Advances in Swine in Biomedical Research, edited by Tumbleson and Schook
Plenum Press, New York, 1996

or quantify amino acid kinetics are ethically unacceptable because treatments which could be deficient in a nutrient could endanger the infant. Therefore, we developed a piglet model to study amino acid kinetics and requirements during TPN. The long term goal of these studies is to make significant improvements in existing parenteral amino acid solutions by matching the amino acid supply to the neonate's requirement for obligatory oxidation, maintenance and growth.

In this review, we include: a description of existing amino acid solutions used in TPN to illustrate the need for in depth studies designed to determine the requirements for individual amino acids during total parenteral nutrition, a discussion of limitations to nutritional studies in human infants which required the development of an animal model, a comparison of piglet and infant metabolism and a description of the advantages and applications of a piglet model. Procedures and protocols for a piglet model for TPN will be described and recent data derived from the use of this model for the development of an optimal amino acid pattern for neonates will be reviewed briefly.

2. WHY CURRENT TPN SOLUTIONS NEED IMPROVEMENT

Significant advancements have been made in TPN since initial studies during the late 1960's and early 1970's illustrated its potential benefits.[3,4] Initial parenteral amino acid solutions were hydrolysates of proteins, usually casein or fibrin, that allowed limited flexibility in formulating TPN solutions. Availability of crystalline amino acids allowed for development of the current generation of amino acid solutions. However, due to a lack of empirical data, the optimal pattern of amino acids for TPN fed LBW infants has yet to be defined. Although amino acid solutions for parenteral feeding of neonates have been formulated and marketed, questions remain as to their suitability.

Several amino acid solutions currently in use (Table 1) are based upon amino acid patterns of reference proteins of high biological value when fed enterally, such as egg albumen (Vamin, Kabi Pharmacia) and human milk protein (Vaminolact, Kabi Pharmacia). Although the concept of a reference protein is sound for enterally fed subjects, this approach may not be justified during TPN feeding. Amino acids administered parenterally bypass splanchnic control mechanisms, including first pass uptake effects at the liver. Amino acid kinetics in LBW were shown to be different depending on the route (oral vs parenteral) of nutrient delivery.[5] These differences in amino acid kinetics and physiological responses to TPN have implications for the overall pattern of amino acids required by the TPN fed neonate.

Other TPN solutions (Table 1) were developed from reference protein patterns by adding or reducing amino acids to address specific abberations in plasma amino acid profiles (Travasol Blend C, Clintec; Aminosyn PF, Abbott Laboratories). Another solution (Trophamine, Kendall-McGaw) was developed using an iterative approach with multiple regression analysis to predict plasma amino acid concentration. The problem with all these approaches is the definition of the "normal" plasma amino acid profile. What are "normal" plasma amino acid concentrations for the TPN fed infant? Should the reference pattern represent oral or intravenous feeding during either the fed, postprandial or fasting state and sampled from either capillary, venous or cord blood? Arguments for and against each position exist.[6] Further significant improvements in TPN profiles are unlikely to result from the "best guess" plasma amino acid approach. The next generation of improvements will require direct empirical experimentation involving quantitation of amino acid and protein kinetics. Our goal is to develop an optimum amino acid pattern for neonatal TPN, with the optimum pattern defined as one which meets the demands for maintenance functions, protein accretion and synthesis of nonprotein metabolites while simultaneously providing no amino acids in

Table 1. The amino acid profile of commercially available parenteral amino acid solutions

Amino Acid	Vamin[1]	Aminosyn[2]	Vaminolact[1]	Aminosyn PF[2]	Travasol Blend C[3]	Trophamine[4]	Primene[1]
	(g/100 g Amino Acids)						
Alanine	4.3	12.9	9.7	7.0	20.7	5.4	7.9
Arginine	4.7	9.9	6.3	12.3	11.2	12.2	8.4
Aspartate	5.9	0	6.3	5.3	0	3.2	6.0
Cysteine	2.0	0	1.5	0	0	0.1	1.9
Glutamate	12.9	0	10.9	8.2	0	5.0	9.9
Glycine	3.0	12.9	3.2	3.9	10.3	3.6	4.0
Histidine	3.5	3.0	3.2	3.1	4.8	4.8	3.8
Isoleucine	5.6	7.3	5.5	7.6	6.0	8.2	6.7
Leucine	7.5	9.5	10.8	11.9	7.3	14.0	9.9
Lysine	5.5	7.3	8.6	6.8	5.8	8.2	10.9
Methionine	2.7	4.0	2.0	1.8	4.0	3.4	2.4
Phenylalanine	7.9	4.7	4.2	4.3	5.6	4.8	4.2
Proline	11.6	8.7	8.6	8.1	6.8	6.8	3.0
Serine	10.7	4.2	5.8	5.0	5.0	3.8	4.0
Threonine	4.3	5.2	5.5	5.1	4.2	4.2	3.7
Tryptophan	1.4	1.6	2.2	1.8	1.8	2.0	2.0
Tyrosine	0.7	0.9	0.8	0	0.4	2.3[5]	0.9
Valine	6.1	8.1	5.5	6.6	5.8	7.8	7.6
Taurine	0	0	0.5	0.7	0	0.2	0.6

Manufacturer: [1]Kabi, [2]Abbott, [3]Clintec, [4]Kendall-McGaw.
[5]Approximately 75% of tyrosine present as N-acetyltyrosine.

excess. Since the empirical methods necessary to achieve this goal are difficult or impossible to use in the infant (see Section 3), we have chosen to use the piglet to develop new methods with clinical applications and to evaluate specific changes in the amino acid profile of TPN.

3. WHY USE A PIGLET MODEL FOR THE HUMAN INFANT?

3.1. Limitations to Nutritional Studies in LBW Infants

Clinical research with neonatal patients is subject to many ethical and practical constraints, which make definitive studies difficult to conduct. Infants under intensive care are a heterogeneous population with a variety of illnesses and often are faced with a number of challenges simultaneously, including infection and recovery from surgery, which complicate metabolic demands and thus interpretation of experimental results. Very low birth weight infants are the infants most likely to require parenteral feeding. Because they are the most fragile with the most immature systems, they are the most vulnerable to metabolic effects of amino acid imbalances. This subgroup, that would benefit most from the precise tailoring of TPN regimens to their metabolic needs, is the most difficult to study.

Experimental design, including the design of TPN regimens, is limited in the clinical setting. Neonatal researchers generally have access only to blood, breath and urine and the sampling of these may be restricted for ethical[7] and technical reasons, thus limiting the extent to which nutritional interventions can be monitored. Many of the precise, sensitive methods required to evaluate the high quality amino acid solutions presently available are not clinically feasible. Refinement of the amino acid profile of neonatal TPN depends on understanding of the precise metabolic effects of changes in amino acid supply during TPN. However, as solutions continue to improve, differences in their effects will become more subtle and difficult to elucidate in the clinical setting. Application of an animal model would permit experimental studies involving a healthy, less heterogeneous population, specifically designed feeding regimens and more invasive and sensitive methods which would not be acceptable clinically. Such a model could provide a more comprehensive and in depth picture of TPN metabolic effects.

3.2. Similarities of Porcine and Human Neonates

The neonatal piglet is considered[8-11] to be a suitable model to study nutritional metabolism in the human neonate. This conclusion is based on the many similarities in anatomy, physiology, metabolism and nutrition between the two species.[8,9] Similarities between neonates of the two species include: rate and pattern of gastrointestinal tract development,[11,12] maturation and function of the kidney,[13,14] body composition,[15] pattern of brain growth and development,[16,17] and respiratory and haematologic systems.[14] Pond and Houpt[9] concluded the nutrient requirement patterns of the pig are more similar to the human requirements than those of any other nonprimate mammal.

The ontogeny of the fetal human and pig are similar. Although the gestation period is approximately 114 days in the pig and 285 days in the human, the two species spend a similar percentage of time at each stage of fetal development, except during late stages to which a greater proportion of the human fetus' gestation is devoted.[8] This is because litter species, like the pig, are born relatively immature compared to single birth species.

The neonatal piglet may be more similar to the human infant born prematurely than the infant born at term[11] in several respects: low birth weight, low fat reserves, low thermoregulatory ability, high metabolic rate[8] and susceptibility to hypoglycaemia.[21] These comparisons support the conclusion that the newborn piglet is an appropriate model for the LBW infant.

The piglet has been used frequently to investigate the effects of TPN on the gastrointestinal tract. The absence of luminal nutrients results in significant atrophy of the gastrointestinal tract, which is not reversed despite adequate feeding by parenteral means.[18] We[19] compared the growth and development of body tissues in piglets fed TPN, sow's milk replacer or sow's milk. Piglets receiving TPN had lower weights for digestive organs, including small and large intestines, and stomach, compared to orally fed piglets. Similar findings have been reported by Burrin et al.,[20] who concluded the atrophy of the gut mucosa could not be reduced by providing the amino acid glutamine, a preferred energy substrate for rapidly dividing cells, to piglets on TPN. As data accumulates,[19] it is apparent that TPN in the piglet model produces similar effects on the gastrointestinal tract as observed in the human infant.[18]

3.3. Similarities in Protein Metabolism Between Species

The piglet is an appropriate model to study questions relating to protein and amino acid metabolism in infant nutrition for several reasons. Protein synthesis rate is higher in younger members of each species.[22-25] The concept of metabolic scaling confers advantage to the pig over other species as a model for the human.[25] Metabolic activity per unit weight is affected by size and is compensated for by scaling. However, size also affects organ distribution of protein synthesis.[26] Therefore, data from the pig are more similar to the human than results scaled up from the rat.[25]

Patterns of amino acid nutrition of pigs and infants are similar, including plasma amino acid concentrations.[27,28] The profile of plasma amino acid concentrations is similar in the piglet and human neonate receiving TPN compared to their mother fed counterparts.[29] Few detailed investigations of the consequences of altering amino acid composition of TPN on amino acid and protein metabolism in the piglet model have been reported. Glutamine supplementation of TPN did not affect intestinal growth or enzyme activity[20] or protein accretion.[30]

Estimated indispensable amino acid requirements (to the extent they are known) are proportionately similar in the orally fed growing piglet[31] and human infant.[32, 33] Dietary requirements for amino acids in the two species, expressed as percent of protein, are shown in Table 2 and compared to the amino acid profile of piglet and human fetal tissue, swine and human milk proteins, egg protein, and two parenteral amino acid solutions. Amino acid profiles for requirements, milk and tissue, are similar between the two species. These data lend support to the use of the piglet for the definition of an optimal amino acid pattern for TPN fed infants.

The concept of optimal amino acid profile for neonatal TPN is analogous to the concept of ideal protein in swine nutrition. Amino acid requirements in the pig have been investigated extensively.[31] Several amino acid profiles, which have been suggested as ideal for the growing pig, are listed in Table 3. The pattern of amino acid requirements for orally fed subjects, as they are currently understood, are similar for the two species when expressed relative to lysine requirement.

Amino acids are required by a growing animal for maintenance processes and tissue protein accretion. The pattern of requirements may be quite different for each component.[34] Consequently, any estimate of optimal amino acid profile represents the sum of these components and depends on their relative contributions to total needs. Thus, the overall

Table 2. Amino acid requirements of the piglet and human infant compared to the amino acid composition of various standards

Amino Acid	Piglet Reqt[1]	Human Reqt[2]	Piglet Tissue[3]	Human FetalTissue[4]	Swine Milk[5]	Human Milk[5]	Egg Protein[6]
	(g/100 g protein)						
Isoleucine*	3.2	3.2	3.6	3.5	4.0	5.3	5.8
Leucine*	4.2	7.3	6.9	7.5	8.9	10.4	9.0
Valine*	3.4	4.2	4.7	4.7	4.6	5.1	7.2
Lysine*	6.0	4.7	6.7	7.2	7.9	7.1	6.7
Methionine*	2.9[7]	2.6[7]	1.6	2.0	2.2	1.6	3.0
Cysteine*	—[7]	—[7]	1.5	n.d.	1.6	2.0	2.1
Phenylalanine*	4.7[8]	5.7[8]	3.7	4.1	4.3	3.7	5.3
Tyrosine*	—[8]	—[8]	2.8	2.9	3.9	4.6	4.3
Threonine*	3.4	4.0	3.9	4.1	3.7	4.4	5.3
Tryptophan*	0.9	0.7	n.d.	n.d.	1.4	1.9	1.8
Histidine*	1.5	1.3	2.5	2.6	2.4	2.3	2.6
Arginine	2.5	n.r.	6.9	7.7	4.4	3.6	6.4
Glycine	n.r.	n.r.	9.7	7.2	3.2	2.2	3.8
Alanine	n.r.	n.r.	6.6	9.0	3.6	4.0	n.d.
Aspartate	n.r.	n.r.	8.0	13.0	7.8	8.6	10.7
Glutamate	n.r.	n.r.	13.8	11.8	20.8	19.0	12.3
Proline	n.r.	n.r.	6.7	8.4	11.7	9.5	4.3
Serine	n.r.	n.r.	4.2	4.1	5.1	6.1	7.7
Total*	28.7	33.7	35.4	38.6	44.9	48.4	53.1

[1]Calculated from NRC,[31] requirement for swine weighing 1 to 5 kg. [2]Caclulated from FAO/WHO,[32] requirement for infants 3 to 4 mo of age. [3]From Aumaitre and Duee,[73] tissue of piglets 7 da of age. [4]Calculated from Widdowson,[35] tissue of fetuses 2.9 to 3.4 kg. [5]From Davis et al.[74] [6]Ciba Giegy 1970. [7]Includes requirement for cysteine. [8]Includes requirement for tyrosine.

* = Indispensable amino acids; n.r. = no requirement established; n.d. = not determined.

Table 3. Comparison of suggested indispensable amino acid requirements patterns (including estimation of maintenance and growth components) in the growing pig and human infant (expressed relative to lysine requirement)

Amino Acid	Human Infant Req[1]	NRC Reqt[2]	Illinois Pattern[3]	Rowett Pattern[4]	Maintenance[5]	Growth[5]	Human Fetal Accretion[6]
	(expressed relative to lysine = 100, by mass)						
Isoleucine	68	54	60	60	44	63	48
Leucine	156	71	100	110	64	115	104
Valine	90	59	68	75	56	78	65
Lysine	100	100	100	100	100	100	100
Methionine+Cysteine	56	52	60	63	136	53	27
Phenylalanine+Tyrosine	121	81	95	120	103	124	98
Threonine	84	59	65	72	147	69	58
Tryptophan	17	15	18	18	31	18	n.d.
Histidine	27	26	32	32	n.d.	n.d.	36
Arginine	n.r.	42	42	42	n.d.	n.d.	107

[1]Calculated from FAO/WHO,[32] requirement for infants 3 to 4 mo of age. [2]Calculated from FAO/WHO,[32] requirement for piglets weighing 1 to 5 kg. [3]Illinois ideal amino acid pattern,[75] determined in pigs weighing 10 kg. [4]From Wang and Fuller,[76] determined in pigs weighing 25 to 50 kg. [5]From Fuller et al.,[34] determined in pigs weighing 35 to 50 kg.[6]Calculated from Widdowson,[35] fetuses weighing 2.9 to 3.4 kg, cysteine not determined.n.r. = no requirement established; n.d. = not determined.

requirement for each amino acid in the piglet would reflect the requirement for growth more strongly than maintenance, due to the rapid growth of the animal. Conversely, the overall requirement for each amino acid in the more moderately growing human neonate would be less influenced by the requirement for growth processes, and more by maintenance. Fuller et al.[34] partitioned requirements for each component by measuring protein accretion at graded intakes of each indispensable amino acid, when it was limiting in the diet. Using data on human fetal accretion rate from Widdowson,[35] we calculated the amino acid requirement for growth for the human neonate (Table 3). The value for the sulphur amino acids is quite low, probably representing only half the total, because cysteine was not measured. With this exception, the amino acid requirement pattern for growth in the human infant appears to be similar to that of the pig. We believe these similarities in amino acid composition of swine and human milk and piglet and fetal tissue support the concept that an ideal amino acid profile for TPN developed in the piglet may similarly represent an ideal pattern for human neonates receiving TPN.

3.4. Practical Advantages and Applications of the Piglet as a Model for the Human Neonate

The logistical advantages to using the piglet model are considerable. It is a more practical model than the nonhuman primate or the rat. Cost of animals and housing is much less for the piglet than the primate. Piglets are readily available from commercial agricultural operations at low cost, and with known and repeatable genetics. The large litter size allows allocation of siblings to different treatments within an experiment. Relative to infants in a clinical setting, neonatal piglets represent a homogenous sampling group with minimal confounding effects of disease or surgical stress. Piglets can be removed from the sow immediately after birth and raised colostrum free under intensive care conditions. The piglet is more tolerant to rearing in an artificial environment without its mother than is the primate. Piglets which have received antibodies in colostrum for one or more days readily adapt to metabolic cages, and require care procedures that are less intensive and costly than those weaned at birth. Surgery can be performed soon after birth for multiple catheter implantation. Advantages of using the piglet over the rat relate to more similar metabolism and ontogeny of the piglet compared to the human infant. In addition, neonatal rats are less practical subjects for TPN. Piglets are small enough to be handled easily, yet are large enough to allow collection of large samples of biological fluids and tissues.

Based on these biological and practical advantages, piglets have been used frequently as a model for the human infant. Models have been developed for protein energy malnutrition[36] and rehabilitation.[37] Piglets have been used to evaluate milk formulas for human infants.[38-41] Based on the similarity of the gastrointestinal tract, the piglet has been used as a model for intestinal enzyme development,[12, 42] gastrointestinal motility[43] and necrotizing enterocolitis.[44]

The neonatal piglet has been used as a model for the TPN fed human infant to study calcium metabolism[45] and the effects of energy source in TPN on various aspects of metabolism.[46-49] TPN piglet models also have been used to study hepatotoxicity of TPN,[50, 51] growth and development of the gut[42,52-56] and the immune system.[57]

The rapid growth rate of the neonatal piglet compared to the human neonate confers both advantages and disadvantages. As discussed above, amino acid metabolism of the piglet will reflect the high growth component, which may result in overestimation of the relative requirement for some amino acids, and underestimation of others. Studies using miniature strains with lower growth rates may illuminate this issue. Conversely, the rapid growth rate of the piglet should confer a greater degree of sensitivity, allowing subtle changes in the amino acid profile to be recognized more readily; such differences are meaningful but difficult to measure in subjects growing more slowly. The piglet model offers the opportunity for a controlled study of the effects of manipulation of growth rate, from a rapid rate representing the goal for healthy infants, to states of malnutrition and growth retardation and illness.

The goal of human neonatal TPN is to achieve a population of healthy infants who are attaining their growth potential. Therefore, an important aspect of a piglet model to study amino acid metabolism in TPN is that the piglets should be healthy and growing at a rate comparable to a normative standard. A comparison of some TPN piglet models (Table 4) shows that piglets receiving TPN, according to established practice, often do not gain weight at rates comparable to piglets remaining with the sow. Therefore, a rigorously defined model, in which growth and biochemical measures of metabolism are acceptable,[58] will provide a framework against which to evaluate perturbations to health and growth. Like all animal models, the piglet model should be used in tandem with a clinical neonatal research program to provide directed improvements in neonatal nutrition.

4. DEVELOPMENT OF THE TPN PIGLET MODEL

Previously, we described in detail surgical procedures, animal housing and TPN formulation of our piglet model.[58] Briefly, male Yorkshire piglets approximately 3 da of age, are obtained from a specific pathogen free (SPF) herd. Upon arrival piglets are weighed, anaesthetized and prepared for surgical insertion of catheters, including external jugular, femoral and carotid catheters. Following surgery, piglets are fitted with an adjustable cotton jacket which has an anchoring button sewn to the fabric. Piglets are placed in individual circular metabolic cages set in banks of 4, allowing aural and visual contact with littermates. A tether and swivel system, attached to the anchor button on the piglet's jacket, allows the piglet some freedom of movement, while providing a means of affixing the TPN infusion sets to preclude their tangling or occlusion. To enrich the environment, toys are placed in cages.

Table 4. Comparison of some piglet models of total parenteral nutrition reported in the literature

Group and representative publication	Amino acid intake (g/kg/da)	Energy intake kcal/kg/da	Initial weight (kg)	Initial age (da)	Study length (da)	Weight gain first wk	Area of investigation and comments
Models using the miniature pig							
Shulman et al.[46]	11 Travasol	170	1.7	10	7	31 g/kg/da	Fuel mix, gut development. Skeletal muscle composition, body composition
Mehrazar and Kim[57]	8.2	165	0.5 to 0.8	3 to 5 da preterm	21	130g	Colostrum deprived model - immune system
Cohen et al.[44]	10	200	n.r.	3 to 6	42	n.r.	- full TPN on da 7 Cholestasis
Models using full sized breeds							
Morgan et al.[42]	8	212	1.5 to 2.0	3	21	31 g/kg/da	Gut development -sow raised group gained 51 g/kg/da
Borum[29]	n.r.	n.r.	n.r.	preterm	7	n.r.	Colostrum deprived model -full TPN on da 6
Wykes et al.[58]	14.6	260	1.7 to 1.9	2 to 3	8	63 g/kg/da	-weight gain began da 6 Growth, biochemistry, body composition, hematology
Draper et al.[45]	12.5	260	2.1	4	7	114 g/d	Bone and calcium metabolism

n.r. = not reported.

4.1. TPN Regimen

The goal in designing TPN solutions was the provision of nutrients at rates sufficient to meet the estimated nutrient requirements of swine weighing 1 to 5 kg.[31] Requirements, as reported, were assumed to be sufficient to meet the needs of a 3 kg piglet, allowing calculation of requirements on a per kg per day basis. The TPN regimen provides 15 g/kg/da of amino acids and 1.1 MJ/kg/da of metabolizable energy, with lipid and glucose each providing 50% of the nonprotein energy component. Two approaches were used for the provision of amino acids in TPN solutions. Initial TPN solutions were formulated using a commercially available general purpose amino acid solution (Vamin™, Kabi Pharmacia).[58] However, use of commercially prepared solutions severely limit experimental treatments. We addressed this issue by using crystalline amino acids to create TPN solutions with amino acid compositions of our own design.[30,59,60]

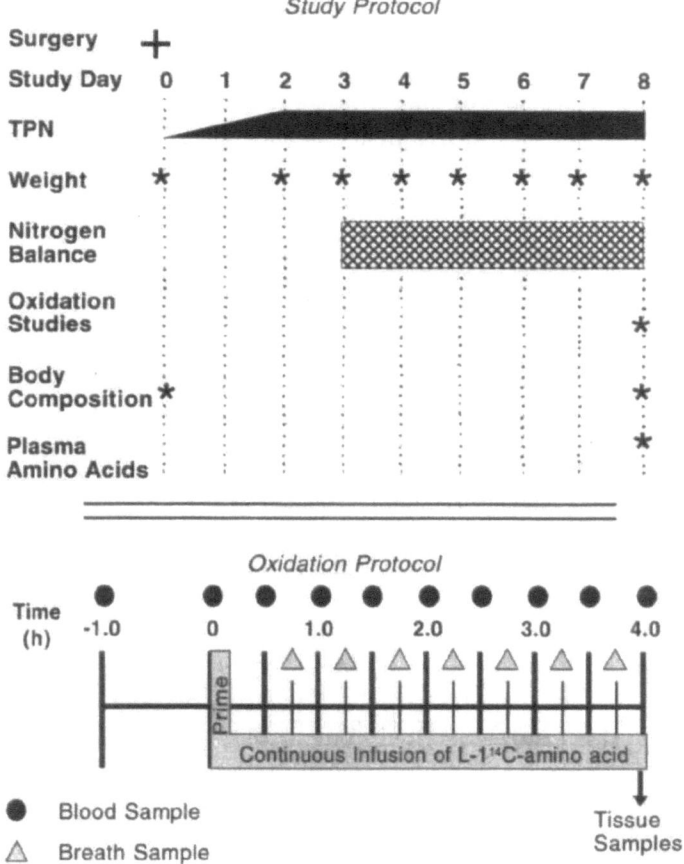

Figure 1. Typical study protocol of experiments designed to investigate amino aid metabolism during TPN in the neonatal piglet.

4.2. Study Protocol

As outlined in Fig. 1, TPN administration is initiated immediately postoperative, at 50% of calculated requirement, with infusion rates adjusted upwards for 2 da, reaching full rates by the morning of the second da. Animals are weighed daily for a period of 8 da postsurgery and TPN infusion rates adjusted accordingly. Beginning on da 2, daily urinary output is collected quantitatively on ice and analyzed with a qualitative reagent strip to monitor urine pH, glucose, ketones, protein, haemoglobin and bilirubin. Aliquots of urine are reserved and stored at -20C for determination of nitrogen and ion balances. Blood can be withdrawn from the sampling catheter for serum biochemistry (ions, protein, glucose, etc.), haematological profiles (complete blood cell counts, packed cell volume, etc.) and plasma amino acid concentrations.[30,58-60]

On the final day of the study, animals are used in amino acid oxidation studies. An L-1[14]C-amino acid is administered via a primed continuous infusion into the infusion catheter (femoral) while the piglet is housed in a covered plexiglass box. Amino acid oxidation rates are determined by the complete collection of $^{14}CO_2$,[59-61] with adjustments for the retention of label in the bicarbonate pool determined during a constant infusion of ^{14}C-sodium bicarbonate 1 da prior to the amino acid oxidation study.[59] Blood samples are withdrawn at 30 min intervals; plasma is collected for analysis of specific radioactivity of infused amino acid (plus metabolites). Animals are killed and body composition analyzed.

Using stochastic modelling techniques, amino acid flux rates are calculated from the plasma specific radioactivity using isotope dilution principles and expressed as a rate per unit of body mass. Oxidation rates are calculated as the product of the flux rate and the percentage of the ^{14}C label transferred from infusate to breath. By measuring the rate of amino acid oxidation, an estimate of net retention of amino acid can be derived by subtracting oxidation rate from rate of intake of the amino acid. This approach offers an unique opportunity to compare kinetic estimates of amino acid retention to those derived from traditional measures of nitrogen retention or protein accretion.

Our experience with this model is that piglets tolerate surgery and the TPN regimen well. Complications involving piglet health and catheter patency have been minimal. Nutrients from TPN are metabolized well: piglets demonstrate normal glycemia, low triglyceride level, low serum urea and high nitrogen retention.[58] Most serum biochemical and hematological results are similar between sow reared and TPN fed piglets, with differences being those normally seen with TPN fed infants.

This piglet model allows for specific investigations requiring invasive procedures (tissue sampling, repeated blood sampling) which would not be ethical in neonatal human subjects. We used this model to study clinically relevant aspects of amino acid metabolism and to develop new methods for the estimation of amino acid requirements in neonates receiving TPN.

5. APPLICATION OF THE MODEL TO AMINO ACID REQUIREMENTS AND METABOLISM DURING TPN

5.1. Aromatic Amino Acid Metabolism

Amino acid requirements of neonates receiving TPN are ill defined, as demonstrated previously (Section 2.). Of particular concern is the concentration of the aromatic amino acids, phenylalanine and tyrosine, due to the relatively poor solubility of crystalline tyrosine. Attempts to meet parenteral aromatic amino acid requirements of neonates by providing large quantities of phenylalanine (8% by weight of total amino acids) have resulted in extremely elevated plasma phenylalanine concentrations.[62, 63] However, failure to compensate for tyrosine insolubility can lead to reduced rates of growth and nitrogen retention. In a series of experiments,[59, 60] we investigated the effectiveness of several tyrosine precursors, as well as 2 amino acid profiles, in our TPN piglet model. Piglets received 1 of 5 TPN regimens, differing only in amino acid profile: 1) Vamin profile, 2) Vaminolact profile, 3) Vaminolact plus phenylalanine, 4) Vaminolact plus glycyl-L-tyrosine or 5) Vaminolact plus N-acetyltyrosine. All 5 regimens contained the same amount of crystalline tyrosine and formulations 1, 3, 4 and 5 contained the same amount of total aromatic amino acids (8.8% phenylalanine plus tyrosine). Complete amino acid profiles of solutions are listed in Table 1. Effects of amino acid profile on nitrogen balance and aromatic amino acid metabolism are reported in Table 5. Piglets receiving TPN solution based, on the amino acid pattern of Vamin, had acceptable nitrogen retention, while those receiving the amino acid pattern of Vaminolact retained less nitrogen. Supplementation of the Vaminolact pattern, with either phenylalanine or glycyl-L-tyrosine, improved nitrogen retention dramatically. However, the provision of N-acetyltyrosine as a source of tyrosine provided no demonstrable benefits compared to piglets given the original Vaminolact based TPN. Phenylalanine oxidation, measured during a 4 hr primed continuous infusion of L-1[14]C-phenylalanine, was higher in the 2 treatment groups receiving high phenylalanine. The oxidation of phenylalanine (as a percentage of the dose oxidized) was more than halved in the 3 treatment groups receiving the lower intake of phenylalanine.

Plasma concentrations of phenylalanine and tyrosine offer significant insights into the metabolism of these amino acids. Delivery of an identical aromatic amino acid pattern, but with other amino acids in a different pattern (Vamin vs VLP), reduced plasma phenylalanine concentrations by more than 5 fold. The amino acid pattern of the Vamin formulation is deficient in at least 1 essential amino acid, potentially lysine,[59] which is limiting protein synthesis. The highest levels of total aromatic amino acids used in these experiments were patterned after the content in human milk protein but probably were supplied in excess of requirements based on phenylalanine oxidation and plasma phenylalanine concentrations.

Glycyl-L-tyrosine can provide tyrosine in vivo; however the rate of inclusion, while consistent with the percentage of tyrosine in human and swine milk proteins (Table 2), was clearly in excess, based on high plasma tyrosine concentrations (Table 5) and the presence of tyrosine catabolites in urine.[60] Furthermore, a total aromatic amino acid content of 8.8% of total amino acids appears to be in considerable excess, although similar to milk fed infants, supporting the need to determine specifically requirements for phenylalanine and tyrosine

Table 5. The effect of aromatic amino acid profile of neonatal TPN solutions on growth, nitrogen retention, phenylalanine (PHE) oxidation and plasma PHE and tyrosine (TYR) concentrations in neonatal piglets[1]

	Vamin	VL[2]	VLP[2]	VLGT[2]	VLNAT[2]
PHE intake (g/kg/da)	1.13	0.59	1.13	0.61	0.61
TYR intake (g/kg/da)	0.10	0.11	0.11	0.71[3]	0.71[3]
Nitrogen retention (%)	82.4	70.0	87.3	84.5	74.0
PHE oxidation (% of dose)	22.8	10.1	24.3	9.0	7.6
Plasma PHE (μmol/l)	2234	156	399	148	118
Plasma TYR (μmol/l)	90	12	79	307	27

[1]From Wykes et al.[59, 60] Presented data represents arithmetic means of treatment groups.
[2]VL = Vaminolact; VLP = Vaminolact plus phenylalanine; VLGT = Vaminolact plus glycyl-L-tyrosine; VLNAT = Vaminolact plus N-Acetyltyrosine.
[3]Presented as a tyrosine equivalent assuming 100% availability

in neonates receiving TPN. Phenylalanine oxidation was sensitive to the level of phenylalanine intake, therefore oxidation techniques for the estimation of amino acid requirements could be adapted to the TPN fed neonate. Amino acid oxidation techniques for the estimation of requirements previously have been confined to the orally fed animal[64]. Development of sensitive and minimally invasive oxidation techniques for the estimation of amino acid requirements during TPN would be a significant contribution to clinical research.

5.2. Determination of Amino Acid Requirements Using Direct and Indicator Oxidation Techniques

As part of our research goals, we wished to establish the indicator amino acid oxidation technique[65-67] as a method for determining amino acid requirements during TPN. This technique usually has monitored phenylalanine oxidation, as an indicator of adequacy, in response to graded additions of another indispensable amino acid.[64] Therefore, before this method could be applied, phenylalanine requirement during TPN had to be determined. To minimize the loss of label from phenylalanine to the tyrosine pool, tyrosine must be supplied in excess of its requirement. However, an extreme excess of tyrosine must be avoided since both phenylalanine and tyrosine share catabolic pathways. In previous studies, we used a level of glycyl-L-tyrosine supplementation at which the tyrosine excess was too great.[60] Therefore, we undertook to measure tyrosine kinetics and to determine the requirement for tyrosine, when supplied as glycyl-L-tyrosine, in TPN fed piglets. The tyrosine requirement thus determined would be used in future experiments designed to determine amino acid requirements, based on the oxidation of L-1^{14}C-phenylalanine, in neonatal piglets receiving TPN.

Changes in nitrogen retention and tyrosine oxidation in piglets receiving TPN containing graded intakes of tyrosine are depicted in Fig. 2. Phenylalanine was provided at the concentration in human milk protein (Table 2). Based on a 2 phase linear regression cross over model, a tyrosine intake of 0.31 g/kg/da, equivalent to 2.1 g/100 g amino acids, was estimated to be the minimal tyrosine necessary to maximize nitrogen retention and minimize tyrosine oxidation, when phenylalanine, total amino acids and available energy were provided at 4.1 g/100 g of amino acids, 15 g/kg/da and 1.1 MJ/kg/da, respectively. However, with the subsequent calculation of the 95% confidence values[64] an adequate supply of tyrosine was concluded to be 2.8% of the total L-amino acid pattern.

This determination of tyrosine requirement provided a level of tyrosine which could be used in subsequent studies designed to determine the phenylalanine requirement by direct oxidation techniques. Piglets received the same TPN for 5.5 da, with phenylalanine provided at 4.2 g/100 g of total amino acids and tyrosine present in excess. Piglets were switched to test phenylalanine intakes 12 hr prior to the initiation of the oxidation study. Phenylalanine oxidation was low and similar for the 3 lowest intakes, and increased linearly thereafter (Fig. 3). Partitioning data between 2 regression equations and determining the cross over yielded a result of 2.9 g/100 g of total amino acids. This value represents the requirement for 50% of the population.[64] To meet the needs of 95% of the population, the upper 95% confidence value was calculated and taken to represent the safe level of phenylalanine intake, corresponding to 3.3 g/100 g of total amino acids. Therefore, a total aromatic amino acid content of 6.1 g/100 g of total amino acids, with phenylalanine providing at least 54%, is necessary to meet the needs of the TPN fed neonatal piglet. These data support earlier observations that an aromatic amino acid content of 8.8 g/100 g of the total L-amino acid profile is excessive for the neonate on TPN[59] and may explain partially the high plasma concentrations of phenylalanine and tyrosine observed in some infants receiving TPN with high aromatic amino acid concentrations.[62, 63]

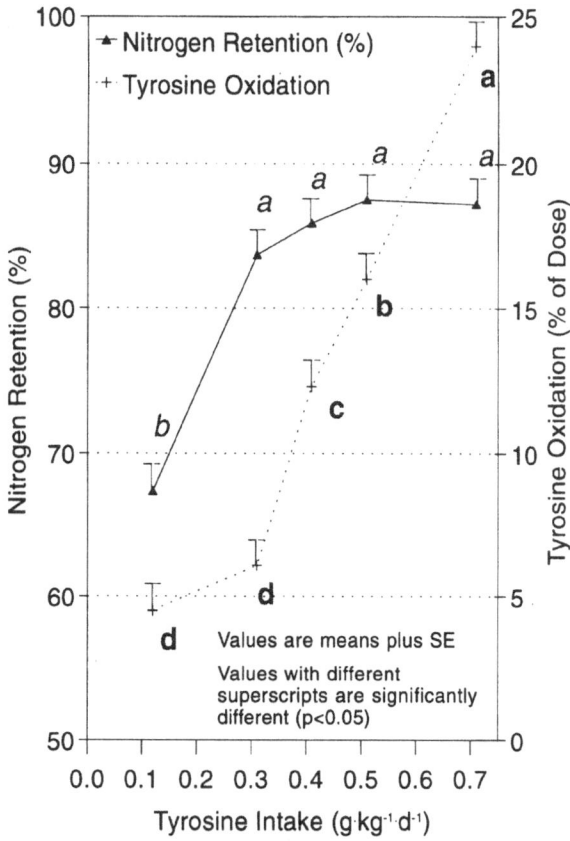

Figure 2. Nitrogen retention and the oxidation of L-[114C]-tyrosine in neonatal piglets (n = 15) receiving TPN with graded intakes of glycyl-L-tyrosine. Tyrosine intakes were 0.11, 0.31, 0.41, 0.51 or 0.71 g/kg/da. From House.[77]

The use of phenylalanine as an indicator amino acid requires that it be provided at safe intake levels to ensure correct response.[64] Therefore, it was necessary to first estimate safe levels of intake for both phenylalanine and total aromatic amino acids. Having achieved this (Figs. 2 and 3), it was possible to adapt the indicator methodology to the case of the neonate receiving TPN. Phenylalanine oxidation was determined in piglets receiving 8.3 g of lysine and 6.1 g total aromatics/100 g of total amino acids with phenylalanine providing 54% of the total (by weight), for 5.5 da. Twelve hr prior to oxidation studies, piglets received 1 of 7 levels of lysine (Fig. 4). Phenylalanine oxidation decreased in a linear fashion as lysine intakes were increased from 0.4 to 0.85 g/kg/da, after which there was little change in the oxidative response (Fig. 4). A safe level of lysine intake, determined as the upper 95% confidence value of the break point estimate, would be met by providing a TPN solution containing 5.6 g/100 g of total amino acids as lysine. This value is well below the lysine concentrations of several parenteral amino acid solutions designed for neonates (Table 1), highlighting potential pitfalls of methods used in formulating these solutions and strengthening our standpoint that amino acid requirements need to be defined specifically for neonates receiving TPN. In addition, the results demonstrate that the indicator method for establishing amino acid requirements can be used during TPN.

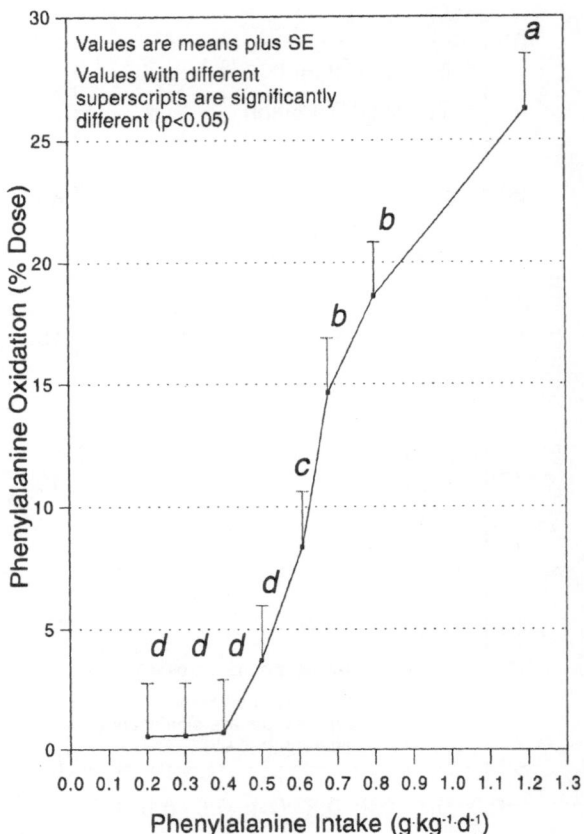

Figure 3. The oxidation of L-[114C]-phenylalanine in neonatal piglets (n = 24) receiving TPN with graded intakes of Phenylalanine. (From House.[77]). Phenylalanine intakes were 0.2, 0.3, 0.4, 0.5, 0.6, 0.7, 0.8 or 1.2 g/kg/da.

5.3. Advantages of the Indicator Amino Acid Oxidation Technique

In our study investigating lysine requirements, we demonstrated the indicator amino acid oxidation technique could be adapted to estimate amino acid requirements during TPN. This finding is significant due to the fact that previous estimations of amino acid requirements using indicator oxidation techniques have been performed in orally fed subjects.[65, 67-72] However, the fact that parenteral nutrients bypass the splanchnic control mechanisms, plus the fact that nutrient delivery is continuous during TPN, could have meant oxidation of the indicator amino acid may not have been sensitive to changes in dietary supply during TPN. As we show, this was not the case.

The indicator oxidation method should improve our knowledge of amino acid requirements of the human neonate for several reasons. First, unlike nitrogen balance studies, the indicator oxidation technique does not require infants to consume diets that are deficient or excessive in an amino acid for lengthy periods; 12 hr is sufficient. However, the choice of this time period was arbitrary, and may be longer than necessary. Second, the end product measure of phenylalanine oxidation, i.e., the percent of the dose oxidized, was a sensitive variable for the estimation of requirements by break point analysis. This means that studies could be performed that would be relatively noninvasive, requiring only breath sampling to measure oxidative response to a particular dietary amino acid intake. This method is therefore

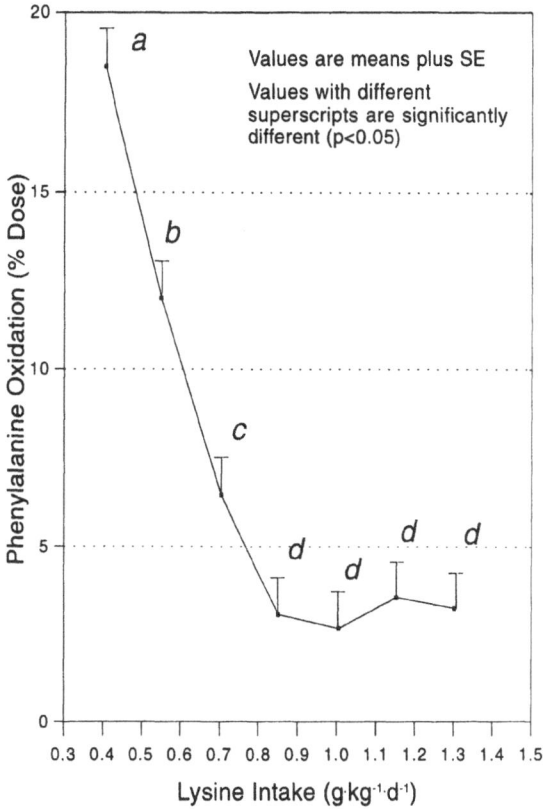

Figure 4. The oxidation of L-[114C]-phenylalanine in neonatal piglets (n = 21)) receiving TPN with graded intakes of lysine. (From House.[77].)Lysine intakes were 0.4, 0.55, 0.7, 0.85, 1.0, 1.15 or 1.3 g/kg/da.

more attractive for studies involving neonates, due to ethical concerns with blood or tissue sampling.

6. CONCLUSIONS

In this review, we describe a model of the TPN fed human neonate, using the metabolically similar newborn piglet. The model achieves the goal of human neonatal management, i.e., healthy neonates with growth and body composition similar to a mother's milk fed counterpart. The model was used to investigate amino acid metabolism and requirements. To our knowledge, our data on tyrosine, phenylalanine and lysine are the first empirical estimates of amino acid requirement during TPN. We also demonstrate conclusively the current approaches to providing aromatic amino acids in commercial TPN solutions are inadequate. Amino acid metabolism during TPN differs from the oral situation and the ideal amino acid pattern must be defined specifically for TPN feeding. The piglet model should continue to offer significant insights into various aspects of protein and amino acid metabolism in the human neonate. A research program using a piglet model, when conducted in tandem with a clinical neonatal research program, provides all aspects of well designed investigation: identification of the problem, rigorous metabolic investigation, clinical metabolic research and clinical implementation.

7. ACKNOWLEDGMENTS

The work described in this review was supported by the Hospital for Sick Children Research Foundation and the Medical Research Council of Canada (MT12928). Amino acids used in the formulation of the TPN solutions were donated by Kabi Pharmacia.

8. REFERENCES

1. Hay, W.W., Jr., 1986, Justification for total parenteral nutrition in the premature and compromised newborn, in: *Total Parenteral Nutrition: Indications, Utilization, Complications, and Pathophysiological Considerations* (E. Lebenthal, ed.), Raven Press, New York, pp. 277-304.
2. Adamkin, D.H., 1986, Total parenteral nutrition in hyaline membrane disease, in: *Total Parenteral Nutrition: Indications, Utilization, Complications, and Pathophysiological Considerations* (E. Lebenthal, ed.), Raven Press, New York, pp. 305-318.
3. Dudrick, S.J., Wilmore, D.W., Vars, H.M., and Rhoads, J.E., 1968, Longterm total parenteral nutrition with growth, development and positive nitrogen balance, *Surgery* 64:134-142.
4. Heird, W.C., 1972, Intravenous alimentation in pediatric patients, *J. Pediatr.* 80:351-372.
5. Wykes, L.J., Ball, R.O., Menendez, C.E., Malyon Ginther, D., and Pencharz, P.B., 1992, Glycine, leucine and phenylalanine flux in low-birth-weight infants during parenteral and enteral feeding, *Am. J. Clin. Nutr.* 55:971-975.
6. Atkinson, S.A., and Hanning, R.M., 1989, Amino acid metabolism and requirements of the premature infant: Is human milk the "gold standard"?, in: *Protein and Non-protein Nitrogen in Human Milk* (S.A. Atkinson, and B. Lonnerdal, eds.), CRC Press, Boca Raton, pp. 187-209.
7. Medical Research Council of Canada, 1987, Guidelines on research involving human subjects, Ministry of Supply and Services, Ottawa, pp. 29-30.
8. Book, S.A., and Bustad, L.K., 1974, The fetal and neonatal pig in biomedical research, *J. Anim. Sci.* 38:997-1002.
9. Pond, W.G. and Houpt, K.A., 1978, *The Biology of the Pig*, Cornell University Press, Ithaca.
10. Miller, E.R., and Ullrey, D.E., 1987, The pig as a model for human nutrition, *Ann. Rev. Nutr.* 7:361-382.
11. Moughan, P.J., and Rowan, A.M., 1989, The pig as a model animal for human nutrition research, *Proc. Nutr. Soc. New Zealand* 14:116-123.
12. Shulman, R.J., Henning, S.J., and Nichols, B.L., 1988, The miniature pig as an animal model for the study of intestinal enzyme development, *Pediatr. Res.* 23:311-315.
13. Terris, J.M., 1986, Swine as a model in renal physiology and nephrology: an overview, in: *Swine in Biomedical Research*, Volume 1 (M.E. Tumbleson, ed.), Plenum Press, New York, pp. 1673-1689.
14. Glauser, E.M., 1966, Advantages of piglets as experimental animals in pediatric research, *Exp. Med. Surg.* 24:181-190.
15. Shulman, R.J., 1993, The piglet can be used to study the effects of parenteral and enteral nutrition on body composition, *J. Nutr.* 123:395-398.
16. Dobbing, J., and Sands, J., 1979, Comparative aspects of the brain growth spurt, *Early Hum. Dev.* 3:79-83.
17. Purvis, J.M., Clandinin, M.T., and Hacker, R.R., 1982, Fatty acid accretion during perinatal brain growth in the pig. A model for fatty acid accretion in the human brain, *Comp. Biochem. Biophys. [B]* 72:195-199.
18. Rossi, T.M., 1986, Effects of total parenteral nutrition on the digestive organs, in: *Total Parenteral Nutrition: Indications, Utilization, Complications and Pathophysiological Considerations* (E. Lebenthal, ed.), Raven Press, New York, pp. 173-184.
19. Adeola, O., Wykes, L.J., Ball, R.O., and Pencharz, P.B., 1995, Comparison of oral milk feeding and total parenteral nutrition in neonatal pigs, *Nutr. Res.* 15:245-265.
20. Burrin, D.G., Shulman, R.J., Storm, M.C., and Reeds, P.J., 1991, Glutamine or glutamic acid effects on intestinal growth and disaccharidase activity in infant piglets receiving total parenteral nutrition. *J. Parent. Ent. Nutr.* 15:262-266.
21. Mount, L.E., and Ingram, D.L., 1971, The pig as a laboratory animal. Academic Press, London.
22. Pencharz, P.B., Parson, H., Motil, K., and Duffy, B., 1981, Total body protein turnover in children: Is it a futile cycle?, *Med. Hypoth.* 7:155-160.
23. Mulvaney, D.R., Merkel, R.A., and Bergen, W.G., 1985, Skeletal muscle protein turnover in young male pigs, *J. Nutr.* 115:1057-1064.

24. Reeds, P.J. and Harris, C.I., 1981, Protein turnover in animals: man in his context, in: *Nitrogen metabolism in man* (J.C. Waterlow, and J.M. Stephen, eds.), Applied Science, London, pp. 391-408.

25. Benevenga, N.J., 1986, Amino acid metabolism in swine: Applicability to normal and altered amino acid metabolism in humans, in: *Swine in Biomedical Research*, Volume 2 (M.E. Tumbleson, ed.), Plenum Press, New York, pp. 1017-1030.

26. Waterlow, J.C., Garlick, P.J., and Millward, D.J., 1987, *Protein Turnover in Mammaliam Tissues and in the Whole Body*, North Holland, Amsterdam.

27. Wu, P.Y.K., Edward, N.B., and Storm, M.C., 1986, The plasma amino acid pattern of normal term breast-fed infants, *J. Pediatr.* 109:347-349.

28. Hagemeier, D.L., Libal, G.W., and Wahlstrom, R.C., 1983, Effects of excess arginine on swine growth and plasma amino acid levels, *J. Anim. Sci.* 57:99-105.

29. Borum, P.R., 1990, The colostrum-deprived TPN piglets as a model to evaluate the effect of different amino acid formulations on neonatal metabolism, *FASEB J.* 4:A507.

30. House, J.D., Pencharz, P.B., and Ball, R.O., 1994, Glutamine supplementation to total parenteral nutrition promotes extracellular fluid expansion in piglets, *J. Nutr.* 124:396-405.

31. National Research Council, 1988, *Nutrient Requirements of Swine*, Ninth revised ed., National Academy Press, Washington, D.C.

32. FAO/WHO Expert Consultation, 1990, *Protein Quality Evaluation*, WHO, Rome.

33. FAO/WHO/UNU Expert Consultation, 1985, *Energy and Protein Requirements*, (WHO Technical Report Series No 724) WHO, Geneva.

34. Fuller, M.F., McWilliam, R., Wang, T.C., and Giles, L.R. 1989, The optimum dietary amino acid pattern for growing pigs, *Br. J. Nutr.* 62:255-267.

35. Widdowson, E.M., 1979, Body composition of the fetus and infant, in: *Nutrition and Metabolism of the Fetus and Infant*, Fifth Nutricia Symposium (H.K.A. Visser, ed.), Martinus Nijhoff, The Hague, pp. 147-157.

36. Widdowson, E.M., 1968, Growth and composition of the fetus and newborn. in: *Biology of Gestation*, (N.S. Assasli, ed.) Academic Press, New York, pp. 1-49.

37. Pond, W.G., Ellis, K.J., and Schoknecht, P., 1992, Response of blood serum constituents to production of and recovery from a kwashiorkor-like syndrome in the young pig, *Proc. Soc. Exp. Biol. Med.* 200:555-561.

38. Newport, M.J., and Henschel, M.J., 1984, Evaluation of the neonatal pig as a model for infant nutrition: effects of different proportions of caesin and whey protein in milk on nitrogen metabolism and composition of digesta in the stomach, *Pediatr. Res.* 18:658-662.

39. Hrboticky, N., MacKinnon, M.J., and Innis, S.M., 1990, Effect of a vegetable oil formula rich in linoleic acid on tissue fatty acid accretion in the brain, liver, plasma and erythrocytes of infant piglets, *Am. J. Clin. Nutr.* 51:173-182.

40. Odle, J., Benevenga, N.J., and Crenshaw, T.D., 1991, Utilization of medium-chain triglycerides by neonatal piglets: chain length of even- and odd-carbon fatty acids and apparent digestion/absorption and hepatic metabolism, *J. Nutr.* 121:605-614.

41. Arbuckle, L.D., and Innis, S.M., 1992, Docosahexaenoic acid in developing brain and retina of piglets fed high or low w-linolenate formula with and without fish oil, *Lipids*, 27:89-93.

42. Morgan III, W, Yardley, J., Luk, G., Niemiec, P., and Dudgeon, D., 1987, Total parenteral nutrition and intestinal development: a neonatal model, *J. Pediatr. Surg.* 22:541-545.

43. Groner, J.I., Altschuler, S.M., and Ziegler, T.R., 1990, The newborn piglet: a model of neonatal gastrointestinal mobility, *J. Pediatr. Surg.* 25:315-318.

44. Cohen, I.T., Nelson, S.D., Moxley, R.A., Hirch, M.P., Counihan, T.C., and Martin, R.F., 1991, Necrotizing enterocolitis in a neonatal piglet model, *J. Pediatr. Surg.* 26:598-601.

45. Draper, H.H., Yuen, D.E., and Whyte, R.K., 1991, Calcium lycerophosphate as a source of calcium and phosphorus in total parenteral nutrition solutions, *J. Parent. Ent. Nutr.* 15:176-180.

46. Shulman, R.J., Fiorotto, M.L., Sheng, H-P, and Garza, C., 1984, Effect of different total parenteral nutrition fuel mixes on the body composition of infant miniature pigs, *Pediatr. Res.* 18:261-265.

47. Fiorotto, M.L., Shulman, R.J., Sheng, H-P, and Garza, C., 1986, The effects of different total parenteral nutrition fuel mixes on skeletal muscle composition of infant miniature pigs, *Metabolism* 35:354-359.

48. Shulman, R.J., Fiorotto, M.L., Sheng, H-P, Finegold, M.J., and Garza, C., 1987 Liver composition and histology in growing infant miniature pigs given different total parenteral nutrition fuel mixes, *J. Pediat. Gastroenterol. Nutr.* 11:275-279.

49. Shulman, R.J., and Burrin, D.G., 1991, Total parenteral nutrition energy composition affects small intestinal disaccharidase activity in the newborn miniature pig, *J. Parent. Ent. Nutr.* 15:560-563.

50. Cohen, I.T., Meunier, K.M., and Hirsh, M.P., 1990, The effects of enteral stimulation on gallbladder bile during total parenteral nutrition in the neonatal piglet, *J. Pediatr. Surg.* 25:163-167.

51. Van Aerde, J., Higa, T., Chan, G., Feldman, M., Lemke, R., and Clandinin, M., 1990, Eicosapentanoic (EPA) and docosahexanoic (DHA) acid prevent cholestatic jaundice in the intravenously fed neonate. *FASEB J.* 4:A507.

52. Remillard, R., Yardley, J., Mitchell, K., Fuerino, F., and Dudgeon, D., 1992, Oral feeding effects in reversing total parenteral nutrition-induced small bowel atrophy in neonatal piglets, *FASEB J.* 6:A1113.

53. Burrin, D.G., Shulman, R.J., Storm, M.C., and Reeds, P.J., 1991, Glutamine or glutamic acid effects on intestinal growth and disaccharidase activity in infant piglets receiving total parenteral nutrition, *J. Parent. Ent. Nutr.* 15:262-266.

54. Guerino, F., Yardley, J.H., and Dudgeon, D.L., 1991, Effects of total parenteral nutrition (TPN) with IV glutamine or limited enteral feeding on development of the neonatal piglet intestinal tract, *FASEB J.* 5:A1451.

55. Shulman, R.J., 1988, Effect of different total parenteral nutrition fuel mixes on small intestinal growth and differentiation in the infant miniature pig, *Gastroenterol.* 95:85-92.

56. Goldstein, R.M., Hebiguchi, T., Luk, G.D., Taqi, F., Guilarte, T.R., Franklin, F.A., Niemiec, P.W., and Dudgeon, D.L., 1985, The effects of total parenteral nutrition on gastrointestinal growth and development, *J. Pediatr. Surg.* 20:785-791.

57. Mehrazar, K., and Kim, Y.B., 1988, Total parenteral nutrition in germ-free colostrum-deprived neonatal miniature piglets: a unique model to study the ontogeny of the immune system, *J. Parent. Ent. Nutr.* 12:563-568.

58. Wykes, L.J., Ball, R.O., and Pencharz, P.B., 1993, The development and validation of a total parenteral nutrition model in the neonatal piglet, *J. Nutr.* 123:1248-1259

59. Wykes, L.J., House, J.D., Ball, R.O., and Pencharz, P.B., 1994, Amino acid profile and aromatic amino acid concentration in total parenteral nutrition: effect on growth, protein metabolism and aromatic amino acid metabolism in the neonatal piglet, *Clin. Sci.* 87:75-84.

60. Wykes, L.J., House, J.D., Ball, R.O., and Pencharz, P.B., 1994, Aromatic amino acid metabolism of neonatal piglets receiving TPN: effect of tyrosine precursors, *Am. J. Physiol.* 267:E672-E679.

61. Ball, R.O., and Bayley, H.S., 1985, Time course of evolution of total and ^{14}C-labelled carbon dioxide by young pigs receiving diets containing ^{14}C-phenylalanine, *Can. J. Physiol. Pharmacol.* 63:1170-1174.

62. Puntis, J.W.L., Edwards, M.A., Green, A., Morgan, I., Booth, I.W., and Ball, P.A., 1986, Hyperphenylalanaemia in parenterally fed newborn babies, *Lancet* ii:1105-1106.

63. Walker, V.A., Hall, M.A., Bulusu, S., and Allan, A., 1986, Hyperphenylalanaemia in parenterally fed newborn babies, *Lancet* ii:1284.

64. Zello, G.A., Wykes, L.J., Ball, R.O., and Pencharz, P.B., 1995, Recent advances in methods of assessing dietary amino acid requirements for adult humans, *J. Nutr.* 125: 2907-2915.

65. Kim, K.I., McMillan, I., and Bayley, H.S., 1983, Determination of amino acid requirements of young pigs using an indicator amino acid, *Br. J. Nutr.* 50:369-382.

66. Ball, R.O., and Bayley, H.S., 1984, Tryptophan requirement of the 2.5-kg piglet determined by the oxidation of an indicator amino acid, *J. Nutr.* 114:1741-1746.

67. Zello, G.A., Pencharz, P.B., and Ball, R.O., 1993, The dietary lysine requirement of young adult males determined by the oxidation of an indicator amino acid, L-[1 ^{13}C]phenylalanine, *Am. J. Physiol.* 264:E677-E685.

68. Kim, K.I., and Bayley, H.S., 1983, Amino acid oxidation by young pigs receiving diets with varying levels of sulphur amino acids, *Br. J. Nutr.* 50:383-390.

69. Kim, K.I., Elliott, J.I., and Bayley, H.S., 1983, Oxidation of an indicator amino acid by young pigs receiving diets with varying levels of lysine or threonine, and an assessment of amino acid requirements, *Br. J. Nutr.* 50:391-399.

70. Ball, R.O., Atkinson, J.L, and Bayley, H.S., 1986, Proline as an essential amino acid for the young pig, *Br. J. Nutr.* 55:659-668.

71. Lazaris-Brunner, G., Pencharz, P.B., and Ball, R.O., 1994, Tryptophan requirement of young women determined by indicator amino acid oxidation, *FASEB J.* 8:A462.

72. Duncan, A.M., Pencharz, P.B., and Ball, R.O., 1995, Lysine requirement of adult males using indicator amino acid oxidation. The effect of a lower protein intake, *FASEB J.* 9:A865.

73. Davis, T.A., Nguyen, H.V., Garcia-Bravo, R., Fiorotto, M., Jackson, E.M., Lewis, D.S., Lee, D.R., and Reeds, P.J., 1994, Amino acid composition of human milk is not unique, *J. Nutr.* 124:1126-1132.

74. Aumaitre, A., and Duee, P.H., 1974, Composition en acides amines des proteines corporelles du porcelet entre la naissance et l'age de huit semaines, *Ann. Zootech.* 23:231-236.

75. Chung, T.K., and Baker, D.H., 1992, Ideal amino acid pattern for 10-kilogram pigs, *J. Anim. Sci.* 70:3102-3111.

76. Wang, T.C., and Fuller, M.F., 1989, The optimum dietary amino acid pattern for growing pigs. I. Experiments by amino acid deletion, *Br. J. Nutr.* 62:77-89.

77. House, J.D., 1995, The determination of amino acid requirements in neonatal piglets receiving total parenteral nutrition, *Ph.D. Thesis*, University of Guelph, Guelph, Ontario, Canada.

64

THE NEONATAL PIGLET AS A MODEL TO STUDY INSULIN LIKE GROWTH FACTOR MEDIATED INTESTINAL GROWTH AND FUNCTION

Sharon M. Donovan,[1][*] Vicki M. Houle,[1] Marcia H. Monaco,[1]
Elizabeth A. Schroeder,[1] Yookyoung Park,[1] and Jack Odle[2]

[1] Food Science and Human Nutrition
College of Agricultural
Consumer and Environmental Sciences
University of Illinois
Urbana, Illinois 61801
[2] Animal Science
College of Agriculture and Life Sciences
North Carolina State University
Raleigh, North Carolina 27695

1. INTRODUCTION

Colostrum and milk contain a variety of hormones and peptides which are known to stimulate cellular growth and differentiation both in vivo and in vitro.[1, 2] The marked enhancement of porcine neonatal intestinal growth and development upon ingestion of milk or colostrum[3,4] has led to the speculation that milkborne hormones and growth factors may be at least partially responsible for this effect.[1,2] Widdowson et al.[3] observed a 61% increase in intestinal mass in piglets which had suckled colostrum for 24 hr compared to piglets fed water. Although a portion of the intestinal mass increase can be attributed to macromolecular absorption of colostral immunoglobulin (IgG),[4] increases in intestinal length and DNA content suggested cellular proliferation also was occurring.[3] Recently, these early observations were validated and extended upon in a series of studies by Burrin et al.[4-6] Piglets fed colostrum or milk for 6 hr postpartum showed a 300% increase in jejunal protein synthesis rate compared to water fed controls. A potential role for nonnutritive components in colostrum was demonstrated in a subsequent study in which piglets were fed either colostrum

[*]Reprint requests to: Dr. Sharon Donovan, Food Science and Human Nutrition, 386 Bevier Hall, 905 S. Goodwin Ave. University of Illinois, Urbana, IL, USA 61801, (217) 333-2289.

Advances in Swine in Biomedical Research, edited by Tumbleson and Schook
Plenum Press, New York, 1996

or a milk replacer with comparable macronutrient composition.[5] In that study, skeletal muscle and jejunal protein synthesis rates were higher in piglets fed colostrum, indicating that colostral components themselves, or endogenous factors which are released in response to colostrum intake, may mediate growth not only of the intestine but of tissues outside the gastrointestinal tract as well. The rapid small intestinal growth observed by Widdowson et al.[3] was associated with functional development in that total small intestinal lactase and acid phosphatase activities were increased in suckled piglets. Burrin et al.[6] observed changes in lactase isoform abundance in newborn pigs fed milk or colostrum for 6 hr compared to water fed controls, which suggests the increase in intestinal lactase activity may arise from alterations in posttranslational processing of lactase in response to feeding.

Based on these previous observations and similarities in gastrointestinal structure and function between human and porcine neonates,[7-9] we chose the piglet model to study the role of milkborne growth factors in neonatal intestinal growth and development. Although it must be kept in mind that the growth factor composition of milk is complex and dynamic, consisting of factors which conceivably could potentiate or inhibit each others actions, we will focus solely upon the biological actions of one class of growth factors in milk, the insulin like growth factors.

2. INSULIN LIKE GROWTH FACTORS (IGF-I AND IGF-II)

Insulin like growth factors (IGF-I and IGF-II) are 7.5 kDa single chain peptides, consisting of 70 and 67 amino acids, respectively. IGF-I and -II exhibit 70% amino acid homology with each other and a 50% homology with proinsulin.[10] The IGF amino acid sequence is highly conserved across species, showing 100% identity between human, porcine, ovine and bovine IGF-I.[11] IGFs may interact with three distinct receptors: type I and type II IGF receptors and, due to their structural resemblance to insulin, the insulin receptor.[12] The type I IGF receptor has a heterotetrameric structure (2 alpha and 2 beta subunits), which is homologous to that of the insulin receptor, and has the highest affinity for IGF-I, followed by IGF-II and insulin. The type II receptor is a monomeric protein with an apparent Mr of 220 kDa which shows highest affinity for IGF-II, then IGF-I, and does not bind insulin. The type I IGF receptor contains a tyrosine binding domain on each beta subunit which is phosphorylated in response to IGF-I binding; whereas, IGF-II binding to the type II receptor activates a calcium permeable cation channel, perhaps through coupling to the guanine nucleotide binding protein (Gi_2).[12] Although the type II receptor has the higher affinity for IGF-II, it appears that the type I receptor mediates the mitogenic responses of both IGF-I and -II.

Type I and type II IGF receptors[13-15] have been detected along the entire length of the neonatal intestine. Autoradiography of rat jejunal segments incubated with [^{125}I]-IGF-I demonstrated receptor binding from the submucosa to the villi; however, receptor density was 5 times higher on crypt cells than on cells located on villi tips.[13,15] In the piglet, specific binding of [^{125}I]-IGF-I to intestinal receptors was highest at birth, declined at 3 and 5 da postpartum and recovered by 21 da postpartum.[14]

Within the body, IGF-I and IGF-II exist almost entirely (99%) in association with binding proteins (IGFBP). Six genetically distinct proteins (IGFBP-1 through -6) have been cloned and sequenced in the rat and human;[16] whereas, less information is available on IGFBP sequences in production species.[17] However, the apparent similarities in serum IGFBP profiles (IGFBP-1, -2, -3 and -4) in porcine and human sera[18,19] and the presence of IGFBP-5[20] and IGFBP-6[21] in other porcine body fluids, suggests that similar IGFBP are present in both species. Separation of sera by gel permeation chromatography under native conditions results in 2 major peaks of IGF binding activity, a 150 kDa peak consisting of a ternary complex of IGFBP-3, IGF and an 85 kDa protein known as the acid labile subunit (ALS).[22] Greater than 95% of serum IGF is associated with the 150 kDa complex. The second peak of IGF binding activity elutes between 30 and 50 kDa and consists of the remaining 5 IGFBP. Association of IGF-I or -II with IGFBP extends their circulating half life from 20 min to approximately 12 and 3 hr for the 150 KDa and the 30 to 50 kDa peak, respectively. In addition to serving as carrier proteins, IGFBP both positively and negatively affect the interaction of IGFs with their target tissues.[23]

3. INSULIN LIKE GROWTH FACTORS IN MILK

IGF-I and IGF-II have been reported in the milk of all species studied.[2,24] Concentrations of IGF-I and -II in porcine, bovine, human and rat colostrum and mature milk are shown in Table 1. As is commonly observed for most bioactive components in milk, IGF-I and -II are present in higher concentration in colostrum than mature milk. Human[25,26] and rat[27] colostrum contain 2 to 3 fold higher IGF-I concentrations than mature milk; whereas, bovine[28,29] and porcine[19,30,31] colostrum contain 10 to 500 fold higher levels of IGF-I than mature milk. Recently, we characterized porcine serum and milk IGF and IGFBP profiles throughout lactation.[19] Serum IGF-II is the predominant IGF in porcine serum and in milk for the first 7 da postpartum, after which, IGF-I and -II levels are comparable.

IGFBP-1, -2, -3 and -4 have been identified in milk by western ligand blotting[32] or radioimmunoassay. Specific IGFBP present in milk, and their relative concentrations, vary among species.[2] In addition, milk IGFBP do not always reflect maternal serum IGFBP profiles, suggesting some regulation of IGFBP entry into milk by the mammary gland. For example, rat milk contains IGFBP-2, which is not present in the maternal rat serum;[27] whereas, human milk only contains IGFBP-2[26] and IGFBP-1[33] although maternal serum also contains IGFBP-3 and -4.[26] Analysis of porcine milk IGFBP by western ligand blotting demonstrated the presence of bands with apparent molecular weights similar to those observed for IGFBP-1, -2, -3 and -4 in porcine serum.[19] Total milk IGF binding activity also was assessed in porcine milk. IGF binding activity was low in prepartum secretions and early colostrum, peaked on da 4 postpartum, followed by a decline throughout lactation.[19] The fact that prepartum secretions and colostrum contained the highest concentrations of IGF-I and IGF-II, while IGF binding activity was at its lowest level, suggests that some IGF-I in

Table 1. Comparison of IGF-I and IGF-II concentrations (mean ± SD) in porcine, rat and human milk

	IGF-I		IGF-II		
	Colostrum[a]	Mature	Colostrum	Mature	Reference
	(μg/L)				
Porcine	220 ± 113	8.4 ± 4.0	NR[b]	NR	30
Landrace	584 ± 95	4 to 20	NR	NR	31
Duroc	1271 ±236	21 to 30	NR	NR	31
York/Duroc	39 ± 22	11.4 ±1.4	82.3 ± 57.5	16.8 ± 5.6	19
Bovine	226 ± 38	24 ± 2.0	467 ± 46	117 ± 10	28
Human	10.9 ± 5.3	7.1 ± 0.4	NR	NR	25
	NR	1.5 ± 0.5	NR	2.7 ± 0.7	26
Rat	27.8 ± 2.0	16.2 ± 2.2	1.2 ± 0.2	<1.0	27

[a]Colostrum 1 to 4 da postpartum; Mature milk > 5 da postpartum.
[b]NR = Not reported.

these fluids may occur in the free state. A relatively high proportion of free IGF-I has been reported previously in human[25] and bovine[29] colostrum. How milk IGFBPs modulate actions of IGF-I and -II within neonatal intestine is unknown. Association of IGF-I with IGFBP-3 does not protect IGF-I from in vitro digestion by adult rat gastrointestinal secretions.[34]

4. ORAL ADMINISTRATION OF IGF-I: STABILITY AND ABSORPTION

Serum IGF-I concentrations tend to be higher in neonatal calves[35] and piglets[4] fed colostrum than mature milk or milk replacers, suggesting that milk borne IGF-I may be absorbed by the neonate. This question was investigated directly by orally administering [^{125}I]-IGF-I to neonatal calves,[36] piglets[37] or rats.[38] Baumrucker et al.[36] orally administered 160 µCi of [^{125}I]-IGF-I to calves within 4 hr of birth. The [^{125}I]-IGF-I was added to 1 l of bovine colostrum, which contained 350 µg/l of endogenous IGF-I and IGFBPs. Blood samples were obtained at various time points for approximately 22 hr postadministration. Approximately 12% of radioactivity in serum pooled during the first 5 hr of the study was immunoprecipitable with an antibody to IGF-I, suggesting that a small percentage of orally administered IGF-I was absorbed in an immunologically intact state.[36] In piglets, the degree of IGF-I absorption was assessed by fitting colostrum deprived newborn piglets with umbilical arterial and venous catheters within 2 hr of birth.[37] Baseline blood samples were drawn, then piglets were given 25 µCi of [^{125}I]-IGF-I (100 ng) in milk replacer by orogastric gavage. The milk replacer was devoid of immunoreactive IGF-I and IGFBPs. Serial arterial and venous blood samples were obtained for 4 hr postadministration. Radioactivity in both venous and arterial blood was detectable within 15 min and was significantly higher in venous than arterial samples until 60 min postfeeding. Total serum radioactivity and radioactivity which was immunoprecipitable with an antibody to IGF-I reached a plateau by 60 min and were maintained throughout the 4 hr study. Approximately 3 to 5% of radioactivity in serum was immunoprecipitable with an antibody to IGF-I. However, it was determined the absorbed [^{125}I]-IGF-I constituted less than 0.05% of the total circulating IGF-I pool (860 ng) of the piglets.[37] To summarize, limited absorption of IGF-I by calves and piglets does occur in the immediate postpartum period; however, the amount of milk borne IGF-I absorbed does not contribute significantly to the circulating IGF-I concentration.

Although IGF-I is absorbed poorly by the neonate, it appears to be sequestered by the gastrointestinal tract and survives therein for at least 30 min postingestion in a biologically active form. Philipps et al.[38] orogastrically administered 4.5 µCi [^{125}I]-IGF-I or [^{125}I]-IGF-II (4 to 7 ng) to rats at 10 to 11 da postpartum. After 30 min, serum and tissues were collected. Confirming results in calves[36] and piglets,[37] less than 4% of the dose was detected in blood and less than 2% in organs other than the gastrointestinal tract. In contrast, 42% of the [^{125}I]-IGF-I dose and 38% of the [^{125}I]-IGF-II dose were recovered within the walls and lumenal contents of the stomach and small intestine. Of the [^{125}I]-IGF-I recovered, approximately 70% in the stomach wall and 10% in the small intestinal wall was found to be intact by size exclusion chromatography. Additionally, the intact [^{125}I]-IGF-I and [^{125}I]-IGF-II recovered from stomach wall retained the ability to bind to tissue type I and type II receptors, respectively,[38] suggesting that orally administered IGF-I retains bioactivity with the neonatal gastrointestinal tract. It is possible the high concentrations of casein present in bovine, porcine and rat milk or their comparable milk replacers are partially protecting IGF-I from digestion.[34]

Table 2. Study designs of experiments investigating the effect of oral administration of insulin like growth factor-I and -II to neonatal animals

| Animal Model | Dose | | Timing of study (da postpartum) | References |
	μg/l	μg/kg BWT/da		
Calf	750[a]	62.5[b]	1 to 7	39-42
Piglet	2,000[c]	440[b]	1 to 2	43
Piglet	10,000-20,000[d]	3,500	1 to 4	44
Piglet	500[d]	200	1 to 14	45

[a]IGF-I added to bovine milk replacer.
[b]Approximate calculations based on food intake and mean body weights reported by the authors.
[c]Studied both IGF-I or IGF-I added to human infant formula.
[d]IGF-I added to porcine milk replacer.

5. INTESTINAL EFFECTS OF ORALLY ADMINISTERED IGF-I

Since 1990, 4 studies have been conducted investigating the role of orally adminis-tered IGF-I or IGF-II in neonatal calves[39-42] and piglets.[43-45] In all studies, recombinant human IGF-I or -II was added to a species specific artificial milk replacer devoid of other bioactive components. Compared to physiological concentrations of IGF-I and -II in bovine and porcine milk (Table 1), doses of growth factors used in these studies range from high colostral to pharmacological concentrations. In addition, with the exception of the study of Houle et al.,[45] all were relatively short term (24 hr to 7 da).

Study designs are summarized in Table 2. The first study was conducted in newborn dairy calves which were fed milk replacer, milk replacer containing 750 μg/l of recombinant human IGF-I (milk replacer+IGF-I) or pooled bovine colostrum for the first 4 feedings; milk replacer was used for the remainder of the study.[39-42] Pooled colostrum contained 500 μg/l of IGF-I and 400 μg/l of IGF-II. To assess gut absorptive development, D-xylose and gamma glutamyltransferase were added to diets at the third feeding. Da 7 postpartum, no differences in body weight, small or large intestinal weight or length[41] or serum IGFBP profiles[42] were observed among groups. Serum hormones were not affected consistently by any of the treatments, although several transient effects on IGF-I, prolactin and insulin were observed.[39] Receptor binding studies with intestinal microsomal membranes demonstrated 50% higher [^{125}I]-IGF-I binding by membranes from calves fed milk replacer+IGF-I than calves fed colostrum or milk replacer alone.[41] When [^{3}H]-thymidine incorporation into intestinal explants was assessed in vitro, incorporation per μg DNA was 2 fold higher in tissue from the formula+IGF group than the other 2 groups.[41] The authors speculated that up regulation of the type I receptor may be responsible for the enhanced thymidine incorporation in intestinal explants from calves which were fed milk replacer+IGF-I. Lastly, the effect of ingestion of milk replacer+IGF-I on blood protein profiles and gut absorptive capacity was assessed.[40] No differences in serum total protein, albumin or globulin concentrations were observed among groups receiving milk replacer with or without IGF-I. No effect on gamma glutamyltransferase absorption was detected, indicating the rate of gut closure was not influenced by oral IGF-I. Oral IGF-I may affect either the development or activity of intestinal sugar transporters, because pharmacokinetic analysis of D-xylose absorption showed that calves fed milk replacer absorbed 17.9% of the dose; whereas, calves fed milk replacer+IGF-I only absorbed 12.5% of the dose.[40]

In the study of Xu et al.,[43] newborn colostrum deprived piglets were fed a human infant formula supplemented with either 2,000 μg/l IGF-I or 2,000 μg/l IGF-II for the first 24 hr postpartum. Burrin et al.[44] fed newborn colostrum deprived piglets a sow milk replacer supplemented with 10,000 to 20,000 μg/l of IGF-I, to achieve an ingested dose of 3500 μg/kg/da for the first 4 da postpartum. The dose was chosen based on previous studies in which systemic administration of of 3500 μg/kg/da stimulated intestinal growth in rats.[34] Houle et al.[45] fed newborn colostrum-deprived piglets a sow milk replacer supplemented with 500 μg/l of IGF-I on da 1 to 14 postpartum. The 500 μg/l concentration was chosen to be in the range of the highest concentrations of IGF-I reported in porcine colostrum[30,31] and resulted in an ingested IGF-I dose of 200 μg/kg/da.

As can be appreciated from Table 3, orally administered IGF-I over a range from 200 to 3500 μg/kg/da exerted limited systemic effects. Serum IGF-I,[44,45] IGF-II[45] and IGFBP profiles[44,45] were unaffected by oral IGF-I, which supports the studies in which limited absorption of [^{125}I]-IGF-I was observed in piglets.[37] In addition, neither final body weight nor body weight gain were affected by oral IGF-I, although there was a tendency for piglets receiving the 3500 μg/kg/da dose of IGF-I to gain more weight than control.[44] Organ weights (g/kg) generally were unaffected, with the exception that pancreas weight was increased by oral IGF-I and IGF-II in the study of Xu et al.[43] The increased pancreas weight in IGF-I treated animals was associated with greater tissue DNA and RNA content compared to piglets fed formula alone. In addition, liver and spleen weight tended to be heavier in piglets fed 3,500 μg/kg/da than control;[44] whereas, organ weights were unaffected by a more physiological 200 μg/kg/da dose.[45]

With respect to the intestine, the magnitude of response to IGF-I was affected both by dose administered and duration of exposure (Table 4). When piglets consumed 3,500 μg IGF-I/kg/da for 4 da, small intestinal weight (g/kg), protein (mg/kg) and DNA (mg/kg) content were elevated compared to piglets receiving milk replacer alone.[44] In contrast, no differences in intestinal weight (g/kg), length (g/kg) or protein content (mg/g tissue) were observed when piglets were consuming IGF-I at a dose of 200 μg/kg/da for 14 da[45] or 440 μg/kg/da for 24 hr.[43] It is likely the 24 hr treatment was insufficient, since an increase in bromodeoxyuridine (BrdU) labelling in crypts of IGF-I and IGF-II treated piglets suggested a longer period of treatment could have resulted in increases in intestinal villus growth.[43] IGF-I and, to a lesser extent, IGF-II were mitogenic in neonatal pig intestine.

Intestinal histomorphology was assessed in all 3 studies. No effect was observed with a 24 hr treatment of either IGF-I or -II.[43] In other studies, only villus height was affected; crypt depth and muscularis thickness did not differ among treatment groups.[44,45] At 3500 μg/kg/da, villus height was greater in both jejunum (74% over control) and ileum (34% over control);[44] whereas, an effect of 200 μg IGF-I/kg/da on villus height was observed only in distal ileum (40 to 60% over control).[45]

Disaccharidase activity was determined by Houle et al.[45] as an index of intestine functional development. Specific activities of sucrase in jejunum and lactase in jejunum and proximal ileum were 2 to 4 fold higher in piglets receiving oral 200 μg IGF-I/kg/da; whereas, no effect on leucine aminopeptidase was observed. These results are consistent with those of Young et al.,[45] in which low doses of IGF-I (1 μg/da) administered orally to neonatal rats increased jejunal brush border lactase and sucrase activity, without stimulating intestinal cellular growth. Although the mechanism underlying the increase in intestinal disaccharidases remains to be elucidated, parallels can be drawn to previous studies in which piglets were fed a formula containing 85 U/l of insulin of insulin.[46,47] Ileal mucosal weight, protein, RNA and DNA content, and lactase and maltase specific activities were greater in piglets fed insulin containing formula than piglets fed formula alone.[46] A subsequent study was conducted in which ileal lactase activity was increased; however, no effect on lactase mRNA or the posttranslational processing of enzymes were found.[47] It is possible that IGF-I increases lactase activity indirectly by increasing cell number. This could be achieved either by enhancing mitotic activity in crypts[43] or by inhibiting apoptosis,[48] thus increasing residence time of mature villus cells.

Table 3. Summary of systemic responses from studies investigating the role of orally administered IGF-I and IGF-II

Ref.	Body wt.	Serum				Stomach	Pancreas	Tissue weights		
		IGF-I	IGF-II	IGFBP	Other			Liver	Kidneys	Spleen
39, 42	No[a]	↑ on da 4	No	No	↓ insulin on da 4 ↑ prolactin first 8 hr	NR	NR	NR	NR	NR
43[b]	No	NR	NR	NR	NR	No[c]	Yes[c]	No	No	No
44	No	No	NR	No	↑ Insulin	No	No	No	No	No
45	No	No	No	No	NR	No	No	No	No	No

[a]Yes = observation was different from control; No = observation was not different from control: NR = not reported.
[b]Similar results were observed for both IGF-I and IGF-II.
[c]DNA content of stomach and pancreas was higher in IGF-I or IGF-II treated than control.

4 broad

Table 4. Summary of small intestinal responses from studies investigating the role of orally administered IGF-I and IGF-II

Ref.	Weight	Length	Protein	DNA	Villus height	Crypt death	Enzymology	Other
41	No[a]	No	No	No	NR	NR	NR	↑ thymidine incrop. ↑ type I receptor binding
43[b]	No	NR	No	No	No	No	NR	
44	No	No	Yes	Yes	Jejunum and ileum	No	NR	↑ BrdU labelling in crypts
45	No	No	No	No	Only ileum	No	↑ lactase, sucrase	

[a]Yes = observation was different from control; No = observation was not different from control; NR = not reported.
[b]Similar results were observed for both IGF-I and IGF-II.

In summary, orally administered IGF-I and IGF-II appear to survive digestion for at least 30 min[38] and exert local effects within intestine. Orally administered IGF-I does not influence circulating concentrations of IGF-I,[44,45] even at pharmacological doses. In addition, the inability of colostral levels of IGF-I (500 μg/l) consumed for 14 da to influence body weight or tissue weights suggests that orally administered IGF-I at a high physiological dose is not a major regulator of neonatal growth.[45] Within intestine, ingestion of 200 μg IGF-I/kg/da for 14 da increased ileal growth and sucrase and lactase activities in piglets without increasing small intestinal weight, length or protein content.[45] However, ingestion of a dose of IGF-I that was 70 times higher (3,500 μg/kg/da) increased the weight and protein and DNA content of the small intestine.[44] Ingestion of formula containing 2000 μg/l of IGF-I or IGF-II increased proliferative activity in crypts (BrdU labelling); however, the study duration was too short (24 hr) to detect differences in intestinal histology.[43] Results of the study in calves suggest the effects of IGF-I in neonatal intestine are due in part to an increase in cellular proliferation ([³H]-thymidine incorporation) and this effect may be mediated by an up regulation of the type I IGF receptor in response to oral IGF-I intake.[41] Regional distribution of IGF receptors within intestine support this hypothesis. IGF-I receptors are highest in crypt cells;[13,15] several investigators have shown enhanced proliferation in crypts in response to oral IGF-I.[41,43]. In addition, IGF receptor binding is highest in ileum[13,15] and ileum was found to be the most responsive region of intestine to oral IGF-I in terms of villus growth,[45] [³H]-thymidine incorporation and type I receptor binding.[41] Also, ileum was most responsive to orally administered insulin,[46] suggesting the potential for a common mechanism of action for insulin acting through the insulin and IGF-I through type I IGF receptor.

6. FUTURE DIRECTIONS

Although work to date on the ability of orally administered IGF-I to exert intestinal effects in the normal neonate is promising, future work is merited in at least 6 areas: 1) determining whether IGF-I at more physiological doses is effective, 2) investigating the role of IGF-II alone or in combination with IGF-I as it is present in milk, 3) assessing the role of IGFBP in milk on mediating IGF-I action within the intestine, since all studies to date have added IGF to a formula devoid of IGFBP, 4) studying whether other bioactive components in milk interact with IGF in either complementary or inhibitory ways, 5) determining underlying mechanisms of IGF-I action on villus growth and enzyme activities and 6) investigating the potential application of orally administered IGF-I during compromised gastrointestinal function. Parenterally administered IGF-I stimulates small intestinal growth when administered to rats in a variety of states which result in compromised intestinal function, including gut resection,[49] dexamethasone treatment[50] or total parenteral nutrition.[51] The ability of enteral IGF-I to enhance small intestinal growth in these clinical situations has not been assessed, although data exists supporting beneficial effects of another milk borne growth factor, epidermal growth factor (EGF). Oral administration, in rats and piglets, of EGF following methotrexate or rotaviral induced intestinal damage enhances the recovery process.[52,53] From the perspective of applicability of basic research on biological actions of growth factors in milk obtained from the piglet model, it is likely potential clinical uses of orally administered growth factors during recovery from gastrointestinal damage holds the greatest promise towards improving human infant health.

7. REFERENCES

1. Grosvenor, C.E., Picciano, M.F., and Baumrucker, C.R., 1993, Hormones and growth factors in milk, *Endocr. Rev.* 14:710-728.
2. Donovan, S.M., and Odle, J., 1994, Growth factors in milk as mediators of infant development, *Ann. Rev. Nutr.* 14:147-167.
3. Widdowson, E.M., Colombo, V.E., and Artavanis, C.A., 1976, Changes in the organs of pigs in response to feeding for the first 24 h after birth. II. The digestive tract, *Biol. Neonate* 28:272-281.
4. Burrin, D.G., Shulman, R.J., Reeds, P.J., Davis, T.A., and Gravitt, K.R., 1992, Porcine colostrum and milk stimulate visceral organ and skeletal muscle protein synthesis in neonatal piglets, *J. Nutr.* 122:1205-1213.
5. Burrin, D.G., Davis, T.A., Ebner, S., Schoknecht, P.A., Fiorotto, M.L., Reeds, P.J., and McAvoy, S., 1995, Nutrient-independent and nutrient-dependent factors stimulate protein synthesis in colostrum-fed newborn pigs, *Pediatr. Res.* 37:593-599.
6. Burrin, D.G., Dudley, M.A., Reeds, P.J. Shulman, R.J., Perkinson, S., and Rosenberger, J., 1994, Feeding colostrum rapidly alters enzymatic activity and the relative isoform abundance of jejunal lactase in neonatal pigs, *J. Nutr.* 124:2350-2357.
7. Aumaitre, A., and Corring, T., 1978, Development of digestive enzymes in the piglet from birth to 8 weeks, *Nutr. Metab.* 22:244-253.
8. Moughan, P.J., Birtles, M.J., Cranwell, P.D., Smith, W.C., and Pedraza, M., 1992, The piglet as a model animal for studying aspects of digestion and absorption in milk-fed human infants, in: *Nutritional Triggers for Health and in Disease*, Volume 37 (A.P. Simonopoulos, ed.), Karger, Basel, Switzerland, pp. 40-113.
9. Donovan, S.M., Zijlstra, R.T., and Odle, J., 1994, The use of the piglet to study the role of growth factors in neonatal intestinal development, *Endocrine Reg.* 28:153-162.
10. Rechler, M.M., and Nisslley, S.P., 1991, Insulin-like growth factors, in: *Peptide Growth Factors and Their Receptors*, Volume 1 (M.B. Sporn, A.B. Roberts, and L.A. Hanson, eds.), Springer-Verlag, New York, pp. 263-368.
11. Tavakkol, A., Simmen, F.A., and Simmen, R.C.M., 1988, Porcine insulin-like growth factor-I (pIGF-I): complementary deoxyribonucleic acid cloning and uterine expression of messenger ribonucleic acid encoding evolutionarily conserved IGF-I peptides, *Mol. Endocrinol.* 2:674-681.
12. Oh, Y., Müller, H. L., Neely, E. K., Lamson, G., and Rosenfeld, R.G., 1993, New concepts in insulin-like growth factor receptor physiology, *Growth Reg.* 3:113-123.
13. Laburthe, M., Rouyer-Fessard, C., and Gammeltoff, S., 1988, Receptors for insulin-like growth factor I and II in rat gastrointestinal epithelium, *Am. J. Physiol.* 254:G457-G462.
14. Schober, D.A., Simmen, F.A., Hadsell, D.L., and Baumrucker. C.R., 1990, Perinatal expression of type I IGF receptors in porcine small intestine, *Endocrinology* 126:1125-1132.
15. Young, G.P., Taranto, T.M., Jones, H.A., Cox, A.J., Hogg, A., and Werther G.A., 1990, Insulin-like growth factors in the developing and mature rat small intestine: receptors and biological actions, *Digestion* 46:240-252.
16. Shimasaki, S., and Ling, N.L., 1991, Identification and molecular characterization of insulin-like growth factor binding proteins (IGFBP-1, -2, -3, -4, -5, and -6), *Prog. Growth Factor Res.* 3:243-266.
17. Walton, P.E., Baxter, R.C., Burleigh, B.D., and Etherton, T.D., 1989, Purification of the serum acid-stable insulin-like growth factor binding protein from the pig (*Sus scrofa*), *Comp. Biochem. Physiol.* 92B:561-567.
18. McCusker, R.H., Campion, D.R., Jones, W.K., and Clemmons, D.R. 1985, The insulin-like growth factor-binding proteins of porcine serum: endocrine and nutritional regulation, *Endocrinology* 125:501-509.
19. Donovan, S.M., McNeil, L.K., Jimenez-Flores, R., and Odle, J., 1994, Insulin-like growth factors and IGF binding proteins in porcine serum and milk throughout lactation, *Pediatr. Res.* 36:159-168.
20. Shimasaki, S., Shimonaka, M., Ui, M., Inouye, S., Shibata, F., and Ling. N.L., 1990, Structural characterization of follicle-stimulating hormone action inhibitor in porcine ovarian follicular fluid: its identification as the insulin-like growth factor binding protein, *J. Biol. Chem.* 265:2198-2202.
21. Shimasaki, S., Gao, M., Shimonaka, M., and Ling, N.L., 1991, Isolation and cloning of insulin-like growth factor binding protein -6, *Mol. Endocrinol.* 4:938-948.
22. Baxter, R.C., 1990, Circulating levels and molecular distribution of the acid-labile (a) subunit of the high molecular weight insulin-like growth factor binding complex, *J. Clin. Endocrinol. Metab.* 70:1347-1353.
23. Bach, L.A., and Rechler, M.M. 1995, Insulin-like growth factor binding proteins, *Diabetes Rev.* 3:38-61.
24. Odle, J., Zijlstra, R.T., and Donovan, S.M., 1996, Intestinal effects of milkborne growth factors in neonates of agricultural importance, *J. Anim. Sci.* 74s.

25. Baxter, R.C., Zaltsman, Z., and Turtle, J.R., 1984, Immunoreactive somatomedin-C/insulin-like growth factor I and its binding protein in human milk, *J. Clin. Endocrinol. Metab.* 58:955-959.

26. Donovan, S.M., Hintz, R.L., and Rosenfeld, R.G., 1991, Insulin-like growth factors I and II and their binding proteins in human milk: effect of heat treatment on IGF and IGF binding protein stability, *J. Pediatr. Gastroent. Nutr.* 13:242-253.

27. Donovan, S.M., Hintz, R.L., Wilson, D.M., and Rosenfeld, R.G., 1991, Insulin-like growth factors I and II and their binding proteins in rat milk, *Pediatr. Res.* 29:50-55.

28. Malven, P.V., Head, H.H., Collier, R.J., and Buonomo, F.C., 1987, Periparturient changes in secretion and mammary uptake of insulin and in concentrations of insulin and insulin-like growth factors in milk of dairy cows, *J. Dairy. Sci.* 70:2254-2265.

29. Vega, J.R., Gibson, C.A., Skaar, T.C., Hadsell, D.L., and Baumrucker, C.R., 1991, Insulin-like growth factor (IGF) and IGF binding proteins in serum and mammary secretions during the dry period and early lactation, *J. Anim. Sci.* 69:2538-2544.

30. Simmen, F.A., Simmen, R.C.M., and Reihnart, G., 1988, Maternal and neonatal somatomedin-C/insulin-like growth factor-I (IGF-I) and IGF binding proteins during early lactation in the pig, *Dev. Biol.* 130:16-20.

31. Simmen, F.A., Whang, K.Y., Simmen, R.C.M., Peterson, G.A., Bishop, M.D., and Irvin, K.M., 1990, Lactational variation and relationship to postnatal growth of insulin-like growth factor-I in mammary secretions from genetically diverse sows, *Domest. Anim. Endocrinol.* 7:199-206.

32. Hossenlopp, P., Seurin, D., Segovia-Quinson, B., Hardouin, S., and Binoux, M., 1986, Analysis of serum insulin-like growth factor binding proteins using western blotting: use of the method for titration of the binding proteins and competitive binding studies, *Anal. Biochem.* 154:138-143.

33. Suikkari, A-M., 1989, Insulin-like growth factor (IGF-I) and its low molecular weight binding protein in human milk, *Eur. J. Obstet. Gynecol.* 30:19-25.

34. Xian, C.J., Shoubridge C.A., and Read, L.C., 1995, Degradation of IGF-I in the adult rat gastrointestinal tract is limited by a specific antiserum or the dietary protein casein, *J. Endocrinol.* 146:215-225.

35. Ronge, H., and Blum, J.W., 1988, Somatomedin C and other hormones in dairy cows around parturition, in newborn calves, and in milk, *J. Anim. Physiol. Nutr.* 60:168-174.

36. Baumrucker, C.R., Hadsell, D.L., Skaar, T.C., Campbell, P.G., and Blum, J.W., 1992, Insulin-like growth factors (IGFs) and IGF binding proteins in mammary secretions: origins and implications in neonatal physiology, in: *Contemporary Issues in Clinical Nutrition, Mechanisms Regulating Lactation and Infant Nutrient Utilization*, Volume 15 (M.F. Picciano and B. Lönnerdal, eds.), Wiley-Liss, New York, pp. 285-308.

37. Donovan, S.M., Chao, J.C.-J., Zijlstra, R.T., and Odle, J., 1996 Orally administered iodinated insulin-like growth factor-I (^{125}I-IGF-I) is poorly absorbed by the neonatal piglet, *J. Pediatr. Gastroent. Nutr.* 23: In Press.

38. Philipps, A.F., Radhakrishna, R., Anderson, G.G., McCracken, D.M., Lake, M., Koldovsky, O., 1995, Fate of insulin-like growth factors I and II administered orogastrically to suckling rats, *Pediatr. Res.* 37:586-592.

39. Baumrucker, C.R., and Blum, J.W., 1994, Effects of dietary recombinant insulin-like growth factor-I on concentrations of hormones and growth factors in the blood of newborn calves, *J. Endocrinol.* 140:15-21.

40. Baumrucker, C.R., Green, M.H., and Blum, J.W., 1994, Effects of dietary rhIGF-I in neonatal calves on the appearance of glucose, insulin, d-xylose, globulins, and gamma-glutamyl transferase in blood, *Domest. Anim. Endocrinol.* 11:393-403.

41. Baumrucker, C.R., Hadsell, D.L., and Blum, J.W., 1994, Effects of dietary insulin-like growth factor I on growth and insulin-like growth factor receptors in neonatal calf intestine, *J. Anim. Sci.* 72:428-433.

42. Skaar, T.C., Baumrucker, C.R., Deaver, D.R., and Blum, J.W., 1994, Diet effects and ontogeny of alterations of circulating insulin-like growth factor binding proteins in newborn dairy calves, *J. Anim. Sci.* 72:421-427.

43. Xu, R-J., Mellor, D.J. Birtles, M.J., Breier, B.H., and Gluckman, P.D., 1994, Effects of oral IGF-I or IGF-II on digestive organ growth in newborn piglets, *Biol. Neonate* 66:280-287.

44. Burrin, D.G., Wester, T.J., Davis, T.A., Amick, S., and Heath, J.P.,1996 Orally administered insulin-like growth factor-I increases intestinal mucosal growth in formula-fed neonatal pigs, *Am. J. Physiol.* 270: R1085-R1091.

45. Houle, V. M., Schroeder, E.A., Park, Y., Odle J., and Donovan, S.M., 1995, Orally administered insulin-like growth factor-I stimulates neonatal piglet intestinal development, *FASEB J.* 9:A580(Abstr.).

46. Shulman, R.J., 1990, Oral insulin increases small intestinal mass and disaccharidase activity in the newborn miniature pig, *Pediatr. Res.* 28:171-175.

47. Shulman, R.J., Tivey, D.R.,Sunitha, I., Dudley, M.A., and Henning, S.J., 1992, Effect of oral insulin on lactase activity, mRNA, and posttranslational processing in the newborn pig, *J. Pediatr. Gastroent. Nutr.* 14:166-172.

48. Sell, C.R., Baserga, R., and Rubin R., 1995, Insulin-like growth factor I (IGF-I) and the IGF-I receptor prevent etoposide-induced apoptosis, *Cancer Res.* 55:303-306.

49. Lemmey, A.B., Martin, A.A., Read, L.C., Tomas, F.M., Owens, P.C., and Ballard, F.J. 1991, IGF-I and the truncated analog des-(1-3)IGF-I enhance growth in rats after gut resection, *Am. J. Physiol.* 260:E213-E219.

50. Read, L.C., Tomas, F.M., Howarth, G.S., Martin, A.A., Edson, K. J., Gillespie C.M., Owens P.C., and Ballard, F.J., 1992, Insulin-lkike growth factor-I and its N-terminal modified analogues induce marked gut growth in dexamethasone-treated rats, *J. Endocrinol.* 33:421-431.

51. Yang, H., Grahn, M., Schlach, D.S., and Ney, D.M., 1994, Anabolic effects of IGF-I coinfusion with total parenteral nutrition in dexamethasone-treated rats, *Am. J. Physiol.* 266:E690-E698.

52. Petschow, B.W., Carter, D.L., and Hutton, G.D., 1993, Influence of orally administered epidermal growth factor on normal and damaged intestinal mucosa in rats, *J. Pediatr. Gastroenterol. Nutr.* 17:49-58.

53. Zijlstra, R.T., Odle, J., Hall, W.F., Petschow, B.W., Gelberg, H.B., and Litov, R., 1994, Effect of orally administered epidermal growth factor on intestinal recovery of neonatal pigs infected with rotavirus, *J. Pediatr. Gastroent. Nutr.* 19:382-390.

65

THE PERINATAL PIG IN PEDIATRIC GASTROENTEROLOGY

Per T. Sangild,[1*] Marian Silver,[2] Mette Schmidt,[1] and Abigail L. Fowden[2]

[1] Clinical Studies
Division of Reproduction
Royal Veterinary and Agricultural University
DK-1870 Frederiksberg C, Denmark
[2] Physiological Laboratory
University of Cambridge
CB2 3EG Cambridge, United Kingdom

1. ABSTRACT

Immature intestinal development is a predisposing factor to enteric disease in the newborn. Hence, in the present study, the effects of colostrum and birth on the development of intestinal enzymes were investigated in the fetal and neonatal pig. In the first experiment, 30 fetal pigs from 10 litters (99 to 102 da of gestation; term = 114 ± 2 da) were prepared with chronic oesophageal and intravascular catheters. Through the oesophageal catheters, fetuses were infused with saline (n = 6), amniotic fluid (n = 8), milk whey (n = 6) or colostrum whey (n = 10) into the fetal stomach *in utero*, 4 times daily for 6 da. In a further 3 fetuses, the oesophagus was ligated to prevent the fetal gut from receiving any fluids. Blood gases, blood haemoglobin and plasma cortisol values were measured to monitor the fetal well being *in utero*. At the end of the experiment, infused and ligated fetuses were removed by caesarean section and small intestine collected for enzyme analyses. Treatment did not affect sucrase or lactase activity but maltase, aminopeptidase and dipeptidylpeptidase IV activities were altered. Mean activities of these enzymes were highest in colostrum infused fetuses and lowest in ligated fetuses. Maltase and aminopeptidase A activities in the colostrum group were 100% higher than in the amniotic fluid group. In the second experiment, 73 newborn pigs from 6 litters were born either by caesarean section or by the vaginal route after induction of parturition with a prostaglandin analogue. Newborn pigs were born prematurely (106 to 108 da of gestation) or close to normal term (113 to 115 da of gestation); small intestine was collected for enzyme analyses at birth or after 2 da of colostrum feeding. Most enzyme activities were lower for newborn, premature piglets than term piglets. Vaginal birth inhibited

---

*Reprint requests to: Dr. Per Sangild, Clinical Studies, Reproduction, Royal Veterinary and Agricultural University, 13 Bülowsvej, DK-1870 Frederiksberg C, DENMARK (45) 35 28 29 71.

Advances in Swine in Biomedical Research, edited by Tumbleson and Schook
Plenum Press, New York, 1996

sucrase activity and stimulated aminopeptidase A activity. For 2 da old pigs, vaginal birth was associated with increased activities of sucrase, lactase, aminopeptidase N and dipeptidylpeptidase IV. Consistent with findings in fetal pigs, colostrum feeding for 2 da induced a large (2 to 3 fold) increase in maltase and aminopeptidase A activities; whereas, the other enzyme activities were similar at birth and at 2 da of age. The fetal and neonatal pig is a useful model to investigate the effects of luminal factors (nutrients, hormones, growth factors) on immature intestine. In addition, time of birth (preterm or term), delivery type (caesarean section or induced vaginal birth) and diet (milk or colostrum) have pronounced effects on the development of intestinal function.

2. INTRODUCTION

Brush border enzymes in the small intestine are nutritionally important for the hydrolysis of carbohydrates and peptides into monosaccharides and small peptides or amino acids. Therefore, development of intestinal hydrolase activities in the fetus is an essential maturational event which ensures the gut can handle oral uptake of nutrients after birth. In the human, serious enteric diseases, such as necrotizing enterocolitis, are more common in preterm than term infants. The effects of different enteral diets on gut maturation and disease have been studied extensively in both preterm and term infants.[1,2] However, little is known about the developmental regulation of brush border enzymes in the human or in species, such as the pig, in which gut development resembles that in the human.[3-5] Hence, in the first part of this study, chronically catheterized fetal pigs were used to study the effects of colostral factors, e.g., nutrients, hormones, growth factors, on the maturation of immature small intestine.

In the preterm human infant, the incidence of necrotizing enterocolitis can be reduced by prenatal glucocorticoid treatment.[6,7] These observations are consistent with the well known maturational effects of cortisol on the gut epithelium and other fetal tissues during the perinatal period.[8-11] In both the pig and human, there is a prepartum rise in fetal cortisol production which correlates well with the prenatal increase in the activity of brush border enzymes.[3,12] In the pig, fetal cortisol levels rise slowly during the last 10 da of gestation and then escalate rapidly during labor and delivery.[13] In the second part of the present study, the role of this final peripartum cortisol surge was examined by comparing intestinal function in neonatal piglets delivered either vaginally or by caesarean section. From these two studies, we should be able to demonstrate whether or not the pig is a feasible model for studying human pediatric gastroenterology. All experiments were approved by relevant statutory licensing authorities.

3. MATERIALS AND METHODS

3.1. Collection and Preparation of Fluids for Infusion or Feeding

A pool of sterile amniotic fluid was collected from a series of different pregnant sows undergoing fetal surgery during late gestation (82 to 102 da of gestation). Fluid was kept in 100 ml aliquots and stored at -20C until use. Sow's milk and sow's colostrum were obtained by hand milking sows from 2 different farms either at 2 to 3 wk of lactation (milk) or immediately after parturition (colostrum). The ejection of milk or colostrum was stimulated by oxytocin (3 IU, IV; Oxytocin, Leo, Ballerup, Denmark) and fluids were kept frozen at -20C until further preparation. Approximately 25 l of sow's colostrum and 5 l of sow's milk were collected for the 2 experiments.

All the globulins and various water soluble regulatory peptides of milk and colostrum can be found in the whey fraction of these fluids. For this reason, and to avoid infusing excessive amounts of digestible fat and protein into fetuses, only the whey fraction of the fluids was used. Fluids were thawed and pooled milk or pooled colostrum centrifuged (5000 g; 4C, 30 min). The top layer, containing fat, and the bottom layer, containing cellular debris and other insoluble material, were discarded. Casein was precipitated at room temperature, pH 6 to 7, by the addition of appropriate amounts of stomach extracts from newborn pigs containing large amounts of the porcine milk clotting enzyme, chymosin.[16,17] This milk clotting procedure reflects the natural conditions in the newborn pig stomach and avoids the possible damage to whey proteins and peptides induced after precipitating casein micelles by creating a highly acidic environment, pH 2 to 3.[16] After 30 min of continuous stirring, precipitate was removed by centrifugation and the milk whey or colostrum whey emptied into 50 ml bottles. Bottles were capped, sealed and frozen (-20C) until the time of the experiment. Colostrum whey contained 80 mg IgG/ml and milk whey less than 0.5 mg IgG/ml. No IgG could be detected in amniotic fluid. Total proportions of protein and lactose were 14.0 and 2.0% in colostrum whey and 2.8 and 5.8% in the milk whey, respectively.

Hand milking of sows and preparation of whey fractions inevitably resulted in some degree of contamination of the fluids with microorganisms from the environment. Therefore, frozen milk whey and colostrum whey were sterilized by gamma irradiation (7.0 kGy at -20C; Riso National Laboratory, Copenhagen, Denmark). The irradiation dose used was the lowest possible to attain 100% sterility of the samples. Irradiation at this dose and temperature induces minimal destruction of biological components (proteins, peptides) by the formation of potentially harmful free radicals. By using this sterilization procedure, excessive antibiotics administration to fluids was avoided. Sterilization by ultrafiltration techniques was not feasible due to the high viscosity of colostrum whey.

3.2. Fetal Pigs

Ten pregnant sows (99 to 102 da of gestation, litter size 8 to 14 fetuses, term 114 ± 2 da) were sedated with 50 mg azeperone IM (Suicalm, Janssen Pharmaceutical, Belgium) and a superficial ear vein catheterized under local anaesthesia (Lignocaine, Willotox, UK) either on the day of surgery or 1 to 2 da previously. General anaesthesia was induced and maintained with sodium pentobarbitone (40 mg/kg, IV) after intubation under halothane. There was no mechanical ventilation of the sows during surgery but pure oxygen was supplied continuously via the endotracheal tube.

The uterus was exposed and dehydration of the uterus during surgery was prevented by placing the exposed area in a sterile plastic bag. The head of a fetus was exposed through a uterine incision.[14] A fetal carotid artery and the oesophagus were exposed through a 2 to 3 cm incision in the neck. A small hole was cut in the oesophagus and a silastic catheter (vinyl tube, 0.86 mm i.d., 1.52 mm o.d., Dural Plastics & Engineering, Auburn, Australia) was inserted into the oesophagus so the catheter tip entered the stomach. A short catheter was also passed into the oesophagus towards the larynx to allow fluid swallowed by the fetus to be returned to the amniotic cavity. A vascular catheter also was inserted into a fetal carotid artery. Antibiotics (50 mg Ampicillin, Penbitrin, Beecham Labs, UK) were injected into the amniotic cavity of each operated fetus.

After catheterization, the fetal skin incision, fetal membranes and uterine incision were closed. This operation was repeated on 2 to 3 fetuses in different areas of the uterus; finally, the sow was prepared with a uterine vein catheter. The maternal vascular catheter and the fetal vascular and oesophageal catheters all were exteriorized on the back of the sow and kept in a small plastic bag. In a further 3 fetuses, the oesophagus only was ligated and not catheterized. As in the other operated fetuses, swallowed amniotic fluid was prevented from passing down into the gastrointestinal tract, but these 3 fetuses received no "artificial diet". After fetal surgery, the uterine incision was closed and the sow was returned to her crate. The total duration of surgery was about 3 hr. A schematic drawing of a chronically catheterized pig fetus is depicted in Fig. 1. Studies were carried out in Cambridge, UK.

Sows were maintained on antibiotics for 3 to 4 da from the day of surgery (20 ml Depocillin IM, Mucofarm, UK) and also were given a daily injection of progesterone (50 mg, IM, Progesterone, Intervet UK, Cambridge, UK) throughout the experimental period. Pregnant sows usually tolerate uterine surgery and chronic experiments on fetuses well.[14,15] However, the present experiments were performed relatively close to normal term and included a large number of blood samplings and infusions. Therefore, progesterone injections were used as a precaution against the possible onset of preterm labor.

Starting 24 hr after surgery (day 0), saline (n = 6), amniotic fluid (n = 8), milk whey (n = 6) or colostrum whey (n = 10) was infused as a 10 ml bolus into the oesophageal catheter in each operated fetus. Fluids were warmed to approximately 35C immediately before infusion and maximal care was taken to assure aseptic conditions around catheters at each infusion. This treatment was continued every 6 hr for 6 consecutive days. During the experiment, fetal well being was monitored by following blood gas values in a blood sample taken every morning from each fetus before the first oesophageal infusion. Blood was collected in ice chilled EDTA containing tubes; plasma was separated by centrifugation and stored at -20C until further analysis.

At the end of the experimental period, 7 da after fetal surgery, infused fetuses were delivered by caesarean section under general anaesthesia as above. Each fetus was removed from the uterus and the gut collected after administration of a lethal dose of anaesthetic (200 mg sodium pentobarbitone). The intestine was dissected free of its mesentery, placed on ice and divided into 3 parts of equal length. The middle 10 cm of each piece was cut open lengthwise, blotted dry on filter paper and stored at -80C until enzyme analysis. The proximal segment (17% down the intestine) always was taken distal to the ligament of Treitz and

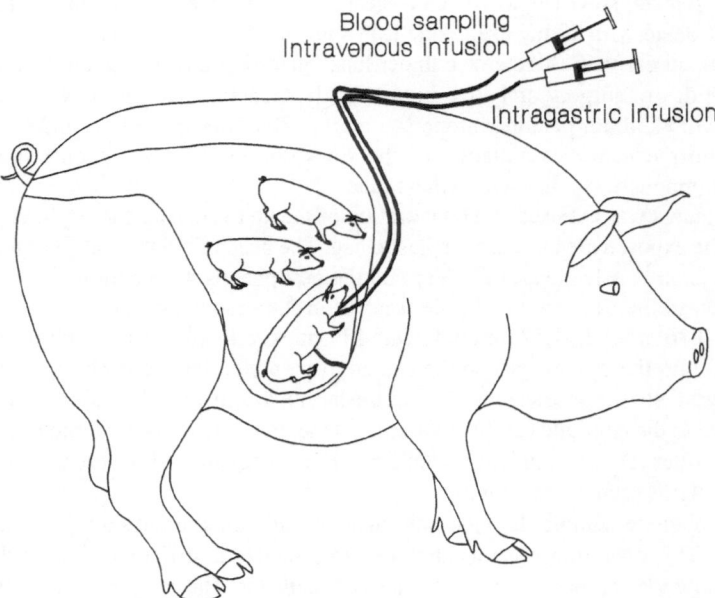

Figure 1. Fetal pigs: The experimental model involved fetuses fitted with chronic vascular and oesophageal catheters. Schematic illustration, see text for details.

defined as proximal jejunum. The distal segment (83% down the intestine) represented the distal jejunum or proximal ileum in the pig.

3.3. Neonatal Pigs

Pregnant sows (106 da of gestation, n = 3, or 113 da of gestation, n = 3) were sedated with azeperone (50 mg, IM, Sedaperone, Janssen Pharmaceutical, Belgium) and anaesthetized with sodium pentobarbitone (10 mg/kg, IV). All, or the majority of, fetuses in the left uterine horn (6 to 9 fetuses/litter, 39 fetuses total) were removed by caesarean section through 2 or 3 incisions in the uterus. All piglets were born less than 45 min after the induction of anaesthesia. After closure of the incisions, the sow was returned to her crate and given 200 μg cloprostenol IM (a $PGF_{2\alpha}$ analogue; Estrumate, Pitman-Moore, Harefield, UK) to induce parturition. Two da later (30 to 45 hr after induction), sows gave birth to the remaining fetuses in each litter (a total of 50 piglets). A venous blood sample was taken at birth from the naval cord in both caesarean delivered and vaginally delivered pigs. After measurements of blood gases and haemoglobin in whole blood, plasma was stored at -20C for further analysis. Induced parturitions were associated with relatively long time spans from delivery of the first to the last piglets (9 to 12 hr). However, sows did not react differently or express excessive pain compared with spontaneous deliveries at term (lasting 4 to 6 hr). Therefore, analgesics for the sow during labour were not used.

The present experimental procedure enabled us to study the effects of delivery type within litters and compare intestinal function in piglets having similar genetic backgrounds and maternal environments. It cannot be excluded that the small differences in fetal age of piglets from the same litters (with vaginally delivered pigs being 30 to 45 hr older than the corresponding caesarean delivered pigs) will have specific effects on enzymic development at this time but these effects should be minimal. Also, vaginally delivered pigs were subjected to the stress of being anaesthetized at the previous caesarean section in combination with the fetal stress caused by induced labor and vaginal birth. A schematic presentation of the experimental protocol is given in Fig. 2. Experiments were carried out in Copenhagen, Denmark.

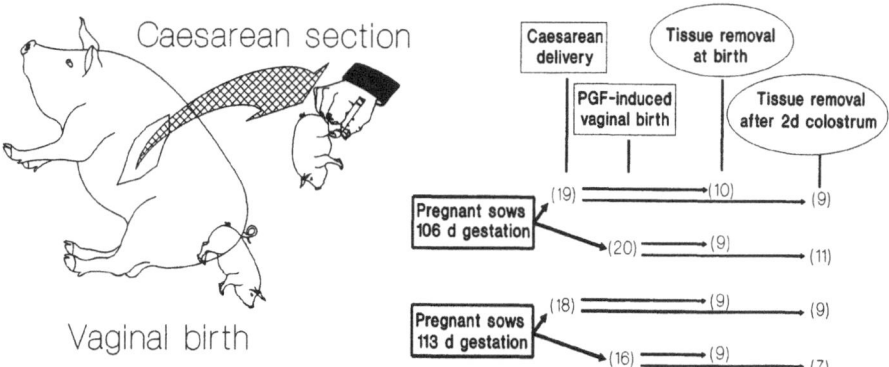

Figure 2. Neonatal pigs: Within litters, fetuses from one uterine horn were first removed by caesarean section. Fetuses in the other uterine horn were subjected to vaginal birth (30 to 45 hr after the caesarean section) following induction of parturition with a prostaglandin $F_{2\alpha}$ analogue. The diagram shows how the 73 liveborn pigs were allocated among different treatment groups; piglet numbers in each group are given in brackets.

Liveborn piglets (n = 73) were killed (by a blow on the head) either immediately after birth (n = 37) or after 2 da of colostrum feeding by stomach tube (15 ml/kg bwt every 3 hr) (n = 36). Intestine was removed and divided as for fetal pigs. Mucosa from pieces of the proximal, middle and distal intestine were scraped off the underlying muscle layers with a glass slide and stored at -80C before further analysis.

3.4. Biochemical Analyses

Blood gases, pH, hemoglobin concentration and oxygen concentration were measured in whole blood (Acid Base Laboratory and Haemooxymeter; Radiometer, Copenhagen, Denmark). Plasma cortisol was measured by radioimmunoassay by the method validated for pig plasma by Silver et al.[18]

Frozen intestinal tissue (fetal pigs) or intestinal mucosa (neonatal pigs) was cut into fine pieces, extracted in 1% Triton X-100 (6 ml/g tissue), homogenized (20,000 rpm, 2 min, 0C; Ultra Turax, IKA-Labortechnik, Stauten, Germany), centrifuged (20,000 g, 4C, 60 min) and the supernatant used for enzyme analyses. Protein concentrations were measured in tissue homogenates by the method of Lowry et al.[19] Disaccharidase activities were measured using sucrose, maltose and lactose as substrates (28 nM) and sodium maleate pH 6.0 (50 mM) as buffer in the reaction mixture (37C, 30 min).[12] Lactase activity is specific for the enzyme lactase:phlorizin hydrolase (EC 3.2.1.23-62); whereas, maltase activity arises from both sucrase:isomaltase (EC 3.2.1.48-10) and maltase:glucoamylase (EC 3.2.1.20). Sucrase activity is specific for sucrase:isomaltase. Peptidase activities were measured using specific nitroanilide substrates and Tris-HCl as buffer.[12] Substrates were specific for aminopeptidase N (EC 3.4.11.2), dipeptidyl peptidase IV (EC 3.4.14.5) or aminopeptidase A (EC 3.4.11.7). Protein contents were expressed as mg/g tissue and enzyme activities as enzyme units (U), where one unit is equal to one μmol substrate hydrolyzed/min at 37C. Values from proximal, middle and distal parts of the intestine were averaged to provide mean protein contents and mean enzyme activities for the intestine as a whole.

3.5. Statistical Analyses

Mean or mean ± SEM values are given throughout. Data were analyzed by analysis of variance (ANOVA) using a general linear model;[20] the LSD test was used to detect significant ($P<0.05$) differences between 2 individual means. For fetal pigs, the model included the linear effects of treatment (ligation, saline, amniotic fluid, milk, colostrum), litter and the treatment x litter interaction. For neonatal pigs, the model included the linear effects of gestational age at delivery (preterm or term), litter (nested within age at delivery), delivery type (caesarean section or induced vaginal birth), age (0 or 2 da) and all possible interactions between these class variables. Linear correlations between parameters, e.g., plasma cortisol vs enzyme activity, were tested by linear regression analyses.

4. RESULTS AND DISCUSSION

4.1. Fetal Pigs: Survival and Health

Of the 30 fetuses implanted with catheters, 28 were alive at the end of the experiment. Acidity, gas pressures, hemoglobin concentration and oxygen saturation in whole blood and cortisol concentration in plasma for these fetuses are shown in Fig. 3. Values have been pooled for the 4 groups of chronically catheterized fetuses.

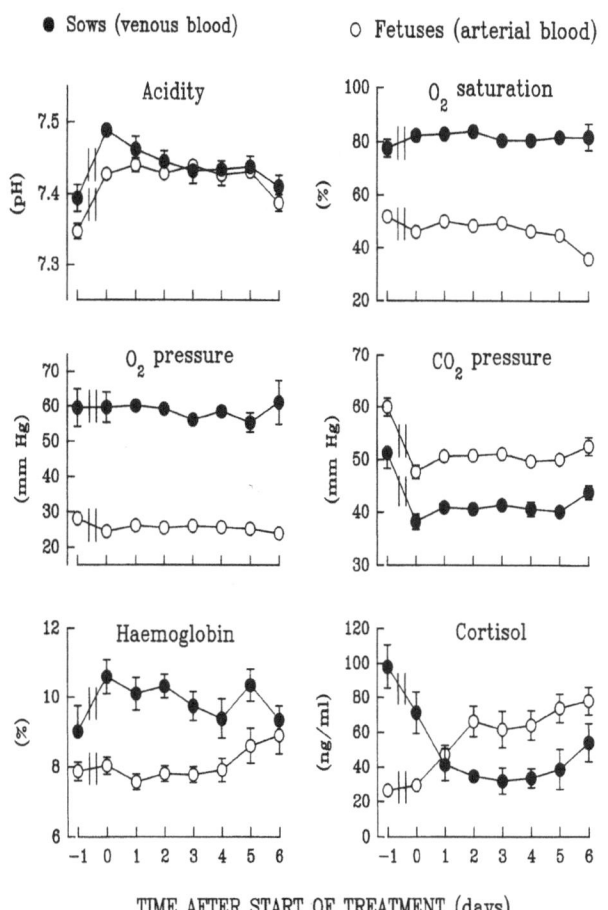

● Sows (venous blood)   ○ Fetuses (arterial blood)

TIME AFTER START OF TREATMENT (days)

Figure 3. Acidity, blood gases and hemoglobin level in whole blood, and concentration of cortisol in plasma from pregnant sows (n = 10) and their chronically catheterized fetuses (n = 28). Starting the day after surgery (day 0), 1 of 4 fluids (saline, amniotic fluid, milk whey or colostrum whey) was infused into the fetal stomach via the oesophageal catheter (10 ml/fetus every 6 hr). The first blood sample (before the line breaks, day -1) was taken from anaesthetized fetuses after surgery, while subsequent samples (after the line breaks, days 0 to 6) were taken from conscious pigs before the first daily oesophageal infusion on each day. For fetal pigs, values from different treatment groups have been pooled.

Anesthesia and fetal surgery resulted in a slight acidosis and hypercapnia for both the sow and her fetuses, as indicated by the lower blood pH values and higher blood CO_2 pressure on da -1 (at the end of surgery, under anaesthesia). Probably, this acidosis could have been prevented by mechanical ventilation of the sow. On the other hand, blood gas values never were critical for survival and always returned to normal levels on the first or second da after surgery. Towards the end of the experiment, mean blood acidity and blood hemoglobin concentration tended to increase in the catheterized fetuses ($P<0.05$ for values on da 6 vs da 1).

Plasma immunoglobulin G (IgG) levels increased from 0 on da 0 to 10.7 ± 0.8 mg/ml on da 6 in fetuses infused with colostrum whey (which contained approximately 80 mg IgG/ml). However, IgG levels also increased slightly in ligated fetuses and in some of the fetuses infused with saline, amniotic fluid or milk whey (from 0 to 2.3 ± 0.6 mg/ml, pooled values across the 4 treatment groups). These unexpected IgG levels in nonIgG infused fetuses may arise either from endogenous IgG production by individual piglets or from maternofetal transfer of IgG. Vascular anastomosis may develop between fetal and maternal circulations in response to fetal surgery and incising the placenta, which completely surrounds the fetus in the pig.[21] Earlier studies[21] also reported a correlation between fetal globulin concentrations and packed cell volume in fetal blood in certain fetuses. This was in agreement with the present observations in that IgG levels in nonIgG infused fetuses correlated with blood hemoglobin values ($r = 0.59$, $P<0.05$). However, further studies are needed to identify the origin of IgG in nonIgG infused fetuses in the present study.

Mean fetal cortisol values increased gradually over the entire experimental period and were higher on da 6 than on da 0 (74 ± 8 ng/ml vs 29 ± 2 ng/ml. There was no effect of treatment on plasma cortisol and the time related increases arose mainly from large changes in a few fetuses in each group with more moderate or no changes in the remaining fetuses. Plasma IgG levels in nonIgG infused fetuses were not correlated with plasma cortisol values. However, the mean cortisol increase in the present chronically catheterized fetuses was higher than that occurring in untreated catheterized fetuses at this gestational age[22] and higher than in fetuses catheterized earlier in gestation.[9] Therefore, we can not exclude the possibility that treatment at the present stage of gestation may lead to some degree of fetal stress for a proportion of the fetuses. This probably could result in precocious organ maturation because cortisol is known to stimulate the maturation of a series of fetal organs in the pig and other species.[8]

4.2. Fetal Pigs: Intestinal Enzymes

Mean values for intestinal protein and enzyme activities in fetal pigs are shown in Table 1. There was no correlation between mean plasma cortisol values and any of the parameters measured in the small intestine. Therefore, the variable increases in plasma cortisol in catheterized fetuses are unlikely to have exerted a major influence on the activity of brush border enzymes.

There was no general effect of treatment on sucrase or lactase activity but there were effects on maltase and peptidase activities. For the latter enzymes, mean values were highest for colostrum infused fetuses and lowest for ligated fetuses. Particularly, maltase and aminopeptidase A activities were much higher (>100%) in fetuses infused with colostrum (1.23 ± 0.22 and 10.16 ± 1.43 U/g, respectively) than in other groups of treated fetuses (0.66 ± 0.05 and 4.63 ± 0.47 U/g, respectively). Intestinal total protein content was higher in fetuses infused with milk whey or colostrum whey than in other fetuses.

Colostrum appears to have a selective (rather than general) effect on intestinal enzyme development in the fetal pig. Further studies are needed to clarify whether the

Table 1. Total protein concentration (mg/g tissue) and activities of 6 intestinal brush border enzymes (U/g tissue) in fetal pigs following treatment *in utero*. ApN/ApA, aminopeptidases N/A; DPP IV, dipeptidylpeptidase IV

Treatment	Protein	Sucrase	Maltase	Lactase	ApN	ApA	DPP IV
Ligated	95	0.042	0.42	10.8	5.26	2.70	1.72
Saline	104	0.036	0.82	10.9	8.27	4.81	2.39
Amniotic fluid	107	0.042	0.59	7.7	7.41	5.87	1.99
Milk whey	128	0.040	0.75	10.9	6.55	3.80	1.73
Colostrum whey	135	0.032	1.23	9.1	10.37	10.16	2.61

colostrum induced increases in maltase and aminopeptidase A activity are associated with an enhanced digestive capacity and maybe a survival advantage in the neonatal pig. As there were no consistent differences in enzyme activities among the groups of ligated, saline infused and amniotic fluid infused fetuses, the swallowing of amniotic fluid by the pig fetus play only a minor role for normal development of brush border hydrolases during the late fetal period. However, it remains possible that ligation of the oesophagus for a longer period would induce changes in both the gastrointestinal tract and other organs, as shown in the fetal lamb.[23]

4.3. Neonatal Pigs: Survival and Health

The 6 sows gave birth to 73 live piglets and 16 stillborn piglets. The proportion of stillborn pigs (18%) is slightly higher than for spontaneous deliveries at term.[24] However, the higher proportion of stillborn piglets was not unexpected because premature deliveries and induced deliveries are known to be associated with more intrapartum piglet losses.[18] A number of the stillborn piglets had low body weights at birth (<0.7 kg); this can be explained by the fact that the mean litter size was relatively high in the present study (15 piglets/litter, range 11 to 17).

Among the 36 piglets not killed at birth but fed colostrum during the first 2 da after birth, 7 died before they reached the age of 2 da. These piglets showed the typical signs of hypoglycemia (low blood glucose levels, weakness, convulsions, hypothermia). These clinical signs and the associated piglet mortality are common observations among neonatal pigs.[24,25] Caesarean delivered pigs, particularly those delivered prematurely, were less active and appeared weaker during the first 24 to 36 hr after birth than pigs delivered after induced vaginal birth. Among the 7 piglets which died during the first 2 da after birth, there were 5 preterm and 2 term piglets; 3 piglets were delivered caesarean while the remaining 4 were delivered vaginally.

Blood acidity, blood hemoglobin and plasma cortisol (Fig. 4) increased with gestational age at delivery and was higher in piglets delivered vaginally than by caesarean section. There were interactions between gestational age at delivery and values for oxygen saturation and blood gases. Elevated CO_2 pressure in vaginally delivered piglets indicated that hypercapnia played a role in the acidosis in these piglets, particularly when pigs were delivered prematurely. As indicated by Randall,[25] lactacidemia also may contribute to acidemia in such piglets.

Blood hemoglobin levels were higher in term than preterm piglets and higher in piglets delivered vaginally than by caesarean section (Fig. 4). The mechanisms by which vaginal birth stimulates blood hemoglobin levels are not known but may be related to both blood volume changes and to specific increases in the synthesis of haemoglobin. Hypoxemia, as well as birth related increases in the circulating levels of cortisol, catecholamines and prostaglandins, are known to stimulate erythropoietin synthesis and thus hemoglobin production.[26] In addition, the stress of birth may stimulate the spleen to release more erythrocytes into the circulation. There was a correlation between plasma cortisol values and blood hemoglobin levels in newborn pigs.

4.4. Neonatal Pigs: Intestinal Enzymes

Activities of lactase and aminopeptidases in newborn pigs increased with advancing gestational age at delivery (Table 2). These increases may have been stimulated by cortisol because earlier studies have shown that exogenous cortisol increases the synthesis of these enzymes and tends to decrease the activity of sucrase in the pig fetus.[12] There were correlations between plasma cortisol values at birth and lactase, aminopeptidase N and

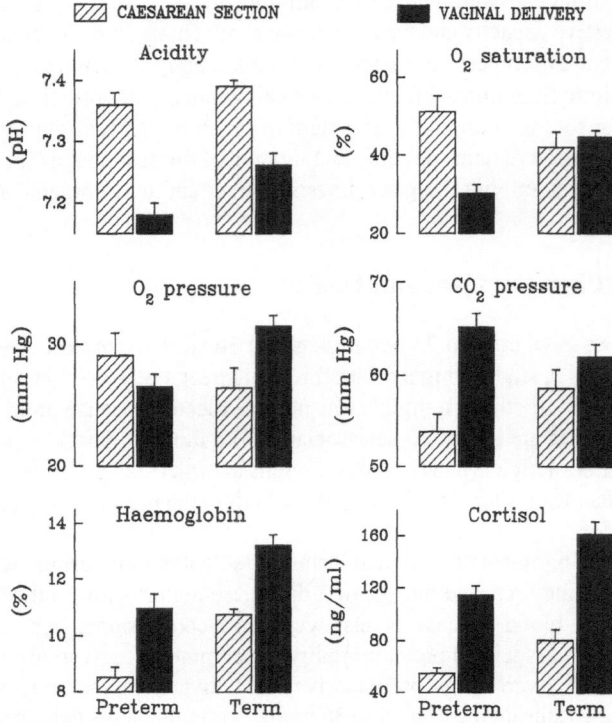

Figure 4. Acidity, blood gases and hemoglobin level in whole blood, and concentration of cortisol in pig plasma at birth (n = 10 to 18 piglets from 6 litters). Pigs were delivered prematurely or at normal term and born after caesarean section or by the vaginal route after induction of parturition with a prostaglandin $F_{2\alpha}$ analogue.

dipeptidylpeptidase IV activities in 0 to 2 da old pigs. In premature newborn piglets, vaginal delivery (elevated plasma cortisol levels) stimulated mucosal protein contents and aminopeptidase A activities at birth and inhibited sucrase activity. In 2 da old pigs, vaginal birth was associated with increased activities of all enzymes, except maltase and aminopepidase

Table 2. Total protein concentration (mg/g tissue) and activities of 6 intestinal brush border enzymes (U/g mucosa) in newborn pigs (0 da) and 2 da old pigs fed colostrum (2 da). Pre, preterm delivery (106 to 108 da gestation); Ter, term delivery (113 to 115 da gestation). CS, caesarean section; VD, vaginal delivery. ApN/ApA, aminopeptidases N/A; DPP IV, dipeptidylpeptidase IV

Treatment	Protein	Sucrase	Maltase	Lactase	ApN	ApA	DPP IV
Pre-CS-0 da	94	0.079	0.54	10.9	9.0	8.8	3.30
Pre-VD-0 da	114	0.043	0.51	9.9	11.0	12.2	3.17
Pre-CS-2 da	224	0.052	1.93	8.9	14.3	26.9	2.37
Pre-VD-2 da	214	0.054	1.90	11.4	16.9	31.2	3.32
Term-CS-0 da	95	0.073	0.45	15.4	15.2	16.7	3.73
Term-VD-0 da	110	0.071	0.45	15.4	16.1	15.9	3.58
Term-CS-2 da	200	0.073	1.35	12.7	15.6	25.6	3.10
Term-VD-2 da	162	0.104	1.21	16.8	19.0	27.2	4.44

A. However, these 2 enzyme activities increased more from birth to 2 da (2 to 3 fold) than any of the other enzymes. The increases apparently were unrelated to delivery type or plasma cortisol and may have resulted from selective effects of factors in colostrum on activities of maltase and aminopeptidase A.

In conclusion, the activity of brush border hydrolases in the pig small intestine is stimulated by vaginal birth and increasing gestational age at delivery. Cortisol may have a role in these effects although the relatively poor correlations between plasma cortisol and enzyme activities in these animals suggest that other factors are involved. Certainly, the large increases in activities of maltase and aminopeptidase A induced by colostrum in both fetuses and neonates demonstrated that colostral factors were involved in regulating specific brush border hydrolases. Although the extent to which these specific enzymes can be used as markers of colostrum ingestion and gastrointestinal competency in human infants remains to be determined, we have demonstrated the perinatal pig is a useful model for studying gastroenterology in the human infant.

5. ACKNOWLEDGMENTS

Supported in part by grants from The Danish Agricultural and Veterinary Research Council, The NOVO Foundation, Denmark, and The Wellcome Foundation, UK.

6. REFERENCES

1. Lebenthal, E., and Leung, Y.K., 1988, Feeding the premature and compromised infant: gastrointestinal considerations, *Pediatr. Clin. N. Am.* 35:215-238.
2. Lucas, A., and Cole, T.J., 1990, Breast milk and necrotizing enterocolitis, *Lancet* 336:1519-1523.
3. Grand, R.J., Watkins, J.B., and Torti, F.M., 1976, Development of the human gastrointestinal tract. A review, *Gastroenterology* 70:790-810.
4. Moughan, P.J., Birtles, M.J., Cranwell, P.D., Smith, W.C., and Pedraza, M., 1992, The piglet as a model for studying aspects of digestion and absorption in milk-fed human infants, *Wld. Rev. Nutr. Diet.* 67:40-113.
5. Henning, S.J., Rubin, D.C., and Shulman, R.J., 1994, Ontogeny of the intestinal mucosa, in: *Physiology of the Gastrointestinal Tract*, 3rd ed., Volume 1 (L.R. Johnson, ed.), Raven Press, New York, pp. 571-610.
6. Bauer, C.R., Morrison, J.C., Poole, W.K., Korones, S.B., Boehm, J.J., Rigatto, H., and Zachman, R.D., 1984, A decreased incidence of necrotizing enterocolitis after prenatal glucocorticoid therapy, *Pediatr.* 73:682-688.
7. Halac, E., Halac, J., Bégué, E.F., Casañas, J.M., Indiveri, D.R., Petit, J.F., Figueroa, M.J., Olmas, J.M., Rodriguez, L.A., Obregón, R.J., Martinez, M.V., Grinblat, D.A., and Vilarrodona, H.O., 1990, Prenatal and postnatal corticosteroid therapy to prevent neonatal necrotizing enterocolitis: a controlled trial, *J. Pediatr.* 117:132-138.
8. Silver, M., 1990, Prenatal maturation, the timing of birth and how it may be regulated in domestic animals, *Exp. Physiol.* 75:285-307.
9. Sangild, P.T., Hilsted, L., Nexø, E., Fowden, A.L., and Silver, M., 1994, Secretion of acid, gastrin and cobalamin-binding proteins by the fetal pig stomach: Developmental regulation by cortisol, *Exp. Physiol.* 79:135-146.
10. Sangild, P.T., Silver, M., Fowden, A.L., Turvey, A., and Foltmann, B., 1994, Adrenocortical stimulation of stomach development in the prenatal pig, *Biol. Neonate* 65:378-389.
11. Sangild, P.T., Weström, B.R., Fowden, A.L., and Silver, M., 1994, Developmental regulation of the porcine exocrine pancreas by glucocorticoids, *J. Pediatr. Gastroenterol. Nutr.* 19:204-212.
12. Sangild, P.T., Sjöström, H., Norén, O., Fowden, A.L., and Silver, M., 1995, The prenatal development and glucocorticoid control of brush-border hydrolases in the pig small intestine, *Pediatr. Res.* 37:207-212.
13. Sangild, P.T., Hilsted, L., Nexø, E., Fowden, A.L., and Silver, M., 1995, Vaginal birth versus elective caesarean section: effects on gastric function in the neonate, *Exp. Physiol.* 80:147-157.

14. Silver, M., 1981, An assessment of the chronically catheterized fetal preparation in sheep and other species, *Placenta* Suppl. 2:89-108.
15. Randall, G.C.B., 1986, Chronic implantation of catheters and other surgical techniques in fetal pigs, in: *Swine in Biomedical Research*, Volume II (M.E. Tumbleson, ed.), Plenum Press, New York, pp. 1179-1185.
16. Foltmann, B., Jensen, A.L., Lønblad, P., Schmidt, E., and Axelsen, N.H., 1981, A developmental analysis of the production of chymosin and pepsin in pigs, *Comp. Physiol. Biochem.* 68B:9-13.
17. Sangild, P.T., Foltmann, B., and Cranwell, P.D., 1991, Development of gastric proteases in fetal pigs and pigs from birth to thirty six days of age. The effect of adrenocorticotropin (ACTH), *J. Devel. Physiol.* 16:229-238.
18. Silver, M., Comline, R.S., and Fowden, A.L., 1983, Fetal and maternal endocrine changes during the induction of parturition with the PGF analogue, cloprostenol, in chronically catheterized sows and fetuses, *J. Devel. Physiol.* 5:307-321.
19. Lowry, O.H., Rosebrough, N., Farr, A., Randall, R.J., 1951, Protein measurement with the Folin phenol reagent, *J. Biol. Chem.* 193:265-275.
20. SAS, 1988, *SAS/STAT^{TM} Users Guide*, 6.03 ed., SAS Institute, Cary, USA.
21. Randall, G.C.B., and Nielsen, K., 1980, Fetal perfusion with maternal blood; a possible sequel to fetal catheterization in the pig, *J. Devel. Physiol.* 2:249-256.
22. Silver, M., and Fowden, A.L., 1989, Pituitary-adrenocortical activity in the fetal pig in the last third of gestation, *Quart. J. Exp. Physiol.* 74:197-206.
23. Trahair, J.F., and Harding, R., 1995, Restitution of swallowing in fetal sheep restores growth after mid gestation oesophageal obstruction, *J. Pediatr. Gastroenterol. Nutr.* 20:156-161.
24. Svendsen, J., 1992, Perinatal mortality in piglets, *Anim. Reprod. Sci.* 28:59-67.
25. Randall, G.C.B., 1992, Perinatal adaptation in animals, *Anim. Reprod. Sci.* 28:309-318.
26. Jain, N.C., 1986, *Schalm's Veterinary Hematology*, 4th ed., Lea and Febiger, Philadelphia, USA.

66

PRENATAL AND PERINATAL DEVELOPMENT OF INTESTINAL TRANSPORT AND BRUSH BORDER HYDROLASES IN PIGS

Randal K. Buddington,[1*] Christiane Malo,[2] and Hongzheng Zhang[1]

[1] Biological Sciences
Mississippi State University
Mississippi State, Mississippi 39762-5759
[2] Membrane Transport Research Group
Physiology
University of Montreal, P.O. Box 6128
Station A, Montreal, Quebec, Canada H3C 3J7

1. ABSTRACT

Although the neonatal intestine is able to process milk and absorb the components, problems of digestion are common. Because of the limited availability of human fetal and neonatal tissues, relatively little is known about functional development of the intestine, particularly during the third trimester when premature births and digestive insufficiencies are likely. Because pigs share with humans similar digestive system structure and functions, we characterized development of hydrolytic and transport activities in pigs, using brush border membrane vesicles (BBMV) and intact tissues prepared from 7, 8, 10 and 12 wk fetuses (43, 49, 61 and 74% of term, respectively), unsuckled neonates and during the first 24 hr of suckling. Intestinal weight increased exponentially between 7 and 12 wk of gestation, more than 7 fold between 12 wk and birth and nearly 2 fold after 24 hr of suckling. Lactase was detected at 7 wk, activity increased dramatically between 10 wk and birth and established a proximal to distal gradient, with a further increase after birth. Sucrase was not detected prenatally, with only low activity during the perinatal period. Active L-leucine uptake was detected at 7 wk, with a decreasing proximal to distal gradient present at birth. D-glucose uptake at 7 wk was low and mostly Na^+ independent but a typical overshoot phenomenon was present at 8 wk. At 12 wk and birth, D-glucose uptake was strictly Na^+

---

[*] Reprint requests to: Dr. R. K. Buddington, Biological Sciences, Mississippi State University, Mississippi State, MS, 39762-5759, (601) 325-7580.

dependent along the entire length of intestine; it declined from proximal to distal. Kinetic analysis of uptake concentration data revealed the presence of both high and low affinity systems at 8 wk of gestation but only a single high affinity Michaelian system at birth and thereafter. The appearance of hydrolytic and transport functions during early gestation of pigs, and the switch in transport systems, is similar to reports for humans. Therefore, the pig may be a valuable model for studying prenatal and perinatal intestinal development and should provide much needed insights about the final trimester of pregnancy.

2. INTRODUCTION

Problems of digestion and nutrition are common in pediatrics, particularly for preterm infants with underdeveloped intestines. However, the limited availability of human tissues, especially from the third trimester and neonatal period, has impeded an understanding about how to treat such infants. As of now, most of what is known about intestinal development, including prenatal phases, is based on studies of rodents. Although these have provided valuable insights about mechanisms and signals involved in intestinal development, there is a need to find more suitable models because of differences from humans with respect to temporal aspects of development, difficulties feeding early life stages and small size. Therefore, progress in this field will be dependent on the availability of more appropriate animal models.

Although standard and miniature swine already are well recognized as a valuable model for studying postnatal intestinal development and the influences of diet,[1,2] less is known about their use for prenatal development. We summarize some of our studies of intestinal development of pigs, with an emphasis on prenatal stages and changes that occur immediately after birth during the first hours of suckling. We include structure and function relations drawing on the wealth of information from classic studies of pig embryology, which also have shown patterns of intestinal development are similar to those of humans. We conclude with a section on the relevance of swine as a model for the human infant, as well as its limitations, and present areas requiring additional attention.

3. MATERIALS AND METHODS

Fetal pigs were obtained from timed pregnancy sows slaughtered at different stages of gestation, i.e., 7, 8, 10, 12 and 15 wk, corresponding to 43, 49, 61, 74 and 92% of term, respectively. Although fetuses were removed from the uterus about 20 min after the sow was killed by electrocution and exsanguination, hearts were observed to be beating. Neonatal pigs were studied either after known periods of suckling (0, 6, 12 or 24 hr) or collected at birth and assigned to 1 of 4 dietary treatments (colostrum, milk replacer, oral electrolyte solution, food deprived) and fed, by feeding tube, hourly for 6 hr before being studied.

The fragile small intestines of 7, 8 and 10 wk fetuses were removed intact, frozen in liquid nitrogen and stored at -80C. Small intestines of 12 and 15 wk fetuses and postnatal pigs were divided into 3 regions of equal length, designated proximal, mid and distal, and portions similarly flash frozen and held at -80C.

A portion of each sample of frozen intestine was transported on dry ice to the University of Montreal where brush border membrane vesicles (BBMV) were prepared by a Mg precipitation technique for measurements of nutrient uptake and enzyme assays. Initial rates, Na^+ dependency and concentration influences of BBMV sugar and amino acid transport were measured using a rapid sampling, rapid filtration apparatus.[3] Frozen tissue from postnatal stages retained at Mississippi State University (MSU) for preparation of

BBMV by a Ca^{++} based technique, were used for additional assays of hydrolase activities. Activities of brush border hydrolases [lactase, sucrase, aminooligopeptidase (AOP), γ-glutamyltranspeptidase (GGT)] were assayed in homogenates and BBMV. Rates of BBMV transport and hydrolase activities were expressed relative to tissue protein concentrations.

Rates of uptake by intact tissues were measured at MSU using intestines from 15 wk fetuses and the same neonatal pigs used to collect mucosa for BBMV studies. Uptakes of glucose and the bile acid taurocholate were evaluated as functions of region and concentration.

Initial rates of uptake by BBMV and intact tissues were estimated by linear regression analysis or by the second degree polynomial, depending on the shape of the curve. Concentration uptake relations for BBMV and intact tissues were defined by nonlinear regression analysis, using model equations for single and multiple transporter types ("Enzfitter", Biosoft, 1987). Analysis of variance was used to search for influences of age, region and diet on intestinal dimensions and functions, with Duncan's and Student-Neuman-Keuhl's test used to identify specific differences.

4. RESULTS

In the following sections, we summarize results for studies of activities of pig intestinal brush border hydrolases and transporters during 1) prenatal development,[3] 2) the first 24 hr after birth and suckling[4] and 3) the first 6 hr after birth when different diets were fed (Buddington, unpublished). We include our preliminary findings for ontogenetic development of taurocholate uptake.

4.1. Prenatal Events

4.1.1. Intestinal Growth and Protein Content. Between 7 wk of gestation and birth, intestinal weight increased 100 fold, with the greatest rate of growth during the last 4 wk of gestation.[3] Concentrations of protein in homogenates, which represent total tissue protein, remained stable during gestation at 70 to 78 mg protein/g intact intestine, with comparable values after birth and onset of suckling. In contrast, BBMV protein concentrations increased during gestation, with the largest increases after 10 wk; about 80% increase between 10 wk of gestation and birth.

4.1.2. Brush Border Hydrolases. Sucrase activity was not detected in homogenates or BBMV prepared from fetal pigs[3] but low activity was present in both fractions at birth.[4] Although lactase was present at 7 wk of gestation (Fig. 1), specific activity did not increase until after 10 wk. GGT also was present during early gestation but specific activity did not increase during the remainder of gestation. When the increase in total protein content during gestation was considered, there were dramatic increases in total intestinal activities (product of specific activity, protein concentration and intestinal weight) of lactase and GGT between 7 wk and birth. The most rapid increases were during the last 4 wk of gestation.

4.1.3. Nutrient Transport.

4.1.3.1. BBMV. Concentrative uptake of glucose and leucine by BBMV was evident by 7 wk of gestation (Fig. 1) but rates and magnitude of accumulation were low relative to later stages of development (Fig. 1). The exception to the prenatal increase was at 10 wk when there was a transient decline in BBMV glucose uptake. Although this remains

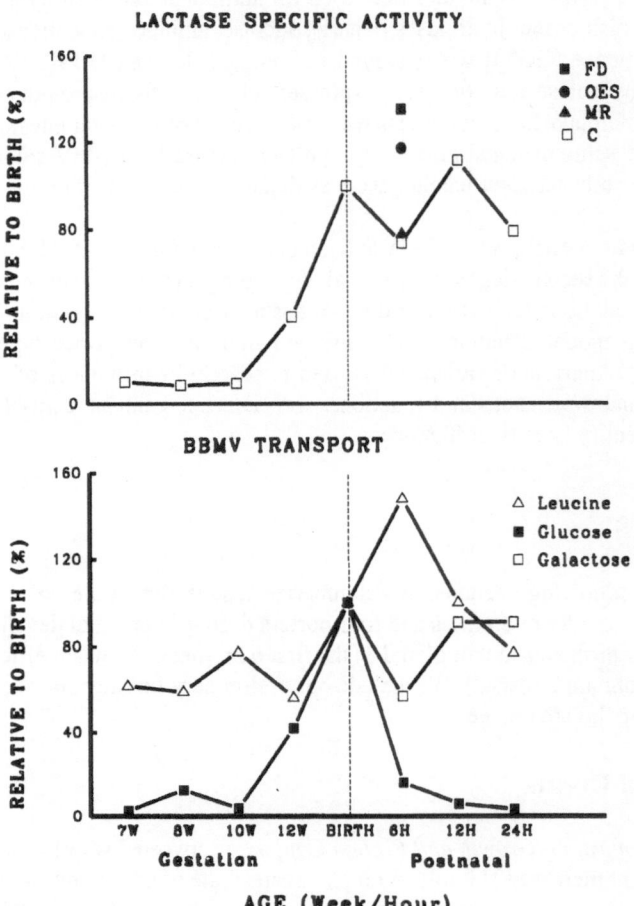

Figure 1. Prenatal and perinatal lactase activity (top panel) and rates of uptake for glucose, galactose and leucine by BBMV (bottom panel) prepared from the entire intestine of 7, 8 and 10 wk fetuses and proximal intestine of 12 wk fetuses and postnatal pigs. Values are expressed as a percentage of activities and rates measured in proximal intestine of neonatal pigs immediately after birth and before suckling. At 6 hr after birth, values are presented for pigs fed colostrum (C, open square), pig milk replacer (MR, filled triangle), oral electrolyte solution (OES, solid circle) or food deprived (FD, solid square).

unexplained, a similar phenomenon has been reported for fetal guinea pigs.[5] The characteristic adult proximal to distal gradient of glucose uptake was present at 12 wk but probably was established earlier in gestation.

Kinetic analysis of BBMV glucose uptake revealed the presence of 2 or more transporter types during early gestation (7, 8 and 10 wk). Uptake via the lower affinity system declined thereafter and by birth and thereafter only a high affinity transporter was detected.

4.1.3.2. Intact Tissues. Intact tissues were capable of active glucose and taurocholate uptake at 92% of gestation, with a declining proximal to distal gradient. Rates of uptake were lower than those measured at birth. Kinetic analysis of taurocholate uptake revealed the majority of uptake was via a nonsaturable pathway. A saturable component with low maximum rates of uptake was present with an apparent affinity comparable to that of weaned and adult pigs.

4.2. Perinatal Development and the Influence of Diet

4.2.1. Intestinal Dimensions and Protein Content. During the first 24 hr after birth and suckling intestinal length, surface area and weight increased 10, 24 and 51%, respectively, with the greatest rate of change during the first 6 hr.[4] The 24 hr increase in intestinal weight was due largely to 72% higher mucosal mass, which coincided with a 156% increase in protein content and was evident by greater tissue thickness. Total brush border membrane protein for the entire length of small intestine also increased but the magnitude of increase was lower and occurred slower (71% at 24 hr).

After 6 hr, pigs fed colostrum had longer, heavier intestines, with more surface area and mucosa, and higher protein content than those fed milk replacer. Intestinal dimensions and protein content were lowest for pigs fed oral electrolyte solution and food deprived and were comparable to those of unsuckled neonates.

4.2.2. Hydrolase Activities and Rates of Nutrient and Bile Acid Uptake. Homogenate and BBMV specific activities for all 4 hydrolases were highest at birth with a transient decline after 6 hr of suckling; activities had recovered at 12 hr and were comparable to those at birth. Apparently the lower values at 6 hr were not caused by a loss of functional hydrolases since total intestinal activities for all 4 hydrolases increased after birth, with more rapid increases for homogenates relative to BBMV. Instead, our findings are indicative of synthesis and insertion of new hydrolases. Presumably, the declines in specific activity were caused by proportionally greater increases in tissue and BBMV protein content, which would dilute enzyme activities. Recoveries for lactase and AOP (24 and 27%, respectively) were higher than for sucrase and maltase (6 and 10%, respectively).

There was a similar reciprocal relationship between rates of BBMV uptake and protein content. However, the decline was more marked for glucose relative to the hydrolases (Fig. 1) and at 24 hr, rates of BBMV glucose accumulation were only 1% of values at birth. The decline was specific for glucose since rates of BBMV uptake did not decline in parallel for the related aldohexose galactose and several amino acids.

Corresponding with the rapid increase in tissue weight, there was a postnatal decline for nutrient uptake by intact tissues when rates were normalized to tissue mass. However, unlike hydrolase activities, total intestinal glucose transport capacities did not increase during the first 24 hr after birth. Also, the decline was less marked than that for BBMV with rates of glucose uptake by intact tissues from proximal intestine at 24 hr averaging 52% of values at birth.

Taurocholate uptake provided a different pattern of postnatal development. Within 12 hr after onset of suckling, the carrier mediated uptake pathway present before and at birth was lost from all 3 regions. Passive influx remained high.

4.2.2.1. Influence of Diet on Perinatal Intestinal Functions. Total intestinal homogenate and BBMV activities for lactase, maltase and AOP at 6 hr were lower in pigs fed colostrum and milk replacer; sucrase activity was not responsive to diet. Dietary influences also were apparent when 6 hr values for the dietary treatments were compared to those of unsuckled neonates. Rates of uptake by BBMV also declined most for pigs fed colostrum, followed by those fed milk replacer, with little or no change for pigs fed an oral electrolyte solution or food deprived. A similar pattern was detected for glucose uptake by intact tissues, but declines associated with feeding colostrum were not as dramatic. Because of greater intestinal growth, colostrum fed pigs actually had higher total intestinal hydrolytic and transport capacities than those of pigs fed the other diets or food deprived.

5. DISCUSSION

We relate our findings about prenatal intestinal functions with information about structural development. In later sections, we address events that occur immediately after birth, the influence of dietary inputs, and other aspects of intestinal development. A final section presents perspectives and unanswered questions.

5.1. Prenatal Development of Intestinal Structure and Functions

Mammalian intestinal development has been separated into 5 phases, each of which is characterized by specific changes in structure and functions.[6] The first 3 occur before birth and prepare the intestine for the transition from placental nutrition to exogenous feeding. The final 2 occur after birth and coincide with the dramatic changes in dietary inputs that occur when neonates begin to suckle and again at weaning when the adult diet is assumed. Despite similar patterns of cell proliferation and differentiation, the timing of developmental events differs among species, with the length of gestation representing an important determinant.[7]

During the first phase, organogenesis, the intestine originates as a folding of the endoderm surrounded by a layer of mesenchyme. These events begin at about 10% of gestation in humans, slightly later in pigs (about 12%), but much later in species with short gestations, such as rats and mice. Cell proliferation occurs throughout the tissue during organogenesis and results in rapid growth and development of a stratified epithelium. Coordination of the early organization may be under control of regulatory peptides that have been detected in developing intestine during this phase.[8]

The second phase starts when villi begin to form. Concurrently, cell proliferation is restricted to presumptive crypts, epithelium lining the villi becomes simple columnar and associated cells differentiate and acquire enterocyte characteristics. The detection of lactase activity and glucose and leucine transport in 7 wk fetal pigs indicates these events occur before 43% of gestation. These same functions are present in 12 wk human fetuses, corresponding with 25 to 30% of gestation. In contrast, intestinal brush border hydrolases and transporters are not detected in rats and mice until the last 10 to 15% of gestation. At 22 wk (53%) of gestation, human fetal intestine is similar morphologically to that of adults. Intestines of piglets at 8 wk (49%) of gestation also are similar morphologically to those of adults. However, adult functional characteristics are not yet fully developed in either species. Furthermore, not all enterocyte functions appear simultaneously, including various amino acid transporters.[9] Notable in pigs is the delayed appearance of sucrase until birth.

The third phase, which coincides with the third trimester in humans, is characterized by continued growth and maturation of the intestine as it prepares for birth and onset of suckling. This is evident in pigs by dramatic increases in intestinal weight, protein content, hydrolase activities and rates of transport that occur during the last 4 wk of gestation. During this time, there is also an apparent shift in relative abundance of different aldohexose transporters. Whereas kinetic analysis revealed the presence of at least 2 different glucose transporters during early gestation, the lower affinity system is lost during later prenatal development. A higher affinity system, with kinetic characteristics similar to that of weaned pigs, is retained. These developmental shifts are similar to the pattern described for humans.[10]

It has been uncertain why intestinal functions develop so early during gestation of some species when nutrients are provided by the placenta and there is virtually no possibility of extrauterine survival. It is now known that fetuses swallow and process substantial volumes of amniotic fluid after the intestine is formed. Although functional demands presumably are less than those after birth intestine plays a critical role in fetal development. This is evident from research with pigs and other species that has shown preventing swallowing of amniotic fluid by esophageal ligation retards fetal growth and development, particularly that of the intestine.[11,12] The dilute nutrients in the amniotic fluid are assimilated into tissue and have been estimated to provide 10 to 14% of the fetus's energy requirements.[13,14] Restoration of intestinal growth and development after ligation is faster when colostrum is infused into the intestine relative to amniotic fluid, milk replacer or other possible nutrient sources.[12] Collectively, these findings have stimulated a search for the identity of nutrients, hormones and/or growth factors that mediate prenatal intestinal development. There are immediate clinical implications of such studies. For example, the resulting information will be useful for studies of transamniotic nutrition to compensate for placental insufficiency. Also, providing nutrients to intestines of premature infants is known to accelerate intestinal development and reduce the period of parenteral feeding. In both cases, there is not enough known to determine optimal mixtures and amounts. However, the large litter sizes and birth weights of pigs and presence of runts in each litter provide opportunities to pursue such studies.

5.2. Postnatal Development and the Influence of Nutrition

Birth and transition from the benign uterus to the external environment causes shifts in functional demands placed on organ systems. This is particularly so for the gastrointestinal tract and lungs which undergo the most dramatic changes in activities as they assume sole responsibility for the provision of nutrients and oxygen. The first swallows of milk trigger surges in intestinal hormones, increase motor activity and elevate oxygen utilization.[15-17] Although these events probably are shared among mammals, there are differences in the structural and functional responses of the intestine to the onset of suckling. As of now, nothing is known about how diet influences postnatal growth and changes in functional characteristics of human intestine.

Whereas intestines of some species, such as pigs and dogs, undergo rapid growth after onset of suckling, those of guinea pigs and cats grow little during the first days after birth, with rodents showing an intermediate pattern.[7] In pigs, the dramatic increase in tissue dimensions and protein content[4,18,19] can be attributed to proliferation of enterocytes, synthesis of new proteins and pinocytosis of milk proteins, particularly immunoglobulins.[20,22]

There is less known about relations between diet and growth and functional characteristics of the intestine. In addition to eliciting changes in activities of hydrolases and transporters, onset of suckling causes increases in concentrations of fatty acid binding proteins[23] and internalization of receptors for epidermal growth factor.[24] Although patterns of change have been established for these and other functions, mechanistic bases have yet to be defined. For example, rapid declines in glucose uptake by intact tissues might be caused by villus swelling, which would limit surface area available for transport. Specifically, transport of nutrients can occur along the entire villus before suckling, not just at the tips as typical for juveniles and adults.[25] However, a reduction in surface area available for absorption does not explain the rapid decline and virtual absence of glucose uptake by BBMV. Since BBMV uptake remained high for other nutrients, notably the related aldohexose galactose, declines are specific for glucose. Using western analysis, we have shown that SGLT-1, the apical Na/glucose cotransporter, is present in BBMV prepared from postnatal pig intestine. Reasons why this transporter apparently loses the ability to concentrate glucose remain unknown but might include postnatal changes in physical and chemical characteristics of apical membrane,[26] which could alter transporter functions such as affinity and turnover rates.[27] The decline and loss of taurocholate uptake also is puzzling. Because the diet of sucklings is high in fat, one would predict that active bile acid uptake would be present to increase efficiency of enterohepatic recycling. Although rates of passive uptake are high, the lack of active uptake would allow an unknown amount of bile acids to enter the colon and be lost.

The responses to colostrum obviously are rapid and lead to changes in gene expression and enterocyte characteristics. Findings comparable to ours are those reported for synthesis of lactase during the first 6 hr after birth.[28] Apparently components of colostrum elicit synthesis and insertion of apical membrane proteins in addition to stimulating intestinal growth. There is a need to understand better the relationship between intestinal growth and changes in functional characteristics, particularly during the first hr after birth and onset of suckling.

5.3. Pre and Perinatal Changes in Other Intestinal Functions

In addition to processing and absorbing dietary components, the intestine serves several other important functions, many of which develop prenatally and undergo dramatic changes after birth and onset of feeding. The enteric immune system provides one example. A fully formed thymus is present in pigs at 30 da of gestation and antibody to sheep red blood cells can be produced by da 74.[29] Development of the enteric immune system probably begins later with maturing lymphocytes migrating to the ileum and upper colon[30] and leading to the formation of nodules, which might be precursors to Peyer's patches. Despite the apparently later start, all components of cellular immunity are present in newborn pig intestine.[31] High densities of intraepithelial lymphocytes develop prenatally, with further increases after birth. Associated with the lack of placental transfer of immunity and dependence on colostrum, enterocytes of newborn pigs are capable of transcytosis and have a well developed apical tubular system (ATS).[18] In comparison, enterocytes of human fetuses, which receive immunity transplacentally, also have an ATS, but it is lost well before birth.[32] The influence of postnatal diet has been investigated for humoral immunity, due to interest in the transfer of maternal immunoglobulins. There is less known for development of enteric cellular immunity, with the limited information available indicating any dietary effects are transient, and may be related more to environment.[31] As a result, it is uncertain if development of the enteric immune system is programmed genetically or is responsive to external influences, such as diet.

Also to be considered is that the gastrointestinal tract represents the largest endocrine organ in the body. Associated hormones regulate numerous digestive and metabolic functions and play an important role in adapting the neonate to the external environment. Despite this, there is little known about prenatal and postnatal development and responses to diet.

5.4. Colonization of the Intestine by the Microflora

Colonization and establishment of normal enteric flora is critical. Throughout life, the enteric microflora plays an important role in health by providing vitamins and nutrients from fermentation of indigestible nutrients and conferring resistance to opportunistic and invasive pathogens.[33] Certain microflora groups also have been implicated in carcinogenesis.[34] The gastrointestinal tract of pigs is sterile at birth, but by 12 hr after onset of suckling, bacterial densities are comparable to those of adults.[35] What differs between newborn pigs and adults are the relative proportions of different bacterial groups. Humans and pigs are similar in that acquisition of adult microflora is a gradual process and takes months to years.

There is an increasing awareness of the complex interplay among developing fecal flora, diet, intestinal functions and health. Although speculative, it is possible that the combination of 1) low gastric acid secretion at birth, 2) perinatal loss of active bile acid uptake and 3) presence of lactose, immunoglobulins and other substances in milk represents an evolutionary adaptation to facilitate colonization and establishment of the normal microflora. Diet is known to influence composition of the microflora.[36] This is evident from differences between breast fed and formula fed infants, and has been attributed to the numerous "bifidogenic" substances in breast milk that promote growth of bifidobacteria and other groups considered to be beneficial. Mode of delivery (Caesarian or vaginal) is also important, apparently by providing different sources of inoculum. In light of these influences, there is interest in understanding how to alter formulas to produce a microflora that more closely resembles that of breast fed infants. The pig provides an excellent opportunity to pursue these studies.

5.5. Conclusions and Perspectives

Prenatal development of intestinal structure is critical for a successful transition to extrauterine life. Although the intestine is not quiescent during gestation and is critical for normal development, there is surprisingly little known about the patterns of development, mechanistic bases and signals triggering changes in structure and functions. For the pig, we suggest it is an appropriate model for studying prenatal and postnatal development of intestinal structure and functions, and influences of dietary inputs. Pigs also satisfy criteria for suitable models of intestinal development and influences of diet. These include 1) long gestations with early appearance of digestive functions, 2) feeding habits similar to those of humans, 3) similar digestive system structure, functions and resident microflora, 4) sufficiently large body and litter sizes and 5) ability to rear easily using milk replacers. Also of interest is the availability of molecular and genetic probes for studies of signals and mechanisms responsible for changes in intestinal structure and functions.

There are key differences between pigs and humans that need to be considered. First, the intestines of human neonates are about 3 times the crown heel length, which is substantially shorter than the 2 to 4 m intestines of much smaller neonatal pigs.[4] Second, whereas human infants develop high sucrase activity early in gestation, this does not occur until after birth in pigs. Third, humans acquire passive immunity transplacentally but pigs do postnatally, via colostrum. Despite these differences, the pig appears to be a suitable model for studying prenatal and perinatal development and the influences of nutrition.

6. REFERENCES

1. Moughan, P.J., Birtles, M.J., Cranwell, P.D., Smith, W.C., and Pedraza, M., 1992, The piglet as a model animal for studying aspects of digestion and absorption in milk-fed human infants, *World Rev. Nutr. Diet.* 67:40-113.
2. Shulman, R.J., Henning, S.J., and Nichols, B.L., 1988, The miniature pig as an animal model for the study of intestinal enzyme development, *Pediatr. Res.* 23:311-315.
3. Buddington, R.K., and Malo, C., 1995, Intestinal brush-border membrane enzyme activities and transport functions during prenatal development of pigs, *J. Pediatr. Gastroenterol. Nutr.* 23: 51-64.
4. Zhang, H., Malo, C., and Buddington, R.K., 1996, Suckling induces rapid intestinal growth and changes in brush border digestive functions of newborn pigs, *J. Nutr.* In Press.
5. Butt, J.H., and Wilson, T.H., 1968, Development of sugar and amino acid transport by intestine and yolk sac of the guinea pig, *Am. J. Physiol.* 215:1468-1477.
6. Menard, D., 1989, Growth-promoting factors and the development of the human gut, in: *Human Gastrointestinal Development*, (E. Lebenthal, ed.), Raven Press, New York, pp. 123-150.

7. Buddington, R.K., 1994, Nutrition and ontogenetic development of the intestine, *Can. J. Physiol. Pharmacol.* 72:251-259.

8. Aynsley-Green, A., Lucas, A., Lawson, G.R., and Bloom, S.R., 1990, Gut hormones and regulatory peptides in relation to enteral feeding, gastroenteritis, and necrotizing enterocolitis in infancy, *J. Pediatr.* 117:S24-S32.

9. Buddington, R.K., and Diamond, J.M., 1989, Ontogenic development of intestinal nutrient transporters, *Annu. Rev. Physiol.* 51:601-619.

10. Malo, C., and Berteloot, A., 1991, Analysis of kinetic data in transport studies: new insights from kinetic studies of Na$^+$-D-Glucose cotransport in human intestinal brush-border membrane vesicles using a fast sampling, rapid filtration apparatus, *J. Memb. Biol.* 122:127-141.

11. Mulvihill, S.J., Stone, M.M., Debas, H.T., and Fonkalsrud, E.W., 1985, The role of amniotic fluid in fetal nutrition, *J. Ped. Surg.* 20:668-672.

12. Wing, S.J., Trahair, J.F., and Sangild, P., 1994, The effects of various enteral diets on the development of the gastrointestinal tract of the fetal pig, *Proc. Aust. Soc. Med. Res.* (Abstr.).

13. Phillips, J.D., Fonkalsrud, E.W., and Mirzayan, A., 1991, Uptake and distribution of continuously infused intraamniotic nutrients in fetal rabbits, *J. Pediatr. Surg.* 26:374-380.

14. Pitkin, R.M., and Reynolds, W.A., 1975, Fetal ingestion and metabolism of amniotic fluid protein, *Am. J. Obstetr. Gynecol.* 123:356-363.

15. Berseth, C.L., 1990, Neonatal small intestinal motility: motor responses to feeding in term and preterm infants, *J. Pediatr.* 117:777-782.

16. Crissinger, K.D., and Burney, D.L., 1991, Postprandial hemodynamics and oxygenation in developing piglet intestine, *Am. J. Physiol.* 260:G951-G957.

17. Lucas, A., 1988, Gut hormones and the adaptation to extrauterine nutrition, in: *Harries Paediatric Gastroenterology*, (P.J. Milla, and D.P.R. Muller, eds.), Churchill Livingstone, New York, pp 302-317.

18. Xu, R.-J., Mellor, D.J., Tungthanathanich, P., Birtles, M.J., Reynolds, G.W., and Simpson, H.V., 1992, Growth and morphological changes in the small and large intestine in piglets during the first three days after birth, *J. Dev. Physiol.* 18:161-172.

19. Cranwell, P.O., 1994, Chapter 5, The development of the neonatal gut and enzyme systems, in: *The Neonatal Pig: Development and Survival*, (M.A. Varley, ed.), Centre for Agriculture and Biosciences (CAB) Int., Wallingford, Oxford, pp. 52-62.

20. Burrin, D.G., Shulman, R.J., Reeds, P.J., Davis, T.A., and Gravitt, K.R., 1992, Porcine colostrum and milk stimulate visceral organ and skeletal muscle protein synthesis in neonatal piglets, *J. Nutr.* 122:1205-1213.

21. Patureau-Mirand, P., Mosoni, L., Levieux, D., Attaix, D., and Bonnet, Y., 1990, Effect of colostrum feeding on protein metabolism in the small intestine of newborn lambs, *Biol. Neonate* 57:30-36.

22. Simmen, F.A., Cera, K.R., and Mahan, D.C., 1990, Stimulation by colostrum or mature milk of gastrointestinal tissue development in newborn pigs, *J. Anim. Sci.* 68:3596-3603.

23. Reinhart, G.A., Simmen, F.A., Mahan, D.C., Simmen, R.C.M., White, M.E., and Trulzsch, D.V., 1992, Perinatal ontogeny of fatty acid binding protein activity in porcine small intestine, *Nutr. Res.* 12:1345-1356.

24. Kelly, D., McFadyen, M. King, T.P., and Morgan, P.J., 1992, Characterization and autoradiographic localization of the epidermal growth factor receptor in the jejunum of neonatal and weaned pigs, *Reprod. Fertil. Dev.* 4:183-191.

25. Smith, M.W., 1988, Postnatal development of transport function in the pig intestine, *Comp. Biochem. Physiol.* 90A:577-582.

26. Omodeo-Sale, F., Lindi, C., Marciani, P., Cavatorta, P., Sartor, G., Masotti, L., and Esposito, G., 1991, Postnatal maturation of rat intestinal membrane: lipid composition and fluidity, *Comp. Biochem. Physiol.* 100A:301-307.

27. LeGrimellec, C., Friedlander, G., El Yandouzi, E.H., Zlatkine, P., and Giocondi, M.-C., 1992, Membrane fluidity and transport properties in epithelia, *Kidney Int.* 42:825-836.

28. Burrin, D.G., Dudley, M.A., Reed, P.J., Shulman, R.J., Perkinson, S., and Rosenberger, J., 1994, Feeding colostrum rapidly alters enzymatic activity and the relative isoform abundance of jejunal lactase in neonatal pigs, *J. Nutr.* 124:2350-2357.

29. Schultz, R.D., Wang, J.T., and Dunne, H.W., 1971, Development of the humoral immune response of the pig, *Am. J. Vet. Res.* 32:1331-1336.

30. Solomon, J.B., 1971, *Foetal and Neonatal Immunology*, North Holland Pub. Co., Amsterdam, 41 pp.

31. Vega-Lopez, M.A., 1991, Immune development in the young pig, PhD Dissertation, University of Bristol, 257 pp.

32. Weaver, L.T., and Walker, W.A., 1989, Uptake of macromolecules in the neonate, in: *Human Gastrointestinal Development*, (E. Lebenthal, ed.), Raven Press, New York, pp. 731-748.
33. Goodman, L.J., 1993, Diagnosis, management, and prevention of diarrheal diseases, *Curr. Opin. Inf. Dis.* 12:88-93.
34. Gorbach, S.L., and Goldin, B.R., 1990, The intestinal microflora and the colon cancer connection, *Rev. Inf. Dis.* 12:S252-S261.
35. Swords, W.E., Wu, C.-C., Champlin, F.R., and Buddington, R.K., 1993, Postnatal changes in selected bacterial groups of the pig colonic microflora, *Biol. Neonate* 63:191-200.
36. Perman, J.A., 1989, Gastrointestinal flora: development aspects and effects of nutrients, in: *Human Gastrointestinal Development*, (E. Lebenthal, ed.), Raven Press, New York, pp. 777-786.

67

CHANGES IN PIG INTESTINAL STRUCTURE AND FUNCTIONS AND RESIDENT MICROBIOTA INDUCED BY ACUTE SECRETORY DIARRHEA

Gayatri Chandra,[1,2] Monika Oli,[1,2] Bryon W. Petschow,[3] and Randal K. Buddington[2*]

[1] Both authors contributed equally.
[2] Biological Sciences
   Mississippi State University
   Mississippi State, Mississippi 39762-5759
[3] Mead Johnson and Company
   Evansville, Indiana 47721

1. ABSTRACT

Diarrhea continues to be an important problem worldwide. Although the losses of fluid and electrolytes during diarrhea are well understood, much less is known about the impact on intestinal structure, function and resident microbiota. Therefore, we studied weanling pigs with acute diarrhea induced by cholera toxin. At peak diarrhea, body weight, small intestinal mass and absorptive surface area, and glucose transport capacities were lower; whereas, hematocrit, stool water content and activities of sucrase, maltase and lactase were higher. Total bacterial densities of lumenal contents from the small intestine, cecum and colon were reduced by up to 90%, especially lactobacilli; whereas, bacteria associated with the small intestinal mucosa were less affected. After 24 hr of oral rehydration therapy, glucose transport and aminooligopeptidase activity remained low and disaccharidase activities were high. Although total bacterial densities had recovered, relative proportions of the different groups were disturbed. Present oral electrolyte solutions address water and ion losses but not changes in intestinal structure, functions and resident microbiota. The pig is a suitable model for studying the impact of diarrhea on the gastrointestinal ecosystem and will be useful for development of improved oral electrolyte solutions that accelerate recovery of the intestine.

[*] Reprint requests to: Dr. R. K. Buddington, Biological Sciences, Mississippi State University, Mississippi State, MS, 39762-5759, (601) 325-7580.

2. INTRODUCTION

Diarrhea continues to be a leading cause of mortality and morbidity worldwide, with infants and young children at greatest risk. It is also one of the most common and devastating conditions afflicting young food animals and has been implicated in food borne diseases and zoonoses.[1] A wide variety of etiologic agents are known to cause diarrhea in humans and animal species. Cholera toxin (CT), the enterotoxin of *Vibrio cholerae*, is the causitive agent of cholera and is recognized for causing a prototype secretory diarrhea. The role of the toxin in activating adenylate cyclase and inducing secretion of ions and loss of water is well understood. The toxin also is used as an important tool for studies of gastrointestinal secretion and regulation of cell functions and subcellular mediators.[2-4]

Whereas the associated disturbances of water and electrolyte balance caused by CT and other diarrheal agents are well established, much less is known about the impact of diarrhea on intestinal structure and functions, and resident microbiota. Corresponding with this, present oral electrolyte solutions (OES) are formulated to address dehydration and ion loss, not damage to the intestine or improvements in bacterial flora composition.

Absorption of water and electrolytes by healthy intestine is associated with carrier mediated transport of nutrients,[5] and, in most cases, adding simple carbohydrates, monosaccharides or amino acids to OES improves absorption of water and electrolytes. What remains uncertain is the relationship between recovery from diarrhea induced changes in intestinal structure and functions and the efficacy of OES. Such information is needed since patient health is dependent on intestinal recovery. Diarrhea induced shifts in densities and relative proportions of bacterial groups residing in different regions of intestine also are understood poorly but are known to be related to health and may be important determinants for risk of secondary infections.

The above highlights the need to use comprehensive, multidisciplinary approaches to understand the impact of diarrhea on intestine. Because many of the methods used in such investigations require invasive procedures these studies require animal models. In light of the relevance of the pig for human gastrointestinal physiology,[6] we evaluated intestinal structure and functions, and resident microbiota in pigs before and during secretory diarrhea induced by cholera toxin and after 24 hr of treatment using OES. We conclude with a section on future research and propose the use of ecological concepts to better understand diarrhea and approaches that will improve the efficacy of OES and accelerate recovery.

3. MATERIALS AND METHODS

3.1. Animals and Their Care

Suckling crossbred standard 20 da old farm pigs were obtained from a commercial producer, transferred to a fully enclosed facility and weaned to a solid diet that was formulated without antibiotics (Baby Pig Pellets; Milk Specialties, Dundee, IL). At 25 da of age, diarrhea was induced using cholera toxin (15 µg/kg). Secretory diarrhea started 3 to 4 hr after administering the toxin and lasted less than 12 hr. Pigs were treated using a commercial OES (Mead-Johnson, Evansville, IN) to replace ions and fluid during and after onset of diarrhea; they were allowed access to solid diet.

3.2. Sampling

Animals were studied in healthy condition, at peak diarrhea and 24 hr after onset of diarrhea and oral rehydration therapy (ORT). Blood samples were collected for routine hematology immediately before the animal was euthanized (Buthanasia; 1 ml/5 kg; IV). The small intestine was removed and its length was measured in a relaxed state, after which it was divided into 3 sections (proximal, mid and distal) of equal length. Twenty cm segments were removed from the midpoint of each section. A portion of each section was used to assess structural characteristics and the impact of diarrhea based on histologic observations of tissue structure and measurements of regional dimensions (weight, surface area without accounting for amplication due to villi and microvilli, mucosal mass). Activities of 4 brush border hydrolases [lactase (LAC), sucrase (SUC), maltase (MAL), aminooligopeptidase (AOP)], and rates of glucose tranport by intact tissues[7,8] were used as indicators of intestinal functions.

Sample sites for bacteriology included lumenal contents and mucosal scrapings of mid small intestine and contents of cecum and distal colon. Microbiology samples were transferred to an anaerobic chamber where serial dilutions were prepared and plated on general and selective media for enumeration of total anaerobes, total aerobes, Enterobacteriaceae, lactobacilli and bifidobacteria.[9-11]

4. RESULTS AND DISCUSSION

In the following sections, we present our results for changes of intestinal structure, functions and resident bacteria caused by CT.

4.1. Intestinal Dimensions, Structure, and Functions

Diarrhea caused a decline in intestinal mass, as well as surface area and length (Fig. 1). Whereas after 24 hr of ORT, surface area was comparable to that of healthy pigs, intestinal weight remained low. There was a corresponding reduction in villus dimensions at peak diarrhea with evidence of some damage at the tips. Our findings conflict with previous reports that intestinal structure is not compromised by cholera infection.[12]

Despite declining intestinal mass, specific activities of the 3 disaccharidases increased and when integrated with intestinal dimensions, capacities of the entire length of small intestine to hydrolyze lactose, maltose and sucrose were highest at peak diarrhea (Fig. 2). In contrast, acute and chronic diarrheas with mucosal damage often are accompanied by sharp declines in brush border disaccharidase activities,[13] which would result in lower hydrolytic capacities of the entire intestine.

Patterns of response differed for rates for glucose transport normalized to tissue weight and AOP specific activity; both were unchanged. As a result, total intestinal capacities to transport glucose and hydrolyze peptides were lower at peak diarrhea. After 24 hr of ORT, total intestinal disaccharidase activities remained high. AOP activity and glucose transport capacities continued to decline (Fig. 2). Implications for recovery and relations with dietary loads that can be processed are unclear.

The reasons for different responses of disaccharidases relative to AOP and glucose transport are unknown. It is possible they are related to changes in enterocyte proliferation and turnover. Because various brush border proteins are not all expressed at the same time as enterocytes migrate up the villi, any distrubance of enterocyte lifespan is likely to influence the relative proportions of different proteins. Additional information about rates of enterocyte proliferation during diarrhea and the recovery process are needed for a better understanding of the impact of diarrhea.

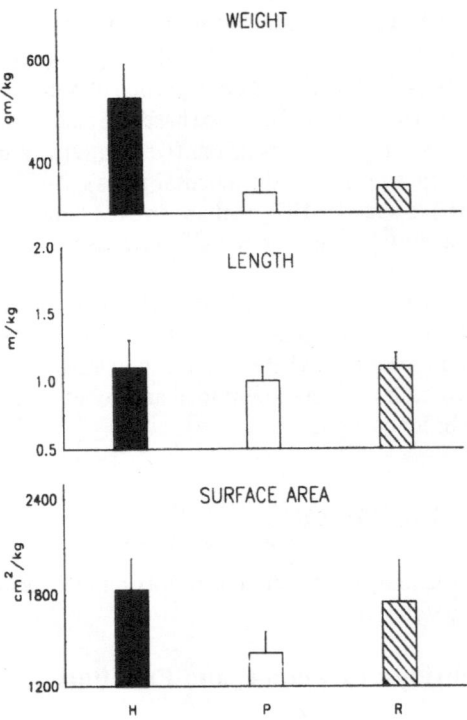

Figure 1. Dimensions of the small intestine before (H = healthy) and during (P = peak) diarrhea and after 24 hr of treatment using OES (R = recovery). Values are expressed relative to body weight.

4.3. Hematology

Changes in hematocrit and serum sodium and glucose concentrations before, during and after diarrhea were monitored to assess dehydration, ion loss and stress. Furthermore, sodium and glucose concentrations might provide insights about coupled transport functions essential for restoration of fluid and electrolyte status. Hematocrit was higher at peak diarrhea and was accompanied by 3.5% higher Na^+ levels (Table 1). After 24 hr of ORT, hematocrit and Na^+ values were closer to normal. Blood glucose values also were higher at peak diarrhea, but declined after 24 hr of treatment using OES.

4.4. Microbiology

The importance of GI bacteria in health and disease was recognized during the last century and has been reviewed extensively.[14,15] The GI tract includes 300 to 500 species of bacteria which coexist in a steady state ecosystem[16,17] and provides the host with vitamins, enables the utilization of indigestible nutrients and confers resistance to opportunistic and invasive pathogens.[18,19] In our present study, we confirm that diarrhea perturbs the ecosystem, as reported by Gorbach et al.[20] Prolonged antibiotic treatment or changes in diet are known to cause shifts in abundance and relative proportions of different bacterial groups.[14]

Microbiology results for the present study are summarized in Fig. 3 and Table 2. In addition to obvious regional differences, 4 points about the impact of diarrhea need to be emphasized. First, bacterial densities in lumenal contents were lower at peak diarrhea. This

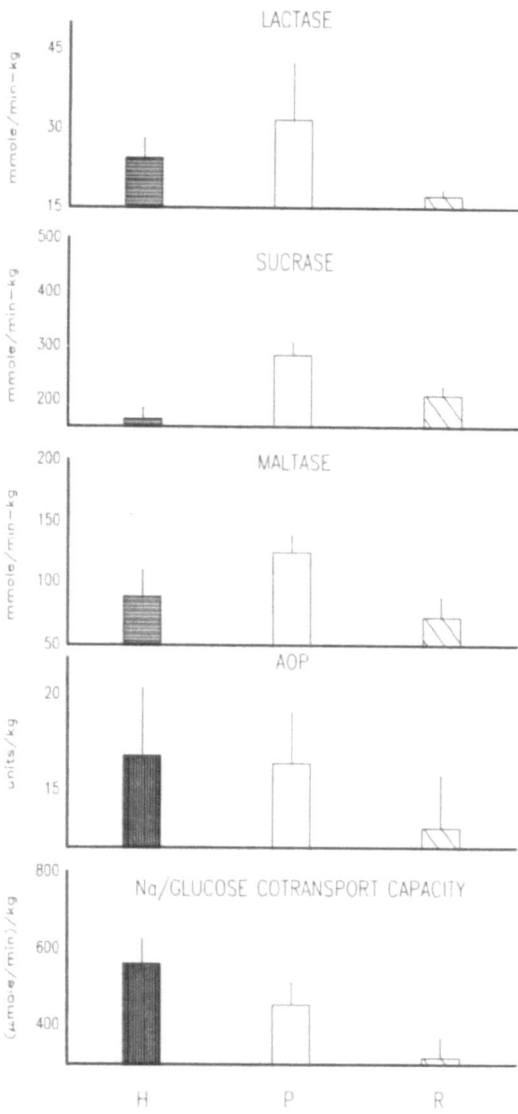

Figure 2. Total intestinal brush border hydrolytic and glucose transport capacities before (H = healthy) and during (P = peak) diarrhea and after 24 hr treatment using OES (R = recovery). Values are expressed relative to body weight.

can be explained by the secretion of fluids, which dilutes lumenal contents and thereby reduces bacterial densities on a wet weight basis. The exception to this was the increase or lack of decline in Enterobacteriaceae. Although the reasons remain uncertain, it might be related to higher lumenal oxygen tension and pH during diarrhea (present studies), which would provide an environment that would favor growth of enterics with short generation times. Responses were even more apparent 24 hr after diarrhea onset.

Table 1. Hematocrit (%), serum sodium (mmol/l) and glucose (mg/dl) for
pig before (healthy) and during (peak) diarrhea and after 24 hr of
treatment using OES (recovery). Values represent means ± SD

	Condition		
	Healthy	Peak	Recovery
Hematocrit	41 ± 2	53 ± 3	38 ± 4
Sodium	143 ± 5	147 ± 5	145 ± 5
Glucose	119 ± 15	177 ± 37	136 ± 17

Second, populations of bacteria associated with mucosa of small intestine were more
resistant to, or not affected by, diarrhea. This probably was related to the lack of serious
mucosal damage. In addition, bacteria which are adherent to the mucosa or present in the
intervillus spaces and crypts probably are protected somewhat from the gut flushing caused
by diarrhea.

Third, although total bacterial densities were near to, or slightly higher than, those
of healthy pigs 24 hr after start of ORT, relative abundances of different groups still were
disturbed. Notable was the continued high densities of aerobes and Enterobacteriaceae.
Fourth, shifts in relative proportions of various groups may have important health relations.
Specifically, *Bifidobacterium* sp. and *Lactobacillus* sp. are considered as beneficial groups
and have been used as sensitive indicators of host health.[18,21] In all sampled sites of healthy
pigs, densities of lactobacilli exceeded those of Enterobacteriaceae (Fig. 3), which includes

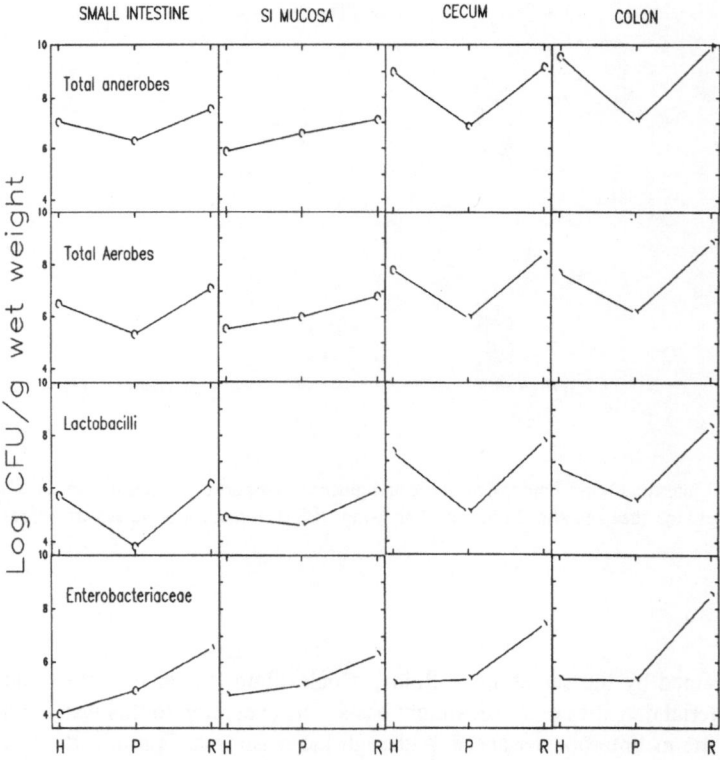

Figure 3. Densities of different bacterial groups in different regions of gastrointestinal tract, before (H =
healthy) and during (P = peak) diarrhea and after 24 hr treatment using OES (R = recovery).

Table 2. Percentage of total anaerobes represented by total aerobes, lactobacilli and Enterobacteriaceae in small intestine (lumen and mucosa), cecum and colon of pigs before (H) and during (P) diarrhea and after 24 hr of treatment using OES (R)

| | | Small Intestine | | | |
		Lumen	Mucosa	Cecum	Colon
Total Aerobes	H	25	40	6.3	1.3
	P	10	25	13	13
	R	32	50	16	7.9
Lactobacilli	H	4.0	10	2.0	0.2
	P	0.3	1.0	1.6	2.5
	R	4.0	6.3	3.2	2.5
Enterobacteriaceae	H	0.08	7.9	0.016	0.006
	P	4.0	3.2	2.5	1.6
	R	7.9	16	1.6	4.0

a large number of pathogens. At peak diarrhea, this relationship was reversed with Enterobacteriaceae exceeding lactobacilli. After 24 hr of ORT, densities of lactobacilli were similar to those of healthy pigs and comparable to those of Enterobacteriaceae. The reciprocal relationship between lactobacilli and Enterobacteriaceae may reflect an inhibitory influence of lactobacilli through production of bactericins, organic acids or other metabolic properties. Therefore, selective stimulation of lactobacilli, with a concurrent reduction of Enterobacteriaceae, may promote health and confer resistance to secondary infections. Bifidobacteria, which are a significant component of the human fecal flora, rarely were detected in samples from any region of gut from experimental pigs. This may have been related to a combination of the recent weaning of the experimental pigs and the diarrhea, both of which are known to reduce densities of bifidobacteria.[17]

5. CONCLUSIONS AND PERSPECTIVES

Acute diarrhea induced by cholera toxin causes changes in intestinal structure and functions and perturbs normal densities, distribution and relative proportions of resident enteric microbiota. Although 24 hr of treatment, using OES, restored fluid and electrolyte status, full recovery of intestinal structure and functions and resident bacteria was not observed by this time. We suggest the pig is an appropriate animal model for studying the impact of intestinal diarrhea. It should prove useful for evaluating treatments and studying new or emerging diarrheal diseases. Efforts to gain additional insights in these areas will be enhanced by further studies in the following areas.

Although both acute and chronic forms of diarrhea are of equal importance to health, there is little known about the impact on the intestine. The pig is an appropriate model for such studies because of its anatomic and physiologic similarities to the human and the need to use invasive techniques. A number of etiologic agents that cause diarrheas in animals are known to have relevance to humans. A partial listing is presented in Table 3. The different agents and associated diarrheas can be classified as to location of diarrhea (small or large bowel), duration (acute or chronic) and how it is mediated (lysis, secretory).

There is increasing interest in the use of compounds that selectively encourage the growth of beneficial bacteria.[18,22] Such "Prebiotics" substances or nutriceuticals also may improve the efficacy of OES. In preliminary studies, adding a short chain fructooligosaccharide to OES accelerated recovery of intestinal structure and functions. In addition, there

Table 3. Selected pathogens that cause diarrhea in humans and/or animals.

Group of pathogen	Name	Infectious for	
		Children	Animals
Virus	Rotavirus	y	y
	Norwalk like viruses	y	
	Corona virus		y
Bacteria	*E. coli* (var. strains)	y	y
	Salmonella and	y	y
	Campylobacter	y	y
	Clostridium difficile	y	y
	Aeromonas, Plesiomonas		y
	Serpulina		y
Protozoans	*Giardia lamblia*	y	
	Cryptosporidium parvum	y	y
	Isospora	y	y

was a selective increase in densities of lactobacilli, with a concurrent decline in Enterobacteriaceae.

In addition to its role in hydrolzying and absorbing foodstuffs, the intestine has extensive immune and endocrine functions. Although cross talk between the immune and endocrine systems is involved in causing certain diarrheas as an adaptive mechanism to rid the intestine of pathogens and toxins, the impact of diarrhea on these functions is unclear. Of importance is the role of the intestine as a barrier to invasion, particularly since disturbances of the intestinal mucosa and associated immune system can compromise animal health and resistance to disease.

Finally, studies of diarrhea can benefit from a multidisciplinary approach. Although the microbial ecosystem has been considered previously, we are unaware of any studies that have considered adequately the complex, but interesting, relations among diet, intestinal structure and functions, and resident microbiota. Evaluating data using ecological concepts, particularly disturbance theory and those derived from stream ecosystems, is likely to provide valuable insights about interactions between diarrheal agents and the host, and subsequent recovery.

6. REFERENCES

1. Holland, R.E., 1990, Some infectious causes of diarrhea in young farm animals, *Clin. Microbiol. Rev.* 3:345-375.
2. Argenzio, R.A., and Whipp, S.C., 1981, Effect of *Escherichia coli* heat-stable enterotoxin, cholera toxin and theophylline on ion transport in porcine colon, *J. Physiol.* 320:469-487.
3. Spangler, B.D., 1992, Structure of cholera toxin and the related *Escherichia coli* heat-labile enterotoxin, *Microbiol. Rev.* 56:622-647.
4. Kaper, J.B., Morris, J.G, Jr., and Levine, M.M., 1995, Cholera, *Clin. Microbiol. Rev.* 8:48-86.
5. Lobley, R.W., 1991, The enterocyte and its brushborder, in: *Molecular Pathogenesis of Gastrointestinal Infections* (T. Wadström, P.H. Mäkelä, A.-M. Svennerholm, and H. Wolf-Watz, eds.), Plenum Press, New York, pp.1-8.
6. Moughan, P.J., Birtles, M.J., Cranwell, P.D., Smith, A., and Pedraza, W.C.M., 1992, The piglet as a model animal for studying aspects of digestion and absoption in milk-fed humans, in: *Nutritional Triggers for Health and Disease. World Rev. Nutr. Diet.* (A.P. Simopoulos, ed.), Karger, Basel, pp. 41-113.
7. Puchal, A.A., and Buddington, R.K., 1992, Postnatal development of monosaccharide transport in pig intestine, *Am. J. Physiol.* 262:G895-G902.
8. Zhang, H., Malo, C., and Buddington, R.K., 1996, Suckling induces rapid intestinal growth and changes in brush border digestive functions of newborn pigs. *J. Nutr.* In Press.

9. Scardovi, V., 1986, Genus *Bifidobacterium* Orla-Jensen, in: *Bergey's Manual of Determinative Bacteriology*, Volume 2 (P.H.A. Sneath, N.S. Mair, M.E. Sharpe, and J.G. Holt, eds.), Waverly Press, Baltimore, pp. 1418-1434

10. Summanen, P., Baron, E.J., Citron, D.M., Strong, C., Wexler, H.M., and Finegold, S.M., 1993, *Wadsworth Anaerobic Bacteriology Manual*, 3rd ed., Star Publ., Los Angeles.

11. Williams, C.H., Witherly, S.W., and Buddington, R.K., 1994, Influence of dietary neosugar on selected bacterial groups of the human fecal flora, *Micr. Ecol. Health Dis.* 7:91-97.

12. Elliot, H.L., Caprenter, C.C.J., and Sack, R.B., 1970, Small bowel morphology in experimental canine cholera: a light and electron microscopic study, *Lab. Invest.* 22:112-120.

13. Nichols, B.F., Carrazza, F., Nichols, V.N., Putman, M., Johnson, P., Rodrigues, M., Quaroni, A., and Shiner, M., 1992, Mosaic expression of brush-border enzymes in infants with chronic diarrhea and malnutrition, *J. Pediatr. Gastroenterol. Nutr.* 14:371-379.

14. Hentges, D.J. (ed), 1983, *Human Intestinal Microflora in Health and Disease*, Academic Press, New York.

15. Simon, L.S., and Gorbach, S.L., 1987, *Physiology of the Gastrointestinal Tract*, 2nd ed. (L.R. Johnson, ed.), Raven Press, New York.

16. Finegold, M., Sutter V.L., Mathiesen, G.E., 1983, Normal indigenous microflora, in: *Human Intestinal Microflora in Health and Disease*, (D.J. Hentges, ed.), Academic Press, New York, pp. 3-31.

17. Tannock, G.W., 1983, Effect of dietary and environmental stress on the gastrointestinal microbiota, in: *Human Intestinal Microflora in Health and Disease* (D.J. Hentges, ed.), Academic Press, New York, pp. 517-541.

18. Gibson, G.R., Beatty, E.R., Wang, X., and Cummings, J.H., 1995, Selective stimulation of bifidobacteria in the human colon by oligofructose and inulin, *Gastroenterology* 108:975-982.

19. Goodman, L.J., 1993, Diagnosis, management, and prevention of diarrheal diseases, in: *Current Opinion in Infectious Diseases*, pp. 88-93.

20. Gorbach, S.L., Banwell, J.G., Jacobs, B., Chatterjee, B.D., Mitra, R., Brigham, K.L., and Neogy, K.N., 1970, Intestinal microflora in Asiatic cholera. II. The small bowel, *J. Inf. Dis.* 121:38-45.

21. Kimura, N., Yoshikane, M., Kobayashi, A., and Mitsuoka, T., 1983, An application of dried bifidobacteria preparation to scouring animals, *Bif. Microfl.* 2:41-55.

22. Kelly, D., Begbie, R., and King, T.P., 1994, Nutritional influences on interactions between bacteria and the small intestinal mucosa, *Nutr. Res. Rev.* 7:233-257.

68

HELICOBACTER PYLORI INFECTION IN GNOTOBIOTIC PIGLETS

A Model of Human Gastric Bacterial Disease

Steven Krakowka[*] and Kathryn A. Eaton

Veterinary Biosciences
College of Veterinary Medicine
The Ohio State University
Columbus, Ohio 43210

1. ABSTRACT

Helicobacter pylori, a gram negative microaerophilic urease positive and motile spiral shaped bacterium is the etiologic agent of human chronic superficial (active) gastritis. In humans, *H. pylori* is the primary cause of gastric ulcers and atrophic gastritis and a probable risk factor for the development of gastric carcinoma and gastric mucosa associated lymphoid tissue (MALT) lymphoma, presumably because of infection induced epithelial cell proliferation and increased mucosal exposure to orally ingested carcinogens in an acid neutral environment in the former and immunogenic overstimulation in the latter. The organism exhibits remarkable genomic variation among bacterial isolates, but genomic homogeneity within infected individuals suggesting that it is not subjected to selectional pressures by the host to ecological competition with other bacterial species. Normal gastric defenses which prevent colonization by most bacteria, do not hinder *H. pylori* and may restrict delivery of specific antibody or immune cells to the gastric lumen.

The gnotobiotic piglet, uniformly susceptible to oral infection from birth onward, is the most widely recognized infection and colonization model of human bacterial gastritis. Human origin virulent *H. pylori*, strain 26695, is adapted to optimal growth in piglets. This motile, urease positive isolate is CagA and cytotoxin (VacA) positive and achieves colonization levels (10^7 to 10^8 colony forming units/g mucosa) within 24 hr after oral inoculation. In addition, other bacterial strains (N6, WV99) but not all (Tx30a) colonize piglets; laboratory passed strains colonize poorly if at all. In pigs, as in humans, infection is restricted to the gastric microenvironment, persists in that location for at least 3 mo (pigs) or yr (humans) in spite of vigorous local and systemic immune responses. The microbe has a

---

[*] Reprint requests to: Dr. Steven Krakowka, Veterinary Biosciences, College of Veterinary Medicine, 1925 Coffey Road, Columbus, OH, 43210, (614) 292-0231.

Advances in Swine in Biomedical Research, edited by Tumbleson and Schook
Plenum Press, New York, 1996

specific affinity for gastric mucus and adheres to gastric mucosal epithelia where it stimulates a multifocal or diffuse mononuclear lymphoplasmacytic inflammatory response with development of lymphoid follicles (pigs and children) and superficial neutrophilic infiltrates plus the lymphoplasmacytic lymphofollicular gastritis in adult humans. In piglets, the gastric inflammatory lesion can be modified to include the neutrophilic component by parenteral but not oral immunization with a killed autologous bacterin prior to infection. Gastric mucosal epithelial lesions, now recognized as diagnostic for *H. pylori* gastric infection, also occur in infected piglets. These include mucus depletion, epithelial cell vacuolation and degeneration, epithelial hyperplasia and necrosis with resultant gastric erosions and ulcerations.

The piglet model permits *in vivo* determination of putative bacterial virulence factors such as motility and/or presence of flagella, vacuolating cytotoxin (VacA) expression, urease enzyme presence and/or activity and infectivity of coccoid forms. In addition, a variety of conventional and experimental pharmaceuticals, antimicrobials and biologicals have been subjected to preclinical evaluation and efficacy in infected piglets. Both microbe related and host associated factors effect the pathogenesis of various components of the disease process. The relative importance of each alone or in combination can be determined in gnotobiotic piglets which possess a susceptible gastric microenvironment devoid of confounding variables such as concurrent commensal microbes, their products or effects on the host, diet, varying *H. pylori* burdens and strain differences, age, preexistent immunity and/or concurrent antimicrobial and antacid therapies.

2. INTRODUCTION

In the last 10 yr, human gastric inflammatory disease and the related entities of nonulcer dyspepsia, atrophic gastritis, gastroduodenal ulcer disease, gastric carcinoma and gastric mucosa associated lymphoid tissue (MALT) lymphoma have received wide attention. The reasons for this burgeoning interest and advances in this area of gastroenterology are 2 fold. Firstly, gastroendoscopic examination and biopsy are available widely and these entities now be may diagnosed readily in an outpatient basis. Secondly, successful isolation and characterization of a novel gastric bacterium, *Helicobacter pylori* (formerly *Campylobacter pyloridis*) in 1983[1] provided the research community with the bacterial agent likely responsible for most of these disorders and has stimulated the discovery of related gastric bacterial pathogens, (genus *Helicobacter*, "*Gastrospirillum*") in other species.

Helicobacter pylori, the type organism for this genus, possesses several unusual characteristics related to its strict gastric ecological niche. All isolates possess the urease enzyme and all are gram negative curved to spiral shaped microaerophiles with multiple polar flagella. All fresh isolates are catalase and oxidase enzyme positive. Bacterial urease is characteristic of this genus. It is a metalloenzyme (nickel) which compromises an estimated 6% of total bacterial protein. Urease is a large (550kDa, 2 subunit) protein that is expressed in bacterial cytosol and is shed into the surrounding microenvironment. Urease genes have been investigated intensively.[2] Two genes code for each subunit (UreA and UreB) and 7 or more may be in the operon. Two are regulatory genes and at least 3 genes/gene products are associated with urease synthesis and nickel transport and incorporation. The 1.6 megabase genome exhibits remarkable individual genomic variability among bacterial isolates. Three copies of 16S ribosomal RNA are present and individual isolates may or may not have plasmids. Isolates from patients are genomically homologous,[3] except for occasional familial clustering, suggesting that the microbe is not subjected to selectional pressures by the host[4-6] nor to ecological competition with other bacterial species. In spite of the daunting genetic individuality of the species, phenotypic characteristics are conserved remarkably. Thus far, the only 2 major features of interest to the pathogenesis of disease are the 87 kDa vacuolation cytotoxin (VacA) protein[7,8] possessed by approximately 50% of isolates examined and possess a *Cag II* genomic region which codes for a 120 kDa protein (cagA) which is possessed by at least 75% of isolates recovered from humans with gastric ulcers. Finally, gastric defenses such as the acid pH of the stomach, the thick mucus layer, rapid transit time(s) and peristalsis "designed" prevent gastric colonization by most bacteria, do not hinder *H.pylori* yet paradoxically may restrict delivery of specific antibody or immune cells to the gastric lumen.

Helicobacter pylori is the etiologic agent of human chronic superficial (active) gastritis.[10] Koch's postulates have been fulfilled in humans[11,12] and animals[13,14] and epidemiology, therapy and pathology studies provide overwhelming evidence of causality. The case for *H. pylori* as the primary cause of nonsteroidal antiinflammatory disease (NSAID) independent gastric ulceration, though slightly less convincing, is suggested strongly by data associating failure to eradicate organisms with a high rate of ulcer recurrence[15-18] and long term cure following successful eradication.[19,20] Similarly, atrophic gastritis, now thought to represent the last stage in the progression of persistent lifelong Helicobacter colonization plus interactions with other environmental factors,[18,21-23] is a known risk factor for development of gastric carcinoma,[22-25] presumably because of infection induced epithelial cell proliferation and increased mucosal exposure to orally ingested carcinogens in an acid neutral environment. Finally, low grade mucosa associated lymphoid tissue (MALT) lymphoma is associated with long term *H. pylori* colonization [26-28] as successful antimicrobial therapy causes tumor regression and "cure". [26]

Studies of the naturally occurring disease in humans have the disadvantages of variable bacterial density and multifocal (patchy) distribution of both microbes and inflammation, important strain differences among colonizing microbes in individuals, unknown duration of infection, discordance among lesions and clinical manifestations in patients, noncompliance with therapy, loss to follow up evaluation, ethical considerations for use of untested and/or unapproved approaches to therapy and concurrent complications such as NSAIDs, alcohol consumption and tobacco use. These difficulties are particularly true during the initial phases of infection where the onset and source of *H. pylori* rarely are determined.

Investigators have long sought to develop animal models of this disease. This search has led to the discovery of a new family of species specific gastric pathogens (the *Helico-bacter* genus) in mustelids,[29] cats and dogs,[30] cheetahs[31,32] and primates,[33] "*Gastrospirillum hominis*" (*Helicobacter heilmannii*)," a related microbe implicated in gastric disease of humans,[34] primates,[35] dogs[36] and swine[37] and a murine microbe (*H. hepaticus*) which colonize the liver[38] as well as other portions of the gastrointestinal tract.[39]

In 1987, we[13] and others[14] reported that gnotobiotic piglets successfully supported gastric colonization when orally inoculated with human origin *H. pylori*. Previously, oral infection by *H. pylori* had been attempted in both adult and neonatal laboratory species including mice, rats, rabbits, guinea pigs and rats[40] including germfree rats and mice and nude mice.[41] All were unsuccessful. Fox and colleagues[42] reported that *H. pylori* has been recovered from domestic cats, suggesting that felids may be a nonhuman reservoir and source to humans. Recently, some isolates have been adapted to growth in mice.[43] The latter, while promising, are compromised by weak and inconsistent colonization levels in the stomach, presence of endogenous, frequently urease positive commensal organisms, lack of adherence to gastric mucosal epithelia, modest or nonexistent mucosal inflammatory response and important dissimilarities in gastric physiology and diet between rodents and humans. Seronegative ferrets, even *H. mustelae* free, can not be infected by *H. pylori* even with porcine infectious strain, 26695 although gnotobiotic dogs support gastric infection by strain 28895.[45] Certain primates harbor *H. nemestremae*[33] and are susceptible to challenge[45] or natural infection with human *Helicobacter* strains [46] but their expense, the sensitivities for use as experimental laboratory animals and frequent contamination with intestinal *Campylobacter* sp. limit their usefulness. Several authors,[47,48] describe successful colonization of barrier born specific pathogen free (SPF) piglets and persistence for as long as 6 mo[48] although formal proof that recovered microbes were identical to parental inoculum was not reported leaving the possibility that these conventional swine became infected naturally with *Gastrospirillum* sp or other urease positive nonhelicobacter sp commonly found in pig stomachs (The Procter and Gamble Co., unpublished, 1990). Others found that infection of young conventional swine, including minipigs, have not yielded reproducible results (D.R. Morgan, unpublished, 1986, Norwich Eaton Pharmaceuticals, Norwich, NY and The Procter and Gamble Co, unpublished, 1990).

In this review, we summarize our laboratory's findings regarding *H. pylori* infection in gnotobiotic piglets and compare our data to that collected in humans. The piglet model provides not only the ideal surrogate to the acute gastric human disease but also is amendable to experimental manipulations, a reliable model for preclinical investigations of antimicrobial therapies in a monogastric omnivore and an excellent model for the study of mucosal defenses and immunity.

3. MATERIALS AND METHODS

3.1. Animals, *in vivo* Sampling Procedures, and Bacterial Inoculation

The surgical procedure[13] used to derive litters of piglets (10 to 12/litter) from date-mated sows was as follows. The sow was restrained in a squeeze chute, tranquilized with ketamine and an epidural administration of 2% (v/v) lidocaine was given to effect anesthesia from the sternum caudally. Hindquarters were elevated and, after preparation of the surgical field, a ventral midline incision is made. The uterus was exteriorized, severed from its abdominal attachments and entered into a transfer tank filled with chlorine disinfectant. The sow was euthanatized immediately by electrocution and exsanguination. Piglets were removed from the uterus, stimulated to breathe and, after resection of the umbilici, transferred into sterile pen/tub isolation units containing 6 separate partitions. Piglets were fed sow milk replacement diet and Similac[R] with iron 3 to 4 times daily. Piglets can be fed this diet until 30 to 35 da of age. Rarely, a few piglets have been weaned onto steam sterilized solid piglet chow and kept within the isolation units for 60 da. Size, space limitations, waste disposal considerations, inevitable contamination of units with steam resistant gram positive spore forming rods in the solid diet and resultant ammonia along with heightened nutritional requirements precluded longer time intervals in isolation units.

Gastric lavage in fasted, ketamine sedated piglets was used to determine infection status in inoculated piglets. For this, a polystyrene feeding tube was introduced into the stomach and 10 ml of Brucella broth was infused slowly by syringe. The abdomen was massaged gently and contents were aspirated, cultured and/or tested for bacterial urease activity. Endoscopic examination and biopsy of ketamine sedated piglets within gnotobiotic environment was performed, if needed in piglets greater than 3 wk of age. For this, sedated piglets were transferred into a sterile 6 ft endoscope isolation unit equipped with endoscopy probe sideports and subjected to endoscopy and biopsy with a flexible pediatric endoscope. Gastric mucosa was examined easily and cardia, fundus, antrum and diverticulum identified and biopsied. A separate attachment permitted intragastric injection of substances into the stomach wall. *In vivo* intragastric pH determinations were made with an endoscopic pH probe. Probes were surface sterilized with ethanol (70%, v/v) and entered into the unit through a double lock side port. Between piglets, probes were removed, disinfected and reintroduced into the unit. Blood samples for serum and *ex vivo* leukocyte studies were collected from jugular veins of sedated piglets.

Gnotobiotic piglets are susceptible to oral inoculation with Brucella broth log phase suspension cultures of *H. pylori*, 26695 from birth onward. For this, fasted piglets were treated orally with 60 mg/kg cimetidine to increase gastric pH, 2 hr prior to oral inoculation with 2 ml broth culture containing approximately 10^9 bacterial colony forming units (cfu). One hundred percent of inoculated piglets became infected;[49] colonization levels in the stomach stabilized at $10^{6-7.5}$ cfu/g mucosa within 24 hr after infection. Except for occasional transient (24 hr) anorexia,[13] infection was asymptomatic clinically.

3.2. Pathology and Microbiology

A standardized procedure for sample collection evaluation of gastric tissues has been developed. On the day of termination, a fasting gastric pH determination was made with the endoscopic probe or with litmus paper on gastric aspirates. In some instances, dividing epithelial (and inflammatory) cells were labeled *in situ* 2 hr prior to termination by a procedure originally developed for a mouse model of Helicobacter sp associated gastric carcinoma. For this, 200 mg/kg bromodeoxiuridine (BrdU) were administered by intraperitoneal injection (Eaton, unpublished, 1994). In some experiments, piglets received intramuscular cimetidine intraperitoneal injection 1 hr prior to termination to inhibit gastric acid secretion, which permitted determination of the distribution of microbes with urea methylcellulose broth. Piglets were sedated heavily with ketamine and removed from the isolation units. Terminal samples for serum and whole blood were collected and the piglet euthanatized via an IV overdose of sodium pentobarbital.

The stomach was exteriorized, ligated at the esophagus and proximal duodenum and removed. Samples of gastric juice from the intact stomach as well as bile and intestinal contents (jejunum/ileum and spiral colon/cecum), were collected in syringes, clarified by low speed centrifugation and frozen (-70C) in labeled vials for future evaluation. The stomach was opened aseptically by dissection along the greater and lesser curvatures and half of the stomach excised for quantitative microbial culture. The remaining half stomach was pinned to a paraffin block in a Petri dish, mucosal surface up. Pertinent gross findings in gastric tissue (submucosal edema, amount of lumenal mucus, presence and extent of lymphoid follicles, and erosions or ulcers if present) also were recorded. Excess mucus was rinsed from the mucosal surface with warm sterile tissue culture phosphate buffered saline (TC-PBS) and sterile urea/ phenol red pH indicator in 1.5% (v/v) methylcellulose was flooded onto the leveled mucosal surface. After 30 to 60 min at 22C, sites of heavy bacterial colonization were identified by urease enzymatic activity (color change from yellow to red) plus generation of CO_2 gas from urea hydrolysis, which remained entrapped as bubbles in the methylcellulose overlay (Fig. 1). These areas were marked on a schematic drawing and unusual features, such as erosions, collections of lymphoid follicles, hemorrhages or ulcers were marked on the sketch and identified *in situ* with pins to facilitate subsequent identification during tissue processing after formalin fixation.

Mucosal punch biopsies were collected aseptically and explanted for *in vitro* immunoglobulin (Ig) production determinations.[50,51] Adjacent mucosal sections were collected, rinsed in saline and embedded in OCT^R mounting medium and snap frozen (-70C) for *in situ* immunohistochemical detection of antigens denatured by cross linking fixatives including epithelial secretory component (SC), interleukin (IL)-8, the leukocyte cluster differentiation (CD) markers CD4, CD8, MHC II with monoclonals, PG130A (monocyte) and DH59B (granulocyte/macrophage) by staining frozen 6 mu sections. Methylcellulose was rinsed away with fresh 10% phosphate buffered formalin and the musosa fixed by emersion of the entire Petri plate in 10 volumes of 10% v/v phosphate buffered formalin for at least 24 hr.

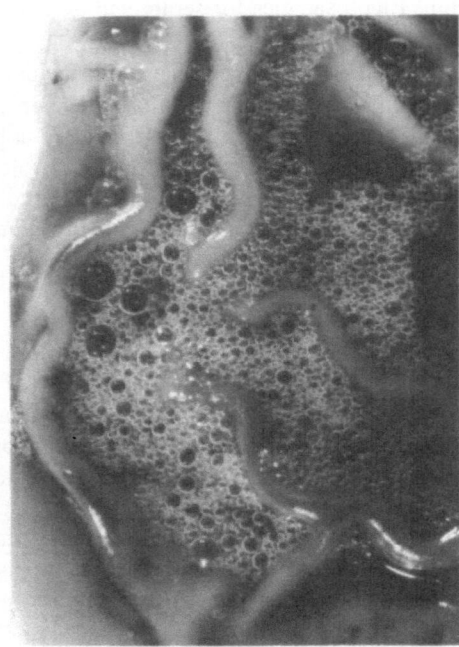

Figure 1. Gross photograph of gastric mucosa from an infected piglet overlaid 1 hr previously with methyl-cellulose containing urea substrate and pH indicator. Note CO_2 immobilized by the methylcellulose.

Gastric samples were processed for histopathology by standard methods and 5 um section replicates were stained with hematoxylin and eosin for morphologic evaluation, periodic acid Schiff (PAS) for carbohydrates, Masson's trichrome for collagen and Warthin-Starry for organisms. Additional unstained deparaffinized tissue section replicates were stained for antigens resistant to formalin fixation associated denaturation such as Ig isotype-specific plasma cells and cytoplasmic alkaline phosphatase as a marker for dendritic and tissue macrophages.[52] Nuclei of BrdU labeled proliferating cells were identified with a murine monclonal antibody to BrdU and biotinylated antimouse IgG avidin/biotin peroxi-dase secondary reagents. For electron microscopy,[53] mucosal fragments were collected, postfixed in osmium tetroxide and processed for ultrastructural examination with a Phillips 300 electron microscope.

3.3. Microbiology

Isolation units were screened for extraneous microbial contaminants after derivation and at intervals thereafter until termination by aerobic and anaerobic culture of swabs from feed pans, feces and the isolation units. Bacterial contaminants, if present, were speciated. For preparation of *in vivo* inocula, Brucella broth, 2.8% w/v Bacto[R] (Difco Laboratories, Detroit, MI), with 10% v/v fetal calf serum cultures in log phase growth, were used. Bacterial enumeration was accomplished by hemocytometer and counts adjusted to 10^9 microbes/2 ml/piglet. Assays for bacterial urease, catalase, oxidase and motility have been described elsewhere.[53-55] For quantitative determination of cfu/g of gastric mucosa recovered from infected piglets, mucosa was stripped free of the underlying tunica muscularis, weighed, homogenized in 10% (w/v) Brucella broth and 10 fold dilutions were plated, in duplicate, onto blood or TVAP (Remel, Lenexa, KS) agar plates and read after 4 da incubation at 37C, 10% CO_2, 5% oxygen and 85% nitrogen in 95% humidity. Porcine resolates were identified as *H. pylori* by gram stain, colony morphology, enzyme profiles and, if necessary, restriction endonuclease digest patterns of isolated bacterial DNA or random primer based PCR to confirm identity to parental inocula.[56]

3.4. *Helicobacter Pylori*: Strains, Naturally Occurring, Chemically Induced, and Isogenic Mutants

Most *in vivo* experiments used *H. pylori*, strain 26695, originally isolated by A. D. Pearson and obtained from scientists at Norwich Eaton Pharmaceuticals, Norwich, NY in 1986. Two substrain variants were used. One was a laboratory passaged strain and the second has been adapted to optimal growth in piglets through 12 serial *in vivo* passes.[49] The latter was used for challenge studies. Both were urease, catalase, oxidase positive motile bacteria which were CagA and VacA positive.

An N-methyl-N'-nitro-N-nitrosoguanidine treated urease activity negative mutant of 26695, otherwise indistinguishable from parental 26695 [56] and cloned nonmotile aflagellate variants of 26695 [57] both were tested for colonizing ability in piglets. Coccoid forms of 26695 were created and tested for *in vivo* infectivity.[58] Strain 60190, also isolated by Pearson, is weakly motile and VacA positive. Strain Tx30a, isolated by G.E. Buck is nonmotile and cytotoxin negative.[53] Wild type strain N6 and the urease negative isogenic mutants, N6 *UreG::km* resistant UreA and *N6 UreB::km* resistant were provided by A. LaBigne. In addition, isogenic mutants of N6, devoid of either *flaA*, *flaB* or both genes, created through allelic exchange by Suerbaum and colleagues,[59] were tested for virulence in piglets. The other porcine virulent strain (WV99) was isolated by T. U. Westbloom.

3.5. Serology, Immunology, and Flow Cytometry

Specific antibody Ig isotype levels in sera, gastric juice, bile and tissue culture fluids from explants[60,62] were accomplished by ELISA with whole, formalin fixed bacteria.[13] The ELISA assays were standardized[51] to a positive control serum. ELISA data also has been generated using crude flagellin and flagellin free outer membrane complex (OMP) as antigens.[63] Whole cell lysates, separated by SDS-PAGE and blotted onto nitrocellulose, identify Ig isotype specific antibody activity to individual bacterial proteins.[64]

Lymphocytes were isolated from heparinized peripheral blood by ficoll hypaque (F/H) centrifugation.[61] Single cell suspensions were prepared from lymphoid tissues by routine methods and further purified by F/H centrifugation. Isolation of gastric lamina propria leukocytes (LPL) was achieved through modifications of several published techniques.[66-68] In brief, thoroughly rinsed isolated mucosa was minced and incubated (25 min at 22C) in buffer containing 1.0 mM dithiothreitol. Tissue fragments were placed in buffer containing 0.75 mM EDTA (20 min at 37C) and washed 3 times. After 2 additional EDTA treatments and washes, fragments were digested with type II collagenase (2 hr at 37C). After digestion, methylcellulose was added to the cell slurry and transferred to a sterile Stomacher bag for mechanical isolation of LPL[67] in a Stomacher model 80 laboratory blender. Recovered LPL were filtered, centrifuged and resuspended in 10 ml of 43% (v/v) percoll (1.057 g/l) and underlaid with 5 ml of 67% (v/v) percoll (1.088 g/l). After gradient centrifugation, interface LPL were collected, washed and assessed for viability and morphology.

Following overnight incubation (37C, 10% CO_2) in lymphocyte culture medium to allow reexpression of cellular markers, LPL were centrifuged through percoll to remove dead cells and debris prior to staining for flow cytometry. Cellular suspensions were stained with optimally diluted monoclonal antibodies to CD4 (PT90A), CD8 (PT81B), MHC Class II (MSA3), monocyte (PG130A) and granulocyte/monocyte (DH59B) or isotype control antibodies (VMRD Inc, Pullman, WA) followed by fluorosceinated flow cytometry grade antiglobulin secondary reagents. Dual labelled (CD4:CD8) positive cells were determined by staining LPL with both monoclonals followed by staining with fluorescein and phyco-erythrin conjugated secondary reagents. Stained cells were fixed in freshly prepared 1.0% v/v paraformaldehyde and processed through an EPICS 753 Flow Cytometer (Coulter Electronics, Kendall, FL).

3.6. Gastrin Determinations

Serum gastrin assays were performed using radioimmunoassay (Becton-Dickinson, Orangeburg, NY) for the detection of human serum gastrin (W. B Green, unpublished, 1994). Gastrin tracer and antibody provided were specific for the major circulating forms of human gastrin, including both the heptadecapeptide, G-17, and G-34. The manufacturer's protocol was followed except for centrifugation of precipitant (22 vs 15 min at 200xg). All samples were analyzed in duplicate or triplicate and a Micromedic Systems 4/200 Automatic Gamma Counter was used to determine the amount of isotope present in standards and samples. Serum gastrins in picograms per ml (pg/ml) were determined from a standard logit log curve.

3.7. Experimental Design for Drug Therapy Trials

One advantage of the *H. pylori* infected piglet was the opportunity to evaluate novel antimicrobials with statistically useful (n = 4 to 5/group) numbers of animals. Many human patients, culture negative after cessation of therapy, revert to culture positive status by the same parental genotype[2] wk or mo later. To replicate this phenomenon in piglets, 2 treatment trial designs have been developed to test intragasric antimicrobial efficacy. All piglets were infected orally with *H. pylori*, 26695 at 3 da of age. Seven da after infection, treatment was initiated and continued 7 da (17 da of age). Piglets were terminated on da 18 (Clearance, design 1) or sedated and lavaged on da 18 to determine interim culture status before termination 14 da later (Eradication, design 2).

4. RESULTS

4.1. Clinical Signs and Gross Pathology

Active infection was asymptomatic clinically except for rare transient anorexia. Otherwise, appetite was unaffected and a febrile response was not observed. Leukocytes from infected piglets responded normally to phytomitogens, a further indication for the lack of detectible systemic effects of infection. In piglets, gross lesions, except for mild extragas-tric lymphoid hyperplasia, were restricted to the stomach.[13] In most cases, detectible lesions were subtle and consisted of excess gastric lumenal mucus, variable submucosal edema, inconsistent elevation of gastric pH from 1.0 to 2.0 (uninfected) to 3.0 to 6.0 (infected). Raised gastric lymphoid follicles (1 to 3 mm diameter) were discerned most easily along the lesser curvature (Fig. 2) from the esophageal glandular cardia to the antrum. Mucosal ulcers at the junction of the antrum and fundus (Fig. 3), if present, occurred only in infected piglets except histologically in one instance. Uninfected control piglets lacked any of these lesions except for rare "stress" erosions in the nonglandular esophageal mucosa [13,69] which occurred in the same frequency (<5%) in infected littermates.

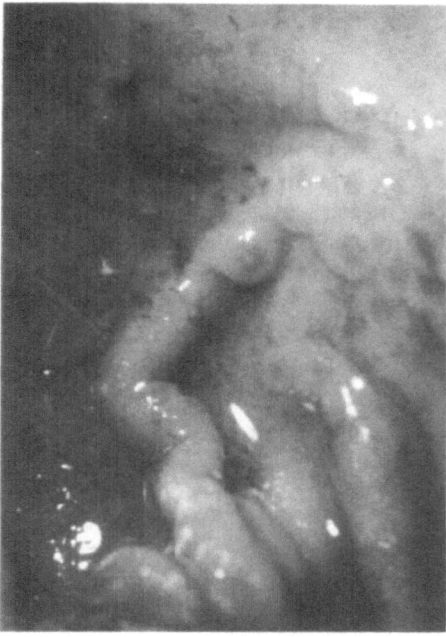

Figure 2. Gross photograph of gastric mucosa from an infected piglet illustrating the appearance of lymphoid follicles in the lesser curvature of the stomach at the junction of the cardia and antrum.

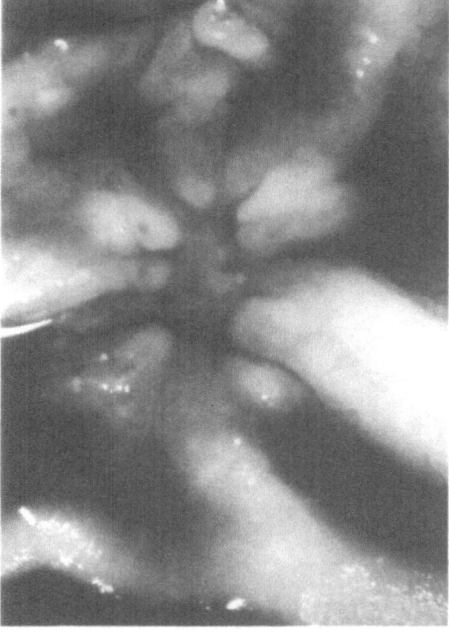

Figure 3. Gross photograph of gastric mucosa from an infected piglet with a grossly visible gastric ulcer. Note that the edges of the ulcer are well defined and raised from the adjacent normal tissue.

4.2. Histologic Findings: The Inflammatory Response

Uninfected gnotobiotic piglets were devoid of visibly detectible leukocytes in the gastric lamina propria and submucosa; less than 1×10^6 LPL have been recovered from whole stomachs of these piglets vs up to 30×10^6 LPL in half stomachs from infected piglets.[61] The hallmark microscopic lesion of *H. pylori* infection was the appearance multifocal to diffuse mucosa oriented lymphocytic to lymphoplasmacytic cellular infiltrates, the appearance of increased numbers of intraepithelial leukocytes (Fig. 4) of undetermined phenotype and superficial gastric mucosa associated lymphoid follicles (Fig. 5). Occasional piglets exhibited focal eosinophilic infiltrates; their role in the disease, if any was unknown. Neutrophils may appear transiently[13] after infection and may be somewhat bacterial strain dependent (Eaton, unpublished, 1995) but, do not dominate the gastric inflammatory response as they do in infected adult humans.[74]

Mucosal immune responses are initiated in Peyer's patches and related organized common mucosa associated lymphoid tissue (MALT) of the intestine. Dogma states the human stomach is devoid of organized lymphoid tissues.[71] This view was reinforced by apparent failure to identify, recognize or distinguish a MALT from gastric inflammation. Yet, this view cannot be reconciled with the fact that the stomach must be the site of induction of immunity to this organism since this microbe is a strict gastric pathogen. Until now, initiation of immunity in the lower gastrointestinal tract with seeding of immunocytes back to the stomach as a response to infection could not be excluded. For several years, we noted the presence of structures anatomically identical to ileal Peyer's patches and colonic lymphoglandular complexes[72] in the esophageal glandular mucosa of the lesser curvature of piglets. Primordial inactive lymphoid aggregates (Fig. 6) were present in microbially sterile

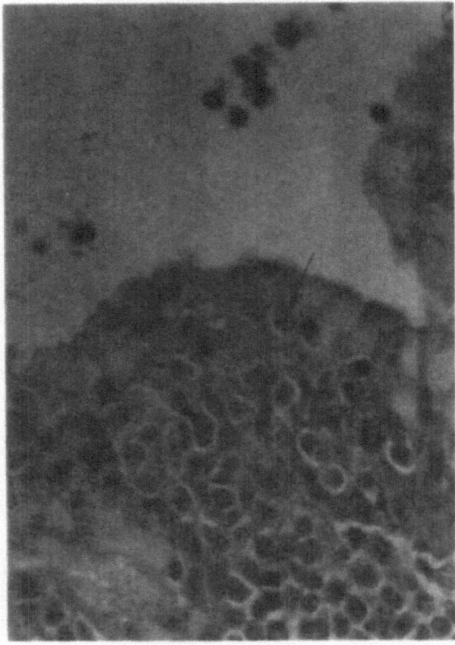

Figure 4. Photomicrograph of gastric mucosal epithelium over an area of inflammation illustrating the presence of intraepithelial leukocytes (arrows).

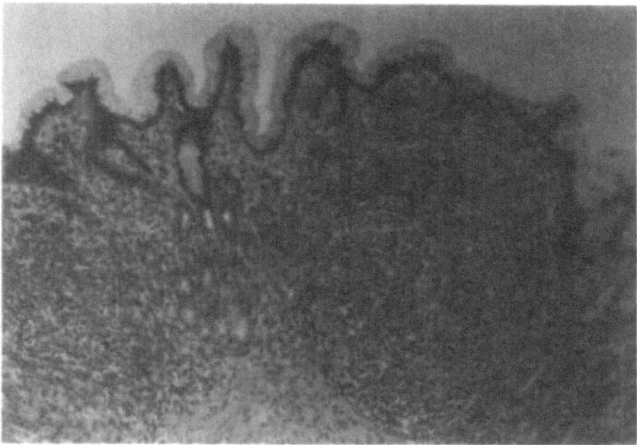

Figure 5. Photomicrograph of gastric mucosa and lamina propria from an infected piglet. Note the close relationship of the developing lymphoid follicles (B cell regions) with the overlying epithelium. These structures correspond to those structures illustrated grossly in Fig. 2. Note also the lymphocytic infiltrates between follicles.

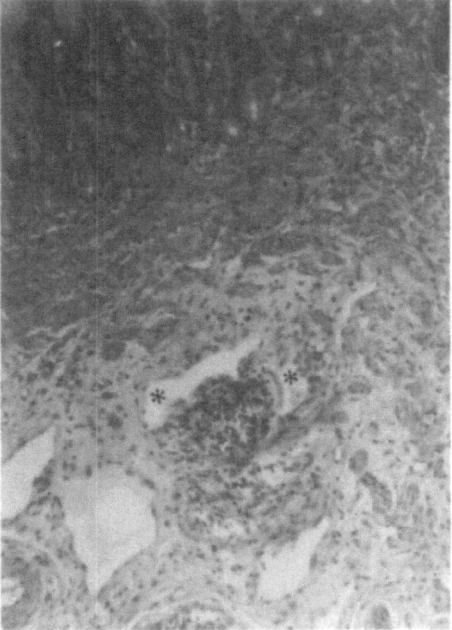

Figure 6. Photomicrograph of the deep gastric lamina propria from the junction of the esophageal and glandular cardia in an uninfected control gnotobiotic piglet, 28 da of age. Note the inactive lymphoid aggregate characteristic of unstimulated gastric MALT and the presence of a fibrous connective tissue capsule around the deep portion. Asterisks denote dilated and acellular lymphatics.

gnotobiotes but their significance was not known. A recently completed morphologic study of these structures in various groups of *H. pylori*, *Staphylococcus* sp. and rotavirus infected gnotobiotic pigs[73] has illuminated their potential importance. They remain inactive and primordial in piglets contaminated with nongastric bacteria and in enteric rotavirus infected piglets. These gastric MALT (Fig. 7) are developed fully in *H. pylori* infected piglets, possessed morphologic features (fibrous encapsulation, efferent lymphatics, location below the muscularis mucosa) which were distinct from bacterial infection/inflammation associated lesions in the superficial mucosa. Their location at the esophageal grove of the gastric glandular mucosa suggest that, after tonsils, gastric MALT are the next site of mucosa involved in antigen processing, regional lymphoid maturation, differentiation and proliferation in the stomach and thus appear to be a component of enteric MALT.

Neutrophilic gastritis ("chronic active gastritis" in humans) was produced regularly in piglets by parenteral immunization of pigs with formalin killed bacteria in incomplete Freund's adjuvant prior to oral infection.[75] In humans, others have suggested that gastric epithelia, stimulated by bacterial infection or associated inflammation, produce and secrete IL-8 which is responsible for local neutrophilia.[76] In piglets, gastric epithelia from infected stomachs stain positively for IL-8 (J. Crabtree, written communication, 1993; Byron, Eaton and Krakowka, unpublished, 1995), yet neutrophils were not present consistently near these cells nor in the inflammatory lesions, even after 30 da of infection.[13] Infected and uninfected control gnotobiotic piglets, developed acute neutrophilic dermatitis and subdermal necrosis when formalin killed, washed *H. pylori* were injected into dermis, indicating that porcine neutrophils were mature and motile chemotactically (Krakowka, unpublished, 1992).

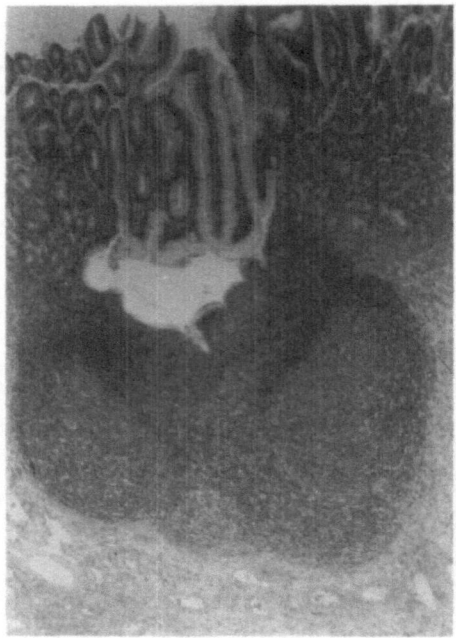

Figure 7. Photomicrograph of the deep gastric lamina propria from the junction of the esophageal and glandular cardia in an infected gnotobiotic piglet given *H. pylori* 28 da previously. Compare to Fig. 6. The structure is fully developed. Note the deep invagination into the lamina propria, the well developed limiting fibrous capsule and the folliclular development within the complex. Efferent lymphatics frequently are filled with lymphocytes.

4.3. Histologic Findings: Gastric Epithelium

H. pylori infection in gnotobiotic piglets resulted in the full spectrum of acute epithelial lesions[26,65] reported in humans including erosions and ulcers. Surface mucus cell stimulation and mucus depletion was appreciated easily in PAS stained histologic sections.[13,70] Cytoplasmic vacuolation (Fig. 8), attributed by some to the *in vivo* action of the bacterial VacA cytotoxin,[53] occurred in close association with mucosal lymphocytic infiltrates and were not detected ultrastructurally in association with adherent bacteria[53,77] nor in Warthin-Starry stained sections which demonstrate adherent organisms.

In situ labeling with BrdU has documented an indirect effect of *H. pylori* infection upon gastric epithelia. In uninfected piglets, the labeling index (number of BrdU labeled epithelial cells/500 epithelial cells) is low (6.6 +/- 1.8%). The labeling index away from areas of inflammation was higher (8.0 +/- 3.3%) but not different from controls. However, the epithelial labeling index adjacent to inflammatory cell follicles and aggregates was high (19.1 +/- 2.7%) and even higher (36.8 +/- 5.1%) in epithelia adjacent to chemically induced ulcers. Finally, gastric epithelia, even over areas of inflammation were MHC class II negative[61] but produced[51] epithelial secretory component (SC).

In addition to the above, close inspection of gastric epithelia revealed small foci of degeneration and necrosis regularly associated with underlying inflammatory cell infiltrates (Fig. 9). These occurred in the presence or absence of recognizable bacteria and were similar morphologically to apoptosis.[78,79] It is logical to expect that some of these lesions progressed from microscopic erosions to grossly visible (Fig. 4) gastric ulcers. In a recent archival study,[70] ulcers and erosions were found in 10/29 *H. pylori*-infected piglets, 1/27 uninfected control gnotobiotes and 0/27 piglets inoculated with noncolonizing strains of *H. pylori*. .

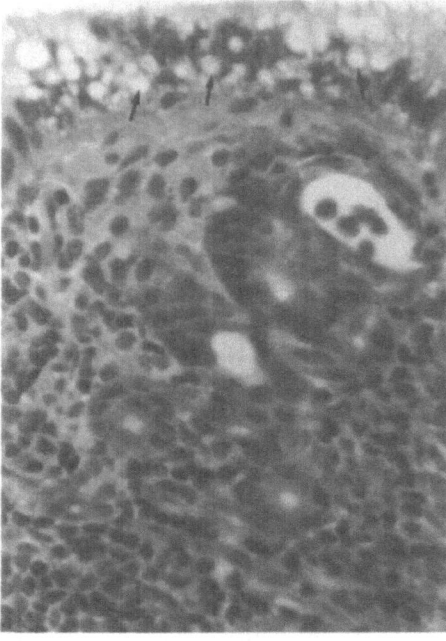

Figure 8. Photomicrograph of gastric mucosal epithelium overlying gastric inflammation in an infected piglet. Note the prominent epithelial vacuoles (arrows).

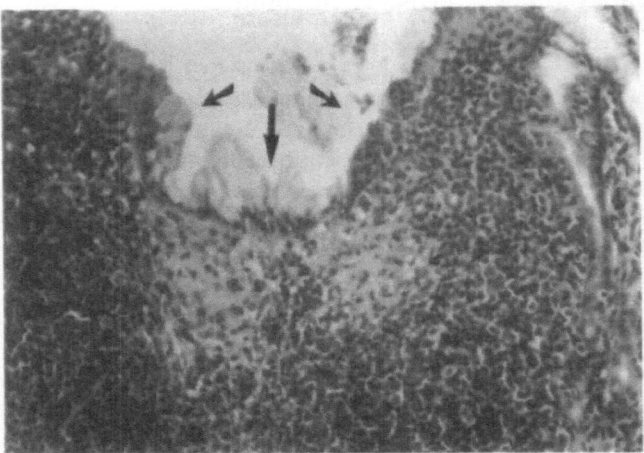

Figure 9. Photomicrograph of gastric mucosal epithelium overlying a region of inflammation from an infected piglet. The arrows illustrate the limits of epithelial loss (erosion). Note also the underlying degeneration and necrosis of the inflammatory cell infiltrate (poorly stained).

4.4. *Helicobacter pylori* Microbiology

Oral inoculation with strains 26695, N6 and WV99 readily colonized the porcine stomach. Most of the quantitative work has been done with 26695; the following data were derived from this strain. *Helicobacter pylori* is a strict extracellular gastric pathogen. Organisms were curved to spiral shaped and located in the extracellar compartment of the stomach, either in the overlying gastric mucus layer (majority) or adherent to gastric epithelia (minority) (Fig. 12). The standard oral inoculation dose/piglet was 10^9 cfu. In our first experiment,[13] cimetidine was used to inhibit gastric acid production to provide optimal conditions for colonization. This pretreatment is not necessary but still is used so that drug efficacy studies can be compared between litters without another variable. We do not know what the minimal infective (cfu) dose for 26695 is but it is likely less than 10^6 as Lambert and colleagues[14] reported success with gnotobiotes at that concentration. The laboratory passaged strain of 26695 colonized poorly ($<10^4$ cfu with $<100\%$ infection rate) but colonization levels increased with *in vivo* pig passage.[49]

Porcine passaged 26695 achieves stable gastric colonization ($10^{6-7.5}$ cfu/gm) within 24 hr; these levels cannot be increased by repeated daily oral reinoculations of infectious inocula. Occasional pigs[70] have cfu as high as 10^{12}/g. The practical limit for reproducible quantitative determinations of bacterial cfu/g as determined by *in vitro* add back experiments of bacteria to sterile gastric mucosal homogenates is 10^3 cfu/g, although fewer organisms can be detected on a "present or absent" scale. Further complicating precise study of bacterial kinetics *in vivo* is the fact that, within the stomach, the distribution of organisms is uneven (Fig. 1) and the intensity varies somewhat from pig to pig in the same litter, even though roughly the same total recoverable cfu are recovered from all. Predilection colonization sites were gastric glandular cardia, lesser curvature and antral mucosa. Rarely are organisms seen in the fundus or pylorus. The sensitivity of the Warthin Starry stain is limited in that organisms are identified easily in pigs with recoverable cfu greater than 10^6, with difficulty at 10^5 cfu and usually falsely negative below 10^4 cfu. Rare microbes ($<10^2$ cfu/g) occasionally are recovered from esophagus and duodenum [13] and occasionally farther down the intestinal tract.[13,55]

Microbes have not yet been recovered from the environment nor from food or milk. In one small gnotobiotic dog study, we demonstrated the infection was transmitted readily among puppies after housing them together in close contact during weaning from milk to semisolid diet.[44] Immunosuppression with cyclosporine (Sandimmune[R], Sandoz Pharmaceuticals, Hanover, NJ) or prednisolone (Depo-Medrol[R], The UpJohn Co., Kalamazoo, MI) did not increase colonization levels; parenteral but not oral immunization prior to inoculation reduced the colonization level by 5 to 10 fold.[75]

Available *in vivo* data indicated that urease is essential for bacterial colonization.[55,56] To determine if urea substrate limitations in diet, in turn, limited colonization levels, the ability of 0.5 g urea (Ureaphil[R], Abbott Laboratories, Chicago, IL) given orally, TID was determined. Infected piglets supplemented with urea alone yielded a mean $10^{7.2}$ cfu/g, (n = 5), compared to piglets treated with urea and cimetidine (mean $10^{7.4}$ cfu/g, n = 5), piglets given cimetidine alone (mean $10^{7.2}$ cfu/g, n = 4) or untreated infected piglets (mean $10^{6.8}$ cfu/g, n = 3). Thus colonization levels in the stomach were not limited by the amount of urease substrate in the gastric microenvironment.

Nicotine is suspected to be a human ulcerogenic cofactor.[82] To test this possibility, we treated 3 wk infected piglets with nicotine impregnated dermal patches (Nicoderm[R], Marion Merrell Dow, Kansas City, MO), 7.0 mg (4 da) and 14.0 mg (4 da). Piglets demonstrated clinical signs of nicotine excess (hyperactivity and slight muscular incoordination) but did not develop ulcers, exhibit increased severity of gastric inflammation nor increased recoverable bacterial cfu ($10^{7.6}$ cfu/g, n = 3). In a pilot experiment, 2 piglets treated daily with Sandimmune[R] (15.0 mg/kg cyclosporine, IP) for 25 da, 22 da after infection did not demonstrate elevated bacterial counts ($10^{7.4}$ cfu/g) but gastric inflammatory cell infiltrates were absent. In contrast, a Depo-Medrol[R] treated piglet (2.0 mg/kg daily, IM) over the same infection/treatment interval yielded $10^{4.6}$ cfu and prominent inflammatory cell infiltrates.

4.5. Virulence Factors of *Helicobacter Pylori*

Since the piglet model is, in essence a short term (45 or less da of infection) one, we concentrated our efforts upon identification of putative bacterial virulence and colonization factors. It provides an unique opportunity to investigate early bacterial colonization events, something not possible in humans. Not all bacterial strains are infectious and known virulent strains, modified to lack suspected virulence/colonization factors can be tested in infected piglets (Table 1). Bacterial urease was required for colonization as was motility.[53] Naturally occurring aflagellate mutant strains (Tx30a) were not infectious and recovered flagellated isolates of 26695 exhibit enhanced motility following *in vivo* porcine passage.[53,54] Isogenic mutants devoid of the flagellin proteins (FlaA and/or FlaB) colonized poorly and did not persist in the gastric compartment (Eaton, unpublished, 1995). Coccoid forms were not infectious.[58]

4.6. Serology and Gastric Mucosal Antibody Responses

The pattern of local antibody production in infected gastric mucosae has been determined by sequential explant biopsy culture and ELISA assay.[60] Humoral immune response to *H. pylori* was initiated in the gastric compartment as an IgA dominant mucosal immune response. IgG and IgM specific Ig also can be detected. Gastric lymph nodes produced high levels of *H. pylori* specific IgA; whereas, IgG production predominated in mesenteric lymph nodes and spleen. Intestinal but not gastric contents contained IgA from

Table 1. Summary of the importance of bacterial virulence/colonization factors in the
establishment of gastric infection by *Helicobacter pylori*

Bacterial factor	*In vivo* efficacy for gastric colonization
Laboratory passaged strains	variable, most are poorly colonizing *in vivo*
Bacterial cytotoxin (vacA)	no effect on colonization
Urease Enzyme	
(a) 26695 nitrosoguanidine induced-	noncolonizing
(b) N6 isogenic mutants (UreG and UreB)	noncolonizing
(c) N6 isogenic mutants (UreG and UreB) in achlohydric stomachs	noncolonizing
Motility	
(a) Aflagellate strain (Tx30a)	noncolonizing
(b) 26695 spontaneous mutants	noncolonizing
(c) N6 isogenic mutants	
FlaA-deficient	poorly colonizing[a]
FlaB-deficient	poorly colonizing[a]
FlaA- and FlaB-deficient	noncolonizing
Coccoid Forms 26695	noncolonizing

[a]Low levels of organisms were recovered from infected piglets on PID 2 and 4 but not PID 10.

PID 14 onward. Local gastric Ig production was dependent upon active bacterial colonization in that gastric explants prepared from piglets successfully treated for infection were Ig negative.[61] Convalescent sera contained high levels of IgG antibody. Western blot analysis of convalescent sera demonstrated many immunoreactive bands including those against flagellins (52 to 54 kDa) and urease subunits (30 to 32 kDa); IgM specific bands were fewer and less well defined.[83] Finally, peripheral blood lymphocytes from infected but not control piglets responded by proliferation to both whole formalin killed organisms and crude outer membrane complex. Peak responses occurred 4 to 6 wk after infection but the significance of this observation is not known.

4.7. Phenotypic Analysis of Gastric Inflammatory Cell Infiltrates

Phenotypic analysis of gastric LPL was accomplished by a combination of flow cytometry and immunocytochemistry.[61] We confirmed that microbially sterile gnotobiotic piglets are devoid of detectible gastric inflammation (0.2 to 0.8×10^6 recoverable LPL/ whole stomach), infected piglets contain significant numbers of recoverable LPL (8.0 to 28.0×10^6 LPL/half stomach) and inflammation develops specifically after gastric colonization with *H. pylori*. Although the flow data (% of total analyzed) indicated the CD8 positive phenotype in uninfected LPL (mean 42.5%, n = 3) was greater than the percent CD8 cells recovered in infected piglets (mean 18.6%, n = 6), the absolute number of CD8 cells (percent x total LPL) demonstrated the CD4 phenotype predominates. In these same animals, 4.5% of LPL were labelled with the CD4 monoclonal, a value similar to the 3.1% positive fraction in *H. pylori* infected LPL. Again, the absolute number calculations revealed that significantly more CD4 positive cells were present in infected piglets. Similar conclusions for the dual labeled LPL (1.9% control, 3.1% infected) and MHC Class II positive (28.7% control, 33.7% infected) cells were reached. Too few cells were recovered from control LPL to analyze the percent of monocytes and granulocytes precluding comparisons to infected values of 5.2% monocytes and 6.3% granulocyte/monocyte.

Immunocytochemical analysis of stained frozen tissue sections confirmed the primacy of the CD8 phenotype *in situ*. The CD8 phenotype was distributed either as a "continuous" cellular layer in the deep lamina propria above the muscularis mucosa, as aggregates, closely associated with peripheral zones of lymphoid follicles, or as prominent perivascular infiltrates in the muscular layers distal from the gastric lumen. In contrast, CD4 cells were uncommon and associated closely with developing lymphoid follicles, presumably B cell origin structures. Macrophages or granulocytes, stained with the monoclonals used, were rare. Inflammatory cells reactive with the MHC class II antibody reagent stained most mononuclear cells and vascular endothelia but not gastric epithelia. Plasma cells, positive for IgA, were abundant and associated closely with follicles and vasculature of lamina propria; IgG positive cells also were present but IgM secreting cells were rare. Follicles were rich in MHC Class II positive leukocytes. Epithelia over areas of inflammation expressed readily detectible SC; whereas, adjacent epithelia were trace positive or negative as was gastric epithelium from uninfected control piglets.[61]

4.8. Serum Gastrin Levels

One of the physiologic features of *H. pylori* associated gastritis in humans is elevated fasting and stimulated serum gastrin levels. Piglets with *H. pylori* gastritis were used to determine serum gastrin levels under a variety of infection and treatment conditions. Gastrin level values ranged from 18 to 270 pg/ml (n = 49) and were correlated positively (R = 0.78) to gastric pH as increased by cimetidine and omeprazole. Serum gastrin levels were determined in fasting and postprandial sera. Thirty min following a meal, fasting serum gastrin levels rose from 46.6 to 85.9 pg/ml in control piglets (n = 3); an identical response pattern was seen in *H. pylori* infected piglets (n = 3). Both sets returned to baseline gastrin levels 9 hr later. There were no differences in serum gastrins in infected vs uninfected piglets, or in female vs male piglets for any of the parameters addressed.

4.9. Evaluation of Experimental Antimicrobials and Biologicals

The infected piglet is ideally suited to the evaluation of standard and innovative approaches to antimicrobial therapy. Using designs outlined above, bench mark antimicrobials have been evaluated in *H. pylori* infected piglets. Novel antimicrobial modalities, including recombinant cytokines, antiadhesins and candidate vaccines, are amendable to evaluation. Summarized in Table 2 are the efficacy of clearance (culture negative at the end of treatment) and eradication (cultural status 2 wk following the last treatment) as well as the doses and route(s) for some commonly employed antimicrobial agents.

5. DISCUSSION

There are numerous points to be made regarding the data summarized above. For simplicity, the discussion is divided into the subheadings of: gastroesophageal ulceration (GEU), a naturally occurring entity of pigs often called "ulcers" but is unrelated to the model system described above, 2) *Helicobacter sp.* infection in other animal species and their significance to *H. pylori* research, 3) gastritis in *H. pylori* infected pigs compared to *H. pylori* infected humans and mice infected with *H. felis*, 4) gastric epithelial damage as a precursor lesion to ulcer, 5) gastric ulcers, 6) chronic manifestations of *H. pylori* ill suited for study in our model, 7) immunoprophylaxis in the model and 8) antimicrobial therapy for *H. pylori*.

5.1. Gastroesophageal Ulceration (GEU) or "Ulcers" in Swine

In modern swine intensive production systems, the development of ulcers and erosions of the nonglandular esophageal (cardiac) gastric mucosa is a common and serious

Table 2. A summary of antimicrobial efficacy of commonly used antimicrobials using the clearance and eradication designs outlined in materials and methods

Antimicrobial	Drug and Dose	Route/Frequency	Clearance[a]	Eradication[a]
Triple Therapy				
bismuth	5.70 mg/kg			
metronidazole	4.40 mg/kg	oral, QID	yes	yes
amoxicillin	6.80 mg/kg			
DeNol[R] (bismuth)	15.00 mg/kg	oral, TID	yes	no
Amoxicillin	5.6-0.01 mg/kg	oral, QID	yes	yes
	0.10 mg/kg	IP, BID	no	not done
Metronidazole	>2.50 mg/kg	oral, QID	yes	yes
	<1.00 mg/kg	oral, BID	no	no[b]
	1.00 mg/kg	IP, BID	no	no
Clarithromycin	5.00 mg/kg	oral, QID	yes	yes
	1.00 mg/kg	oral, QID	no	no
Ciprofloxacin	7.30 mg/kg	oral, QID	yes	yes
	1.90 mg/kg	oral, QID	yes	no
Tetracycline	7.10 mg/kg	oral, QID	yes	no
Erythromycin	6.60 mg/kg	oral, QID	yes/no[c]	yes/no[c]
Nitrofurantoin	5.00 mg/kg	oral, QID	yes	no
Acid Inhibitors Drugs:				
a) Ranitidine	1.60 mg/kg	oral, QID	no	no
b) Cimetidine	75.00 mg/kg	IM, TID	no	not done
Proton Pump Inhibitor(Omeprazole)	5.00 mg/kg	oral, QID	no	no
Syn. Prostaglandin E-1(Misoprostol/HPMC	0.100 mg/kg	oral, QID	no	not done

[a]Clearance defined as culture negative for *H. pylori* immediately after cessation of therapy; eradication defined as culture negate for *H. pylori*, 2 wk after cessation of therapy.
[b]Organisms were susceptible to Metronidazole *in vitro* prior to innoculation; organisms recovered from Metronidazole treated piglets were Metronidazole resistent *in vitro*.
[c]Erythromycin at this dose was partially effective for clearance (50%) and eradication (25%).

problem.[84] A prevalence of GEU of 5 to 100% is reported and death losses from fatal hemorrhages of 3% or more are reported (R.A. Argenzio, personal communication, 1995). Economic losses, due to subclinical hemorrhage and associated secondary disorders of anemia, anorexia and weight loss, are not known but likely substantial.[84]

In pigs, the nonglandular (esophageal) stratified squamous epithelium is 10 to 15 cells thick, extends down into the body of the stomach and is separated sharply from the glandular stomach.[85] Most investigators believe that increases in total gastric luminal acid content (gastric hyperacidity) is the proximate cause of GEU.[84-86] Hyperacidity associated lesions[84,85] in squamous epithelia range from epithelial parakeratosis (acute cellular swelling secondary to disruptions of sodium ion dysregulation)[85] and loss of superficial keratinized layers with affinity for bile salts and subsequent sloughing (preerosive) to erosions (loss of epithelia in the stratus spinosum or gastric erosions) to ulceration (penetration to or through underlying lamina propria).

Pathways for initiation and progression of GEU are not elucidated fully although reduced dietary bulk and fiber content, genetics, stressful husbandry conditions, gastric hyperacidity and resultant hypergastrinemia secondary to gastritis, acidic metabolite prone diet and microbial commensals or *H. heilmannii*[37,87,88] all are incriminated in the disorder. Perhaps, parietal cell resistance to antiacid negative feedback stimuli contribute to excess gastric lumenal hydrogen ion concentrations associated with the latter? Since 1986, 1482 piglets from 136 litters have been utilized in our studies. In these piglets, GEU occurred in

<5% of animals; lesions were mild and have never progressed to deep ulceration or hemorrhage and no fatalities attributable to GEU have occurred. (Krakowka, unpublished, 1995).

5.2. Helicobacter Like Species in Animals: Their Significance for Use in *H. pylori* Infection Models of Human Disease

Almost all domestic or wild animal species examined, including pigs, possess their own endogenous Helicobacter like microbial species. *Helicobacter sp.* infections occur in ferrets (*H. mustelae*),[29] in cheetahs (*H. acinonyx*) coinfected with *Gastrospirillum* ("*H. heilmannii*") sp.[31], in cats (*H. felis*) and dogs[30], mice (*H. hepaticus*),[38] swine[37,87,88] as (*Gastrospirillum* /*H. heilmannii*) and primates[33,46] as (*H. nemestremae* and/or *H. pylori*), the latter being indistinguishable from human isolates.

In 1989, scientists at The Procter and Gamble Company recovered and characterized 3 bacterial species from the stomachs of conventional minipigs which, by the technology available at the time were determined to be *Helicobacter sp.* In an unpublished pilot experiment, we tested the colonizing potential of 2 of these isolates in a litter of 8 gnotobiotes, 2 groups of 4 each. One group (A) received an inoculum containing a urease positive isolate plus a urease negative isolate and the other (group B) received the urease negative isolate alone. The entire GI tract was cultured (13) and 6 gastric sites (2 cardia, 1 fundus , 2 antrum and 1 pylorus) were subjected to quantitative analysis of bacterial cfu.

In group A pigs, only urease positive isolate was recovered from the stomach (3/4); less than 200 cfu were recovered from the pharynx (2/4), esophagus (2/4) and duodenum (3/4). Urease negative coinoculate was not recovered from the stomach (0/4) but instead was recovered from the lower bowel (4/4). In piglets of group B, given urease negative isolate alone, this organism was recovered from 1/4 piglets in one site in the stomach and again from the lower bowel. Neither isolate in both groups of pigs was recovered from other tissues (lung, liver, spleen and mesenteric lymph nodes) and infections were asymptomatic clinically. Subsequent 16S RNA ribotyping of these isolates confirmed the urease positive isolate was a *Helicobacter* sp., not *H. pylori*) and the urease negative isolate was likely a *Campylobacter*-like species (R.A. Leunk, personal communication, 1995).

Every investigator who proposes to use human origin *H. pylori* as the test bacterium in their chosen animal species must exclude the possibility that their test animals are absolutely devoid of endogenous "look a like" *Helicobacter* sp. organisms. Their presence fatally confounds all interpretations of lesions, immunity, vaccination as protection from infection as well as tested antimicrobials proposed for use in *H. pylori* infected humans.

5.3. Gastric Inflammation: Comparisons among Humans Infected with *H. pylori*, Mice Infected with *H. felis* and Pigs Infected with *H. pylori*

An important aspect of model development was to determine similarities/differences between humans bacterial gastritis and piglets infected with the human microbe. A brief discussion of the murine/*H. felis* model is included as mice are the immunological tools of biology, widely used in Helicobacter research and data generated in this system are often "validated" by the assumption that findings in mice are applicable to humans.

5.3.1. Gastric Inflammation in Humans. In humans, the spectrum of microscopic inflammatory lesions attributable to this bacterium varies[89,90] Histologic lesions may include lymphocytic infiltrates and lymphoid follicle development alone[90-92] and/or chronic active inflammation characterized by neutrophilic infiltrates and gland abscesses. CD4 helper T cell phenotype predominates in submucosal lamina propria; this is thought to be consistent with the prominent B cell (antibody precursor cellular sites) lymphoid follicles. Suppressor (CD8) phenotype cells are uncommon and most are located in intraepithelial sites.[92] These data are consistent with the idea that gastritis in humans associated with H. pylori infection is a TH2 dominated (antibody directed) mucosal immune response.[93] Activated mucosal TH1 lymphocytes, through release of IFN-gamma, TNF and other cytokines are epitheliotoxic,[94] stimulate goblet cell mucus production and secretion and promote crypt cell proliferation.[83,84] Phagocytic cells also are present in lamina propria and they have been implicated as a phagocytized source[96] of bacterial immunogen (urease) and as sources of the repertoire of cytokines typical of mucosal macrophages. Finally, gastric epithelia from infected humans are strongly MHC Class II positive[97] suggesting these cells, under appropriate circumstances, can substitute for antigen presenting phagocytic or B cells.

Neutrophilic infiltrates are the often cited histological hallmark of human H. pylori gastritis.[50,89,90,92,98,99] Neutrophils, their hydrolytic enzymes, released cytokines and reactive oxygen intermediates are implicated logically in epithelial damage associated with the human disease[101-104] as is the case with NSAIDs and experimental gastric ulcerogenesis.[104]

What are the possible sources of neutrophil chemotaxins? Bacterial chemotactic factors such as urease,[98] unidentified bacterial proteins[96,105] lipopolysaccharide,[106] are one possibility. Host origin inflammatory mediators also are thought to augment neutrophilic gastritis. Accumulation of neutrophils in gastric mucosae has been attributed to production of neutrophil chemotactic and activating cytokines by T cells[107] and recently, infection induced IL-8 secretion by gastric epithelia.[76,108-111]

5.3.2. Gastric Inflammation in Mice Infected with H. felis. This model, because of its inexpense and universal availability, is under extensive investigation. It is not our intention to provide a detailed description of each facet of this interesting interaction between a foreign microbe in a species with a basic gastric physiology and anatomy markedly dissimilar to humans. The gastric inflammatory response in conventional mice colonized with H. felis,[112] demonstrate that CD4 positive cells predominate in gastric mucosal lamina propria and most of the detectible CD8 cells are, like humans, intraepithelial.[113] The presence and distribution of MHC Class II positive epithelia in these mice parallels that of humans in that both enterocytes and gastric epithelia are positive.[112-115]

5.3.3. Gastric Inflammation in H. pylori Infected Gnotobiotic Piglets. A study[61] was designed to determine the distribution, numbers, anatomical arrangement and cellular phenotype(s) of gastric inflammatory cell infiltrates in H. pylori infected piglets and to compare these findings to those in H. pylori naive piglets and also to comparable data in humans and mice summarized above. Uninfected gnotobiotes or piglets contaminated with nonhelicobacter sp are devoid of visually detectable inflammatory cell/immune cell infiltrates and this fact makes attempts to recover LPL from lamina propria of controls for analysis beyond flow cytometry difficult.

In H. pylori infected piglets, both flow cytometry and immunocytochemistry demonstrated the bulk of the invading inflammatory cells in gastric lamina propria of H. pylori infected piglets were of T cell origin. The infected gnotobiotic piglet differs from humans in that the bulk of infiltrating leukocytes are of the suppressor (CD8) phenotype and only a few intraepithelial leukocytes are observed. In pigs, like humans, CD4 bearing leukocytes are associated closely with developing lymphoid (B cell) follicles. Monocyte/macrophages are present in locations and numbers consistent with their presumed antigen presenting and processing functions. Similar to mice and humans, prominent porcine MHC Class II antigen is detected in organized follicles, disorganized mononuclear cell infiltrates and vascular endothelia.

Gastric neutrophilia in infected piglets is transient[13] and possibly bacterial strain dependent (Eaton, unpublished, 1995). The immediate question in light of the above is why is this so? Diffusible[98] bacterial urease (or any other microbe derived chemotaxin for that matter) has been proposed as the pivotal chemotaxin *in vivo*. Absence of neutrophilic inflammation in piglets infected with urease producing bacteria for 4 wk or more indicate that absorption of bacterial products alone is insufficient to produce local neutrophilia. Local neutrophillic infiltrates are produced when *H. pylori* organisms are inoculated subcutaneously into naive or infected animals (Krakowka, unpublished, 1992), indicating that an intrinsic porcine neutrophil dysfunction can be excluded as an explanation for the paucity of neutrophils in the gastric compartment. Further, neutrophilic leukocytosis induced by recombinant G-CSF injections did not result in neutrophilic gastritis (Krakowka, unpublished, 1992). It is likely that porcine LPL produce endogenous chemoattractants and cytokines similar to those produced in humans and mice, particularly since all of the equivalent cell type(s) believed responsible for cytokine production are present in the porcine gastric microenvironment. Finally, a popular explanation for gastric neutrophilia is that infection induced production/secretion of gastric epithelial IL-8, the definitive neutrophil chemotaxin, is responsible for neutrophilic infiltrates. Infected piglet tissues have been stained for IL-8 and were positive (J. Crabtree, written communication, 1993). We reproduced these results using a commercially available monoclonal antibody known to react with porcine IL-8. Uninfected control pig gastric epithelia are IL-8-negative (J. Byron, unpublished, 1995). Thus, the absence of neutrophils in porcine tissues constitutes the best *a priori* evidence that additional unknown mechanisms for gastric neutrophilia are operable and must be identified.

Finally, the pig differs from humans and mice in MHC Class II antigen distribution in nonlymphoid tissues. Porcine gastric epithelia, even directly overlying inflammatory cell accumulations in the stomach[61] like porcine enterocytes elsewhere,[116] are MHC Class II-negative. The distribution of leukocytes in the stomach is similar to that recently described for the distribution of intestinal leukocyte population(s) in conventional swine.[117] MHC data in swine suggests the concept of enteric epithelium mediated antigen presentation[118] may not be applicable to all species.

5.4. Gastric Epithelial Damage Precedes Development of Gastric Ulcers Associated with *H. pylori* Infection in Both Humans and Piglets

In *H. pylori* infected humans, morphologic studies document progression of epithelial degeneration characterized by loss of apical mucus, cuboidal morphological conversion of epithelia, vacuolar cytoplasmic degeneration, focal cellular necrosis to microerosions and gastric ulceration.[119,120] In the piglet model, a similar spectrum of gastric epithelial precursor lesions occur. Using the Sydney classification system[121] as described by Wyatt,[120] nonulcerating fully developed gastric lesions in the porcine stomach would be scored as antral dominant chronic gastritis with grade 0 neutrophils, grade 2 to 3 mononuclears and grade 1 to 2 epithelial change (combinations of atrophy, degeneration, necrosis and proliferation). Chronic epithelial changes have not been systematically investigated in piglets.

An ill defined reciprocal relationship exists between mucosal epithelial surfaces such as the gastric mucosa and the underlying submucosal inflammatory cell infiltrates. We correlated gastric epithelial cell vacuolation more closely to underlying inflammation[53] rather than to the direct effects of the VacA cytotoxin.[8] Thus, vacuoles may be a consequence of inflammation/cytokine secretion.[17,83,84] Moreover we can intensify the gastric inflammatory cell response by parenteral immunization with whole formalin killed microbes.[75] This maneuver should intensify gastric epithelial lesions in infected piglets. The effects of immunization upon gastric mucosal epithelia, particularly upon IL-8 and secretory component synthesis, cellular proliferation and mucus secretion by epithelia overlying collections of infiltrating lymphocytes and neutrophils,[122,123] remain to be investigated. However, it is tempting to speculate that parenteral immunization "directs" immune responses toward the cytotoxic TH1 pathway rather than to the down regulated TH2 pathway characteristic of naturally developing immune responses on mucosal surfaces;[124] TH1 cytokines are known to be cytotoxic to intestinal epithelia.[125]

5.5. Gastric Ulceration: a Comparable Disease of *H. pylori* Infection in Humans and Piglets

Volumes have been written about human ulcer disease. In fact, "anti ulcerogenic" medications are the most widely consumed class of drugs/medications in the world. In humans, ulcers occur with *H. pylori* associated gastritis,[1,126-130] NSAID use and in patients with gastrin hypersecretory disorders. Ulcers are a bacterial infectious disease in over 90% of the human experience. However, only a few infected people (<10%) will ever develop gastric ulcers as a consequence of infection. Perhaps the more important question to ask is, "Why don't more members of the *H. pylori* infected population develop this manifestation of infection?"

Marshall[1] was the first to suggest that ulcers may a consequence of *H. pylori* associated gastritis. In humans, direct scientific proof that *H. pylori* is ulcerogenic is lacking. Epidemiologic data and concurrence studies document the coexistence of infection and gastric ulcers.[127,131,132] The strongest evidence that infection, inflammation and ulceration are linked is derived from antimicrobial trials in *H. pylori* infected patients with endoscopically confirmed ulcer disease.[128,133-135] Successful bacterial eradication is accompanied by reduced recurrence of ulcers 1 yr later; failure to eradicate the bacterium results in high ulcer relapse rates.[120,133,137] There are correlations among duration of infection, intensity of the inflammatory response, estimated bacterial burden in the stomach and propensity to ulcerogenesis.[138-141,143]

In spite of this strong association, ulcers occur in only a small fraction of infected individuals which suggests that infection alone is not responsible for the ulcerogenesis.[128,144]. It is believed that infection with this bacterium accounts for virtually all of the NSAID negative ulcerative lesions in the upper gastrointestinal tract in humans and these ulcers develop from antecedent *H. pylori* induced gastric inflammatory disease and epithelial degeneration.[89,126,131,145-148] Cofactors suspected as enhancing ulcerogenesis include ill defined psychological disorders, stress and alcohol and tobacco consumption[82] or dietary habits of high salt and low fiber. The importance of the host is exemplified by the tendency of ulcers to exhibit familial linkage in humans.[139-141]

It is compelling to view the pathophysiology of ulcer disease in a context similar to that of delayed wound healing. From this view, the role of *H. pylori* in promoting ulcerogenesis assumes a dual character. On the one hand, focus upon the aggressive tendencies of the bacterium may elucidate the mechanism(s) whereby the mucosal barrier is disrupted. On the other hand, the possibility that *H. pylori* exerts its full "ulcerogenic potential" only when infection interferes with the local wound healing process also must be considered. Bacteria (or their products) themselves may contribute to delayed wound healing through this mechanism.[90,149,150] Perhaps the critical series of events lie not in the production of damage but rather in how the mucosa responds to it? This hypothesis is amendable to testing in the piglet model.

Unlike humans and their environment, gnotobiotic piglets, even though genetically disparate, are well defined. Gastric tissue from sterile gnotobiotic piglets or from piglets inoculated with various noncolonizing strains and mutants of *H. pylori* lack histologic evidence of gastric inflammatory cell infiltrates; gastric mucosal surfaces are intact and epithelia are normal in appearance and staining properties.[13,61]. Thus, the gnotobiote can confirm or deny most ulcerogenic hypotheses by including a defined microbe free environment, absence of other known or suspected external cofactors such as risk factors of age, gender, duration of infection, NSAID usage, complex dietary composition including high (or low) fiber content, alcohol, caffeine and tobacco consumption[80,82] and even the confounding effects of other microbial contaminants or their products which might possess intrinsic coulcerogenic potential. In pigs, no obvious differences in size, gender, the nature (lymphocytic vs neutrophilic), severity or distribution (focal vs regionally diffuse) of the inflammatory cell infiltrates between ulcer bearing and ulcer free piglet are apparent.

Gnotobiotic *H. pylori* infected piglets developed erosions and ulcers, the latter detectible as early as 14 da after infection.[70] In that study, incidence of detectible epithelial lesions was higher (10 of 29, 34%) than that known for humans but the porcine lesion in most instances is less severe and has not been life threatening for at least the first 45 da of life. Though not yet proven, it seems that ulcers develop in sites of epithelial damage. In several instances, microulcers appear to develop within or over developing inflammatory foci; in others, there is no obvious relationship between ulceration and physical presence of gastric microbes. Replicate sections stained with Masson's trichrome stain for collagen as an index of fibroplasia has revealed no deposition of collagenous tissue within ulcer craters, suggesting that ulcers observed were of recent origin. This suggests that ulcerogenesis is a dynamic process in that ulcers may develop and then resolve in the absence of clinical signs and without obvious long term structural damage to gastric mucosa. It seems that local gastric defenses are sufficient to protect mucosa from most instances of damage.

5.6. Pathways of Ulcerogenesis Amendable to Dissection in *H. pylori* Infected Piglets

Three independent or covariate factors appear to be operable. Firstly, the bacterium or its products alone are ulcerogenic. Secondly, various facets of the inflammation/immune response to the bacterium or its products are ulcerogenic. Thirdly, combinations of both are operable in weighted fashion in each individual such that a unifying hypothesis regarding ulcer development will be difficult to construct. The role of inflammation in inducing preulcerogenic gastric epithelial damage is discussed above.

Bacterial infection alone, absent of nonhost influences, may be intrinsically ulcerogenic. In gnotobiotic pigs, devoid of all known confounding variables, *H. pylori* infection alone is ulcerogenic and duration of infection is not important.[70] Bacteria were demonstrated along lateral margins of ulcer craters adjacent to epithelia as both individual adherent organisms or as mucus entrapped colonies. While these data are direct proof against the necessity for the presence of ulcerogenic cofactors, contributions of the host inflammatory/immune response cannot be excluded.

The bacterium alone concept cannot be separated from host responses since it is known that these responses potentiate ulcerogenesis. For example, NSAID induced ulcers can be prevented by suppression of neutropoiesis.[104] In humans, ulcers are more likely to occur in areas of neutrophilic gastritis; others[149] reported no correlation between the severity of neutrophilic infiltrates and ulcer development. Instead, severity of lymphofollicular gastritis is linked most directly to ulcers.[149] Urease enzyme is antigenic, absorbed intact or as antigenic determinants into damaged gastric lamina propria and is implicated as a major target of host inflammatory responses.[150] Finally, ulcers may arise as a consequence of failure of the wound healing process in the presence of infection associated epitheliotoxic events. All of these are testable hypotheses in the piglet model.

With justification, investigators have focused upon bacterial factors suspected of being ulcerogenic principles. Bacterial sonicates are epitheliotoxic in rodents. *In vivo* and *in vitro* studies have focused upon epitheliotoxic (ulcerogenic) bacterial products and metabolites.[96,151-153] Epidemiologic data suggest *in vivo* expression of VacA cytotoxin, CagA gene or its product and/or "high" bacterial loads in the stomach[139] promote ulcerogenesis. Proof that cagA gene or its product is essential for ulcerogenesis awaits construction and *in vivo* testing of virulent isogenic mutants as was accomplished recently for urease genes of strain N6 *H. pylori*.[55] Many strains secrete an 87 kDa cytotoxin designated by its cloned gene, VacA,[9] which causes vacuolation of cultured epithelial cells *in vitro*[7,8] and is produced *in vivo*.[154] A second 120 kDa cytotoxin associated glycoprotein (CAG, gene designated CagA) is expressed *in vivo* and its presence/expression is associated with ulcerogenesis.[156] Bacterial supernatants containing these proteins retard epithelial cell proliferation [150] *in vitro* and may delay ulcer healing by lavage *in vivo*.[149]

In addition to its essential, but still unelucidated role in gastric colonization, bacterial urease has received attention as the bacterial ulcerogenic entity. Initially, it was hypothesized that urease served to protect bacteria during transit through gastric acid to its preferred niche, the extracellular gastric mucus layer, by creating an ammonia "cloud" of neutral pH via released products of urea hydrolysis.[157] This sequence has been reproduced *in vitro*.[8] Urease, by virtue of its ability to hydrolyse urea to release ammonia, produces ammonia, an epitheliotoxic metabolite.[157] It follows that urease negative mutants should not be ulcerogenic. However, these mutants will not even colonize the stomach, even in a pH neutral gastric environment,[55] and supplemental urea in diet did not promote ulcerogenesis. Further, if the urease associated pH protection via generation of ammonia hypothesis is true, uremic patients should be at risk for ulcerogenesis; they are not. *In vitro*, urease negative mutants exhibit growth curves identical to their wild type counterparts and they adhere to gastric epithelial explants in a fashion identical to parental strains[158] indicating that urease does not function as an essential bacterial adhesin.

This organism's preferred ecologic niche is gastric mucus. Active motility facilitates penetration and residence in this space and the microbe has been shown to inhibit or alter mucus production[13,159,160] and produce/secrete mucolysins.[159] It is difficult to believe that therapies designed to prevent colonization of this environment, by providing excess free mucin binding moieties, would be successful, given the intimate but not obligatory relationship between mucus, gastric epithelia and bacteria.

As first demonstrated in patients with gastrin secreting neoplasms and the Zollinger-Edison Syndrome, excessive acid production by parietal cells is ulcerogenic. Thus, it is logical to expect that ulcerogenesis in *H. pylori* associated disease should be accompanied by disrupted, i.e., elevated, gastrin regulatory pathways. In support, human infections frequently are accompanied by hypochlorhydria. In piglets, gastrin levels, at least that detected in serum, is an epiphenomenon unrelated to initiation of ulcerogenesis.

5.7. Chronic Manifestations of Infection Ill Suited to Detailed Study in Gnotobiotic Piglets: Gastric Carcinoma and MALT Lymphoma

Like ulcers, Marshall[1] was the first to suggest that gastric cancer may be a consequence of *H. pylori* associated gastritis. Correa and colleagues provided[161-163] crucial support for this idea through a combination of careful epidemiologic and pathologic studies. Epidemiologic studies[162,164-166] have provided compelling evidence for this association, particularly with the intestinal type gastric carcinoma, a well described entity epidemiologically linked to gastric inflammation of progressive epithelial preneoplastic changes (loss of normal gastric glandular structure) atrophy to intestinal metaplasia to regenerative hyperplasia with dysplasia to neoplasia.[162,163] Recently, Fox and colleagues[38] associated murine *Helicobacter hepaticus* with hepatocellular carcinoma. The role of *H. pylori*, and the associated inflammatory/immune responses in this carcinogenic process, is unknown but is an area of active investigation. The gnotobiotic piglet infected with human *H. pylori*, is the only reliable nonprimate animal species model available. This model is useful in 1) evaluation of bacterial colonization/virulence factors, 2) host immune responses in a genetically outbred omnivore species and 3) therapeutic approaches to antimicrobial investigation. However, it is a short term model poorly suited for long term carcinoma research.

In humans, gastric low grade B cell MALT lymphoma also is associated with *H. pylori* colonization.[26-28] Although, in the end, tumors are likely to represent the full spectrum of low malignancy lymphatic neoplastic disease, it appears that most gastric MALT neoplasms are B cell origin tumors which, if Ig positive, may secrete microbe specific antibody. Even more intriguing is the finding that the T cell infiltrates in these neoplasms are *H. pylori* specific cells which secrete B cell growth/proliferation cytokines. In a portion of these patients, successful eradication of the microbe has resulted in MALT regression and "cure". However, the same limitations for application of the piglet model to this aspect of *H. pylori* disease, unless confined to the earliest of preneoplastic events, are identical to those of carcinoma.

5.8. Immunoprophylaxis and the Gnotobiotic Piglet Model

Currently, there is interest in creating a "vaccine" for this disease although there is controversy regarding who should be immunized and what are the risk factors.[167] The logical recipient of such a product would be third world subjects in situations where the infection is endemic and acquired at an early age. However, limited financial resources available in many of these developing countries and more pressing health concerns preclude effective delivery of an immunogen to a population which may be of benefit only beyond the 4th or 5th decade of life. Moreover, in the absence of a readily defined population of humans who have overcome the infection without antimicrobial therapy, it is difficult to design a vaccination protocol or product which mimics a naturally occurring successful phenomenon. It is cost prohibitive to pursue extensive investigations into defining the best candidate protective immunogen (urease, flagellin, heat shock protein), proper adjuvant, dose and formulation in piglets in the absence of novel potentially successful vaccination strategy or an unique protective immunogen.

In spite of these caveats, published results using *H. felis* infected mice[168,169] or ferrets[170] as test subjects have demonstrated the theoretical feasibility of this approach. Realistically, the modest successes engendered thus far have employed cholera toxin as the orally administered adjuvant at levels toxic to many species except rodents or have utilized parenteral immunization without accounting for the likelihood of inducing severe inflammatory lesions in immune gastric mucosa.[75] Nonetheless in piglets, it is possible to test directed mucosal immunization schemes by taking advantage of the presence of the gastric mucosa associated lymphoid tissue (MALT) system recently identified by our laboratory.[73]

5.9. Antimicrobials and the Gnotobiotic Piglet Model

Summarized in Table 2 are data obtained with *H. pylori* 26695 in gnotobiotic piglets. These experiments represent approximately 500 animals divided for treatment into statistically defensible groups, always including a group of infection alone controls which receive no treatment or the vehicle alone if appropriate. The posttreatment observation interval of 2 wk is not ideal and does not approach the 3 to 6 mo interval recommended in human clinical trials.

6. ACKNOWLEDGMENTS

The authors wish to acknowledge the excellent technical assistance of Judith Younger, Bryan Kessler, Susan Ringler, Nancy J. Hughey, Steven Spencer and Michelle Roberts Trumble of The Ohio State University and Drs. R. A. Leunk, C. Catrenich, D. R. Morgan and Mr. John Berman of The Procter and Gamble Co. Dr. W. B. Green performed the gastrin immuno-assays as a portion of a Master's Degree Dissertation, 1994. Supported in part by research and training grants, DK39570-01A3, A107938-02, R29DK45340, R01 CA67498, NIH, PHS and grants from The Procter and Gamble Co.

7. REFERENCES

1. Marshall, B.J., and Warren, J.R., 1984, Unidentified curved bacilli in the stomach of patients with gastritis and peptic ulceration, *Lancet* i:1311-1314.

2. Hazell, S.L., 1991, Urease and catalase as virulence factors of *Helicobacter pylori*, in: *Helicobacter Pylori*, (H. Menge, ed.), Springer-Verlag, Berlin, 3-14.

3. Labigne, A., Cussac, V., and Courcoux P., 1991, Shuttle cloning and nucleotide sequences of *Helicobacter pylori* genes responsible for urease activity, *J. Bacteriol.* 173:1920-1931.

4. Marshall, D.G., Chua, A., Kneeling, P.W.N., Sullivan, D.J., Coleman, D.C., and Smyth, C.J., 1995, Molecular analysis of *Helicobacter pylori* populations in antral biopsies from individual patients using randomly amplified polymorphic DNA (RAPD) fingerprinting, *FEMS, Immunol. Med. Microbiol.* 10:3-4.

5. Blaser, M.J., 1992, Hypotheses on the pathogenesis and natural history of *Helicobacter pylori*-induced inflammation, *Gasteroenterol.* 102:720-727.

6. Nwokolo, C.U., Bickley, J., Attard, A.R., Owen, R.J., Costas, M., and Fraser, I. A., 1992, Evidence of clonal variants of *Helicobacter pylori* in three generations of a duodenal ulcer disease family, *Gut* 133:1323-1327.

7. Leunk, R.D., Johnson, P.T., David, B.C., Kraft, W.G., and Morgan, D.R., 1988, Cytotoxic activity in broth culture filtrates of *Campylobacter pylori*, *J. Med. Microbiol.* 26:93-99.

8. Cover, T.L., Puryear, W., Perez-Perez, G.I., and Blaser, M.J., 1991, Effect of urease on HeLa cell vacuolation induced by *Helicobacter pylori* cytotoxin, *Infect. Immun.* 59:1264-1270.

9. Tummuru, M.K.R., Cover, T.L., and Blaser, M.J., 1993, Cloning and expression of a high-molecular-mass major antigen of *Helicobacter pylori*: Evidence of linkage to cytotoxin production, *Infect. Immun.* 61:1799-1809.

10. Stolte, M., and Eidt S., 1989, Lymphoid follicles in antral mucosa: Immune response to *Campylobacter pylori*? *J. Clin. Pathol.* 42:1269-1271.

11. Marshall, B.J., Armstrong, J.A., McGechie, D.B., and Glancy, R.J., 1985, Attempt to fulfill Koch's postulates for pyloric campylobacter, *Med. J. Aust.* 142:436-439.

12. Morris, A.J., Ali, R., Nicholson, G.I., Perez-Perez, G.I., and Blaser, M.J., 1991, Long-term follow up of voluntary ingestion of *Helicobacter pylori*, *Ann. Int. Med.* 114:662-663.

13. Krakowka, S., Morgan, D.R., Kraft, W.O., and Leunk, R., 1987, Establishment of gastric *Campylobacter pylori* infection in the neonatal piglet, *Infect. Immun.* 55:2789-2796.

14. Lambert, J.R., Borromeo, M., Pinkard, K.J., Turner, H., Chapman, C.B., and Smith, M.L., 1987, Colonization of gnotobiotic piglets with *Campylobacter pyloridis* - An animal model, *J. Inf. Dis.* 155:1344.

15. Prichard, P.J., and Yeomans, N.D., 1991, *Helicobacter pylori* and duodenal ulcer, *J. Gastroent. Hepatol.* 6:177-178.

16. Peterson, W.L., 1991, *Helicobacter pylori* and peptic ulcer disease, *N. Eng. J. Med.* 324:1043-1047.

17. Sipponen, P., 1991, *Helicobacter pylori* and chronic gastritis: An increased risk of peptic ulcer? A review, *Scand. J. Gastroenterol.* 26:6-10.

18. Moss, S., and Calam, J., 1992, *Helicobacter pylori* and peptic ulcers: The present position, *Gut* 33:289-292.

19. Oderda, G., Vaira, D., Ainley, C., Holton, J., Osborn, J., Altare, F., and Ansaldi, N., 1992, Eighteen month follow up of *Helicobacter pylori* positive children treated with amoxycillin and tinidazole, *Gut* 33:1328-1330.

20. Forbes, G.M., Glaser, M.E., Cullen, D.J.E., Warren, J.R., Christianses, K. J., Marshall, B.J., and Collins, B.J., 1994, Duodenal ulcer treated with *Helicobacter pylori* eradication: Seven-year follow-up, *Lancet* 343:258-260.

21. Sipponen, P., and Hyvarinen, H., 1993, Role of *Helicobacter pylori* in the pathogenesis of gastritis, peptic ulcer and gastric cancer, *Scand. J. Gastroenterol.* 28(Suppl 196):3-6.

22. Nomura, A., Stemmermann, G.N., Chyou, P.-H., Kato, I., Perez-Perez, G.F., and Blaser, M.J., 1991, *Helicobacter pylori* infection and gastric carcinoma among Japanese Americans in Hawaii, *N. Engl. J. Med.* 325:1132-1136.

23. Parsonnet, J., Vandersteen, D., Goates, J., Sibley, R.K., Pritikin, J., and Chang, Y., 1991, *Helicobacter pylori* infection in intestinal and diffuse-type gastric adenocarcinomas, *J. Natl. Canc. Inst.* 83:640-643.

24. Forman, D., Newell, D.G., Fullerton, F., Yarnell, J.W.G., Stacey, A.R., Wald, N., and Silas, F., 1991, Association between infection with *Helicobacter pylori* and risk of gastric cancer: Evidence from a prospective investigation, *Br. Med. J.* 302:1302-1305.

25. Sipponen, P., 1992, *Helicobacter pylori* infection - A common worldwide environmental risk factor for gastric cancer? *Endos.* 24:424-426.

26. Hussell, T., Isaacson, P.G., Crabtree, J.E., Spencer, J., 1993, The response of cells from low-grade B-cell gastric lymphomas of mucosa-associated lymphoid tissue to *Helicobacter pylori*, *Lancet* 342:7571-7574.

27. Rodriguez, L.V., Hansen, S., Gelb, A., Friedman, G., Warnke, R.A., Jellum, E., and Parsonnet, J., 1993, *Helicobacter pylori* and gastric lymphoma: A nested case-control study, *Acta Gastroenterol. Belgium* 56(suppl):47.

28. Eidt, S., Stolte, M., and Fischer, R., 1994, *Helicobacter pylori* gastritis and primary gastric non-Hodgkins lymphomas, *J. Clin. Pathol.* 47:436-439.

29. Fox, J.G., Cabot, E.B., Taylor, N.S., and Laraway, R., 1988, Gastric colonization of *Campylobacter pylori* subsp. *mustelae* in ferrets, *Infect. Immun.* 56:2994-2996.

30. Lee, A., Hazell, S.L., O'Rourke, J., and Kouprach, S., 1988, Isolation of a spiral-shaped bacterium from the cat stomach, *Infect. Immun.* 56:2843-2850.

31. Eaton, K.A., Radin, M.J., Kramer, L., Wack, R., Sherding, R., Krakowka, S., and Morgan, D.R., 1993, Epizootic gastritis in cheetahs associated with gastric spiral bacilli, *Vet. Pathol.* 30:55-63.

32. Eaton, K.A., Radin, M.J., Kramer, L., Wack, R., Sherding, R., Krakowka, S., and Morgan, D. R., 1991, Gastric spiral bacilli in captive cheetahs, *Scand. J. Gastroenterol.* 26:38-42.

33. Euler, A.R., Zarenko, G.E., Moe, J.B., Ulrich, R.G., and Yagi, Y., 1990, Evaluation of two monkey species (*Macaca mulatta* and *Macaca fascicularis*) as possible models for human *Helicobacter pylori* disease, *J. Clin. Microb.* 28:2285-2290.

34. McNulty, C.A.M., Dent, J.C., Curry, A., Uff, J.S., Ford, G.A., Gear, G.W.L., and Wilkinson, S.P., 1989, New spiral bacterium in gastric mucosa, *J. Clin. Pathol.* 42:585-591.

35. Reed, K.D., and Berridge, B.R., 1988, Campylobacter-like organisms in the gastric mucosa of Rhesus monkeys, *Lab. Anim. Sci.* 38:329-331.

36. Henry, G.A., Long, P.H., Barns, J.L., and Charbonneau, D.L., 1987, Gastric spirillosis in Beagles, *Am. J. Vet. Res.* 48:831-836.

37. Mendes, E.N., Queroz, D.M.M., Rocha, G.A., Moura, S.B., Leite, V.H.R., and Fonseca, M.E.F., 1990, Ultrastructure of a spiral micro-organism from pig gastric mucosa (*"Gastrospirillum hominis"*), *J. Med. Microbiol.* 33:61-66.

38. Fox, J.G., Dewhirst, F.E., Paster, B.J., Yan, L., Taylor, N.S. Collins, M.J., Gorelick, P.L., and Ward, J.M., 1994, *Helicobacter hepaticus* sp. nov., a microaerophilic bacterium isolated from livers and intestinal mucosal scrapings from mice, *J. Clin. Microbiol.* 32:1238-1245.

39. Fox, J.G., Yan, L.L., Dewhirst, F.E., Paster, B.J., Shames, B., Murphy, J.C., Hayward, A., Belcher, J.C., and Mendes, E.N., 1995, *Helicobacter bilis* sp nov, a novel Helicobacter species isolated from bile, livers, and intestines of aged, inbred mice, *J. Clin. Microbiol.* 33:225-454.

40. Goodwin, C.S., Armstrong, J.A., and Marshall, B.J., 1986, *Campylobacter pyloridis*, gastritis and peptic ulceration, *J. Clin. Pathol.* 39:353-365.

41. Cantovna, M.T., and Balish, E., 1990, Inability of human clinical strains of *Helicobacter pylori* to colonize the ailmentary tract of germfree rodents, *Can. J. Microbiol.* 36:237-241.

42. Handt, L.K., Fox, J.G., Dewhirst, F.E., Fraser, G.J., Paster, B.J., Yan, L.L., Rozmiarek, H., Rufo, R., and Stalis, I.H., 1994, *Helicobacter pylori* isolated from the domestic cat: Public health implications, *Infect.Immun.* 62:2367-2374.

43. Marchetti, M.B., Arico, D., Burroni, N., Figura, R., Rappouli, R., and Ghira, P., 1995, Development of a mouse model of *Helicobacter pylori* infection that mimics human disease, *Science* 267:1655-1658.

44. Radin, M.J., Eaton, K.A., Krakowka, S., Morgan, D.R., Lee, A., Otto, G., and Fox, J.G., 1990, *Campylobacter pylori* gastric infection in gnotobiotic Beagle dogs, *Infect.Immun.* 58:2606-2612.

45. Baskerville, A. and Newell, D.G., 1988, Naturally occurring chronic gastritis and *C.pylori* infection in the Rhesus monkey: A potential model for gastritis in man, *Gut* 29:465-472.

46. Dubois, A., Fiala, N., Heman-Ackah, L.M., Drazek, E.S., Tarnawski, A., Fishbein, W.N., Perez-Perez, G.I., and Blaser, M.J., 1994, Natural gastric infection with *Helicobacter pylori* in monkeys: A model for spiral bacteria infection in humans, *Gasterenterol.* 106:1405-1417.

47. Engstrand, L., Gustavsson, S., Jorgensen, A., Schwan, A., and Scheynius, A., 1990, Inoculation of barrier born pigs with *Helicobacter pylori*: A useful animal model for gastritis type B, *Infect. Immun.* 58:1763-1768.

48. Engstrand, L., Rosberg, K., Hubinette, R., Berglindh, T., Rolfsen, W., and Gustavsson, S., 1992, Topographic mapping of *Helicobacter pylori* colonization in long-term-infected pigs, *Infect. Immun.* 60:653-656.

49. Akopyants, N.S., Eaton, K.A., and Berg, D.E., 1994, Adaptive mutation and cocolonization during *Helicobacter pylori* infection of gnotobiotic piglets, *Infect Immun.* 63:116-121.

50. Sobala, G.M., Crabtree, J.E., Dixon, M.F., Schorah, C.J., Taylor, J.D., Rathbone, B.J., Heatley, R.V., and Axon, A.T.R., 1991, Acute *Helicobacter pylori* infection: Clinical features, local and systemic response. Gastric mucosal histology and gastric juice ascorbic acid concentrations, *Gut* 32:1415-1418.

51. Green, W.B., Eaton, K.A., and Krakowka, S., 1995, Tissue distribution of specific immunoglobulin synthesis in the gnotobiotic piglet model of *H. pylori*: infection. Am. of Gastroenterol. 89 (217):51339.

51. Taylor, D.N., and Blaser, M.J., 1991, The epidemiology of *Helicobacter pylori* infection, *Epidem. Rev.* 13:42-59.

52. Hutter, C., and Poulter, L.W., 1992, The balance of macrophage subsets may be customised at mucosal surfaces, *FEMS Microbiol. Immunol.* 105:309-316.

53. Eaton, K.A., Morgan, D.R., and Krakowka, S., 1989, *Campylobacter pylori* virulence factors in gnotobiotic piglets, *Infect. Immun.* 57:1119-1125.

54. Eaton, K.A., Morgan, D.R., and Krakowka, S., 1992, Motility as a factor in the colonization of gnotobiotic piglets by *Helicobacter pylori*, *J. Med. Micro.* 37:123-127.

55. Eaton, K.A, and Krakowka, S., 1994, Effect of gastric pH on urease-dependent colonization of gnotobiotic piglets by *Helicobacter pylori*, *Infect. Immun.* 62:3604-3607.

56. Eaton, K.A., and Krakowka, S., 1991, Essential role of urease in the pathogenesis of gastritis induced by *Helicobacter pylori* in gnotobiotic piglets, *Infect. Immun.* 59:2470-2475.

57. Eaton, K.A., Morgan, D.R., and Krakowka, S., 1992, Motility as a factor in the colonization of gnotobiotic piglets by *Helicobacter pylori*, *J. Med. Micro.* 37:487-494.

58. Eaton, K.A., Catrenich, C.A., Makin, K.M., and Krakowka, S., 1995, Virulence of coccoid and bacillary forms of *Helicobacter pylori* in gnotobiotic piglets, *J. Inf. Dis.* 171:459-462.

60. Green, W.B., Eaton, K.A., and Krakowka, S., 1996, Porcine gastric mucosa associated lymphoid tissue (MALT): Stimulation by gastric infection with *Helicobacter pylori*, Vet. Immun. Immunopathol. In Press.

61. Krakowka, S., Ringler, S., Eaton, K.A., Green, W.B., and Leunk, R., 1996, Analysis of the gastric inflmmatory response in gnotobiotic piglets infected with *Helicobacter pylori*, *Vet. Immun. Immunopathol.* 52: 159-173.

62. Eaton, K.A., Morgan, D.R., and Krakowka, S., 1990, Persistence of *Helicobacter pylori* in conventionalized piglets, *J. Inf. Dis.* 161:1222-1301.

63. Krakowka, S., Eaton, K., and Morgan, D.R., 1991, *Campylobacter pylori* gastritis in gnotobiotic piglets: Persistence of gastric infection after conventionalization, in: *Campylobacter*, (G.M. Ruiz-Palacios, E. Calva, and B.R. Ruiz-Palacios, eds.), pp. 350-353.

64. Krakowka, S, Eaton, K.A., and Morgan, D.R., 1991, Gastritis induced by *Helicobacter pylori* in gnotobiotic piglets, *Rev. Inf. Dis.* 13:S681-685.

65. Bull, D.M., and Bookman, M.A., 1977, Isolation and functional characterization of human intestinal mucosal lymphoid cells, *J. Clin. Invest.* 59:966-974.

67. Gibson-D'Ambrosio, R.E., Samuel, M. and D'Ambrosio, S.M., 1986, A method for isolating large numbers of viable disaggregated cells from various human tissues for cell culture establishment, *In Vitro Cell. Devel. Biol.* 22:529-534.

68. Wilson, A.D., Stokes, C.R., and Bourne, F.J., 1986, Responses of intraepithelial lymphocyte to T-cell mitogens: A comparison between murine and porcine responses, *Immunol.* 58:621-625.

70. Krakowka, S., Eaton, K.A., and Rings, D.M., 1995, Occurence of gastric ulcers in gnotobiotic piglets colonized by *Helicobacter pylori, Infect. Immun.* 63:2352-2355.

71. Owen, D.A., 1986, Normal histology of the stomach, *Am. J. Surg. Pathol.* 10:48-61.

72. Genta, R.M., Lew, G.M., and Graham, D.Y., 1993, Changes in the gastric mucosa following eradication of *Helicobacter pylori, Mod. Pathol.* 60:281-289.

73. Mooney, C., Keenan, J., Munster, D., Wilson, I., Allaedyce, R., Bagshaw, P., Chapman, B., and Chadwich, V., 1991, Neutrophil activation by *Helicobacter pylori, Gut* 32:853-857.

74. Krakowka, S., Morgan, D.R., Eaton, L.A., and Radin, M.J., 1991, Animal models of *Camplylobacter pylori* gastritis, in: *Helicobacter pylori*, (H. Menge, M. Gregor, G.N.J. Tytgat, B.J. Marshall, and C.A.M. McNulty, eds.), Springer-Verlag, Berlin, pp. 74-82.

75. Eaton, K. A., and Krakowka, S., 1992, Chronic active gastritis due to *Helicobacter pylori* immunized gnotobiotic giglets, *Gastroenterol.* 103:1580-1586.

76. Crabtree, J.E., Covacci, A., Farmery, S.M., Xaing, Z., Tompkins, D.S.S., Perry, S., Lindley, I.J.D. and Rappuoli, R.R., 1995, *Helicobacter pylori* induced interleukin-8 expression in gastric epithelial cells is associated with CagA positive phenotype, *J. Clin. Pathol.* 48:41-45.

77. Rudmann, D.G., Eaton, K.A., and Krakowka, S., 1992, Ultrastructural study of *Helicobacter pylori* adherence properties in gnotobiotic piglets, *Infect. Immun.* 60:2121-2124.

79. Gorczyca, W., Tuziak, T., Krakm, A., Melamed, M.R., and Darzynkiewicz, Z., 1994, Detection of apoptosis-associated DNA strand breaks in fine-needle aspiration biopsies by *in situ* end labeling of fragmented DNA, *Cytometry* 15:169-175.

80. Marotta, R.B., and Floch, M.H., 1991, Diet and nutrition in ulcer disease, *Med. Clin. of N. Amer.* 75:967-979.

81. Mertz, H.R., and Walsh, J.H., 1991, Peptic ulcer pathophysiology, *Med. Clin. N. Am.* 75:799-814.

82. Friedman, G.D., Siegelaub, A.B., and Seltzer, C.C., 1974, Cigarettes, alcohol, coffee and peptic ulcer, *N. Eng. J. Med.* 290:469-473.

83. Krakowka, S., Eaton, K.A., and Morgan, D.R., 1991, Gastritis induced by *Helicobacter pylori* in gnotobiotic piglets, *Rev. Inf. Disease* 13:S681-S685.

89. Dixon, M., Wyatt, J., Burke, D., and Rathbone, B., 1988, Lymphocytic gastritis-relationship to *Campylobacter pylori* infection, *J. Pathol.* 154:125-132.

90. Dixon, M.F., 1994, Pathophysiology of *Helicobacter pylori* infection, *Scand J. Gastroenterol.* 29(Suppl 201):7-10.

91. Nagata, S., and Golstein, P., 1995, The fas death factor, *Science* 67:1449-1445.

92. Bertram, T., Krakowka, S., and Morgan, D., 1991, Gastritis associated with infection by *Helicobacter pylori*: Comparative pathology in humans and swine, *Rev. Infect. Dis.* 13:S714-722.

93. Valnes, K., Brandtzaeg, P., Elgjo, K., and Stave, R., 1986, Quantitative distribution of immunoglobulin-producing cells in gastric mucosa: Relation to chronic gastritis and glandular atrophy, *Gut* 27:505-514.

94. Mathews, J.B., and Garner, A., 1991, Review article: Stomach wars - a mucosal defense initative, *Ailment. Pharmacol. Therap.* 5:1311-1314.

95. Pruul, H., Lee, P.C., Goodwin, C.S., and Macdonald, P.J., 1987, Interaction of *Campylobacter pyloridis* with human immune defense mechanisms, *J. Med. Microbiol.* 23:233-238.

96. Mai, U., Perez-Perez, G., Allen, J., Wahl, S., Blaser, M., and Smith, P., 1992, Surface proteins from *Helicobacter pylori* exhibit chemotactic activity for human leukocytes and are present in gastric mucosa, *J. Exptl. Med.* 175:517-525.

97. Pavila, P., Hume, D.A., Van De Pol, E., and Doe, W.F., 1993, Dendritic cells, the major antigen-presenting cells of the colonic lamina propria, *Immunol.* 78:132-141.

98. Blaser, M., 1992, Hypotheses on the pathogenesis and natural history of *Helicobacter pylori*-induced inflammation, *Gastroenterol.* 102:720-727.

99. Buck, G., 1990, *Campylobacter pylori* and gastroduodenal disease, *Clin. Microbiol. Rev.* 3:1-12.

100. Madan, E., Kemp, J., Westblom, T.U., Chaffin, C., and Foster, A., 1990, Histologic characteristics of *Campylobacter pylori* (*Helicobacter pylori*) mediated gastritis, *Ann. Clin. Lab. Sci.* 20:329-336.

101. Nielsen, H., and Andersen, L., 1992, Chemotactic activity of *Helicobacter pylori* sonicate for human polymorphonuclear leucocytes and monocytes, *Gut* 33:738-742.

102. Reymunde, A., Deren, M.D., Nachamkin, I., Oppenheim, D., and Weinbaum, G., 1993, Production of chemoattractant by Helicobacter pylori, *Digest. Dis. Sci.* 38:1697-1701.

103. Rautelin, H., Blomberg, B., Fredlund, H., Jarnerot, G., and Danielsson, D., 1993, Incidence of *Helicobacter pylori* activating neutrophils in patients with peptic ulcer disease, *Gut* 34:599-603.

104. Wallace, J.L., Keenan, G.M., and Granger, N., 1990, Gastric ulceration induced by non-steroidal anti-inflammatory drugs is a neutrophil-dependent process, *Am. J. Physiol.* 259:462-467.

105. Norgaard, A., Andersen, L.P., and Nielson, N.D.H., 1995, Neutrophil degranulation by *H. pylori*, *Gut* 36:354-357.

106. Moutiala, A., Helander, L., Pyhala, L., Kosunen, T., and Moran, A., 1992, Low biological activity of *Helicobacter pylori* liposaccharide, *Infect. Immun.* 60;1714-1716.

107. MacDonald, T.T., and Spencer, J., 1990, The role of activated T cells in transformed intestinal mucosa, *Digest.* 46:290-296.

108. Crabtree, J.E., Peichl, P., Wyatt, J.I., Stachl, U., and Lindley, I.J. D., 1993, Gastric interleukin-8 and IgA autoantibodies in *Helicobacter pylori* infection, *Scand. J. Immunol.* 37:65-70.

109. Davies, G.R., and Crabtree, J.E., 1994, *Helicobacter pylori*: Trick or treat? *J. Roy. Soc. Med.* 87:436-439.

110. Fan, X.G., Chua, A., Fan, X.J., and Keeling, P.W.N., 1995, Increased gastric production of interleukin-8 and tumor necrosis factor in patients with *Helicobacter pylori* infection, *J. Clin. Pathol.* 48:133-136.

111. Noach, L.A., Bosma, N.B., Jansen, J., Hoek, F.J., vanDeventer, S.J.H., and Tytgat, G.N.J., 1994, Mucosal tumor necrosis factor-a, interleukin IB and interleukin-8 production in patients with *Helicobacter pylori* infection, *Scand. J. Gastroenterol.* 29:425-429.

112. Fox, J., Blanco, M., Murphy, J., Taylor, J.N., Lee, A., Kabok, Z., and Pappo, J., 1993, Local and systemic immune responses in murine *Helicobacter felis* active chronic gastritis, *Infect. Immun.* 61:2309-2315.

113. Papadimitriou, C., Elli, I.-V., Tsianos E., and Moutsopoulos H., 1988, Epithelial HLA-DR expression and lymphocyte subsets in gastric mucosa in type B chronic gastritis, *Vir. Arch. Pathol. Anat.* 413:197-204.

114. Engstrand, L., Scheynius, A., Pahison, L., Grimelius, L., Schwan, A., and Gustavsson, S., 1989, Association of *Campylobacter pylori* with induced expression of class II transplantation antigens on gastric epithelial cells, *Infect. Immun.* 57:827-832.

115. Valnes, K., Huitfeldt, H.S., and Brandzaeg, P., 1990, Relation between T cell number and epithelial HLA class II expression quantified by image analysis in normal and inflamed human gastric mucosa, *Gut* 31:647-652.

116. Weiner, H.L., Friedman, A., Miller, A., Khoury, S.J., Al-sabbagh, A., Santos, L., Sayegh, M., Nussenblatt, R.B., Trentham, D.E., and Hafler, D.A., 1994, Oral tolerance: Immunologic mechanisms and treatment of animal and human organ-specific autoimmune diseases by oral administration of autoantigens, *Ann. Rev. Immunol.*12;809-837.

117. Vega-Lopez, M., Telemo, E., Bailey, M., Stevens, K., and Stokes, C., 1993, Immune cell distribution in the small intestine of the pig: Immunohistochemical evidence for an organized compartmentalization in the lamina propria, *Vet. Immunol. Immunopathol.* 37:49-60.

118. Kaiserlain, D., and Vidal, K., 1993, Antigen presentation by intestinal epithelial cells, *Immunol. Today* 14:115.

119. Dixon, M.F., Wyatt, J.L., Baarke, D.A., and Rathbone, B.J., 1988, Lymphocytic gastritis -relationship to *Campylobacter pylori* infection, *J. Pathol.* 154:125-132.

120. Wyatt, J.I., 1995, Histopathology of gastroduodenal inflammation: The impact of *Helicobacter pylori*, *Histopathol.* 26:1-15.

121. Price, A.B., 1991, The Sydney system: Histological division, *J. Gastroenterol. Hepatol.* 6:209-222.

122. Campbell, P.A., 1990, The neutrophil, a professional killer of bacteria may be controloled by T cells, *Clin. Exp. Immunol.* 79:141-143.

123. Goodman, 1992, Molecular cloning of porcine alveolar macrophage-derived neutrophil chemotactic factors I and II: Identification of porcine IL-8 and another intercrine-alpha protein, *Biochem.* 31:10483-10490.

124. Deem, R.L., Shanahan, F., Targan, S.R., 1991, Triggered human mucosal T cells release tumour necrosis factor alpha and interferon-gamma which kill human colonic epithelial cells, *Clin. Exp. Immunol.* 83:79-84.

125. Karttunen, R.G., Andersson, K., Poikonen, T.U., Kosunen, T., Kartunen, K., Juuniten, K., and Niemala, S., 1990, *Helicobacter pylori* induce lymphocyte activation in peripheral blood cultures, *Clin. Exp. Immunol.* 82:485-488.

126. Alper, J., 1993., Ulcers as an infectious disease, *Science* 260:159-160.

127. Morfitt, D.C., and Pohlenz, J.F.M., 1989, Porcine colonic lymphoglandular complex: Distribution, structure and epithelium, *Am. J. Anat.* 184:41-51.

128. Graham, D.Y., 1991, *Helicobacter pylori*: Its epidemiology and its role in duodenal ulcer disease, *J. Gastroenterol. Hepatol.* 6:105-113.

129. Lee, A., 1993, *H. pylori*-initiated ulcerogenesis: Look to the host, *Lancet* 341:280-281.

130. Peterson, W.L., 1991, *Helicobacter pylori* and peptic ulcer disease, *N. Eng. J. Med.* 324:1043-1048.

131. Blaser, M.J., and Parsonnet J., 1994, Parasitism by the "slow" bacterium *Helicobacter pylori* leads to altered gastric homeostasis and neoplasia, *J. Clin. Invest.* 94:4-8.

132. Buck, G.E., 1990, *Campylobacter pylori* and gastroduodenal disease, *Clin. Microbiol. Rev.* 3:1-12.

133. Graham, D.Y., Lew, G.M., Klein, P.D., Evans, D.G., Evans, J.D., Jr., Saeed, Z.A., and, Malaty, H.F., 1992, Effect of treatment of *Helicobacter pylori* infection on long-term recurrence of gastric or duodenal ulcer; a randomized controlled study, *Ann. Int. Med.* 116:705-708.

134. Hassal, E., and Dimmick, J.E., 1991, Unique features of *Helicobacter pylori* disease in childern, *Dig. Dis. Sci.* 36:417-423.

135. NIH consensus statement, 1994, *Helicobacter pylori* in peptic ulcer disease, Feb. 7.

136. Hentschel, E.G., Brandstatter, B., Dragosics, A.M., Hirschl, H., Nemec, K., Schutze, M., and Wurzer, H., 1993, Effect of ranitidine and amoxicillin plus metronidazole on the eradication of *H. pylori* and recurrence of duodenal ulcer, *N. Eng. J. Med.* 328:308-312.

137. Marshall, B.J., Warren, J.B., and Blincow, E.D., 1988, Prospective double-blind trial of duodenal ulcer relapse after eradication of *Campylobacter pylori*, *Lancet* ii:1437-1442.

138. Eaton, K.A., and Krakowka, S., 1995, Avirulent, urease-deficient *Helicobacter pylori* colonizes gastric epithelial explants *ex vivo*, *Scand. J. Gastroenterol.* 30:434-437.

139. Katz, J., 1991, The course of peptic ulcer disease, *Med. Clin. N. Am.* 75:831-840.

140. Lee, A., 1994, The microbiology and epidemiology of *Helicobacter pylori* infection, *Scand. J. Gastroenterol.* 29(Suppl 201):2-6.

141. O'Connor, H.J., 1994, The role of *Helicobacter pylori* in peptic ulcer disease, *Scand. J Gastroenterol.* 29(Suppl 201):11-15.

142. Taylor, D., and Blaser, M., 1991, The epidemiology of *Helicobacter pylori* infection, *Epidemiol. Rev.* 13:42-59.

143. Lee, A., Fox, J., and Hazell, S., 1993, Pathogenicity of *Helicobacter pylori*: A perspective, *Infect. Immun.* 61:1601-1610.

144. Leung, K., Hui, M., Chan, P.K., and Thomas, T.M.M., 1992, *Helicobacter pylori*-related gastritis and gastric ulcer. A continumum of progressive epithelial degeneration, *Am. J. Clin. Pathol.* 6:569-574.

145. Clearfield, H.R., 1991, *Helicobacter pylori*: Aggressor or innocent bystander? *Med. Clin. N. Am.* 75:815-829.

146. Louw, J.A., van Rensburg, C., Zak, J., Adams. G., and Marks, I.N., 1993, Distribution of *Helicobacter pylori* colonization and associated gastric inflammatory changes: Difference between patients with duodenal and gastric ulcers, *J. Clin. Pathol.* 46:754-756.

147. Moss, S., and Calam, J., 1992, *Helicobacter pylori* and peptic ulcers: The present position, *Gut* 33:289-292.

148. Sipponen, P., and Hyvarinen, H., 1993, Role of *Helicobacter pylori* in the pathogenesis of gastritis, peptic ulcer and gastric cancer, *Scand. J. Gastroenterol.* 28(Suppl 196):3-6.

149. Bui, H.X., del Rosario, A., Sonbati, H., Lee, C.Y., George. M., and Ross, J.S., 1991, *Helicobacter pylori* affects the quality of experimental gastric ulcer healing in a new animal model, *Exp. Mol. Pathol.* 55:261-268.

149. Genta, R., Hamner, H.W., and Graham, D.Y., 1993, Gastric lymphoid follicles in *Helicobacter pylori* infection: Frequency, distribution, and response to triple therapy, *Human Pathol.* 24:577-583.

150. Mai, U., Perez-Perez, G., Allen, J., Wahl, S., Blazer, M., and Smith P., 1992, Surface proteins from *Helicobacter pylori* exhibit chemotactic activity for human leukocytes and are present in gastric mucosa, *J. Exptl. Med.* 175:517-525.

151. Chang, K.Y., Wyle, F., Stachura, J., and Tarnawski, A., 1993, *Helicobacter pylori* toxin delays healing of experimental gastric ulcers and inhibits cell proliferation at the ulcer margin, *Gastroenterol.* 104:52(Abstr.).

152. Rautelin, H., Blomberg, B., Fredlund, H., Jarnerot, G., and Danielsson, D., 1993, Incidence of *Helicobacter pylori* activating neutrophils in patients with peptic ulcer disease, *Gut* 34:599-603.

153. Reymunde, A., Deren, M.D., Nachamkin, I., Oppenheim, D., and Weinbaum, G., 1993, Production of chemoattractant by *Helicobacter pylori*, *Digest. Dis. Sci* 38:1697-1701.

154. Cover, T.L., Cao, P., Murthy, U.K., Sipple, M.S., and Blaser, M. J., 1992, Serum neutralizing antibody response to the vacuolation cytotoxin of *Helicobacter pylori*, *Clin. Invest.* 90:913-918.

155. Nielsen, H., and Andersen, L., 1992, Chemotactic activity of *Helicobacter pylori* sonicate for human polymorphonuclear leucocytes and monocytes, *Gut* 33:738-742.

156. Brandtzaeg, P., Bjerke, K., Kett, K., Kvale, D., Rognum, D.O., Scott, H, Sollid, I.M., and Valnes, K., 1987, Production and secretion of immunoglobulin in the gastrointestinal tract, *Trans. Ann. Allerg.* 59:21-39.

157. Megraud, F., Neman-Simha, V., and Brugmann, D., 1992, Further evidence of the toxic effect of ammonia produced by *Helicobacter pylori* urease on human epithelial cells, *Infect. Immun.* 60:1858-1863.
159. Sidebotham, R.L., Batten, J.J., Karim. Q.N., Spencer, J., and Baron, J. H., 1991, Breakdown of gastric mucus in presence of *Helicobacter pylori*, *J. Clin. Pathol.* 44:52-57.
160. Sidebotham, R.L., and Baron, J.H., 1990, Hypothesis; *Helicobacter pylori*, urease, mucus, and gastric ulcer, *Lancet* 355:193-195.
161. Correa, P., 1992, Human gastric carninogenesis: A multistep and multifactorial process - First American Cancer Society Award Lecture on Cancer epidemiology and Prevention, *Cancer Res.* 52:6735-6740.
162. Correa, P., Fox, J., Fontham, E., Ruiz, B., Lin, Y., Taylor, N., Mackinley, D., de Lima, E., Portilla, H. and Zarama, G., 1990, *Helicobacter pylori* and gastric carcinoma, *Cancer* 66:2569-2674.
163. Correa, P., and Shiao, Y.-H., 1994, Phenotypic and genotypic events in gastric carcino-genesis, *Cancer Res.* 54:1941s-1943s.
164. Fox, J.G., Correa, P., Taylor, N.S., Zavala, D., Fontham, E., Janney, P.H., Rodrigeuz, E., Hunter, F., and Diavolitis, S., 1989, *Campylobacter pylori*-associated gastritis and immune response in a population at increased risk of gastric carcinoma, *Am. J. Gastroenterol.* 32:775-781.
165. Forman, D, 1991, *Helicobacter pylori* infection: A novel risk factor in the etiology of gastric cancer, *J. Natl. Cancer Inst.* 83:1702-1703.
166. Parsonnet, J., Vandersteen, J.D., Goates, J., Sibley, R.K., Sibley, J., and Chang, Y., 1991, *Helicobacter pylori* infection in intestinal- and diffuse-type gastric adenocarcinomas, *J. Natl. Cancer Inst.* 93:640-643.
167. Engstrand L., 1995, Potential animal models of *Helicobacter pylori* infection in immunological and vaccine research, *FEMS Immunol. Med. Microbiol.* 10:265-270.
168. Michetti, P., Corthesy-Theulaz, I., Davin, C., Haas, R., Vaney, A-C., Heitz, M., Bille, J., Kraehenbuhl, J-P., Saraga, E., and Blum, A. L., 1994, Immunization of BALB/c mice against *Helicobacter felis* infection with *Helicobacter pylori* urease, *Gasterenterol.* 107:1002-1011.
169. Pappo, J., Thomas, W.D., Kabok, Z., Taylor, N.S., Murphy J., and Fox, J.G., 1995, Effect of oral immunization with recombinant urease on murine *Helicobacter felis* gastritis, *Infect. Immun.* 63:1246-1252.
170. Czinn, S., and Nedrud, J.G., 1991, Oral immunization against *Helicobacter pylori*, *Infect. Immun.* 59:2359-2364.

69

AN EXPERIMENTAL ROTAVIRAL ENTERITIS MODEL WITH NEONATAL PIGS

Guillermo G. Gomez,[1][*] Edward J. Rozhon,[2] Richard A. Goforth,[1] and Oulayvanh Thirakoune[1]

[1] Animal Science
Center for Gastrointestinal Biology and Disease
North Carolina State University
Raleigh, North Carolina 27695-7626
[2] Shaman Pharmaceuticals, Inc.
South San Francisco, California 94080-4812

1. ABSTRACT

Forty colostrum deprived newborn piglets were reared individually with an automatic feeding device (Autosow®) from 1 until 20 da of age. Twenty piglets were used to study the effect of rotaviral enteritis on growth, mortality, food scores, feces consistency and fecal virus shedding; another 20 piglets were used to assess its effect on intestinal damage during the postinfection period. Piglets were fed a liquid diet only, which contained 20% dry matter and had a composition similar to sow's milk. At 8 da of age, each piglet was infected, per os, with 2×10^8 porcine rotavirus particles. The inoculation of rotavirus produced vomiting in ~40% of piglets within 24 to 48 hr after ino culation. During da 6 postinfection (PI), piglets gained little or lost weight and were off food. Growth resumed after 6 da PI, but mortality increased from 15 to 70% at the end of the trial (12 da PI). Diarrhea developed from 2 da PI and reached its peak on da 5 PI. High levels of rotavirus shedding in feces were evident on the day after inoculation and continued for at least 10 da PI. At da 4 PI, rotaviral infection produced a drastic reduction in jejunal and ileal lactase activity and villi height, while crypt depth was increased considerably, when compared to values obtained before infection. At the end of the trial, ileal lactase activity was approximately half that found in jejunum; however, both jejunal and ileal lactase activity, as well as villi height, were higher than their levels at 4 da PI, but still lower than those found before infection. Protein concentration in mucosal homogenates was not affected by rotaviral infection. On the basis of these results and others previously reported, and because of their intestinal anatomy and physiology

---

[*] Reprint requests to: Dr. Guillermo Gomez, Animal Science, North Carolina State University, Raleigh, NC, 27695-7626, (919) 515-2019.

Advances in Swine in Biomedical Research, edited by Tumbleson and Schook
Plenum Press, New York, 1996

similarities to that of human infants, neonatal pigs are a useful animal model for studies designed to evaluate prevention and treatment studies for rotaviral enteritis.

2. INTRODUCTION

Gastrointestinal viral infections are common in young animals and humans and often are associated with enteritis and diarrhea.[1,2] Rotavirus infection is a major cause of severe diarrhea in children[3] as well as of neonatal diarrhea in several animal species worldwide.[4] The main characteristics of rotavirus infections have been reviewed extensively.[5,6] The need for an animal model in which to study the pathogenesis, as well as the prevention and treatment of rotavirus enteritis, has long been recognized.[1] Animal models using gnotobiotic as well as conventional piglets,[7-9] rabbits[10,11] and mice[12,13] have been studied for their relevance to acute infantile diarrhea. Our involvement with gastrointestinal research has led to the development of an experimental rotaviral enteritis model with neonatal pigs. The objectives of this study were to assess the effect of rotaviral enteritis on growth, mortality, food scores, feces consistency and fecal rotavirus shedding, as well as on intestinal damage during the PI period. Assessment of intestinal damage included villi height, crypt depth, mucosal protein concentration and lactase activity, both at midjejunum and midileum of rotavirus infected piglets.

3. MATERIALS AND METHODS

3.1. Experimental Animals and Environmental Conditions

The protocol of this research was approved by the NCSU Institutional Animal Care and Use Committee. Gestating sows were obtained from the NCSU Swine Farm and transferred to an isolated farrowing facility, 5 da before farrowing. Piglets farrowed by 6 crossbred sows were used in this study. Crossbred pigs carrying no known or defined pathogens were farrowed in an antiseptically clean (washed 3 times daily) stall after 4 to 5 da of repeating bathing and sanitizing of sows, before delivery, with an iodinated detergent (Wescodyne®, American Sterilizer Company, Medical Products Division, Erie, PA). Newborn piglets were not allowed to nurse and were transferred to an isolated room containing an automatic feeding device (Autosow®, BioAg Associates, Raleigh, NC). The temperature of the room was maintained at 32C during the first wk and lowered to 27 to 29C throughout the remainder of the trial period. The ambient relative humidity in the room varied between 55 and 75%. Lights were on from 0730 until 2145 hr.

3.2. Feeding Protocol and Basal Diet

The Autosow® is a machine containing individual cages (length, 0.50 m; width, 0.30 m; and height, 0.40 m) which regularly dispenses aseptically, small volumes of liquid diet according to piglet weight. Piglets were fed a liquid diet only and did not have access to drinking water. The diet reservoir was refrigerated; therefore, dietary bacterial growth, if any, was minimal. After each feeding, food pans were washed, under pressure, with a chlorinated detergent (Tri-Foam™, Diversey, Wyandotte, MI). Details of this device have been reported previously.[14] Piglets were fed a daily volume of diet that was 30% of body weight, i.e., a 1 kg piglet was fed 300 ml diet/da; food volume/da was divided in 8 equal portions (2 hr feeding interval; feeding schedule from 0730 to 2130 hr). This daily volume to weight ratio is near optimum with regard to weight gain and food efficiency when diets are made from milk solids and have a dry matter (DM) content of 20%.[14,15] Detailed composition of the basal diet has been described elsewhere.[16] The approximate theoretical dietary energy supply was 986 Kcal (4.1 MJ)/l, with caloric contributions of 28, 50 and 20%, on a DM basis, for protein, lactose and fat, respectively. During the first 3 da, piglets were fed basal diet containing 27 mg of IgG (Porcine plasma immunoglobulins; American Protein, Corp., Ames, IA). On the 3rd da of life, each piglet was injected IM with 100 mg of hematinic iron (Iron Dextran Complex, Fermenta Animal Health, Kansas City, MO).

3.3. Rotavirus Inoculation

At 8 da of age, after a 4 to 5 hr fasting period, each piglet was administered 2×10^8 rotavirus particles in 15 ml of diet followed by 15 ml of regular diet. The concentration of rotavirus was determined by measuring the absorbance at 260 nm, where it was determined previously that one unit of absorbance at this wavelength was equivalent to 8.4×10^8 rotavirus particles.[17] Aliquots were obtained from a stock of bacteria free fecal supernatant from rotavirus infected piglets.

3.4. Experimental Protocol

Piglets were weighed every other day and diet volume for each piglet was adjusted according to body weight. Food scores were recorded according to the following scale: 1 = eating normally, 2 = off food and 3 = not eating. Feces consistency or diarrhea scores were based on the following scale: 1 = normal, solid feces, 2 = soft, looser than normal stools and 3 = liquid diarrheal feces. Food and feces scores were recorded twice daily (midmorning and midafternoon) throughout the trial. Before rotavirus inoculation, rectal swabs of piglets with feces consistency of 2 or 3 were taken, placed in tubes containing 2 ml of 0.01 mol/l phosphate buffered saline (PBS), pH 7.5, and assayed within a few hours after collection for hemolytic *Escherichia coli* and rotavirus to determine if these pathogens were the cause of diarrhea. To assess virus shedding, on the day of rotavirus inoculation and every day thereafter, rectal swabs of each piglet were taken daily and assayed for rotavirus. Rectal swabs were processed for bacteriological culturing of hemolytic *E. coli* using blood agar with 5% sheep blood. After 24 hr incubation at 37C, cultures were evaluated for hemolysis. A commercial kit (Virogen Rotatest®, Wampole, Cranbury, NJ), based on a rapid latex particle agglutination slide test, was used for detection of rotavirus in fecal specimens. Rotavirus positivity was scored from 0 to 4 based on the following scale for speed of agglutination: 0 = none in 4 min, 1 = in 3 min, 2 = in 2 min, 3 = in 1 min and 4 = in <1 min.

3.5. Measurement of Intestinal Parameters

Piglets killed to ascertain the effect of rotavirus infection on intestinal damage were sedated with an IM injection of a mixture of 0.8 ml ketamine hydrochloride (Ketaset®, 100 g ketamine/l, Fort Dodge Laboratories, Fort Dodge, IA) and 0.2 ml of xylazine (Rompun®, 20 g xylazine/l, Miles, Shawnee Mission, KS) and killed with an intracardiac lethal dose of an euthanasia solution (Somlethol, Pentobarbital sodium, J.A. Webster, Sterling, MA). Piglets were fasted for at least 10 hr (from 2130 until 0730 hr) before being killed. Immediately after a piglet was killed, the abdomen was opened and the entire intestine, from the pylorus to the rectum, was removed rapidly, placed on ice and dissected from its mesentery. Jejunum and ileum were defined as the segment of the small intestine between the peritoneal inflection and the ileocecal junction. This segment was divided in half, and the proximal and distal halves were designated as jejunum and ileum, respectively. At approximately midjejunum and midileum, 2 adjacent segments, one slightly longer than 10 cm and another ~3 cm long, were taken. Intestinal contents, if any, were squeezed gently from the long segment, and a 10 cm piece was measured under uniform tension of a 10 g weight. Each 10 cm segment was opened longitudinally on an iced glass plate and the mucosa was scraped, using a glass slide, and weighed. The scraped mucosa was transferred into a Kontes tissue grinder and homogenized with a Teflon pestle in 4 volumes (v/w) of iced 0.9 % NaCl solution. Homogenate tubes were placed on ice and within 30 min were transferred into a freezer. Homogenates were kept frozen until assayed for protein and lactase activity. Homogenates were thawed, centrifuged in a refrigerated centrifuge at 3,000 rpm and supernatants were used for these assays. Protein determination was by the method of Lowry et al.[18] and protein concentration expressed as mg/protein/ml homogenate. Lactase activity was assayed by the method of Dahlqvist[19] and expressed as μmoles of lactose hydrolyzed/min/g protein.

Each 3 cm segment was submerged individually in and flushed with cold PBS, pH 7.4, and then submerged in fresh chilled fixative (FEA: formalin, 95% ethanol and glacial acetic acid solution). After 24 hr fixation, each segment was kept in 70% ethanol until it was dehydrated and embedded in paraffin wax. Measurements of villous morphology were made on jejunal and ileal sections stained with hematoxylin and eosin according to conventional techniques.[20] Morphometric measurements were performed by one person, using light microscopy with a computer assisted morphometric system (BioScan Optimetric, BioScan Inc., Edmonds, WA). Height and crypt depth of 5 representative and well oriented villi were measured.

3.6. Statistical Analyses

Values are reported as means ± SEM. Intestinal data were analyzed as a completely random design, using an individual piglet as the experimental unit, following the general linear procedures of SAS. Duncan's multiple range test[21] was used to identify significant (P<0.05) differences among observations during the PI period.

4. RESULTS

Before rotavirus inoculation (8 da of age), average body weight of piglets used to assess the effect of rotaviral infection on growth and mortality was 2.21 ± 0.08 kg (n = 20). The 20 piglets used to assess intestinal parameters weighed 2.23 ± 0.10 kg. During the 6 da PI, piglets gained little or lost weight (2.16 ± 0.11 kg; n = 17) and practically all piglets were off food (feces score of 2) between 2 and 6 da PI (Fig. 1). Growth resumed 6 da PI; however, the mortality rate increased from 15% on 6 da PI to 70% at the end of the trial (ET) (12 da PI). Most of the mortality (45%, from 15 to 60%) occurred between 6 and 8 da PI.

The inoculation of rotavirus produced vomiting in ~40% of piglets within 24 to 48 hr after rotavirus ingestion and before the onset of diarrhea. Diarrhea developed from 2 da PI (feces consistency of 2.0 ± 0.2, n = 20), reached its peak at 5 da PI (2.7 ± 0.2, n = 17) and progressively declined thereafter (Fig. 2). Fecal rotavirus shedding was evident on the day after inoculation (agglutination score of 3.4 ± 0.1, n = 20). High levels of virus shedding in feces continued throughout most of the PI period until 10 da after rotavirus inoculation (2.9 ± 0.1, n = 7; Fig. 2).

In Table 1 are data on jejunal and ileal villi height and crypt depth as well as lactase activity in mucosa homogenates of piglets killed before infection (BI)(8 da of age), on the 4th da after rotavirus inoculation (PI) and at ET (12 da PI). At 4 da PI, rotaviral enteritis

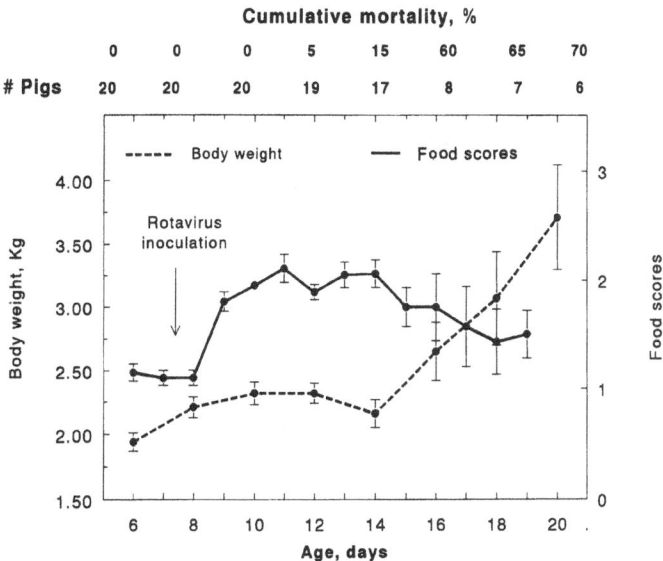

Figure 1. Effect of rotavirus infection on growth, food scores and mortality.

produced a reduction of jejunal and ileal villi height, while crypt depth was increased. At ET, both jejunal and ileal villi were higher than at 4 da PI but still lower than villi of piglets before infection; crypt depth, particularly at midjejunum, remained unchanged.

The blunting of jejunal and ileal villi was accompanied by a reduction of lactase activity in mucosal homogenates to 10% of the activity present BI (Table 1). At ET, lactase activity in jejunal mucosa homogenates was twice that in ileal homogenates (130 ± 16 vs 68 ± 9, n = 8). However, lactase activity, particularly at midileum, was lower than that of piglets BI (Table 1). Protein concentration in mucosal homogenates was not affected by rotaviral infection (16.2 ± 0.5 and 15.4 ± 0.5 mg/ml, n = 23, for midjejunum and midileum homogenates, respectively).

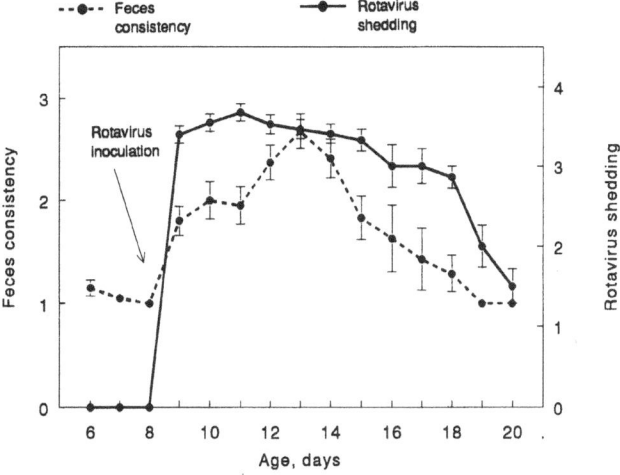

Figure 2. Effect of rotavirus infection on feces consistency and rotavirus shedding.

Table 1. Effect of rotavirus enteritis on villi height, crypt depth and lactase activity

Infective stage*	Age days	n	Midjejunum**			Midileum**		
			VH	CD	Lactase	VH	CD	Lactase
BI	8	6	771 ± 51^a	135 ± 10^b	193 ± 31^a	980 ± 99^a	138 ± 8^c	195 ± 25^a
PI	12	9	305 ± 28^c	210 ± 13^a	18 ± 3^c	319 ± 19^c	226 ± 5^a	17 ± 3^c
ET	20	8	616 ± 53^b	209 ± 4^a	130 ± 16^b	519 ± 64^b	194 ± 8^b	68 ± 9^b

* BI = before infection; PI = 4 da postinfection; ET = end of trial.

** Values in a column with unlike superscripts are different (P<0.01).

5. DISCUSSION

Numerous scientists have demonstrated the usefulness of neonatal pigs as an animal model for rotaviral enteritis. Most of these investigators used gnotobiotic pigs[7,8,22-27] while a few studies were performed with either conventional (sow reared and early weaned)[8,28] or colostrum deprived, artificially reared,[29] piglets. Unfortunately, the experimental conditions used have been variable, in most cases the PI period has been relatively short, the number of animals studied has been limited and the effect of rotavirus infection on subsequent growth has not been examined fully.

Human rotavirus isolates obtained from human infants with acute gastroenteritis[7,22-25,28] and porcine rotavirus isolates,[8,26,27,29] as well as calf and foal rotaviruses,[23] have been used as inocula to infect pigs. The concentration of rotavirus in inocula was not always given but in several studies[25,26,28,29] in which concentration was mentioned, it varied between 10^6 and 10^8 particles/pig. Oral administration was the most common route of rotavirus inoculation but in a few studies, intranasal[24,26] or nasogastric[28] administration were used. The age of piglets at infection varied from 12 to 14 hr[7] to 28 da[24] of age; in the majority of studies, piglets were infected between 2 and 7 da of age.[8,22,23,25,26] Detailed information about the feeding regimen used prior to and after infection is limited.[8,27,29]

The onset of diarrhea in practically all studies occurred within 2 to 3 da after rotavirus inoculation. The duration of diarrhea, however, was quite variable. Middleton et al.,[25] using conventional piglets which were removed from the sow at 2 to 3 da of age and inoculated per os at 6 da of age with human rotavirus, reported that diarrhea began 32 hr after virus ingestion and lasted for only 24 hr. Furthermore, they reported that diarrhea in germfree piglets was difficult to assess since these animals produced loose stools. In another study,[24] none of the gnotobiotic piglets intranasally inoculated with human rotavirus developed any clinical signs of infection during the following 3 to 4 wk; however, piglets infected with porcine rotavirus had profuse diarrhea as early as 18 hr after viral infection. Others reported diarrhea duration varying from 3 to 6 da[29] to 10 to 18 da[7,26] after rotavirus inoculation. In all cases, severity of diarrhea diminished with time after rotavirus inoculation.

Large amounts of virus in feces were found shortly before or at the onset of diarrhea;[7,8,22,24,25,29] the amount of virus shed in infected piglet feces exceeded the quantity of virus ingested.[25] Rotavirus was detected in infected piglets feces regardless of whether or not they had diarrhea.[23]

Despite the variability of experimental conditions described above, experimentally induced rotavirus enteritis in neonatal pigs, particularly when porcine rotavirus isolates were used,[8,26,27,29] has been characterized clinically by anorexia, diarrhea, occasional vomiting and high titers of rotavirus shedding in feces. The experiment reported herein provides additional detailed information, particularly on the extent of anorexia (off food), severity of diarrhea and extent of fecal virus shedding, as well as on the effect of rotavirus infection on growth and mortality rate during the PI period.

Under our experimental conditions, the onset of diarrhea occurred within 48 hr after virus ingestion, profuse diarrhea lasted for 4 to 5 da (from 2 to 7 da PI) and diarrhea progressively ceased thereafter (Fig. 2). Furthermore, high levels of rotavirus shedding were found in feces of infected piglets until 10 da PI (Fig. 2), even after diarrhea in most pigs has ceased. The relatively long period of anorexia (off food from 2 to 6 da PI, Fig. 1) along with the diarrhea produced by rotavirus infection (Fig. 2) resulted in a loss of body weight of infected piglets by 6 da PI (Fig. 1). Thereafter, infected piglets resumed growth, but the cumulative mortality increased, notably between 6 and 8 da PI (Fig. 1). In one study,[29] none of the colostrum deprived, artificially reared piglets died, while in other reports,[8,22] 25 to 30% of gnotobiotic rotavirus infected piglets died by 4 to 5 da after virus ingestion. The high dose of rotavirus inoculum (2×10^8 rotavirus particles per piglet) used appears to be the main factor responsible for the high cumulative mortality of this study. In several experiments carried out in our laboratory, colostrum deprived piglets reared under similar conditions and infected at 5 to 6 da of age with 10^6 to 10^7 rotavirus particles have shown mortality rates between 40 and 50% by the end of the experimental periods (10 to 12 da PI)(Gomez, unpublished). An experiment in which newborn piglets were allowed to nurse their dams for 24 hr and then artificially reared and infected at about the same age and with a similar dose of rotavirus inoculum as in this study showed a cumulative mortality of 30% (Gomez, unpublished). Protection of neonatal pigs against rotavirus is known to be related to maternal antibodies in sow's milk.[30] Nursing pigs receiving maternal lactation antibodies have reduced disease signs, but antibody may not prevent completely infection or disease.[31] Artificially reared piglets are therefore at high risk for infection and subsequent severe rotaviral disease.[32]

Young conventional pigs infected with human rotavirus have shown viral invasion of villous epithelium, lesions on jejunal mucosa and depression in mucosal disaccharidase activity.[28] Viral induced histopathological changes have been restricted to the small intestine and were more pronounced within jejunum and ileum.[27] Severe stunting of villi has been reported in studies with gnotobiotic pigs infected with either porcine rotavirus[26,27] or neonatal calf rotavirus,[33] conventional pigs infected with human rotavirus[28] and colostrum deprived piglets infected with porcine rotavirus.[29] Villous atrophy in pigs with clinical signs of rotaviral enteritis appeared to be more severe between 24 and 48 hr after diarrhea onset.[27] In addition, crypt hyperplasia has been found in most studies[26-28] but crypt depth was not affected in infected piglets studied by Rhoads et al.[29] Histological data obtained in our study confirm the severe blunting of both jejunal and ileal villi at 4 da PI (Table 1), when villi height was reduced to 40 and 33% of that found in pigs before viral infection. At 12 da after viral inoculation, jejunal and ileal villi were still shorter that villi of piglets before infection.

Rotaviral induced villous atrophy appears to be the result of a rapid and extensive infection of the differentiated epithelial cells that subsequently leads to an accelerated desquamation of these cells from the villi.[27] The decrease in body weight following infection and the mortality of pigs presumably are the result of gut damage. As a consequence of gut damage from rotaviral infection, depression of disaccharidases, particularly lactase, has been reported in children[2] and piglets.[28,29] The decrease in jejunal and ileal lactase activity at 4 da PI (Table 1) confirms these reports. Diarrhea in rotavirus infected animals appears to be caused by repopulation of the damaged mucosa with immature cells that cannot absorb nutrients as well as mature cells.[28]

Although rabbits[10,11] and mice[12,13] also have been used as animal models for studies related to rotaviral enteritis, neonatal piglets appear to be the most suitable species for this type of research, mainly because of the similarities in intestinal anatomy and physiology to that of human infants. Results reviewed above clearly indicate that differences in lesions severity may be due to factors such as age and immunological status of animals at infection, infective dose used, strain and infectivity of the virus, feeding regimen used prior to and after infection and stage of disease at which samples are studied. We suggest the established methodology to artificially rear neonatal pigs with an automatic feeding device (Autosow®) offers an interesting experimental animal model to further study rotaviral enteritis.

6. ACKNOWLEDGMENTS

The authors thank Shaman Pharmaceuticals, Inc., for partial funding; Ms. Gema Gomez, for editing the manuscript; Ms. Noris Carbajal, for preparation of graphs; Milk Specialties and Central Soya, for supplying feed ingredients and vitamin premix, respectively; and Wampole Laboratories, for donating the Virogen Rotatest® kit.

7. REFERENCES

1. Greenberg, H.B., Wyatt, R.G., Kalica, A.R., Yolken, R.H., Black, R., Zapikian, A.Z., and Chanock, R.M., 1981, New insights in viral gastroenteritis, *Perspect. Virol.* 11:163-187.
2. Wolf, J.L., and Schreiber, D.S., 1982, Viral gastroenteritis, *Med. Clin. N. Am.* 66:575-595.
3. LeBaron, C.W., Lew, J., Glass, R.I., Weber, J.M., and Ruiz-Palacios, G.M., 1990, Annual rotavirus epidemic patterns in North America. Results of a 5-year retrospective survey of 88 centers in Canada, Mexico, and the United States, *J. Am. Med. Assoc.* 264:983-988.
4. Jawetz, E., Melnick, J.L., and Adelberg, E.A., 1987, Reoviruses, rotaviruses & other human viral infections, in: *Review of Medical Microbiology*, Appleton and Lange, Norwalk, CT, pp. 513-516.
5. Flewett, T.H., and Woode, G.N., 1978, The rotaviruses, brief review, *Arch. Virol.* 57:1-23.
6. McNulty, M.S., 1978, Rotaviruses, *J. Gen. Virol.* 40:1-18.
7. Torres-Medina, A., Wyatt, R.G., Mebus, C.A., Underdahl, N.R., and Kapikian, A.Z., 1976, Diarrhea caused in gnotobiotic piglets by the reovirus-like agent of human infantile gastroenteritis, *J. Infect. Dis.* 133:22-27.
8. Tzipori, S., and Williams, I.H., 1978, Diarrhea in piglets inoculated with rotavirus, *Aust. Vet. J.* 54:188-192.
9. Lecce, J.G., King, M.W., and Dorsey, W.E., 1978, Rearing regimen producing piglet diarrhea (Rotavirus) and its relevance to acute infantile diarrhea, *Science* 199:776-778.
10. Connner, M.E., Estes, M.K., and Graham, D.Y., 1988, Rabbit model of rotavirus infection, *J. Virol.* 62:1625-1633.
11. Hambraeus, A.M., Hambraeus, L.E.J., and Wadell, G., 1989, Animal model of rotavirus infection in rabbits - protection obtained without shedding of viral antigen, *Arch. Virol.* 107:237-251.
12. Coelho, K.I.R., Bryden, A.S., Hall, C., and Flewett, T.H., 1981, Pathology of rotavirus infection in suckling mice: A study by conventional histology, immunofluorescence, ultrathin sections, and scanning electron microscopy, *Ultrastruct. Pathol.* 2:59-80.
13. Ijaz, M.K., Dent, D., Haines, D., and Babiuk, L.A., 1989, Development of a murine model to study the pathogenesis of rotavirus infection, *Exp. Mol. Pathol.* 51:186-204.
14. Coalson, J.A., and Lecce, J.G., 1973, Herd differences in the expression of fatal diarrhea in artificially reared piglets weaned after 12 hours vs. 36 hours of nursing, *J. Anim. Sci.* 36:1114-1121.
15. Lecce, J.G., 1969, Rearing colostrum-free pigs in an automatic feeding device, *J. Anim. Sci.* 28:27-33.
16. Gomez, G.G., Sandler, R.S., and Seal, Jr., E., 1995, High levels of inorganic sulfate cause diarrhea in neonatal piglets, *J. Nutr.* 125:2325-2332.
17. Keljo, D.J., and Smith, A.K., 1988, Characterization of binding of Simian rotavirus SA-11 to cultured epithelial cells, *J. Ped. Gastroenterol. Nutr.* 7:249-256.
18. Lowry, O.H., Rosebrough, N.J., Farr, A.L., and Randall, R.J., 1951, Protein measurement with the folin phenol reagent, *J. Biol. Chem.* 193:262-275.
19. Dahlqvist, A., 1964, Method for assay of intestinal disaccharidases, *Anal. Biochem.* 7:18-25.
20. Luna, L.G., 1968, *Manual of Histologic Staining Methods of the Armed Forces Institute of Pathology*, Armed Forces Institute of Pathology, Washington, DC, pp. 38-39.
21. Steel, R.G.D., and Torrie, J.H., 1980, *Principles and Procedures of Statistics - A Biometrical Approach*, 2nd ed., McGraw-Hill, New York.

22. Torres-Medina, A., Wyatt, R.G., Mebus, C.A., Underdahl, N.R., and Kapikian, A.Z., 1976, Patterns of shedding of human rotavirus-like agent in gnotobiotic newborn piglets with experimentally-induced diarrhea, *Intervirology* 7:250-255.

23. Tzipori, S.R., Makin, T.J., and Smith, M.L., 1980, The clinical response of gnotobiotic calves, pigs and lambs to inoculation with human, calf, pig and foal rotavirus isolates, *Aust. J. Exp. Biol. Med. Sci.* 58:309-318.

24. Bridger, J.C., Woode, G.N., Jones, J.M., Flewett, T.H., Bryden, A.S., and Davies, H., 1975, Transmission of human rotaviruses to gnotobiotic piglets, *J. Med. Microbiol.* 8:565-569.

25. Middleton, P.J., Petric, M., and Szymanski, M.T., 1975, Propagation of infantile gastroenteritis virus (Orbi-group) in conventional and germfree piglets, *Infect. Immun.* 12:1276-1280.

26. Davidson, G.P., Gall, D.G., Petric, M., Butler, D.G., and Hamilton, J.R., 1977, Human rotavirus enteritis induced in conventional piglets, *J. Clin. Invest.* 60:1402-1409.

27. Crouch, C.F., and Woode, G.N., 1978, Serial studies of virus multiplication and intestinal damage in gnotobiotic piglets infected with rotavirus, *J. Med. Microbiol.* 11:325-334.

28. Theil, K.W., Bohl, E.H., Cross, R.F., Kohler, E.M., and Agnes, A.G., 1978, Pathogenesis of porcine rotaviral infection in experimentally inoculated gnotobiotic pigs, *Am. J. Vet. Res.* 39:213-220.

29. Rhoads, J.M., Keku, O.E., Quinn, J., Woosely, J., and Lecce, J.G., 1991, L-Glutamine stimulates jejunal sodium and chloride absorption in pig rotavirus enteritis, *Gastroenterology* 100:683-691.

30. Hess, R.G., and Bachmann, P.A., 1981, Distribution of antibodies to rotavirus in serum and lacteal secretions of naturally infected swine and their suckling pigs, *Am. J. Vet. Res.* 42:1149-1152.

31. Debouck, P., and Pensaert, M., 1983, Rotavirus excretion in suckling pigs followed under field circumstances, *Ann. Rech. Vet.* 14:447-448.

32. Lecce, J.G., and King, M.W., 1978, Role of rotavirus (Reo-like) in weanling diarrhea of pigs, *J. Clin. Microbiol.* 8:454-458.

33. Hall, G.A., Bridger, J.C., Chandler, R.L., and Woode, G.N., 1976, Gnotobiotic piglets experimentally infected with neonatal calf diarrhoea reovirus-like agent (Rotavirus), *Vet. Pathol.* 13:197-210.

70

INFLUENCE OF DIETARY PROTEIN AND FAT SOURCES ON THE LEVELS OF BLOOD HOMOCYSTEINE IN A PIG MODEL

Preliminary Observations

Ghulam Sarwar,[*] Nimal Ratnayake, Robert W. Peace, and Herbert G. Botting

Nutrition Research Division
Food Directorate
Health Canada
Tunney's Pasture, Banting Building (PL: 2203C)
Ottawa, Ontario, Canada K1A OL2

1. ABSTRACT

A pig growth study was conducted to investigate the effects of 2 protein sources (casein and soy protein isolate (SPI); varying in methionine/cysteine ratios) and 2 fat sources (lard and fish oil; varying in n-3 polyunsaturated fatty acids) on the levels of homocysteine and lipids in blood and tissues. The 4 experimental diets (casein-lard, casein-fish oil, SPI-lard, SPI-fish oil) were formulated to contain 22% protein (N X 6.25, from casein or SPI), 10% fat (10% lard or 5% fish oil plus 5% lard) and required levels of minerals and vitamins. Male, weanling, 9 kg Yorkshire pigs (6/diet) were fed the 4 experimental diets for a period of 6 wk. Blood samples were collected before the test and after 2, 4 and 6 wk of test. Samples of tissues (liver, heart and kidneys) were collected before the test and after 6 wk of test.

The average fasting baseline value for total (both free and protein bound) homocysteine in plasma of pigs was 11 ± 1 umol/l. Plasma homocysteine increased after feeding test diets. At 2 wk of test, pigs fed SPI diets had lower levels of plasma homocysteine (20 to 23 umol/l) than those fed casein diet (30 to 31 umol/l). Differences in plasma homocysteine levels of pigs fed fish oil diets compared to those fed lard diets were, however, small. Pigs fed SPI diets also had lower levels of total serum cholesterol (1.26 to 2.11 mmol/l) than those fed casein diets (1.70 to 2.68 mmol/l). Similarly, pigs fed fish oil diets had lower levels of

---

[*] Reprint requests to: Dr. G. Sarwar, Nutrition Research Division, Food Directorate, Health Canada, Tunney's Pasture, Banting Building (PL: 2203C), Ottawa, Ontario, Canada K1A OL2, (613) 957-0933.

serum cholesterol (1.26 to 1.70 mmol/l) than those fed lard diets (2.11 to 2.68 mmol/l). Differences in serum triglycerides of pigs fed experimental diets were small. Analyses for blood samples obtained after 4 and 6 wk of test, and for tissue samples obtained after 6 wk of test were not completed at the time of the preparation of this report.

2. INTRODUCTION

A moderate increase in plasma homocysteine is considered a common and independent risk factor for premature cardiovascular disease.[1-3] Large intakes of foods high in methionine, coupled with marginal deficiencies of folate, pyridoxine or cobalamin, may result in elevated levels of plasma homocysteine. On the other hand, fish oils (high in n-3 polyunsaturated fatty acids) have been shown to be associated with lower plasma homocysteine levels.[4]

The purpose of this study was to evaluate the usefulness of a pig model for studying the influence of diet on homocysteine levels in blood and tissues. The effects of feeding 2 protein sources (casein and SPI; varying in cysteine/methionine ratios) and 2 fat sources (lard and fish oil-lard; varying in n-3 polyunsaturated fatty acids) on homocysteine levels in plasma and tissues of pigs were studied.

3. MATERIALS AND METHODS

The 4 experimental diets (Table 1) contained 22% protein (N X 6.25) from SPI or casein, 10% fat (5% fish oil + 5% lard or 10% lard) and required levels of

Table 1. Composition of experimental diets fed to pigs

Ingredient (g/kg)	SPI-Fish oil	SPI-Lard	Casein-Fish oil	Casein-Lard
Soybean protein isolate[a]	244.45	244.45	-	-
Casein (ANRC)[b]	-	-	244.45	244.45
Ground yellow corn	100.00	100.00	100.00	100.00
Sucrose	32.50	32.50	32.50	32.50
Molasses	10.00	10.00	10.00	10.00
Cellulose [c]	50.00	50.00	50.00	50.00
Fish oil[d]	50.00	-	50.00	-
Lard	50.00	100.00	50.00	100.00
Mineral mixture[a]	44.00	44.00	44.00	44.00
Vitamin mixture[f]	10.00	10.00	10.00	10.00
Corn starch	409.05	409.05	409.05	409.05

[a]Isolated soy protein (066-906), ADM, Decatur, IL. [b]Animal Nutrition Research Council Reference Protein, ICN, St-Laurent, Quebec, Canada. [c]Alfafloc, Teklad Test Diets, Madison, WI. [d]Menhaden oil (supplied by Zapata Protein Inc., Reedvile, VA.[e]Teklad Mineral Mix (Swine, TD 94135), supplied the following (g/kg mixture): calcium phosphate, diabasic, 287.0; calcium carbonate, 278.3; potassium phosphate, monobasic, 239.2; potassium citrate, monohydrate, 52.2; sodium chloride, 83.0; magnesium oxide, 18.5; ferric citrate, 16.1; zinc oxide, 2.72; manganese sulfate, 0.81; cupric sulfate, 0.6; chromium potassium sulfate, 0.21; potassium iodate, 0.0074; sodium selenate, 0.0146; sucrose, finely ground 21.338.[f]Teklad Vitamin Mix (Swine, TD 94142), supplied the following (g/kg mixture): vitamin A palmitate (500 000 U/g), 1.0; vitamin D_3 (500 000 U/g), 0.1; vitamin E acetate (500 U/g), 5.0; menadione sodium bisulfate, 0.91; vitamin B_{12} (0.1 % titration), 3.0; biotin, 0.02; calcium pantothenate, 2.73; choline bitartarate, 244.0; folic acid, 0.2; niacin, 4.0; pyridoxine HCl, 0.7; riboflavin, 1.0; thiamine HCl, 1.5; sucrose finely ground, 735.84.

minerals and vitamins.[5] Three young (8 to 10 kg) male Yorkshire pigs were allotted randomly to each diet within each of 2 consecutive complete replicates, in a temperature (24 to 25C) and humidity controlled housing facility. Pigs were allowed access to allotted feeders in individual stall enclosures within 12 pig pens for approximately 1 hr twice daily. Water was available ad libitum except during feeding. Records of daily food consumption and weekly body weights were kept for a period of 6 wk. Samples of blood were obtained from each pig (after an overnight fast) at 2, 4 and 6 wk of test. For collection of tissues (liver, heart and kidneys), animals were sacrificed after 6 wk of test. To obtain blood and tissues at zero time, 5 additional piglets were sacrificed, after an overnight fast, at the start of the experiment. Animals were immobilized with 10 to 15 mg/kg ketamine hydrochloride before collecting blood from the jugular vein. Ketamine was given in the muscle of the hind leg using a 3 or 5 cc syringe and a butterfly with 21 ga X 3/4"length needle. Animals were subjected to Isoflurane anesthesia (2 to 4%) before collection of tissues. Blood samples were collected into 2 sets of tubes, SST Vacutainer tubes (Becton-Dickinson, Rutherford, NJ) and heparinized tubes. The blood was allowed to clot, and tubes were centrifuged 10 min at 1600 X g to obtain serum for serum cholesterol and triglyceride analysis. Heparinized tubes containing blood were placed on ice and plasma separated by centrifuging at 3000 X g for 15 min. Plasma samples were analyzed immediately or stored at -80C until analyzed for total homocysteine. Subsamples of plasma were deproteinized with acetonitrile;[6] deproteinized plasma samples were stored at -80C until analyses for free amino acids. Tissues (liver, heart and kidneys) were weighed and a portion of each tissue was frozen under liquid nitrogen and stored at -80C until analysis. Health Canada's guide for the care and use of laboratory animals was followed and the protocol was approved by the animal care committee.

Total homocysteine (both protein bound and free) in plasma was determined by the reversed phase HPLC method of Fiskerstrand et al.[7] using a Waters NOVA-PAK-C18 15 cm long column at 25C. The method included precolumn derivatization with monobromobimane and fluorescence detection of derivatives. Baseline separation of homocysteine, cysteinylglycine and glutathione standards, as monobromobimane derivatives was completed in 15 min (Fig. 1). Similarly, separation of homocysteine, cysteinylglycine and glutathione, as monobromobimane derivatives, in pig plasma was obtained in 15 min (Fig. 2).

Free amino acids in deproteinized plasma were determined using precolumn phenylisothiocyanate derivatization and liquid chromatography.[6] Levels of total serum cholesterol and triglycerides were determined using a CX-5 Autoanalyzer (Beckman, Brea, CA). SPI and casein were hydrolysed in duplicate with 6N HCl at 110C for 22 hr for the determination of all total amino acids except sulfur amino acids and tryptophan.[8] Hydrolysates for determination of methionine as methionine sulfone and cystine/cysteine as cysteic acid were prepared by performic acid oxidation of protein followed by 6N HCl hydrolysis.[8] Hydrolysis with 4.2N NaOH was used to recover tryptophan quantitatively.[8] All amino acids, except tryptophan, in hydrolysates were determined by liquid chromatography of precolumn phenylisothiocyanate derivatives.[9] Tryptophan in alkaline hydrolysates was determined by a simple liquid chromatographic method requiring no derivatization.[9] Fatty acid compositions of dietary oils were determined by capillary gas chromatography analysis of the fatty acid methyl ester derivatives using a SP-2560 capillary column. Methyl ester derivatives were prepared by direct methylation using boron trifluoride:methanol.[10] Peak identity was established by cochromatography with fatty acid methyl ester standards (GLC 68, NU Check Prep Elysian, MN).

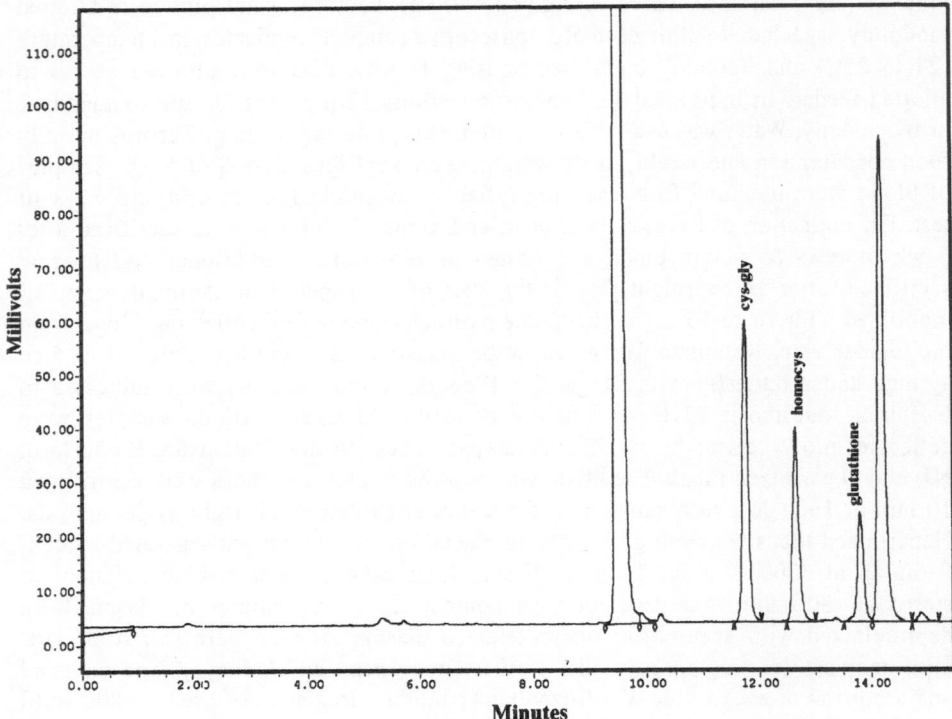

Figure 1. Chromatogram of homocysteine, cysteinylglycine and glutathione standards as monobromobimane derivatives.

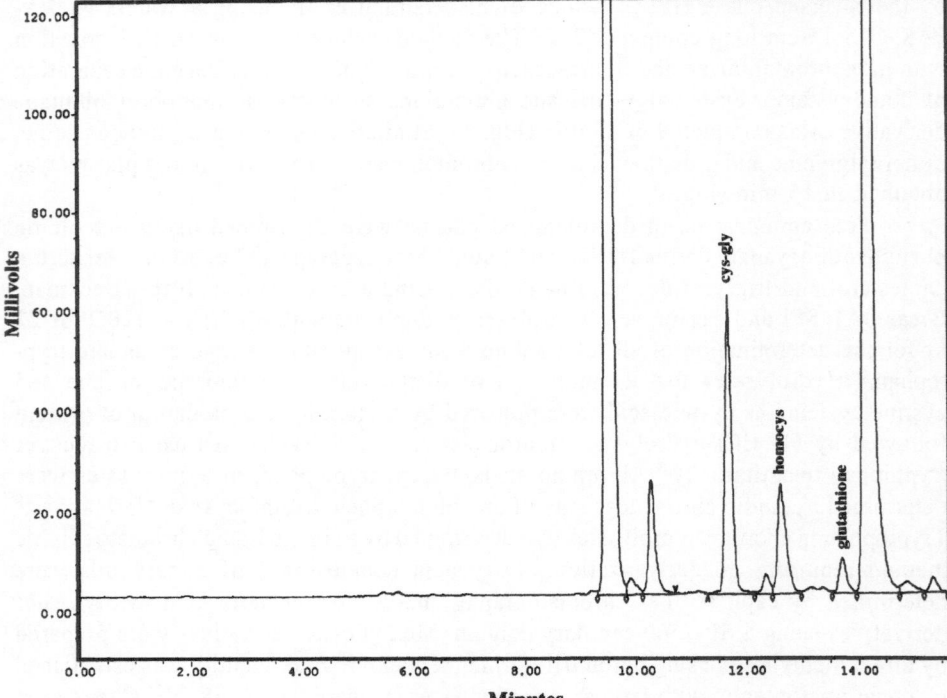

Figure 2. Chromatogram of homocysteine, cysteinylglycine and glutathione (as monobromobimane derivatives) in pig plasma.

Table 2. Growth of pigs fed experimental diets for 6 wk

Diet	Food consumed[a]	Weight gain [a]	Gain/food[a]
	kg/6 wk		Ratio
SPI-Fish oil	31.01 ± 1.86	17.27 ± 0.69	0.55 ± 0.02
SPI-Lard	35.36 ± 1.77	20.00 ± 0.90	0.56 ± 0.02
Casein-Fish oil	37.69 ± 2.03	28.63 ± 1.43	0.76 ± 0.03
Casein-Lard	38.26 ± 1.72	26.06 ± 1.04	0.68 ± 0.02

[a]Mean ± SEM (n=6).

4. RESULTS AND DISCUSSION

Data on 6 wk food consumption, weight gains and gain/food ratios are shown in Table 2. After 6 wk of test, pigs fed SPI diets had lower food consumption, weight gains and gain/food ratios compared to those fed casein diets. Since gain/food ratios of pigs fed SPI diets were higher than the ratio (0.47) recommended by the NRC committee on Nutrient Requirements of Swine[5] for 10 to 20 kg pigs, SPI diets met the nutrient requirements for pigs. Lower growth performance of pigs fed SPI diets, compared to those fed casein diets, may have been due to the presence of antinutritional factors, such as trypsin inhibitors and lysinoalanine in SPI.

Blood analyses have been completed only for samples obtained at zero time and at 2 wk of test. Average baseline fasting value for total homocysteine in plasma of pigs was 11 ± 1 umol/l. Similar normal values (10 to 12 umol/l) for fasting total homocysteine in human plasma have been reported.[11,12] After 2 wk of test (Table 3), plasma homocysteine increased after feeding test diets containing 22% protein from SPI or casein. Pigs fed SPI diets had lower levels of plasma homocysteine (20 to 23 umol/l) than those fed casein diets (30 to 31 umol/l). Levels of plasma homocysteine appeared to be related positively to levels of dietary methionine (Table 4; 1.47 and 3.03 g/100 g protein in SPI and casein, respectively) and plasma methionine (Table 3; 39 to 43 and 75 to 89 umol/l in pigs fed SPI and casein diets, respectively).

Homocysteine is an intermediate in the transsulfuration pathway by which conversion of methionine to cysteine occurs. Methionine not incorporated into proteins may be catabolized through transamination but most is converted to S-adenosylmethionine. S-adenosylhomocysteine, the demethylated product of S-adenosylmethionine, is hydrolyzed further to adenosine and homocysteine; this reaction is the only known source of homocysteine in vertebrates.[13] Homocysteine is catabolized to cysteine through 2 vitamin B_6 dependent reactions that complete the transsulfuration pathway. Remethylation of homocysteine to methionine is catalyzed either by 5-methyltetrahydrofolate homocysteine methyltransferase

Table 3. Concentrations of total homocysteine and free methionine, cystine and taurine in plasma of pigs fed the experimental diet for 2wk

Diet	Homocysteine[a]	Methionine[a]	Cystine[a]	Taurine[a]
	μmol/l			
SPI-Fish oil	23 ± 2	43 ± 2	26 ± 1	40 ± 2
SPI-Lard	16 ± 2	39 ± 2	26 ± 1	38 ± 2
Casein-Fish oil	29 ± 2	89 ± 4	27 ± 1	30 ± 2
Casein-Lard	27 ± 2	75 ± 3	28 ± 1	36 ± 2

[a]Mean ± SEM (n=6).

Table 4. Amino acid profiles of soybean protein isolate (SPI) and casein

Amino acid (g/100 g protein)	SPI	Casein
Arginine	7.84	3.71
Histidine	2.51	2.97
Isoleucine	5.08	5.36
Leucine	8.15	10.16
Lysine	6.08	8.44
Methionine	1.47	3.02
Cyst(e)ine	1.15	0.47
Phenylalanine	5.42	5.47
Tyrosine	3.80	6.04
Threonine	3.98	4.64
Tryptophan	1.30	1.31
Valine	5.12	6.85
Alanine	4.17	3.30
Aspartic acid	11.05	7.71
Glutamic acid	18.79	24.00
Glycine	3.93	2.00
Proline	5.21	11.72
Serine	5.12	6.10
Methionime/cysteine ratio[a]	0.91	5.22

[a]Molar basis.

(methionine synthase) or betaine homocysteine methyltransferase; both these reactions conserve methionine. Methionine synthase requires 5-methyltetrahydrofolate as methyl donor and vitamin B_{12} as cofactor. Folate and vitamin B_{12} dependent remethylation of cysteine explains the clinically well established relation between folate and B_{12} status.[13] Homocysteine is an important regulatory branch point metabolite that connects the metabolism of methionine, cysteine, vitamin B_{12}, reduced folates and vitamin B_6. Homocysteine has been shown to cause damage to endothelial cells, possibly by generation of reactive oxygen species such as hydrogen peroxide.[13] Oxidized low density lipoproteins (LDL), but not native LDL, is highly cytotoxic to endothelial cells. Endothelial lesions induced by oxidized LDL may contribute to atherogenesis.[13]

Differences in plasma homocysteine levels of pigs fed fish oil diets, compared to those fed lard diets, were small (Table 3), suggesting no effect of difference in n-3 polyunsaturated fatty acids contents of the experimental diets. Fish oil diets contained a higher level of total n-3 polyunsaturated fatty acids compared to lard diets (17.3 vs 0.5%, Table 5).

Epidemiologists have demonstrated an inverse relation between fish consumption and mortality from coronary heart disease.[14] High fish consumption is considered responsible for the low prevalence of atherosclerosis and myocardial infarction in Greenland Eskimos, despite high dietary fat and cholesterol intake.[15] The n-3 fatty acids of fish oils, especially 20:5n-3 and 22:6n-3 fatty acids, are believed to be responsible for the antiatherogenic effect of high fish consumption.[16,17] Protection against coronary heart disease afforded by a diet rich in fish may be due to lowering of serum homocysteine levels by the n-3 polyunsaturated fatty acids of fish oils.[4] Administration of fish oil supplement, 12 g/da for 3 wk, was reported to cause a decrease in serum homocysteine in hyperlipemic men, while consumption of the same amount of olive oil supplement had no such effect.[4] On the other hand, a semipurified diet with corn oil resulted in lower serum homocysteine levels in rabbits compared to the same diet with butter.[18] Corn oil is a rich source of linoleic acid, an n-6 polyunsaturated fatty acid precursor of arachidonic acid; olive oil contains an abundant amount of oleic acid, a monounsaturated fatty acid; butter contains predominantly saturated and monounsaturated fatty acids. Therefore, the lowering of serum homocysteine by poly-

Table 5. Fatty acid composition (%) of the 4 experimental diets

Fatty acids	SPI-fish oil	SPI-Lard	Casein-Fish oil	Casein-Lard
14:0	4.1	1.5	4.1	1.5
16:0	20.2	23.8	20.2	23.8
18:0	10.1	15.7	10.1	15.7
ΣSat. FA	35.9	41.9	35.9	41.9
16:1	6.2	2.1	6.2	2.1
18:1	27.7	43.1	27.7	43.1
20:1	1.3	1.0	1.3	1.0
ΣMono. FA	35.5	46.6	35.5	46.6
18:2n-6	6.4	11.1	6.4	11.1
20:4n-6	0.4	0.8	0.4	0.8
22:5n-6	0.1	0.0	0.1	0.0
Σn-6 PUFA	7.8	11.6	7.8	11.6
18:3n-3	0.8	0.5	0.8	0.5
20:5n-3	7.3	0.0	7.3	0.0
22:5n-3	1.3	0.0	1.3	0.0
22:6n-3	5.0	0.0	5.0	0.0
Σn-3 PUFA	17.3	0.5	17.3	0.5
Σ PUFA	28.7	11.6	28.7	11.6

Fish oil = Menhaden oil (Supplied by Zapata Protein, Reedville, VA). Lard was purchased locally.

unsaturated n-3 fatty acids of fish oil in humans and by the polyunsaturated n-6 fatty acids of corn oil in rabbits may provide an explanation for the antiatherogenic effect of dietary polyunsaturated fatty acids.[4] In the present investigation, fish oil diets contained higher levels of n-3 polyunsaturated fatty acids than lard diets; however, the latter diets contained higher levels of n-6 polyunsaturated fatty acids than the former diets (Table 5) which may have modified the influence of n-3 polyunsaturated fatty acids on blood homocysteine levels.

After 2 wk of study, pigs fed SPI diets had lower levels of total serum cholesterol (1.26 to 2.11 mmol/l) than those fed casein diets (1.70 to 2.68 mmol/l). Similarly, pigs fed fish oil diets had lower levels of serum cholesterol (1.26 to 1.70 mmol/l) than those fed lard diets (2.11 to 2.68 mmol/l). Differences in serum triglycerides of pigs fed the 4 experimental diets were, however, small (0.24 to 0.26 mmol/l). Casein is more cholesterolemic and atherogenic than soy protein.[19] Differences in amino acid profiles of proteins may be partly responsible for the alteration of plasma cholesterol in rats fed different animal and plant proteins.[20-23] Casein contained higher levels of glutamic acid, methionine and tyrosine (cholesterol raising amino acids) than SPI, while SPI contained higher levels of arginine and glycine (cholesterol lowering amino acids) than casein.

5. ACKNOWLEDGMENTS

The authors are grateful to J. Matte, K. Kittle, N. Beausoleil and S. Stals for their technical assistance and to Dr. J. Fournier for his advice and supervision during collection of blood and tissue samples.

6. REFERENCES

1. Clarke, R., Daly, L., Robinson, K., Naughten, E., Cahalane, S., Fowler, B., and Graham, I., 1991, Hyperhomocysteinemia: an independent risk factor for vascular disease, *N. Engl. J. Med.* 324:1149-1155.

2. McCully, K.S., 1993, Chemical pathology of homocysteine 1. Atherogenesis, *Ann. Clin. Lab. Sci.* 23:477-493.

3. Ueland, P.M., and Refsum, H., 1989, Plasma homocysteine, a risk factor for vascular disease: plasma levels in health, disease, and drug therapy, *J. Lab. Clin. Med.* 114:473-501.

4. Olszewski, A.J., and McCully, K.S., 1993, Fish oil decreases serum homocysteine in hyperlipemic men, *Coronary Art. Dis.* 4:53-60.

5. National Research Council, 1988, *Nutrient Requirements of Swine*, 9th revised ed., National Academy Press, Washington, D.C., pp. 50-53.

6. Sarwar, G., and Botting, H.G., 1990, Rapid analysis of nutritionally free amino acids in serum and organs (liver, brain and heart) by liquid chromatography of precolumn phenylisothiocyanate derivatives, *J. Assoc. Off. Anal. Chem.* 73:470-475.

7. Fiskerstrand, T., Refsum, H., Kvalheim, G., and Ueland, P.M., 1993, Homocysteine and other thiols in plasma and urine: automated determination and sample stability, *Clin. Chem.* 39:263-271.

8. Association of Official Analytical Chemists, 1990, *Official Methods of Analysis*, 15th ed., A.O.A.C., Arlington, VA, sections 982.30, 985.28 and 988.15.

9. Sarwar, G., Peace, R.W., and Botting, H.G., 1988, Complete amino acid analysis in hydrolysates of foods and feces by liquid chromatography of precolumn phenylisothiocyanate derivatives, *J. Assoc. Off. Anal. Chem.* 71:1172-1175.

10. Ratnayake, W.M.N., Hollywood, R., O'Grady, E., and Pelletier, G., 1993, Fatty acids in some common food items in Canada, *J. Am. Coll. Nutr.* 12:651-660.

11. Anderson, A., Brattstrom, L., Israelsson, B., Isaksson, A., Hamfelt, A., and Hultberg, B., 1992, Plasma homocysteine before and after methionine loading with regard to age, gender, and menopausal status, *Eur. J. Clin. Invest.* 22:79-87.

12. Stampfer, M.J., Malinow, M.R., Willett, W.C., Newcomer, L.M., Upson, B., Ullmann, D., Tishler, P.V., and Hennekens, C.H., 1992, A prospective study of plasma homocyst(e)ine and risks of myocardial infarction in US physians, *J. Am. Med. Assoc.* 268:877-881.

13. Ueland, P.M., Refsum, H., and Brattstrom, L., 1992, Plasma homocysteine and cardiovascular disease, in: *Atherosclerotic Cardiovascular Disease, Hemostasis, and Endothelial Function*, (R.B. Francis, Jr., ed), Marcel Dekker Inc, New York, pp. 183-236.

14. Kromhout, D., Bosschieter, E.B., and Coulander, C.D.L., 1985, The inverse relation between fish consumption and 20-year mortality from coronary heart disease, *N. Engl. J. Med.* 312:1205-1209.

15. Kromann, N., and Green, A., 1980, Epidemiological studies in the Upernavik district, Greenland: incidence of some chronic diseases 1950-1974, *Acta. Med. Scand.* 208:401-406.

16. Dyerberg, J., Bang, H.O., and Hjorne, N., 1975, Fatty acid composition of the plasma lipids in Greenland Eskimos, *Am. J. Clin. Nutr.* 28:958-966.

17. Leaf, A., and Weber, P.C., 1988, cardiovascular effects of n-3 fatty acids, *N. Engl. J. Med.* 318:549-557.

18. McCully, K.S., Olszewski, A.J., and Vezeridis, M.P., 1990, Homocysteine metabolism in atherogenesis: effect of the homocysteine thiolactonyl derivatives, thioretinaco and thioretinamide, *Atherosclerosis* 83:197-206.

19. Kritchevsky, D., 1995, Dietary protein, cholesterol and atherosclerosis: a review of the early history, *J. Nutr.* 125:589S-593S.

20. Jaques, H., Deshaies, Y., and Savoie, L., 1986, Relationship between dietary proteins, their in vitro digestion products, and serum cholesterol in rats, *Atherosclerosis* 61:89-98.

21. Sautier, C., Dieng, K., Flament, C., Doucet, C., Suquet, J.P., and lemonnier, D., 1983, Effect of whey protein, casein, soya bean, and sunflower proteins on the serum, tissue and faecal steroids in rats, *Br. J. Nutr.* 49:313-319.

22. Sautier, C., Flamant, C., Doucet, C., and Suquet, J.P., 1986, *Nutr. Rep. Int.* 34:1051-1061.

23. Sugiyama, K., and Muramatsu, K., 1990, Significance of the amino acid composition of dietary protein in the regulation of plasma cholesterol, *J. Nutr. Sci. Vitaminol.* 36:105S-110S.

71

LUNG EICOSANOID PRODUCTION IN NEONATAL PIGS FED FORMULA SUPPLEMENTED WITH n-3 AND n-6 FATTY ACIDS

Margaret C. Craig-Schmidt[*] and Meng-Chuan Huang

Nutrition and Food Science
Auburn University
Auburn, Alabama 36849

1. ABSTRACT

The neonatal pig was used to investigate the influence of n-3 and n-6 fatty acids on lung eicosanoid metabolism. The effects of feeding fish oil, a source of eicosapentaenoic acid (20:5n-3) and docosahexaenoic acid (22:6n-3), were compared to those of feeding canola oil, a source of linolenic acid (18:3n-3). Male littermates from 5 sows were assigned at 1 da of age to 1 of 4 dietary treatments: 20% corn oil, 2% corn oil + 18% coconut oil, 2% corn oil + 18% fish oil or 2% corn oil + 18% canola oil. Animals were bottle fed for 28 da. Additionally, 5 piglets served as a naturally reared group. Production of thromboxane A_2 and prostacyclin by minced, incubated lung tissue was determined by radioimmunoassay of the inactive metabolites. Compared to corn oil fed animals, production of thromboxane by incubated lung was decreased in fish oil fed animals to 32% and in canola oil fed animals to 60%. Similar inhibition of prostacyclin was observed. Among the 5 groups, the greatest eicosanoid production was observed in lung of coconut oil fed and naturally reared animals. Dietary canola oil, containing an omega-3 fatty acid of shorter chain length than that in fish oil, inhibited lung eicosanoid production but to a lesser extent than fish oil. In another experiment, more physiologically relevant concentrations of very long chain polyunsaturated fatty acids were used. We evaluated if supplementation of infant formula with microbial sources of arachidonic acid (20:4n-6), docosahexaenoic acid or both at concentrations only slightly greater than those found in human milk would modulate eicosanoid production in neonatal pig lung. For 25 da, piglets (n = 5/group) received 1 of 4 diets: standard diet containing a fat blend similar to conventional infant formula, diet containing 0.9% of total

---

[*] Reprint requests to: Dr. Margaret Craig-Schmidt, Nutrition and Food Science, Auburn University, Auburn, AL 36849, (334) 844-3263.

Advances in Swine in Biomedical Research, edited by Tumbleson and Schook
Plenum Press, New York, 1996

fatty acids as arachidonate, diet containing 0.7% docosahexaenoate or diet containing 1.0% arachidonate plus 0.8% docosahexaenoate. Inclusion of arachidonate in the diet increased both thromboxane and prostacyclin production by 25 to 35%. Although docosahexaenoic acid supplementation resulted in the least eicosanoid production among treatments, suppression was observed for only thromboxane when supplementation with both fatty acids was compared to supplementation with arachidonate alone. Thus, eicosanoid production increased in response to small amounts of dietary arachidonate and tended to decrease in response to dietary docosahexaenoate. Eicosanoid production by neonatal pig lung can be manipulated by dietary n-3 and n-6 fatty acids.

2. INTRODUCTION

Eicosanoids are powerful regulatory compounds derived from arachidonic acid (AA, 20:4n-6) by the action of the enzyme cyclooxygenase.[1] These compounds, which include prostaglandins, thromboxane A_2, prostacyclin and leukotrienes, play important roles in late gestation and in the neonate. There is evidence that prostacyclin serves as a key mediator of pulmonary vasomotor tone during the perinatal period.[2,3] A developmental increase in pulmonary artery cyclooxygenase-I (COX-I) gene expression[4] and the subsequent rise in pulmonary prostacyclin production[5,6] are believed to be important in decreasing pulmonary vascular resistance, leading to successful cardiopulmonary transition at the time of birth. Further, impaired prostacyclin status is associated with several pulmonary neonatal diseases such as idiopathic respiratory distress syndrome,[7] sudden infant death syndrome[8] and persistent pulmonary hypertension.[9]

On the other hand, increased amounts of some eicosanoids are associated with some disease states.[1,10-12] An eicosanoid, thromboxane A_2, is a vasoconstrictor and platelet aggregator, yet shares the same substrate with the vasodilator, prostacyclin. The effects of thromboxane A_2 oppose those of prostacyclin in regulating pulmonary vascular integrity,[13] and its potent vasoconstrictory and bronchoconstrictory effects have been implicated in allergen mediated bronchoconstriction in asthmatics[14] and patients suffering from other respiratory stress.[15]

The biosynthesis of eicosanoids is dependent upon an adequate supply of dietary essential fatty acids, including linoleic acid (18:2n-6). Arachidonic acid, which serves as the immediate substrate for eicosanoid biosynthesis, is derived from dietary linoleic acid through the action of desaturases and elongases. The parent n-3 series fatty acid is α-linolenic acid (18:3n-3), with eicosapentaenoic acid (EPA, 20:5n-3) and docosahexaenoic acid (DHA, 22:6n-3) serving as its 20 and 22 carbon long chain polyunsaturated fatty acid derivatives.[1] The families of the n-6 and n-3 series are not interconvertible and compete for the same desaturation and elongation system.[16,17]

Marine oil rich in n-3 polyunsaturated fatty acids, such as EPA and DHA, is known to lessen the effects of several diseases by decreasing proinflammatory eicosanoid production.[10,11,18] Some of the beneficial effects of n-3 fatty acids may be exercised via competition with n-6 fatty acids. Increasing dietary n-3 fatty acids results in changes in the phospholipid fatty acid composition of cell membranes which subsequently exerts an influence on the type and the amount of eicosanoids generated from arachidonic acid. Eicosanoid production responsive to dietary polyunsaturated fatty acids of the n-6 and n-3 series has been reported in rodents,[19-22] chicken,[23] human[24] and pigs.[18,25] Little is known about the effects of physiological concentrations of dietary AA and DHA on eicosanoid metabolism in the neonate. These "preformed" long chain polyunsaturated fatty acids are present in small amounts in human milk and are needed during the perinatal period for optimal neural development and visual acuity.[26-28] Conventional infant formula do not contain AA and

DHA; addition of these fatty acids to infant formula may be beneficial. However, before arachidonic acid can be added safely to infant diets even in small amounts, e.g., less than 1% of total fatty acids, neonatal eicosanoid metabolism effects must be elucidated.

Because of the physiological similarities between piglets and human infants, neonatal pigs are a suitable animal model to study lipid metabolism of term gestation infants.[29] Therefore, the overall objective of a series of experiments in our laboratory has been to use the neonatal piglet to investigate the influence of n-3 and n-6 fatty acids on lung eicosanoid production. In the first experiment, we investigated the influence of dietary n-6 and n-3 fatty acids at concentrations higher than normally would be consumed; whereas, in a second experiment, the influence of dietary n-6 and n-3 long chain polyunsaturated fatty acids at physiologically relevant concentrations was studied. The results of these experiments have possible implications for treatment of lung disease in the infant and for the composition of infant formula.

3. METHODS

3.1. Experimental Design

In Experiment I,[30] the effects of feeding fish oil, a source of n-3 fatty acids with 20 and 22 carbons, were compared to those of feeding canola oil, a source of the shorter chain n-3 precursor fatty acid, linolenic acid (18:3n-3). These two n-3 containing diets were compared to two n-6 containing treatments. Corn oil was used as a dietary source rich in n-6 fatty acids; coconut oil was chosen because it contained linoleic acid in approximately the same concentration as fish oil, but minimal amounts of n-3 fatty acids. Basal amounts of essential fatty acids, specifically linoleic acid (18:2n-6), were supplied by adding corn oil to the coconut oil, fish oil or canola oil. Effects of these dietary treatments on lung eicosanoid production were studied.

In Experiment II,[31] more physiologically relevant concentrations of very long chain polyunsaturated fatty acids were studied. Microbial sources of AA and DHA were incorporated into the diet at concentrations only slightly greater than those normally present in human milk. These 2 long chain polyunsaturated fatty acids were added separately or in combination to investigate the possible effects of adding DHA and AA to infant formula on eicosanoid production.

3.2. Animals

In each Experiment, 25 male piglets (Yorkshire x Landrace sows bred to Duroc boars) of normal gestation were obtained at 1 da of age from the Swine Nutrition Unit of the Alabama Agricultural Experimental Station, Auburn University, Auburn, AL. Littermates, weighing at least 1.36 kg from each of 5 litters, were assigned randomly to receive 1 of 4 formulas (n = 5). Additionally, 1 piglet from each of the 5 litters remained with the sow to serve as a naturally reared group (SOW). In Experiment II, the size of the remaining litter was standardized to 5 by cross fostering if necessary. The 20 formula fed neonatal pigs were housed individually in metal cages with 1 cm^2 plastic mesh bottoms in temperature controlled rooms and artificially fed for 28 da (Experiment I) and 25 da (Experiment II). During the first 2 wk, heat was provided with heat lamps attached to each cage. Passive immunity was provided by administration of bovine immunoglobulin concentrate (LitterMaker™, Protein Technology, Petaluma, CA) to the formula for the first 72 hr of feeding. Animals were weighed at 4 da intervals. Piglets were fed milk based liquid formula by bottle 5 times a da during the first wk and 4 times a da after da 8. All animal protocols were approved by the Auburn University Institutional Animal Care and Use Committee.

3.3. Diets

Formulas were prepared using a nonfat sow's milk replacer (Pet-Ag, Elgin, IL) at 20% solids by weight of the total (Table 1). Fat blends were added such that the formula contained 20% of the solids as fat. Thus, 1000 g of formula contained 160 g nonfat milk replacer, 40 g oil and 800 g water.

In Experiment I, 4 oils were used in preparation of experimental diets: corn oil, coconut oil, fish oil and canola oil. As shown in Table 2, all oils were balanced with respect to cholesterol and antioxidants as analyzed by the National Marine Fisheries Service (Charleston, SC). The fatty acid compositions of oil mixtures used in experimental diets are shown in Table 3. Thus, dietary treatments for Experiment I were: 18% coconut oil + 2% corn oil, 20% corn oil, 18% fish oil + 2% corn oil and 18% canola oil + 2% corn oil.

Dietary treatments for Experiment II were: STD, a basal diet; STD + AA, the standard diet modified to contain AA; STD + DHA, the standard diet modified to contain DHA; and STD + BOTH, the standard diet modified to contain AA and DHA. In the SOW group, piglets were reared naturally. The fatty acid composition of the standard (STD) diet was similar to that of some fat blends currently used in infant formula. Microbial sources of lipids were added to the STD diet such that long chain polyunsaturated fatty acids were present in the following concentrations as percent of total fatty acids: STD+AA, 0.9% as AA; STD+DHA, 0.7% as DHA; and STD+BOTH, 1.0% as AA plus 0.8% as DHA. Fats were balanced with respect to antioxidant (carotenoid) content. Complete fatty acid compositions of fats in dietary treatments and sow's milk are shown in Table 4.

Formula diets were prepared within 24 hr of feeding and stored at 4C. Formulas were warmed in a water bath immediately before they were fed; consumption was recorded at

Table 1. Nutrient and Energy Composition of Pig Milk Replacer

Ingredient	Percentage
Dried Skim Milk	82.810
Ca-Na Caseinate	15.963
MR Vitamin Premix	0.499
Dicalcium Phosphate	0.399
Iron Sulfate, 31% Fe	0.014
Zinc Methionine	0.015
Manganese Sulfate	0.002
Copper Sulfate	0.002
Calcium Iodate	0.0003
MR Choline Chloride, 50%	0.301
Crude Protein (%)	42.5
ME (Kcal/kg)	3370

[1]Vitamins and minerals in 1 kg of nonfat pig milk replacer: 110,000 I.U. vitamin A; 9,900 I.U. vitamin D_3; 198 I.U. vitamin E; 22 mg riboflavin; 84 mg pantothenic acid; 62 mg niacin; 110 µg vitamin B_{12}; 1.32 mg menadione; 1.25 mg folic acid; 6.82 mg thiamine; 0.506 mg pyridoxine; 2691 mg biotin; 10.6 g choline; 9 g phosphate; 1 g magnesium; 13.6 g potassium; 50.6 mg iron; 13.2 g copper; 63.8 mg zinc; 6.8 mg manganese; 1.76 mg iodine.

Table 2. Antioxidant Contents of Oils Used in Experimental Diets (Experiment I)

Diet	Antioxidant			
	α-Vit E[1]	γ-Vit E[2]	TBHQ[3]	CHOL[4]
	mg/g Oil			
18% Coconut Oil + 2% Corn Oil	1.0	1.2	0.191	2.0
20% Corn Oil	0.9	1.1	0.186	2.1
18% Fish Oil + 2% Corn Oil	1.1	1.1	0.199	2.2
18% Canola Oil + 2% Corn Oil	0.9	1.2	0.192	2.1

[1]α-Tocopherol.
[2]γ-Tocopherol.
[3]Tertiary Butyl hydroquinone.
[4]cholesterol.

every feeding. Piglets were weighed at 4 da intervals. In Experiment II, collection of sow's milk was performed on da 12 of lactation.

3.4. Analysis of Lung Eicosanoids

Approximately 0.4 to 0.5 g of fresh lung was rinsed with saline, weighed to the nearest 0.01 mg, minced in a defined manner (100 strokes of a scalpel blade) and placed into an Erlenmeyer flask containing 10 ml preincubated Kreb's-Ringer bicarbonate buffer. Lung tissue was incubated in an atmosphere of 95% oxygen and 5% carbon dioxide with shaking at 37C in a water bath. The reaction was stopped by addition of 1N HCl to obtain a pH of 3.5 to 4.0. Using 20 ml of ethylacetate, eicosanoids were extracted twice from the incubation media. Ethylacetate fractions were pooled, evaporated to dryness and then redissolved in ethanol. Samples were stored at -80C until analysis. Immediately prior to analysis by

Table 3. Fatty Acid Composition of Oil Blends Used in Experimental Diets (Experiment I)

FattyAcid	Treatment diets			
	18% Coconut+ 2% Corn	20% Corn	18% Fish+ 2% Corn	18% Canola+ 2% Corn
	% of total fatty acids			
8:0	3.26	-	-	-
10:0	4.40	-	-	-
12:0	41.77	-	-	-
14:0	13.60	0.03	5.44	0.05
16:0	7.48	9.90	13.48	4.02
16:1	0.01	0.14	7.67	0.18
18:0	2.75	1.44	1.99	1.35
18:1 n-9	13.87	25.13	11.04	47.64
18:2 n-6	8.90	60.99	7.27	23.29
18:3 n-3	0.11	1.08	1.01	8.68
18:4 n-3	-	-	3.18	-
20:4 n-6	-	-	0.39	-
20:5 n-3	-	-	14.34	-
22:5 n-3	-	-	2.02	-
22:6 n-3	-	-	9.41	-
Other FA[1]	3.85	1.29	21.96	4.79

[1]Individual fatty acids less than 2%.

Table 4. Fatty Acid Composition of Experimental Diets (Experiment II)

Fatty Acid	Treatment Diets				
	STD	STD+DHA	STD+AA	STD+BOTH	SOW
	% Total fatty acids				
<18C	36.7	38.3	37.5	37.8	52.0
18:0	6.8	6.8	7.5	6.6	3.8
18:1n-9	39.7	37.7	38.0	37.0	28.1
18:1n-7	1.1	1.1	1.2	1.1	1.7
18:2n-6	12.8	12.6	12.1	12.7	12.7
18:3n-3	1.4	1.4	1.3	1.4	0.6
20:0	0.2	0.2	0.2	0.3	-
20:1n-9	0.2	0.2	0.2	0.2	-
20:4n-6	-	-	0.9	1.0	0.7
22:0	0.2	0.2	0.2	0.2	-
22:6n-3	-	0.7	-	0.8	0.04

radioimmunoassay (RIA), samples were evaporated and redissolved in gel phosphate buffered saline. Extraction efficiency was measured by addition of 1000 cpm [^{3}H] prostaglandin $F_{2\alpha}$ (Dupont NEN, Boston, MA) prior to extraction. The average extraction efficiency was 93%.

Eicosanoids were analyzed using [H^3] RIA kits (Dupont NEN, Boston, MA) for assay of 6-keto-prostaglandin $F_{1\alpha}$ (6-keto-PGF$_{1\alpha}$), the stable metabolite of prostacyclin, and thromboxane B_2 (TXB$_2$), the stable metabolite of thromboxane A$_2$. All reagents were reconstituted with distilled water, stored in the refrigerator and used within 1 mo. All samples and standards were assayed in duplicate. The assay consisted of 100 µl of lung extract, 100 µl of antibody, 100 µl of tracer and enough assay buffer (50 mM phosphate buffer, pH 6.8) to give a total volume of 500 µl. Mixtures were incubated for 18 hr at 4C. Then, 500 µl of charcoal suspension (0.5 mL of 0.5% Norit A in 10 mM phosphate buffer, pH 6.8, with 0.5% Dextran T-70) was added to each tube. Samples were placed in an ice bath for 10 min and centrifuged for 15 min at 4C at 1300 x g. Supernatant radioactivity was determined using a Packard Liquid Scintillation Analyzer (1600TR Model, Packard Instrument, Downers Grove, IL). Data reduction was performed with Riasmart (Packard Instrument, Downers Grove, IL).

3.5. Statistical Analyses

Data were analyzed by the Department of Research Data Analysis of the Alabama Agricultural Experiment Station of Auburn University. Differences in weight gain and formula consumption were tested by Repeated Measures Analysis of Variance. Differences among treatments in fatty acids and eicosanoid data among treatments were tested by analysis of variance (ANOVA) procedures. Differences among treatment means were determined using Duncan's Multiple Range Test (Experiment I) or Least Significance Difference (LSD) analysis (Experiment II) (P=0.05). The Statistical Analysis System Computer Program (SAS/STAT, SAS Institute, Cary, NC) was used to analyze the data.

Table 5. Body Weight and Weight Gain of Neonatal Pigs

Treatment	BW_i^3	BW_f^4	Weight Gain
		Kg	
Coconut + Corn	$1.72^a \pm 0.19$	$10.80^a \pm 0.44$	$9.08^a \pm 0.27$
Corn	$1.81^{ab} \pm 0.07$	$11.66^a \pm 0.51$	$9.85^a \pm 0.51$
Fish + Corn	$1.73^a \pm 0.14$	$10.76^a \pm 0.93$	$9.03^a \pm 0.82$
Canola + Corn	$1.76^{ab} \pm 0.11$	$11.88^a \pm 0.17$	$10.12^a \pm 0.18$
Sow Group[5]	$2.11^b \pm 0.13$	$9.06^b \pm 0.63$	$6.95^b \pm 0.75$

[1]Data are presented as mean ± SEM (n=5 for each group).
[2]Means in the same column not sharing a common superscript are different ($p<0.05$).
[3]Initial body weight.
[4]Final body weight.
[5]Naturally reared group.

4. RESULTS

4.1. Body Weight and Formula Consumption of Piglets

In Experiment I, both the final body weight and weight gain of the naturally reared group was less than that of the piglets bottle fed experimental diets (Table 5). This was true in spite of the fact that the naturally reared group had the greatest initial body weight among all treatments. Formula consumptions were not different among the 4 formula fed groups (Table 6).

Initial mean body weights (1.7 to 1.8 kg) among the 5 treatment groups in Experiment II were not different (Fig. 1). After 25 da of dietary treatment, both the final weight (9.8 to 10.4 kg) and weight gain (8.0 to 8.6 kg) of the 4 dietary treatment groups (STD, STD+AA, STD+DHA, STD+BOTH) were greater than the naturally reared group (SOW), which had a final weight and weight gain of 7.8 ± 0.3 kg and 6.1 ± 2.7 kg, respectively. No differences in formula consumption or in final body weight due to diet were observed among the 4 dietary treatment groups during the experimental period (Figs. 1 and 2).

4.2. Thromboxane and Prostacyclin

In Experiment I, production of TXB_2 and 6-keto-$PGF_{1\alpha}$ was decreased in the 2 groups fed diets containing n-3 fatty acids (fish and canola) compared to the naturally reared group (Figs. 3 and 4). Production of TXB_2 by incubated lung was decreased in fish oil fed animals to 32% and in canola oil fed animals to 60% of that of corn oil fed animals. Similar reductions

Table 6. Formula Consumption by Neonatal Pigs[1,2]

Week	18% Coconut + 2% Corn	20% Corn	18% Fish+ 2% Corn	18% Canola+ 2% Corn
	liters of milk per day per piglet			
1st	0.42 ± 0.04	0.40 ± 0.02	0.40 ± 0.03	0.44 ± 0.03
2nd	0.82 ± 0.03	0.88 ± 0.09	0.84 ± 0.09	0.94 ± 0.05
3rd	1.38 ± 0.07	1.50 ± 0.11	1.37 ± 0.14	1.57 ± 0.04
4th	1.98 ± 0.06	2.05 ± 0.07	1.73 ± 0.21	2.07 ± 0.04

[1]Data are presented as mean ± SEM (n=5 for each group).
[2]No significant difference for means of formula consumption among dietary groups was observed ($p>0.05$).

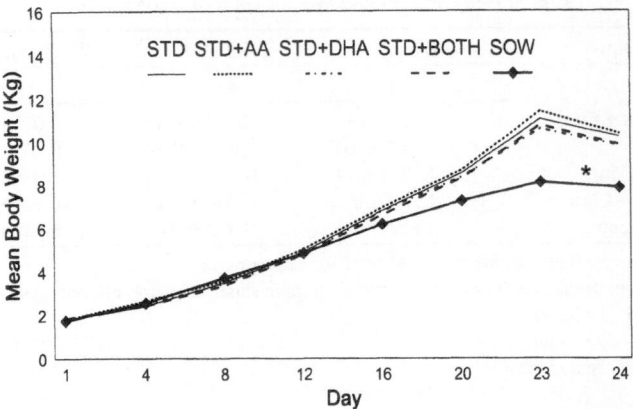

Figure 1. Body weights of neonatal pigs (Experiment II). Data are expressed as mean body weight (kg/pig) for each group of piglets (n=5). * The overall weight gain of the naturally reared group (SOW) was less (p<0.05) than the other 4 dietary treatment groups.

occurred for 6-keto-PGF$_{1\alpha}$ production in response to feeding diets containing n-3 fatty acids, i.e., decreases to 16% with fish oil feeding and to 59% with canola oil. Animals fed fish oil as a source of n-3 polyunsaturated fatty acids, with chain lengths greater than 20 carbons, exhibited less eicosanoid production than animals fed canola oil containing linolenic acid (18:3n-3). This difference in effect between the C20-22 and C18 n-3 fatty acids was significant for 6-keto-PGF$_{1\alpha}$ but not for TXB$_2$. Fish oil fed animals exhibited less eicosanoid production than coconut oil fed animals, as did corn oil fed animals compared to the naturally reared animals.

In Experiment II, addition of arachidonic acid to the diet increased both TXB$_2$ and 6-keto-PGF$_{1\alpha}$ production by 25 to 35% (Figs. 5 and 6). A difference was observed both when STD + AA was compared to STD and when STD + BOTH was compared to STD + DHA. Although docosahexaenoic acid supplementation resulted in the least eicosanoid production among treatments, suppression was observed for only TXB$_2$, when supplementation with

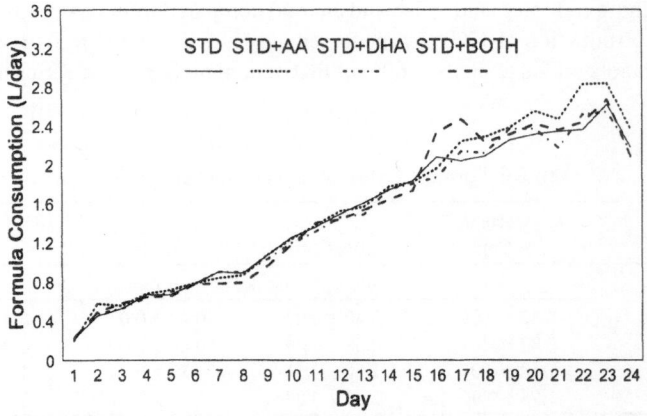

Figure 2. Formula consumption by neonatal pigs (Experiment II). Data are expressed as the mean formula consumption l/pig/da for each group of piglets (n=5). Overall formula consumption among the 4 dietary treatment groups was not different.

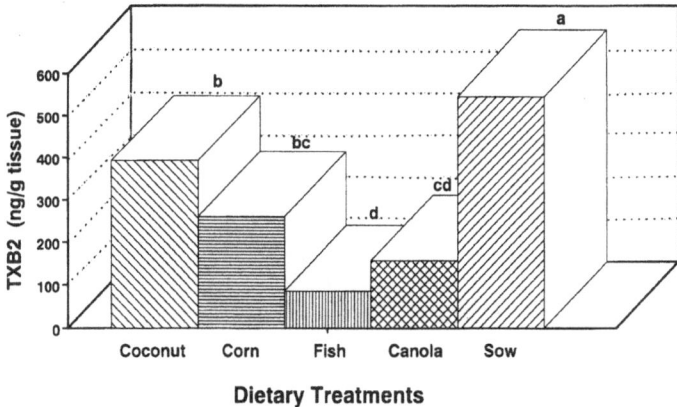

Figure 3. Effect of dietary n-6 and n-3 polyunsaturated fatty acids on TXB_2 production by incubated pig lung.[30] Treatments not sharing a common superscript are significantly ($p<0.05$) different.

both fatty acids was compared to supplementation with arachidonate alone. TXB_2 production in naturally reared pigs (SOW) lung was similar to that of the STD + AA group. Production of 6-keto-$PGF_{1\alpha}$ in the SOW group was similar to that of the STD + AA group as well as to the STD + BOTH treatment.

5. DISCUSSION

5.1. Body Weight and Formula Consumption of Piglets

Final body weights and formula consumptions for bottle fed piglets in both experiments were similar among all groups, indicating dietary treatment did not adversely affect the piglets. This finding was particularly important in the case of fish oil fed animals in light of reports that feeding fish oil can result in decreased food consumption and body weights, attributable to rancidity of marine oil.[32] The antioxidants added to oils in the current

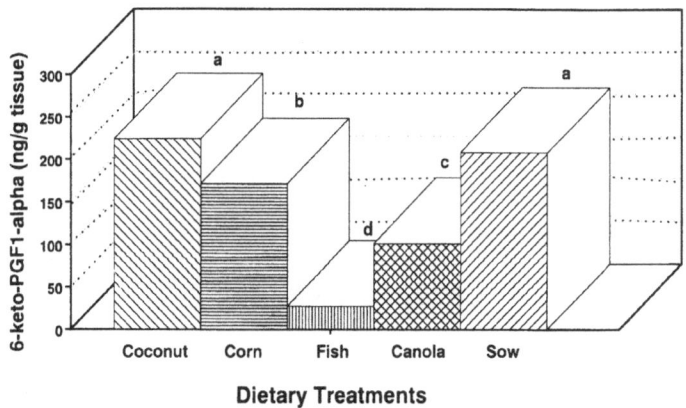

Figure 4. Effect of dietary n-6 and n-3 polyunsaturated fatty acids on 6-keto-PGF_1 alpha production by incubated pig lung.[30] Treatments not sharing a common superscript were significantly ($p<0.05$) different.

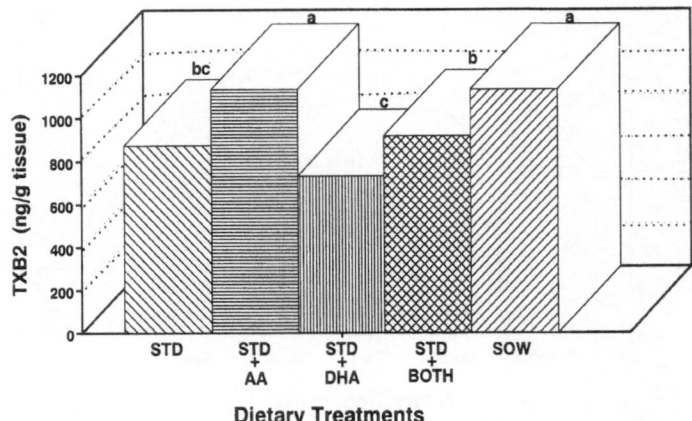

Figure 5. Effect of dietary arachidonic acid and docosahexaenoic acid on TXB_2 production by incubated pig lung.[31] Treatments not sharing a common superscript were significantly (p<0.05) different.

experiment, as well as daily preparation of diets, may have protected the fish oil from oxidation and thus prevented any weight loss in the fish oil fed animals in comparison to the other groups.

Bottle fed animals exhibited greater weight gain than naturally reared animals in both experiments. The lower weight gain in naturally reared animals was observed only in the latter part of the experimental period. Sow milk production becomes limiting during this period; in commercial swine operations, solid food or creep feeding is provided to supplement sow's milk.[33] However, creep feeding was withheld in the experimental protocol because of the potential for confounding effects.

5.2. Effects of n-6 Fatty Acids on Eicosanoid Production

5.2.1. Effects of Dietary Linoleic Acid (18:2n-6). Even though dietary linoleic acid, after elongation and desaturation to arachidonic acid, serves as the precursor to eicosanoids,

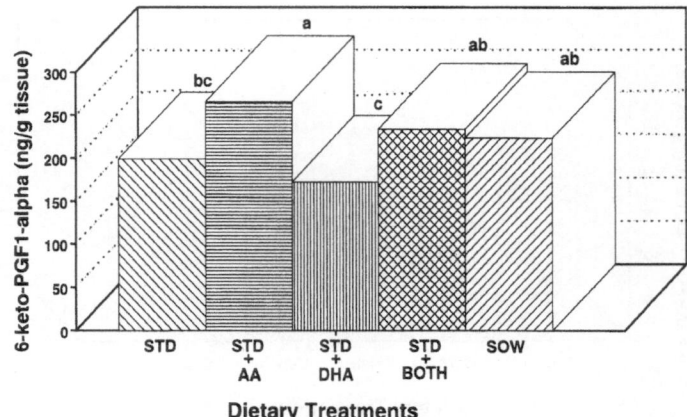

Figure 6. Effect of dietary arachidonic acid and docosahexaenoic acid on 6-keto-PGF_1 alpha production by incubated pig lung[31]. Treatments not sharing a common superscript are significantly different (p<0.05).

the relationship between dietary linoleic acid and eicosanoid production is not a direct linear relationship. Mathias and Dupont[34] have shown that thromboxane production during clotting of whole blood from rats peaks at less than 5% of calories as linoleic acid; increases in dietary linoleic acid after this point resulted in decreased thromboxane production. This offers a possible explanation for why piglets fed corn oil exhibited less thromboxane and prostacyclin production compared to coconut oil fed and naturally reared animals. The 18% coconut oil + 2% corn oil diet contained 3.5% of calories as linoleic acid, compared to 24% of calories as linoleic acid in the 20% corn oil diet. Based on analysis of milk samples collected in Experiment II and literature values for the nutrient content of sow's milk,[35] it is estimated that sow's milk contained 6.9% of calories as linoleic acid. Thus, greater eicosanoid synthesis occurred when linoleic acid was in the range of 3 to 7% of calories, compared to 24% of calories. Our results in pigs are therefore consistent with the results of Mathias and Dupont[34] in rats.

5.2.2. Effects of Dietary Arachidonic Acid (20:4n-6). Arachidonic acid serves as the primary precursor for eicosanoid synthesis and the production of eicosanoids is dependent upon the availability of phospholipid AA.[19,21,23] The effects of the addition of AA to the diet in small, but physiologically significant, amounts has not been well studied. A clear effect of AA supplementation, even in very small amounts, was observed in Experiment II. Inclusion of 0.9% of total fatty acids (0.36% kcal) as AA in the formula elevated 6-keto $PGF_{1\alpha}$ and TXB_2 production by incubated lung. Production of both eicosanoids was 33% greater in the STD+AA than in the STD treatment. Similarly, addition of AA in combination with DHA resulted in a 25 to 35% increase when compared with the treatment to which only DHA was added. Eicosanoid production in the naturally reared group was similar to that of the STD + AA group. This result was consistent with the fact that sow's milk contained very small amounts of n-3 fatty acids and AA in amounts comparable to the STD + AA diet. Thus, in all groups in which the diet contained AA, increased eicosanoid production was observed.

In the current study, AA was supplied as the microbial oil and thus was in the triglyceride form. The increase in eicosanoid metabolism observed in the current experiment is in agreement with other studies which used ethyl esters as a source of supplemental AA and which used greater concentrations of AA than were used in our study. Seyberth et al.[24] reported the major urinary prostaglandin E metabolite in man was increased by 47% in 3 volunteers after 2 to 3 wk administration of ethylarachidonate. More recently, Whelan et al.[19] investigated the influence of addition of moderate amounts of AA as the ethyl ester to the diet on eicosanoid metabolism. There was increased production of TXB_2 by platelets following stimulation with thrombin in animals fed supplemental AA. Moreover, Li et al.[36] reported that production of 6-keto $PGF_{1\alpha}$ and TXB_2 in peritoneal macrophages in vivo stimulated with opsonized zymosan was elevated in AA supplemented animals compared to the control. The enhancement effect of dietary AA on eicosanoid synthesis also has been observed in rat aorta. Mann et al.[21] reported greater 6-keto $PGF_{1\alpha}$ production by animals fed a high fat diet supplemented with AA as the ethyl ester. The urinary metabolite of prostacyclin also increased in the AA-supplemented group.

5.3. Effects of Dietary n-3 Fatty Acids

5.3.1. Effects of Canola Oil Versus Marine Oil as a Source of n-3 Polyunsaturated Fatty Acids. Dietary canola oil, containing an omega-3 fatty acid of shorter chain length than that in fish oil, inhibited lung eicosanoid production but to a lesser extent than fish oil (Experiment I). In earlier work from our laboratory,[23] we found inhibition of eicosanoid production in chicken lung by dietary linolenic acid. Decreased n-6/n-3 ratios in the diet were reflected in decreases in the arachidonic/linoleic acid ratio of serum phospholipids as well as in lung eicosanoid production. In rat hepatic microsomes, inhibitory effects of dietary n-3 fatty acids have been observed on desaturase enzyme activities.[37,38] One possible explanation of this inhibitory effect is that 18:3n-3 depletes 20:4n-6 by inhibiting delta-6-desaturase, thereby preventing the conversion of linoleic acid to AA and subsequently decreasing the availability of AA for eicosanoid biosynthesis.

Inhibitory effects on eicosanoids by marine oil feeding have been observed by a number of investigators.[39] In studies using pigs, Fritsche et al.[25] reported that dietary supplementation with menhaden oil during late gestation and lactation can increase the content of n-3 fatty acids in nursing piglet's immune cells and result in reduced release of eicosanoids. Increased consumption of marine oil or supplementation of the diet with the ethyl ester of EPA is reported to diminish eicosanoid production in rodents.[19,21,36] Moreover, Surette et al.[22] reported that rats fed diets supplemented with 2.5% n-3 polyunsaturated fatty acids from marine sources decreased TXB_2 production by peritoneal macrophages compared to animals fed a control diet. The n-3 polyunsaturated fatty acids, DHA and EPA, can be incorporated efficiently into membrane phospholipids at the expense of AA and act as a competitive inhibitor of cyclooxygenase. Generally, EPA, as opposed to DHA, has been regarded as the major inhibitor of eicosanoid synthesis;[40] however, until recently there were few scientists who investigated the effects of DHA independent of EPA.

   5.3.2. Effects of Dietary Supplemental DHA. In Experiment II, we investigated the effect of a small amount of DHA (0.7% to 0.8% of total fatty acids) on eicosanoid production by the neonatal pig lung. Supplementation with DHA alone tended to decrease 6-keto-$PGF_{1\alpha}$ production (STD+AA vs STD+BOTH; STD vs STD+DHA) but the suppression was not significant. On the other hand, the suppression of TXB_2 due to DHA addition was more pronounced and was significant when the STD+BOTH was compared to STD+AA. Thus, at concentrations used in the present experiment, DHA appeared to be a weak inhibitor of eicosanoid production. In contrast, Ikeda et al.[20] reported that DHA fed as the ethyl ester at 9.5% of total fatty acids is a strong inhibitor of prostacyclin production by rat aorta. DHA concentrations used in that experiment were much greater than those used in the current experiment. In human subjects, neither purified DHA nor EPA diet was observed to alter thromboxane synthesis.[41]

5.4. Conclusions and Implications

   In conclusion, eicosanoid production by neonatal pig lung can be manipulated by dietary n-3 and n-6 fatty acids. Dietary canola oil, containing an omega-3 fatty acid of shorter chain length than that in fish oil, inhibited lung eicosanoid production but to a lesser extent than fish oil. Dietary linoleic acid, at relatively low concentrations, resulted in greater eicosanoid production than dietary linoleic acid at greater concentrations. Eicosanoid production increased in response to physiological concentrations (such as those in breast milk) of dietary arachidonic acid and tended to decrease in response to dietary docosahexaenoic acid. The neonatal pig can be used to study dietary manipulation of lung eicosanoids; our results have implications for the composition of infant formula, as well as for possible treatment of respiratory problems affecting the human infant.

6. REFERENCES

1. Sardesai, V.M., 1992, Biochemical and nutritional aspects of eicosanoids, *Annu. Rev. Nutr.* 3:562-569.

2. Green, R.S., Rojas, J., and Sundell, H., 1979, Pulmonary vascular response to prostacyclin in fetal lambs, *Prostaglandins* 18:927-934.
3. Lock, J.E., Olley, P.M., and Coceani, F., 1980, Direct pulmonary vascular response to prostaglandins in the conscious newborn lamb, *Am. J. Physiol.* 238:H631-H638.
4. Brannon, T.S., North, A.J., Wells, L.B., and Shaul. P.W., 1994, Prostacyclin synthesis in ovine pulmonary artery is developmentally regulated by changes in cyclooxygenase-1 gene expression, *J. Clin. Invest.* 93:2230-2235.
5. Leffler, C.W., and Hessler, J.R., 1981, Perinatal pulmonary prostaglandin production, *Am. J. Physiol.* 241:H756-H759.
6. Leffler, C.W., Hessler, J.R., and Green, R.S., 1984, The onset of breathing at birth stimulates pulmonary vascular prostacyclin synthesis, *Pediatr. Res.* 18:938-942.
7. Kappa, P., Koivisto, M., Viinikka, L., and Ylikorkala, M., 1982, Increased plasma immunoreactive 6-keto-prostaglandin $F_{1\alpha}$ levels in newborns with idiopathic respiratory distress syndrome, *Pediatr. Res.* 16;827-829.
8. Seto, D.S.Y., Matsuo, M., Hokama, J.L.R.Y., Shirai, L.K., and Hokama, Y., 1983, Prostacyclin formation in lung tissues of infants in sudden infant death syndrome (SIDS), non-SIDS and adults, *J. Med.* 14:451-459.
9. Philips, J.B., and Lyrene, R.K., 1984, Prostaglandins, related compounds and perinatal pulmonary circulation, *Clin. Perinatol.* 11:565-579.
10. Kinsella, J.E., Lokesh, B., and Stone, R., 1990, Dietary n-3 polyunsaturated fatty acids and amelioration of cardiovascular disease: possible mechanisms, *Am. J. Clin. Nutr.* 52:1-28.
11. Simopoulos, A.P., 1991, Omega-3 fatty acids in health and disease and in growth and development. *Am. J. Clin. Nutr.* 54:438-463.
12. Pauly, T.H., Aziz, S.M., Horstman, S.J., and Gillespie, M.N., 1992, Impact of prostaglandin and thromboxane synthesis blockade on deposition of group B Streptococcus in lung and liver of intact piglet, *Pediatr. Res.* 31:14-17.
13. Barnard, J.W., Ward, R.A., Adkins, K.W., and Taylor, A.E., 1992, Characterization of thromboxane and prostacyclin effects on pulmonary vascular resistance, *J. Appl. Physiol.* 72:1845-1853.
14. Hardy, C.C., Bradding, P., Robinson, C., and Holgate, S.T., 1988, Bronchoconstrictor and antibronchoconstrictor properties of inhaled prostacyclin in asthma, *J. Appl. Physiol.* 64:1567-1574.
15. Slotman, G.J., Burchard, K.W., Yellin, S.A., and Williams, J.J., 1986, Prostaglandin and complement interaction in clinical acute respiratory failure, *Arch. Surg.* 121:271-274.
16. Brenner, R.R., and Peluffo, R.O., 1966, Effect of saturated and unsaturated fatty acids on the desaturation in vitro of palmitic, stearic, oleic, linoleic, and linolenic acids, *J. Biol. Chem.* 241:5213-5219.
17. Holman, R. T., 1986, Nutritional and biochemical evidences of acyl interaction with respect to essential polyunsaturated fatty acids, *Prog. Lipid Res.* 25:29-39.
18. Murray, M.J., Svingen, B.A., Holman, R.T., and Yaksh, T.L., 1991, Effect of a fish oil diet on pigs' cardiopulmonary response to bacteremia, *J.P.E.N.* 15:152-158.
19. Whelan, J., Surette, M.E., Hardard'ottir, I., Lu, G., Golemboski, K.A., Larsen, E. and Kinsella, J.E., 1993, Dietary arachidonic acid enhances tissue arachidonate levels and eicosanoid production in Syrian hamsters, *J. Nutr.* 123:2174-2185.
20. Ikeda, I., Wakamatsu, K., Inayoshi, A., Imaizumi, K., Sugano, M., and Yazawa, K., 1994, α-Linolenic, eicosapentaenoic and docosahexaenoic acids affect lipid metabolism differently in rats, *J. Nutr.* 124:1898-1906.
21. Mann, N.J., Warrick, G.E., O'Dea, K., Knapp, H.R., and Sinclair, A.J., 1994, The effect of linoleic, arachidonic acid and eicosapentaenoic acid supplementation on prostacyclin production in rats. *Lipids* 29:157-162.
22. Surette, M.E., Whelan, J., Lu, G., Hardard'ottir, I, and Kinsella, J.E., 1995, Dietary n-3 polyunsaturated fatty acids modify Syrian hamster platelet and macrophage phospholipid fatty acyl composition and eicosanoid synthesis: a controlled study, *Biochim. Biophy. Acta* 1255:185-191.
23. Craig-Schmidt, M.C., Faircloth, S.A., and Weete, J.D., 1987, Modulation of avian lung eicosanoids by dietary omega-3 fatty acids. *J. Nutr.* 117:1197-1206.
24. Seyberth, H.W., Oelz, O., Kennedy, T., Sweetman, B.J., Danon, A., Frolich, J.C., Heimberg, M., and Oats, J.A., 1975, Increased arachidonate in lipids after administration to man: Effects on prostaglandin biosynthesis, *Clin. Pharmacol. & Ther.* 18:521-529.
25. Fritsche, K.L., Alexander, D.W., Cassity, N.A., and Huang, S.- C., 1993, Maternally-supplied fish oil alters piglet immune cell fatty acid profile and eicosanoid production, *Lipids* 28:677-682.
26. Innis, S.M., 1991, Essential fatty acids in growth and development, *Prog. Lipid Res.* 30:39-103.
27. Carlson, S.E., 1995, The role of PUFA in infant nutrition, *Inform* 6:940-946

28. Craig-Schmidt, M.C., Stieh, K., and Lien, E.L., 1996, Retinal fatty acids of piglets fed docosahexaenoic and arachidonic acids from microbial sources, *Lipids* 31:53-59.
29. Innis, S.M., 1993, The colostrum-deprived piglet as a model for study of infant lipid nutrition, *J. Nutr.* 123:386-390.
30. Craig-Schmidt, M.C., Johnson, J.A., Chaung, H.C., Powe, T.A., 1992, Effect of dietary fish oil vs. canola oil on lung eicosanoids of the neonatal pig, in: *Third International Congress on Essential Fatty Acids and Eicosanoids* (R. Gibson, and A. Sinclair, eds), Adelaide, South Australia.
31. Huang, M.C., and Craig-Schmidt, M.C., 1996, Arachidonate and docosahexaenoate added to infant formula influence fatty acid composition and subsequent eicosanoid production in neonatal pigs. J. Nutr. 126: 2199-2208.
32. Ip, C., Ip, M.M., and Sylvester, P., 1986, Relevance of trans fatty acids and fish oil in animal tumorigenesis studies, in: *Dietary Fat and Cancer,* Volume 222 (C. Ip, D.F. Birt, and A.E. Rogers, eds.), Liss, New York, pp. 283-294.
33. English, P., Smith, W., and MacLean, A., 1979, *The Sow-Improving her Efficiency, 3rd ed,* Farming Press, Ipswich, Suffolk, Great Britain.
34. Mathias, M.M., and Dupont, J., 1985, Quantitative relationships between dietary linoleate and prostaglandin (eicosanoid) biosynthesis, *Lipids* 20:791-801.
35. Pond, W.G., and Houpt K.A., 1978, *The Biology of the Pig,* Cornell University Press, New York.
36. Li, B., Birdwell, C., and Whelan, J., 1994, Antithetic relationship of dietary arachidonic acid and eicosapentaenoic acid on eicosanoid production in vivo, *J. Lipid Res.* 35:1869-1877.
37. Garg, M.L., Thomson, A.B.R., Clandinin, M.T., 1988, Effect of dietary cholesterol and/or omega-3 fatty acids on lipid composition and Δ^5-desaturase activity of rat liver microsomes, *J. Nutr.* 118:661-668.
38. Garg M.L., Sebokova, E., Thomson, A.B.R., Thomas, M.T., 1988, Δ^6-desaturase activity in liver microsomes of rats fed diets enriched with cholesterol and/or omega-3 fatty acids, *J. Biochem.* 249:351-356.
39. Nettleton, J.A., 1995, *Omega-3 Fatty Acids and Health,* Chapman & Hall, New York.
40. Croft, K.D., Beilin, L.J., Legge, F.M., and Vandongen, R., 1987, Effects of diets enriched in eicosapentaenoic or docosahexaenoic acids on prostanoid metabolism in the rat, *Lipids* 22:647-650.
41. Von Schacky, C. and Weber, P.C., 1985, Metabolism and effects on platelet function of the purified eicosapentaenoic and docosahexaenoic acids in humans, *J. Clin. Invest.* 76:2446-2450.

72

GENETICALLY DIVERSE PIG MODELS IN NUTRITION RESEARCH RELATED TO LIPOPROTEIN AND CHOLESTEROL METABOLISM

Wilson G. Pond[*] and Harry J. Mersmann

USDA/ARS, Children's Nutrition Research Center
Department of Pediatrics
Baylor College of Medicine
Houston, Texas 77030

1. ABSTRACT

The pig is a widely accepted animal model for nutrition research related to growth and development, atherogenesis, obesity, and lipoprotein and cholesterol metabolism. We review current knowledge of lipoprotein metabolism in pigs and describe research in populations of genetically diverse swine. Specifically, we report the results of experiments showing that dietary cholesterol during neonatal life affects atherogenesis and indices of central nervous system development and lipid metabolism differentially in genetically obese and lean pigs and in pigs selected genetically for low or high plasma total cholesterol.

2. INTRODUCTION

The pig has been used for approximately 40 yr as a model organism for research into cholesterol and lipoprotein metabolism with application to atherogenesis. In addition to developing spontaneous atherogenic lesions, young prepubertal pigs develop fatty streaks and early atherosclerotic lesions within several wk when fed diets high in fat and cholesterol. There were reviews of much of this work about a decade ago;[1,2] whereas, an excellent summarization of the history of the use of pig atherogenic models was presented 6 yr ago.[3] In addition to use as an atherogenesis model, the pig continues to be a model for other aspects of cardiovascular research because of a number of similarities to the human. The anatomy

---

[*] Reprint requests to: Dr. Wilson Pond, USDA/ARS, Children's Nutrition Research Center, 1100 Bates Street, Houston, TX, 77030, (713) 798-7055.

Advances in Swine in Biomedical Research, edited by Tumbleson and Schook
Plenum Press, New York, 1996

of the coronary arteries and the pattern for development of collateral circulation after permanent ischemic episodes are 2 important similarities. Many aspects of various pig cardiovascular models were reviewed a decade ago.[4] We do not attempt a comprehensive review of porcine cholesterol and lipoprotein metabolism but will briefly survey general aspects including a few references to indicate current endeavors in the field; the major emphasis is on results obtained in our laboratory with genetically obese and lean pigs, and pigs genetically selected for high and low plasma cholesterol.

3. CHOLESTEROL AND LIPOPROTEIN METABOLISM

In humans, cholesterol is obtained partially from animal products in the diet and synthesized partially de novo. Synthesis is from the 2 carbon moiety, acetate, to yield the 6 carbon moiety, 3-hydroxy-3-methylglutaryl-CoA (HMG-CoA). After decarboxylation, the remaining 5 carbon moieties are assembled into a 30 carbon moiety and formed into the cholesterol ring structure with 4 rings and 27 carbons.[5] Cholesterol synthesis is regulated primarily by the enzyme HMG-CoA reductase that catalyzes an early (6 carbon molecule) committed step toward cholesterol synthesis; the reductase is regulated highly and activity is decreased in the presence of cholesterol, to achieve a balance between endogenous synthesis and dietary intake.[4] In most mammals, liver is a major site of cholesterol synthesis, with considerable activity in intestinal mucosa and in organs that synthesize steroid hormones, such as adrenal cortex and gonads. The central nervous system (CNS) has the capacity to synthesize cholesterol as well. The cholesterol molecule is essential for homeostasis of mammals; cholesterol is the precursor for synthesis of steroid hormones and for bile acids, it is a major component of several lipoproteins and it is an integral part of biological membranes. Sources of cholesterol are endogenous synthesis plus a dietary contribution ranging from zero in strict herbivores to considerable amounts in carnivores and some omnivorous humans.

Whole body cholesterol input (synthesis + diet + reuptake from the gut) equals output (endogenous products + deposition + excretion). Approximations of input and output of cholesterol metabolism in an adult omnivorous human follow (adapted from Fig. 5 of Bietz and Knight[6] and Fig. 4-1 of Marinetti[7]). There are 100 g cholesterol in the total body. Diet provides 400 mg (300 to 500 mg); whereas, 900 mg is synthesized endogenously. Cholesterol removal from the body is great via secretion into the gut lumen, mostly as bile acids and sterols; fecal sterols are to a large extent reabsorbed, so net excretion is approximately 1100 mg/da, balanced by accretion rate. Some 200 mg/da enters various body pools such as skin, steroid hormones and cell membranes. Thus, the 5 major facets of cholesterol metabolism are dietary intake, endogenous synthesis, production of various products such as hormones and membranes, bile acid synthesis and fecal sterol reabsorption.

Cholesterol is transported in plasma as a component of various lipoprotein molecules. Dietary cholesterol is absorbed, then exits the intestinal cell mostly as chylomicron particles. These large particles contain primarily triglyceride, but are also the carrier for entrance of dietary cholesterol, as cholesterol esters, into the lipoprotein transformation mechanisms. Chylomicrons are reduced to chylomicron remnants after removal of triglycerides; remnants reenter the liver. Liver synthesizes and secretes a smaller triglyceride and cholesterol ester rich particle, the very low density lipoprotein (VLDL), that gives rise to intermediate and low density lipoproteins (LDL) after removal of triglyceride. Apolipoprotein B-100 (apo B-100) is present on the surface of LDL particles which allows them to bind to the LDL receptor, present in various peripheral tissues, including liver. Uptake of LDL allows cholesterol to be transferred to tissues for use in biosynthetic mechanisms and as structural components of cell membranes. The high density lipoprotein (HDL) particle is synthesized primarily in liver; it has a major role in recirculating cholesterol back to liver from peripheral tissues. In this brief discussion of lipoprotein metabolism, we did not attempt to elicit the complexities of lipoprotein metabolism, including interchanges of individual apolipoproteins and lipid moieties between lipoprotein particle classes; numerous reviews of lipoprotein metabolism are available.[7-9]

4. ATHEROGENESIS

Development of atherogenic lesions in arteries of humans is a gradual process that begins very early in those consuming the typical U.S. diet, that is high in fat and cholesterol.[10] Pigs naturally develop atherosclerosis; aortic fatty streaks are observed at an early age when diets are high in fat (particularly saturated fat) and cholesterol.[2-4] The full blown lesion that produces partial occlusion of arteries does not appear until experimental diets are fed for many mo but the pathological state is hastened by denuding of arterial intima.[11] There is evidence for inheritance of the propensity toward atherogenesis in pigs.[12] Pigs also have been used to investigate regression of atherosclerotic plaques.[13,14]

5. PIG MODELS

Not only has the pig been accepted as a relatively good model for the human cardiovascular system, but it is readily available, large enough to use for many types of intervention experiments, comparatively easy to care for and handle, even in biomedical settings, and it can be obtained in any of several miniature versions. The pig is particularly interesting as a nutritional model because it is one of the few common and available mammals that is omnivorous. Canine and feline species are carnivorous; whereas, many small laboratory species are herbivorous. The sheep, a suitably sized mammal used extensively for fetal and neonatal research, is an herbivorous ruminant; whereas, the rabbit is, in effect, a ruminant. Although pigs routinely are fed grain based diets in the U.S. commercial swine industry, their natural diet is composed of both plant and animal products. In fact, the commercial swine production industry has fed diets containing fish meal, meat scraps, whey and/or blood; the use of these ingredients is based on availability and cost. There are many reviews of the pig as a model for humans[15-19] including the particular applications to cardiovascular research.[3,4]

5.1. Merits of the Pig Model

Porcine cholesterol synthesis in vivo appears to occur primarily in the liver (>60%) with 30% in adipose tissue and <5% in intestine;[20] these early studies have technical flaws but more sophisticated approaches using labelled water are not available. Sites of cholesterol synthesis vary considerably among other species; intestine contributes 10 to 25% and the liver contributes 10 to 50% of whole animal synthesis.[21,22]

Thirty years ago, it was established that the lipoprotein pattern of pig serum was similar in many ways to that of human serum.[23] Many researchers since that time have confirmed and expanded this information; comparison of the serum lipoprotein pattern among several mammalian species indicates the pig is perhaps most similar to the human.[24] Human serum has more HDL than LDL; the majority of cholesterol is carried in LDL, as in the pig.[24] Although there are considerable similarities in apolipoprotein structures of the pig and human,[25] there are differences, as expected.[26-28] The complexity of lipoprotein metabolism, involving multiple tissues, receptors, lipoproteins and apolipoproteins, essentially ensures divergence in details of metabolism across species. For example, compared to the human, the pig has 1) a broader density range for LDL, 2) 2 distinct LDL particles and 3) no plasma cholesterol ester transfer activity (with consequent low synthesis of LDL from VLDL[29], yet transfer of HDL cholesterol esters to LDL).[30] In some species, apolipoprotein B (apo B) is edited by liver to produce apolipoprotein B-48 in addition to apo B-100 (from 18% editing in the dog to 70% editing in the mouse).[31] In the human, pig, cow, sheep and cat, the liver produces only apo B-100.[31] All models have limitations. As pointed out recently,[32] dietary cholesterol intake to down regulate LDL receptors 50% in chow fed hamsters equates to 2000 to 5000 mg/da in the human. If a 20 kg pig eats 1 kg diet containing 0.1% cholesterol, this equals 1000 mg/da for 20 kg body weight or 3500 mg/da when extrapolated to a 70 kg human. Obviously, many of our models should be scrutinized carefully when extrapolating to the human.

Porcine serum lipids and lipoproteins have been quantified during the fetal, neonatal and growing periods,[33-35] as well as in pregnancy.[36] The newborn pig has been used extensively as a model to examine the effects of the lipid composition of formula on lipid and cholesterol metabolism, particularly by Innis and coworkers.[37,38]

5.2. Types of Pig Models

5.2.1. Nutritional Models. There have been many diverse studies of nutritional influences on cholesterol and lipoprotein metabolism in swine. The literature was surveyed rather comprehensively a decade ago;[1] the references cited are to indicate these types of studies continue today. Recently, scientists have examined various dietary fat sources with a wide array of fatty acid compositions, including the currently popular fish oils,[39-42] protein sources, particularly comparisons of plant and animal proteins,[43-48] minerals,[49] fiber sources[50] and other dietary factors, including pharmacologically active materials.[51] The pig is a favorable model for these types of approaches because of its omnivorous nature; however, many investigators do not take advantage of this fact to feed pigs diets that contain animal proteins (a major component of human diets in the United States, Canada and Europe).

A particular note should be made of the extensive studies of Lee, Kim, Thomas and coworkers[2] who used the pig as a model for nutritionally induced atherogenesis for many years. Some of the studies from that laboratory involved feeding experimental diets for extended periods to allow development of full blown atherosclerosis. Scientists in that laboratory not only investigated the effects of various diet components but also devoted considerable effort toward understanding development of the atherosclerotic lesion and factors modulating regression of the lesion. Metabolic aspects of cholesterol metabolism were investigated,[52,53] including early publication of hepatic activities of the rate limiting enzymes for cholesterol (HMG-CoA reductase)[54,55] and bile acid (cholesterol-7-alpha-hydroxylase)[55] synthesis. The investiagtors have continued to be active.[56,57]

5.2.2. Genetic Models. Four models will be discussed: 1) the von Willebrand pig, 2) genetically selected lean and obese pigs, 3) hyperapo B lipoproteinemic pigs selected by Rapacz and 4) the hyper and hypocholesterolemic pigs selected by Pond and coworkers

5.2.2.1. The von Willebrand Pig. This model has been used to some extent in atherosclerosis research. This pig has a genetic defect in that it produces no von Willebrand factor and thus has extended bleeding time. The lesser adherence of platelets in these pigs is presumed to be a factor in the lesser atherogenesis observed.[58] However, some have demonstrated a lesser development of atherosclerosis in these pigs,[59,60] while others have not.[60,61] Perhaps some differences between studies may relate not to the von Willebrand factor, but to the presence of particular apo B genotypes. Pigs with the type 5 apo B allele have much higher serum cholesterol concentrations and develop more atherosclerosis than those with the type 1 or type 8 allele, regardless of the presence or absence of the von Willebrand disease.[60]

5.2.2.2. Obese Pigs. Obesity generally is viewed as a factor in cardiovascular disease and more specifically as one of the multiple contributors to atherosclerosis development. Genetically lean and obese pigs were developed by Hetzer and Harvey[62] beginning in the 1950's. Pigs were selected for thin and thick backfat at 80 kg body weight; selection in a Duroc line continued for 18 generations; whereas, that in a Yorkshire line continued for 14 generations. In the late 1970's at the US Meat Animal Research Center, obese Duroc were crossed with obese Yorkshire pigs to produce the current 50% Duroc-50% Yorkshire obese pigs. Similarly, lean Duroc were crossed with lean Yorkshire pigs to produce the current 50% Duroc-50% Yorkshire lean pigs. The pigs represent a multigene obesity (as are most human obesities) that already is expressed at or before birth as evidenced by several endocrine and cellular measurements. Phenotypic obesity is not evident grossly until 4 to 6 wk postpartum. The known growth, metabolic and endocrine properties of lean and obese pigs have been summarized.[63] These pigs have been used for various nutritional studies of atherogenesis and cholesterol metabolism by Pond and coworkers over the last several years.

A modified sow milk replacer diet containing no cholesterol (Table 1) was used as the basal diet in all neonatal pig experiments reported from this laboratory. Crystalline USP grade cholesterol (ICN Nutritional Biochemicals, Cleveland, OH) was added to this basal diet at levels ranging from 0.20 to 1.00% of the dry diet in individual experiments.

5.2.2.2.1. Experiment 1[64] The purpose was to determine the effect of dietary cholesterol and genetic background on plasma lipid and lipoprotein metabolism in newborn pigs. Sixteen 1 da old pigs (8 lean, 8 obese) were fed a sow milk replacer diet (Table 1) containing 0 (LC; 4 lean, 4 obese) or 0.5% (HC; 4 lean, 4 obese) cholesterol to da 32; from da 33 to 56, all pigs were fed the same diet containing 1.0% cholesterol. Blood was collected at 4 da intervals from da 28 to 56 for measurement of plasma total cholesterol (TC), HDL-cholesterol (HDL-C), triglycerides (TG), apolipoprotein A-I (apo A-I), apo B and LDL and HDL particle size distribution profiles. Mean concentrations of TC from da 28 to 32 were 2.43, 2.50, 3.16 and 8.74 mmol/l for lean/LC, lean/HC, obese/LC and obese/HC pigs, respectively. Corresponding values for HDL-C were 1.29, 1.15, 1.30 and 1.41 mmol/l, respectively. TC in obese pigs responded more dramatically than in lean pigs to the HC diet; whereas, HDL-C tended to decrease in response to HC in lean pigs but increase in obese pigs. TG concentration was unaffected by dietary cholesterol in either genotype. After all pigs were changed to the 1% cholesterol diet on da 33, plasma TC increased in all groups; mean values for the 10 sampling times from da 33 to 56 were 13.9, 10.0, 9.7 and 21.8 mmol/l for lean/LC, lean/HC, obese/LC and obese/HC, respectively. Corresponding values for HDL-C were 1.78, 1.71, 1.48 and 1.91 mmol/l, respectively. There was an effect of genotype and a diet x genotype interaction on plasma TC and HDL-C concentrations, resulting from a greater increase in TC and HDL-C in obese than in lean pigs in response to the extra high (1%) dietary cholesterol challenge. The higher plasma HDL-C in obese than in lean pigs would suggest protection against atherogenesis.

Plasma apo B levels responded to dietary cholesterol differently in lean and obese pigs; obese pigs fed LC and lean pigs fed LC or HC had similar plasma apo B levels before the 1% cholesterol challenge. Apo B levels in obese pigs fed HC, in contrast, were nearly 4 fold greater. After the challenge, apo B increased 2 to 6.5 fold in all groups. Again, a diet x genotype interaction was observed, in which apo B in lean pigs previously fed cholesterol (lean/HC) increased less than in lean pigs not previously fed cholesterol (lean/LC); whereas,

Table 1. Composition of the no cholesterol dry diet[1]

Ingredient	$(g/100g)^2$
Coconut oil	9.50
Corn oil	1.00
Soybean Flour	10.00
Whey protein concentrate	5.00
Calcium caseinate	15.50
Whey (grade 1)	50.04
Dicalcium phosphate	1.70
Propylene glycol	6.00
Soweena Pig Krave Extra	0.20
Vitamin Premix 181[3]	0.25
Mineral Premix 181[3]	0.63
Choline chloride	0.05
Antioxidant/mold inhibitor	0.10
Neomycin sulfate	0.03

[1]Modified Soweena (coconut and corn oil replacing tallow) from Merrick Foods (Union Center, WI). Each dry diet contained 13.65 joules of metabolizable energy per kilogram. Propylene glycol improves pelleting properties in cases requiring a pelletted diet; Soweena Pig Krave Extra is a nonnutritive sweetener to enhance diet palatability; BHA and propionic acid provide antioxidant and mold inhibition, respectively; neomycin sulfate minimizes diarrhea.
[2]Trace amounts of cholesterol possibly were provided by whey and whey protein concentrate, but the quantity was considered insignificant (< 1 mg/kg).
[3]Vitamin Premix 181 contains (per g): 8.3 mg retinyl acetate, 55 mcg cholecalciferol, 44 mg all cis-$\propto$ tocopheryl acetate, 11 mg vitamin B-12, 11 mg riboflavin, 17.6 mg D-calcium pantothenate, 14 mg niacin, 11 mg menadione, 22 mcg biotin, 66 mg ascorbic acid. Mineral Premix 181 contains (per g): 844 µg iodine, 47.9 µg selenium, 2.3 mg cobalt, 39.9 mg copper, 24.1 mg iron, 76 mg magnesium, 3.5 mg manganese, 23.3 mg zinc.

apo B in obese pigs previously fed cholesterol (obese/HC) was much higher than in those not previously fed cholesterol (obese/LC). In contrast, the apo A I level was similar in all groups during the 1% cholesterol feeding period, although before the challenge, apo A I levels in obese/HC pigs were 20% greater than in those in other groups, all of which were similar. The LDL particle size distribution profiles were affected by both diet and genotype. Before the 1% cholesterol challenge, lean/LC, lean/HC and obese/LC pigs had similar bimodal profiles (major peaks around Rf 0.45 and 0.55); peaks corresponding to smaller LDL (Rf 0.55) predominated in LC pigs; whereas, peaks corresponding to larger sized LDL predominated in HC pigs. In contrast, LDL peak size in obese/HC pigs was much larger and had major trimodal peaks near Rf 0.32, 0.38 and 0.42 plus large components that produced a broad peak near Rf 0.13. Challenge with the 1% cholesterol diet shifted LDL size distribution profiles toward larger components in all 4 groups.

Dietary cholesterol in the suckling age pig markedly affects lipoprotein metabolism and the response is modulated by genetic background of the animal. The greater rise in plasma TC, and other differential changes in lipoprotein profiles in response to a dietary cholesterol challenge in obese compared to lean pigs, suggests the possibility of a greater susceptibility to atherogenesis in the former. Experiment 3 was designed to test that hypothesis.

5.2.2.2.2. Experiment 2[65] The purpose was to determine the effects of age from 3 to 42 da and the role of dietary cholesterol on plasma lipids and activity of hepatic HMG-CoA reductase (the rate limiting enzyme in synthesis of cholesterol from acetate) in young obese and lean pigs and in crossbred contemporary pigs. Stage of development was a dominant factor in regulating liver HMG-CoA reductase activity in the pig (Table 2). Both dietary and endogenously synthesized cholesterol apparently are used predominantly for tissue building in the very young pig. HMG-CoA reductase activity increased considerably at da 13 even in pigs fed high cholesterol diets.

5.2.2.2.3. Experiment 3[66] The purpose was to determine the effect of dietary cholesterol and genetic background on plasma lipids and atherogenesis in suckling age pigs. Genetically lean (n = 7) and obese (n = 7) pigs were removed from their dams at 1 da of age and assigned randomly within genetic group in a 2 x 2 factorial arrangement of genetic group and diet to receive either the no cholesterol (NC) diet (Table 1) or the same diet with 1.0%

Table 2. Effect of age, genetic background and dietary cholesterol on body weight, plasma total cholesterol and liver HMG-CoA reductase in young pigs

		Genetic background		
Age, da	Diet chol, %	Crossbred(A)	Lean(L)	Obese(O)
		Plasma total cholesterol,[a] mmol/l		
3	0	3.19(0.61)	4.86(0.85)	2.95(0.96)
13	0	1.66	2.06	2.11
13	1	2.22	2.79	2.45
25	0	2.07	2.29	3.04
25	1	3.44	4.67	6.35
Pooled SD		0.9	0.8	0.8
		HMG-CoA reductase,[b] pmol/mg/min		
3	0	18.8(6.5)	13.0(12.2)	13.1(1.6)
13	0	78.9	48.4	81.2
13	1	45.5	40.1	42.3
25	0	17.7	39.8	26.7
25	1	7.6	13.0	6.9
Pooled SD		2.5	1.9	0.9

[a]Data indicated as mean with the pooled SD except for da 3 data, which are means with individual SD because these data were not included in the statistical model.In A pigs, there were significant age, diet and age x diet (plasma cholesterol increased more at older ages in pigs fed cholesterol) and age x gene (increase in plasma cholesterol at older ages was greater in obese than in lean pigs) effects. Gene effects were $P = 0.1$.
[b]Data indicated as mean with the pooled SD except for the da 3 data, which are means with individual SD because these data were not included in the statistical model. In A pigs, there was a significant age effect; whereas, in the obese and lean pigs there were age and diet effects. The age x gene effect was $P = .07$ (reductase was not elevated as much in 13 da lean pigs as in obese pigs and it remained at a higher level in 25 da Lean pigs than in Obese pigs). From McWhinney et al.[65]

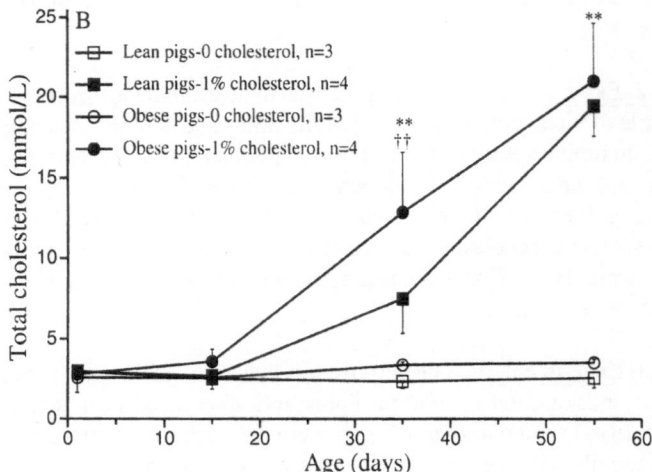

Figure 1. Plasma total cholesterol (TC) for lean pigs fed diet containing 0 cholesterol or 1% cholesterol and obese pigs fed 0 cholesterol or 1% cholesterol. Values are means ± SD. **Significant difference ($P<0.01$) due to 0 cholesterol vs 1% cholesterol diet. †† Significant difference ($P<0.01$) between lean and obese pigs. From Hackman et al.[66]

cholesterol added by weight (HC). Body wt was recorded every 7 to 10 da and blood was sampled from the anterior vena cava on da 1, 15, 35 and 55 for determination of TC, TG and HDL-C. Whole body in vivo cholesterol synthesis rate was estimated by measuring erythrocyte (RBC) deuterium labeled cholesterol enrichment following deuterium oxide administration. Deuterium oxide was administered subcutaneously on da 35 to 38. In vivo cholesterol fractional synthesis rate (FSR) was estimated by enrichment of deuterium in RBC cholesterol isolated from blood sampled on da 35 to 38.[67] Liver was used to determine activity of HMG-CoA reductase. The heart with left ventricle was opened longitudinally and the aorta to the bifurcation of the iliac arteries were used for histopathology. Plasma TC (Fig. 1), HDL-C (Fig. 2) and LDL-C + VLDL-C increased in both genetic groups fed cholesterol but obese pigs responded with a more rapid increase in TC and HDL-C and a higher concentration of HDL-C and a lower ratio of TC to HDL-C (Figure 3) than in lean pigs. As expected, FSR was reduced by dietary cholesterol; obese and lean pigs responded similarly (Fig. 4). Likewise, HMG-CoA reductase activity was depressed by dietary cholesterol in both lean and obese pigs to a similar degree (Table 3). Intimal lesions (plaques) were observed in the aorta of all 4 lean pigs fed HC and in 3 of 4 obese pigs fed HC; no lesions were found in lean or obese pigs fed NC. Microscopically, plaques developed in tunica intima; the intima was up to 0.5 mm thick and consisted of degenerate fibers with often preserved nuclei. Lean pigs fed HC had more severe plaques ($P<0.05$) than obese pigs fed HC based on plaque size (Fig. 5).

Atherogenic lesions can be produced by 8 wk of age in pigs by high dietary cholesterol and genetic differences exist in susceptibility to atherogenesis. The less severe plaque produced in obese than in lean pigs by dietary cholesterol may be related to the protective effect of the higher HDL-C and lower TC/HDL-C ratio in obese pigs in the presence of severe hypercholesterolemia. The data confirm the earlier observation[71] that obesity per se is not necessarily related to an increased propensity to atherosclerosis.

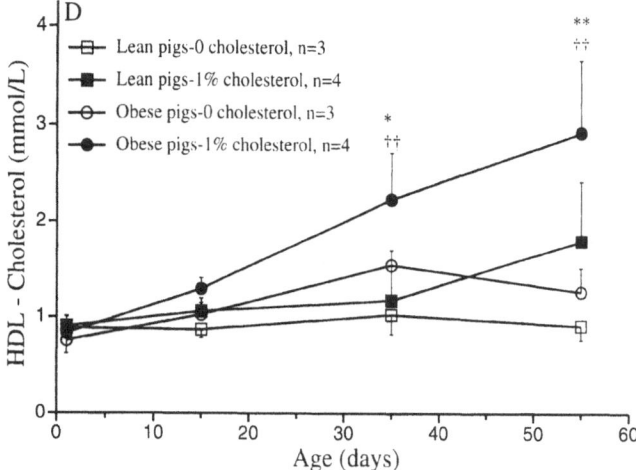

Figure 2. Plasma HDL-cholesterol for lean pigs fed diet containing 0 cholesterol or 1% cholesterol and obese fed 0 cholesterol or 1% cholesterol. Values are means ± SD. *Significant difference ($P<0.05$) due to 0 cholesterol vs 1% cholesterol **Significant difference ($P<0.07$) due to 0 cholesterol vs 1% cholesterol. †† Significant difference ($P < 0.01$) between lean and obese pigs. From Hackman et al.[66]

5.2.2.3. Hyperapo B Lipoproteinemic Pigs. Rapacz and coworkers began to examine the variation in immunological properties of porcine serum proteins more than 25 yr ago. Initial observations indicated 4 phenotypes associated with lipoproteins of a density characteristic of LDL; 2 different dominant genes controlled expression of the phenotypes. Eventually this work led to preparation of 16 distinct immunoreagents to identify 16 distinct variations in protein structure (16 epitopes). Classification of porcine genotypes (*Lpb*'s) for apolipoprotein B in thousands of pigs indicated several prominent genotypes in the general

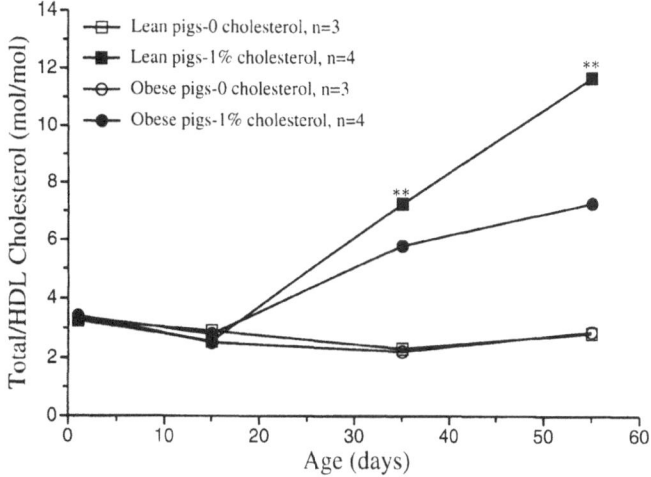

Figure 3. Ratio of plasma total cholesterol (TC) to plasma HDL cholesterol (HDL-C) for lean pigs fed 0 cholesterol (chol) or 1% cholesterol and obese pigs fed 0 cholesterol or 1% cholesterol. ** Significant difference ($P<0.01$) due to 0 cholesterol vs 1% cholesterol. From Hackman et al.[66]

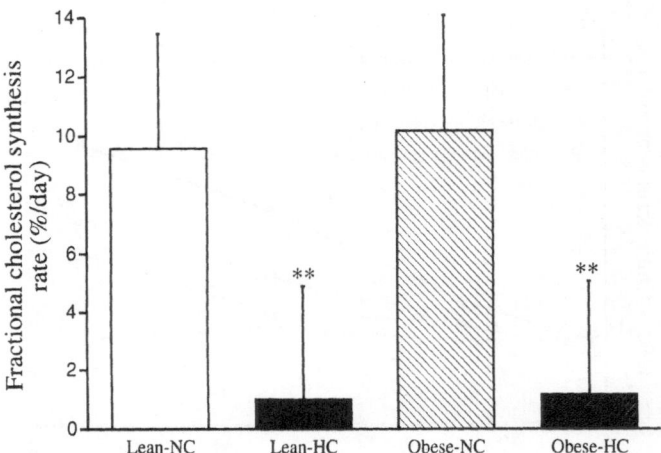

Figure 4. Fractional cholesterol synthesis rates from da 35 through da 38. Values are means ± SD. NC, no cholesterol diet; HC, high cholesterol diet. Synthesis rates are lower in pigs fed HC than in those fed NC diet (**, *P*<0.01). From Hackman et al.[66]

pig populations with several others restricted to particular breeds. The type 5 apo B allele (*Lpb* 5) was associated strongly with hypercholesterolemia and spontaneous atherogenesis in pigs, even when fed low fat, no cholesterol diets;[3] pigs with the highest LDL and cholesterol were homozygous for *Lpb* 5 (as well as *Lpu* 1 and *Lpr* 1).[68] These pigs do not have a defective LDL receptor but the LDL is defective in its binding to the normal LDL receptor and catabolism of LDL from *Lpb* 5 pigs is slow;[69]; both factors would be expected to contribute to hypercholesterolemia. The activity of the enzyme, lecithin:cholesterol acyltransferase, is low in the serum of *Lpb* 5 pigs and there is decreased HDL-C and decreased apolipoprotein A I.[70] These defects begin to account for the hypercholesterolemia and accelerated atherogenesis. Recently, using controlled breeding, it has been demonstrated that propensity toward hypercholesterolemia and atherogenesis is not limited to the homozygous *Lpb* 5/*Lpu* 1/*Lpr* 1 pigs; obviously many genes can be involved in producing the pathological state.[71]

5.2.2.4. Hyper and Hypocholesterolemic Pigs. Beginning in 1987, Pond and co-workers[72,73] began to select pigs for high and low plasma TC. Selection was at 56 da of age with pigs fed a diet containing <4% fat and no cholesterol. Progress in changing mean plasma cholesterol of the LG and HG population from 1987 (base population) through 1994 (generation 7) is summarized in Table 4. This population of pigs consists of a line with high

Table 3. HMG-CoA reductase activity (pmoles/mg protein/min) in liver at necropsy in lean and obese pigs fed either a no cholesterol or 1% cholesterol diet[a]

Lean NC	Lean HC	Obese NC	Obese HC
(n = 3)	(n = 4)	(n = 3)	(n = 4)
7.07 ± 2.87[b]	1.53 ± 0.87	9.47 ± 3.86	3.15 ± 2.80

[a]NC = no cholesterol diet, HC = 1% cholesterol diet.
[b]HC < NC (*P*<0.002) From Hackman et al.[66]

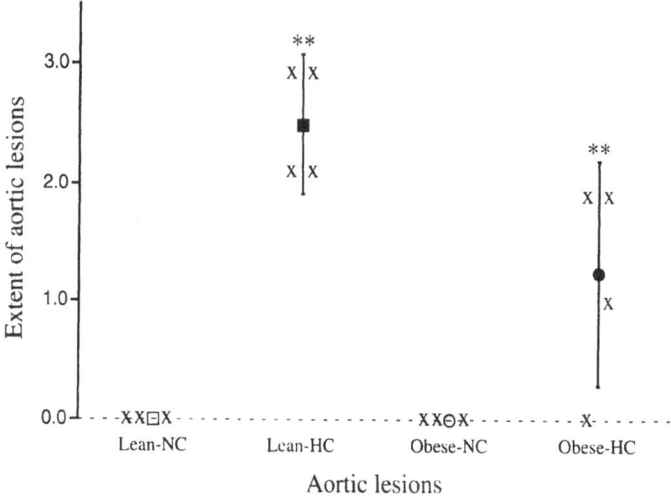

Figure 5. Extent of aortic atherosclerosis (0 = absent, 1 = mild, 2 = moderate, 3 = severe). X, individual pigs; □, ■,○,• are means ± SD; NC, no cholesterol diet; HC, high cholesterol diet; atherosclerosis is more extensive in pigs fed HC vs NC diet (**, $P<0.01$). $P = 0.08$ for lean vs obese genetic groups. From Hackman et al.[66]

plasma TC (mean of about 125 mg/dl = HG) and one with low plasma TC (mean of about 70 mg/dl = LG).

In generation 3, there was a positive correlation in LG pigs[73] at 8 wk between plasma TC and body weight (r = 0.470, $P<0.05$ for males and r = 0.215, $P<0.05$, for females) (Fig. 6). This relationship has persisted through generation 7. Selection for low or high plasma cholesterol has been found[74] to be associated with correlated responses in other traits, e.g., greater ovulation rate and litter size in LG than in HG pigs; those unexpected findings are not addressed further here.

We have been exploring the cardiovascular and plasma lipid response of these genetically diverse pigs to neonatal dietary cholesterol intake. Research with genetically high and low plasma TC pigs in the third generation of selection[75] showed that when they were allowed to suckle for the first 4 wk of postnatal life, both HG and LG pigs had increased plasma cholesterol after being fed a high cholesterol, high fat diet during the latter half of the growing finishing period (12 to 25 wk of age). However,

Table 4. Total plasma cholesterol (TC) at 8 wk of age for 7 generations of pigs selected for high or low plasma cholesterol (Mean ± SD)[a]

Year	Generation	HG(mg/dL)	LG(mg/dL)
1988	1	87 ± 16(201)[b]	74 ± 17(201)
1989	2	108 ± 17(199)	78 ± 14(201)
1990	3	107 ± 15(173)	65 ± 12(219)
1991	4	136 ± 30(63	81 ± 18(75)
1992	5	130 ± 20(92)	82 ± 16(93)
1993	6	118 ± 18(67)	78 ± 31(63)
1994	7	117 ± 18(86)	64 ± 13(69)

[a]HG = pigs selected for high plasma cholesterol at 56 da of age, LG = pigs selected for low plasma cholesterol at 56 da. From Young et al.[73]
[b]# of animals in parentheses.

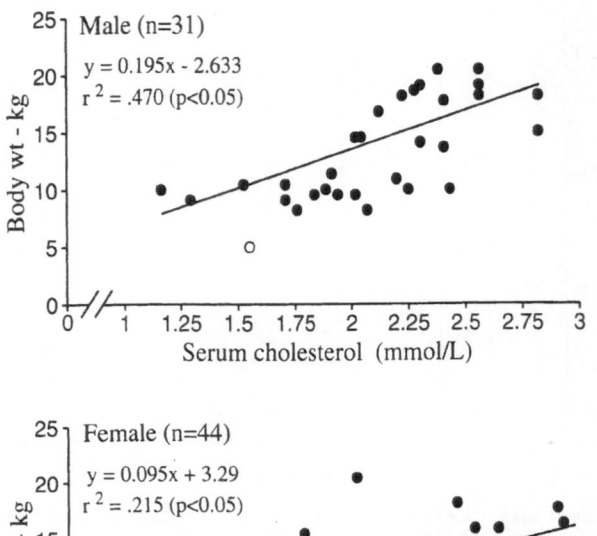

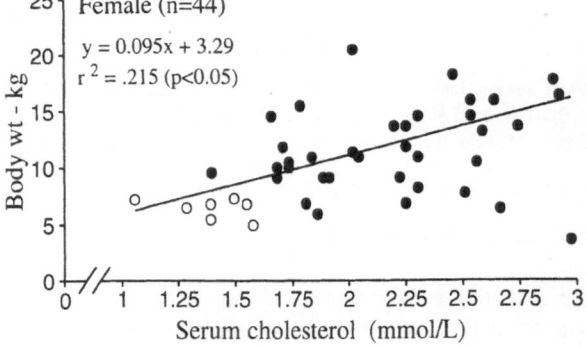

Figure 6. Relationship between body wt and serum cholesterol in male and female pigs with genetically low serum cholesterol. The open circle (O) represents the 8 pigs selected for Experiment 5 and the solid circle (•) represents the remainder of pigs in study. From Schoknecht et al.[77]

there was no difference between genetic or dietary groups in organ or carcass fat or cholesterol content, except for liver, which had higher concentrations of lipid and cholesterol in HG than in LG pigs. There were no genetic x diet interactions in any of these traits and there was no coronary artery or aortic plaque formation in any group.

5.2.2.4.1. Experiment 4[76] The purpose was to determine whether dietary cholesterol deprivation during neonatal life would affect subsequent lipid metabolism or atherogenesis in HG and/or LG pigs fed a high fat, high cholesterol diet during the postweaning period from 8 to 24 wk of age. Twenty four neonatal pigs (12 HG and 12 LG) were assigned at 1 da of age to a sow milk replacer diet (Table 1) containing 0 (NC; n = 6 HG and 6 LG) or 0.5% (HC; n = 6 HG and 6 LG) cholesterol. These diets were fed for 4 wk. Pigs were fed and housed as described for the obese and lean pigs in Experiments 1, 2 and 3. From 4 to 8 wk of age, all pigs were fed the NC diet. From 8 to 24 wk, all pigs were fed a high fat (14%), high cholesterol (1156 mg/kg) diet. Plasma TC, HDL-C and TG concentrations were measured monthly from 4 to 24 wk. All pigs were killed at 24 wk to obtain measurements of carcass traits, tissue cholesterol concentration and plaque formation in the aorta. Overall means for plasma TC were 94, 133, 135 and 180 mg/dl for LG-NC, LG-HC, HG-NC and HG-HC groups, respectively. At 4 wk, LG and HG pigs fed NC had plasma TC concentrations of 88 and 160 mg/dl, compared with 267 and 363 mg/dl, in LG and HG pigs fed HC. At 8 wk (after being fed the NC diet for 4 wk), corresponding plasma TC concentrations were 72, 110, 75 and 107 mg/dl, respectively. There was an effect of diet and genotype on TC at 4 wk and an effect of genotype but not of diet at 8 wk, but no diet x genotype interaction at either time. Concentrations of cholesterol at 24 wk of age in brain, liver and kidney fat but not in muscle or subcutaneous fat, were greater in LG and HG pigs fed HC than in those fed NC. Genotype had no effect on tissue cholesterol except for cerebrum, in which HG pigs tended to have more cholesterol than did LG pigs. Aortic plaques similar to those observed in obese and lean pigs fed HC[66] were produced in both LG and HG pigs. The frequency and severity of the lesions were greater in LG-NC pigs than in other groups (Table 5). This unexpected finding may be related to the lower HDL-C in LG than in HG pigs and the failure of HDL-C to be increased as much by dietary cholesterol in LG as in HG pigs. The results must be confirmed before inferences are drawn.

5.2.2.4.2. Dietary cholesterol and central nervous system (Cns) development in pigs differing in genetic background. Based on the observed relationship between plasma TC and body weight at 8 wk of age in the LG and HG pig populations (Fig. 6),[73] we became interested in the relationships among growth, plasma TC concentration and CNS development. An experiment was designed to address this question.

5.2.2.4.2.1. Experiment 5[77] Eight LG pigs in the lowest quintile of plasma TC from generation 3 of selection for high or low plasma TC, which had body wt less than half that of normal pigs (lowest decile of body wt, Fig. 6) were observed to be unusually lethargic and inactive. These small for age pigs were assigned to a diet containing 0 (n = 4) or 0.2% (n = 4) cholesterol for 4 wk. Mean growth rate was 24% greater in cholesterol supplemented pigs. They were killed to measure tissue cholesterol concentrations. Liver, kidney, adipose, muscle and ileum free and esterified cholesterol concentrations were similar in the 2 groups; whereas, in cerebrum, free cholesterol was higher in supplemented pigs than in those not fed cholesterol (372 vs 307 ± 31 μmol/g). Esterified cholesterol was unaffected by diet (0.52 vs 0.67 ± 0.10 μmol/g. The observed correlation between plasma TC and body wt gain in this pig population selected for low plasma TC[73] and the increase in cerebrum free cholesterol concentration in response to dietary cholesterol in the cohort representing the lowest quintile of plasma TC, suggested the possibility that exogenous cholesterol may be associated with improved growth and brain cholesterol accretion in the neonatal pig from some genetic lines.

Table 5. Aorta atherosclerotic plaques at age 6 mo[a] in pigs fed 0 or 0.5% cholesterol from 1 to 28 da of age and a high fat (14%) high cholesterol (1156 mg/kg) diet from 56 da to 6 mo of age

	HG-HC[b]	HG-HC[c]	LG-HC[d]	LG-NCe
N	6	6	6	6
N with plaque	3	2	2	5
mean Score[f]	2.8	2.7	2.7	4.1

[a]All pigs were fed no cholesterol fat from 27 to 56 da of age.[b]HG-HC = genetically high plasma cholesterol, fed 0.5% cholesterol.[c]HG-NC = genetically high plasma cholesterol, fed zero cholesterol.[d]LG-HC = genetically low plasma cholesterol, fed 0.5% cholesterol.[e]LG-NC = genetically low plasma cholesterol, fed zero cholesterol.[f]Score calculated as product of plaque length x plaque width in mm. Plaques occurred at level of exit of carotid artery.

5.2.2.4.2.2. Experiment 6[77] The purpose was to test the effect of dietary cholesterol deprivation in HG and LG lines of pigs on growth and tissue cholesterol accretion. The same sow milk replacer formula was used as in the neonatal studies with obese and lean pigs described previously (Table 1). Neonatal HG and LG female pigs were fed for 5 wk the NC diet (n = 8 HG and 8 LG) or the same diet supplemented with 0.2% USP cholesterol (HC). Because of the importance of cholesterol in myelin formation and the observation of lethargy noted in Experiment 5 among small for age pigs in the LG population, a simple behavior test was devised to determine relationships between cholesterol ingestion and exploratory behavior in HG and LG neonates. The test consisted of placing the pig in a kennel, putting the kennel in the corner of an observation room in which a grid had been painted on the floor, opening the kennel door and leaving the room. Movements of the pig were observed for 10 min via closed circuit television camera. Three behavioral indices were quantified: 1) time elapsed before the pig first left the kennel, 2) total time spent in kennel and 3) point total based on above criteria and number of grid squares traversed.[77] Behavioral testing was done on d 5, 10, 15, 20 and 25. Pigs were killed on da 35. Dietary cholesterol tended to increase daily body wt gain in both HG and LG pigs. LG pigs fed NC scored lower on 1 component and numerically lower on 2 components of the exploratory behavior test than pigs in all other groups (Fig. 7). LG pigs fed 0.2% cholesterol (LG-HC) had scores similar to those HG pigs fed 0 cholesterol (HG-NC) or 0.2% cholesterol (HG-HC), that is, LG-NC < LG-HC = HG-NC = HG-HC. Free cholesterol in the cerebrum of HG and LG pigs at 35 da of age was increased by dietary cholesterol (Table 6). One can infer the LG-HC pigs (normal exploratory behavior and increased cerebral cholesterol) may have had more mature and /or functionally more competent neurological development in the CNS than did LG-NC pigs.

In Experiment 4,[76] LG and HG pigs fed 0.5% dietary cholesterol from 1 to 28 da of age had higher cerebrum cholesterol concentrations at age 6 mo than pigs which had been deprived of cholesterol during that postnatal period, even though all pigs consumed a high fat, high cholesterol diet from 2 to 6 mo. Neonatal cholesterol ingestion had a persistent, long term effect on CNS development. Mean cerebrum cholesterol concentrations at 6 mo of age in HG pigs which had been fed 0.5% cholesterol, HG pigs fed 0 cholesterol, LG pigs fed 0.5% cholesterol and LG pigs fed 0 cholesterol from da 1 to da 28 were, respectively (in mmol/g, ± SEM), 0.78 ± 0.11, 0.63 ± 0.12, 0.67 ± 0.09 and 0.60 ± 0.07. There was an effect of diet and a trend toward an effect of genetic background but no diet x gene interaction.

The relationship of nutrition of the neonate to CNS functional development is well recognized, but the role of neonatal cholesterol ingestion in this relationship is less clear. The specific role of dietary lipids has been addressed recently;[78,79] however, cholesterol was not considered. Research with the neonatal pig, whose brain growth spurt occurs during late prenatal and early postnatal life in parallel with the ontogeny of brain development in the human infant,[80] may provide new insights into the role of dietary cholesterol and fatty acid composition in CNS development in the neonate. Genetic background appears to be important in modulating the response of neonatal pigs to dietary cholesterol deprivation, because genetically obese pigs failed to respond to dietary cholesterol during neonatal life,[64,81] in contrast to the response we observed in the HG and LG pigs.

5.2.2.4.3. restriction fragment lenth polymorphisM (RFLP) analysis of the cholesterol-7∝-hydroxylase (CYP7) and low density lipoprotein receptoR (LDLR) gene loci in hg and lg pigs. Cholesterol homeostasis involves a balance among cholesterol synthesis, absorption, transport via lipoproteins and uptake via lipoprotein receptors by liver and other tissues, and degradation and excretion as bile acids. Cholesterol synthesis, transport and degradation may be approximated by measurement of key enzymes or receptors, namely HMG-CoA reductase, low density lipoprotein receptor (LDLR) and cholesterol-7∝-hy-

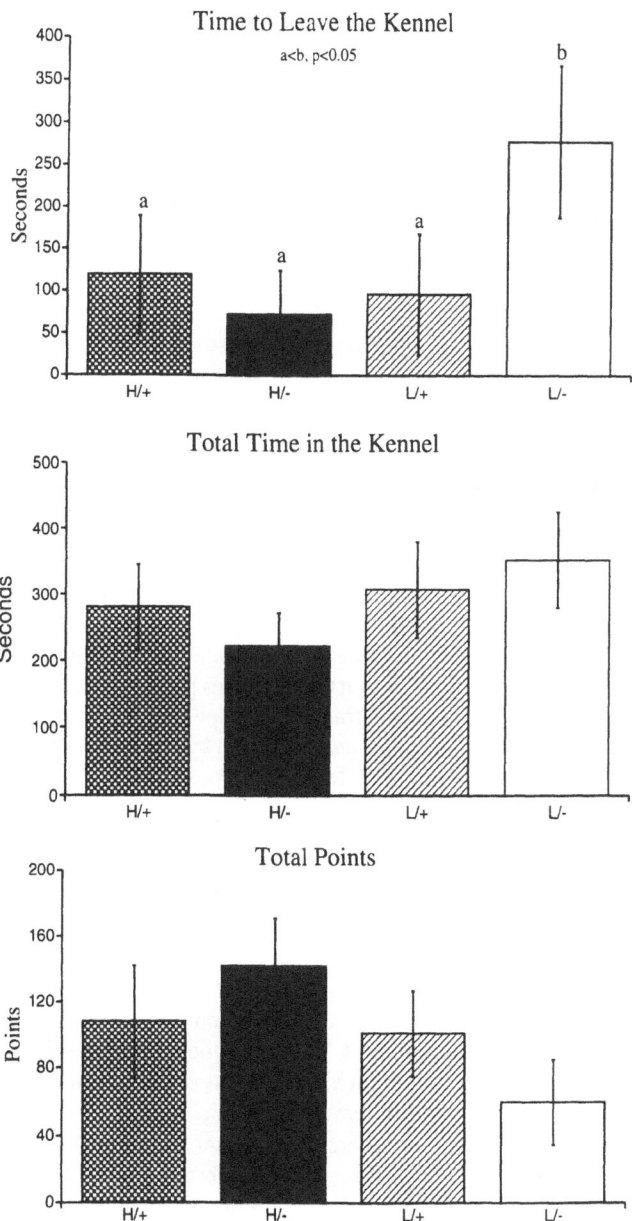

Figure 7. Behavioral indices for piglets of Experiment 6 with genetically high (H) or genetically low (L) serum cholesterol receiving a cholesterol free milk replacer (H/-, L/-; n = 8/group). Top: time taken by pigs to leave kennel (a < b, $P<0.05$); middle: total time spent in kennel; bottom: total points. From Schoknecht et al.[77]

Table 6. Esterified and free cholesterol concentrations and contents of cerebrum in the genetically high cholesterol (HG) and genetically low cholesterol (LG) piglets with or without dietary cholesterol supplementation (n = 4/group)[1]

	HG-HC[a]	HG-NC[b]	LG-HC[c]	LG-NC[d]	SD	P value + Gene	Diet
Cerebrum cholesterol							
Ester concentration, μmol/g	1.0	0.83	0.96	0.75	0.18	0.45	0.06
Ester content, μmol	37.8	30.0	35.4	25.9	2.3	0.38	0.04
Free concentration, mmol/g	0.38	0.33	0.36	0.31	0.04	0.44	0.06
Free content, mmol	13.9	11.8	13.4	11.0	1.1	0.47	0.02

From Schoknecht, et al.[7]

[7][a]HG-HC = genetically high plasma cholesterol, fed 0.5% cholesterol.
[b]HG-NC = genetically high plasma cholesterol, fed zero cholesterol.
[c]LG-HC = genetically low plasma cholesterol, fed 0.5% cholesterol.
[d]LG-NC = genetically low plasma cholesterol, fed zero cholesterol.

droxylase (CYP7), respectively. There is no evidence to suggest that differences between HG and LG pigs in plasma TC concentration are related to HMG-CoA reductase activity. Schoknecht et al.[77] found no difference between HG and LG pigs at 35 da of age in HMG-CoA reductase activity in liver or jejunum, although dietary cholesterol suppressed activity, as expected. Also, TaqI RFLP analysis (A. Davis and M. Wheeler, personal communication) revealed no polymorphism for the HMG-CoA reductase gene locus in HG and LG pigs. RFLPs at the porcine CYP7 gene locus[82] and at the LDLR locus[83] have been demonstrated; a sample of 26 HG and 25 LG pigs from the 7th generation of selection for high or low plasma TC was used for RFLP analysis of both of these genes.[84] RFLP analysis of the LDLR locus with TaqI revealed 2.1 and 0.5 kb alleles present in both HG and LG pigs; the allele frequencies of the 2.1 and 0.5 kb fragments, respectively, were 66 and 34% in LG pigs and 30 and 70% in HG pigs. There was an association between LDLR allele frequencies and plasma TC at 8 wk of age (the 0.5 kb fragment appeared to be associated with high plasma TC and the 2.1 kb fragment with low plasma TC). RFLP analysis of the CYP7 locus with TaqI revealed the presence of 5.0 and 2.8 kb fragments. Only the 2.8 kb homozygous genotype was observed in HG pigs; whereas, LG pigs exhibited the heterozygous as well as the 5.0 kb homozygous genotype. Allele frequencies of the 2.8 and 5.0 kb fragments were, respectively, 26 and 74% in LG pigs and 100 and 0% in HG pigs. High or low plasma TC was associated with 2.8 kb and 5.0 kb CYP7 alleles, respectively. The relationship between plasma total cholesterol and CYP7 alleles in these pigs is shown in Fig. 8. The identification of individual HG pigs homozygous for the 2.8 kb allele and of individual LG pigs homozygous for the other allele (5.0 kb) for the CYP7 gene locus will enable the creation of homozygous populations for this gene. Such populations should be useful for studies of cholesterol homeostasis related to bile acid metabolism.

In summary, this research underscores the impact of genetic background on the response of the neonatal pig to manipulation of cholesterol intake. The effect of dietary cholesterol deprivation during the suckling period on the development of the CNS and on atherogenesis appears to be modulated by genetic background in swine. The relevance of the relationships between genetic background and response to dietary manipulation in the pig to the optimal nutritional environment of the human infant for normal long term neural development and cardiovascular function is unknown but this information is of interest to those developing human infant feeding recommendations.

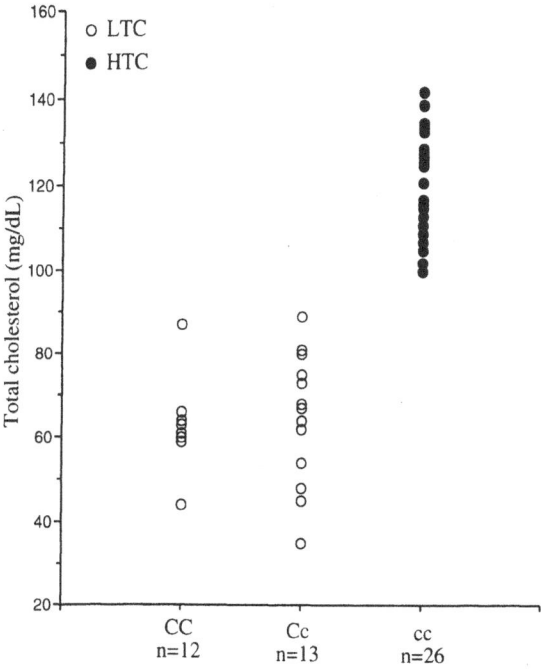

Figure 8. RFLP genotype for CYP7 gene locus compared to plasma total cholesterol in generation 7 of pigs selected for high (HTC) or low (LTC) total cholesterol genotype (CC = 5.0 kb, cc = 2.8 kb, Cc = heterozygote). From Davis et al.[82]

6. REFERENCES

1. Pond, W.G., 1986, Nutrition and the cardiovascular system of swine, in: *Swine in Cardiovascular Research*, Volume II (H.C. Stanton, and H.J. Mersmann, eds.), CRC Press, Boca Raton, pp. 1-31.

2. Lee, K.T., Kim, D.N., and Thomas, W.A., 1986, Atherosclerosis in swine, in: *Swine in Cardiovascular Research*, Volume II (H.C. Stanton, and H.J. Mersmann, eds.), CRC Press, Boca Raton, pp. 33-47.

3. Rapacz, J., and Hasler-Rapacz, J., 1989, Animal models: the pig, *Monogr. Hum. Genet.* 12:139-169.

4. Stanton, H.C., and Mersmann, H.J. (eds.), 1986, *Swine in Cardiovascular Research,* Volumes I and II, CRC Press, Boca Raton.

5. Mead, J.F., Alfin-Slater, R.B., Howton, D.R. and Popjak, G., 1986, Biosynthesis of cholesterol and related substances, in: *Lipids: Chemistry, Biochemistry, and Nutrition* (J.F. Mead, R.B. Alfin-Slater, D.R. Howton, and G Popjak, eds.), Plenum Press, New York, pp. 295-367.

6. Beitz, D.C., and Knight, T.J., 1994, Fats and cholesterol, role in human nutrition, *Encyclo. Agri. Sci.* 2:139-153.

7. Marinetti, G.V., 1990, *Disorders of Lipid Metabolism*, Plenum Press, New York.

8. Mead, J.F., Alfin-Slater, R.B., Howton, D.R., and Popjak, G. (eds), 1986, *Lipids: Chemistry, Biochemistry, and Nutrition,* Plenum Press, New York.

9. Vance, D.E., and J.E. Vance (eds.), 1991, *Biochemistry of Lipids, Lipoproteins and Membranes,* Elsevier, London.

10. Committee on Diet and Health, Food and Nutrition Board, Commission on Life Sciences, National Research Council. 1989, *Diet and Health: Implications for Reducing Chronic Disease Risk.* National Academy Press, Washington, pp 159-258.

11. Nunes, G.L., Sgoutas, D.S., Redden, R.A., Sigman, S.R., Gravanis, M.B., King, S.B., and Berk, B.C., 1995, Combination of vitamins C and E alters the response to coronary ballon injury in the pig, *Arteriosclerosis, Thromb. Vasc. Biol.* 15:156-165.

12. Sprecher, D.L., Kruth, H.S., Pierce, J.E., Lakatos, E., and Papadopoulos, N., 1987, A familial basis for the heterogeneity in coronary atherosclerotic disease, *Atherosclerosis* 65:167-172.

13. Kobari, Y., Koto, M., and Tanigawa, M., 1991, Regression of diet-induced atherosclerosis in Gottingen miniature swine, *Lab. Anim.* 25:110-116.

14. Sassen, L.M.A., Lamers, J.M.J., Sluiter, W., Hartog, J.M., Dekkers, D.H.W., Hogendoorn, A., and Verdouw, P.D., 1993, Development and regression of atherosclerosis in pigs, *Arteriosclerosis Thromb.* 13:651-660.

15. Pond, W.G., and Houpt, K.A., 1978, *The Biology of the Pig*, Comstock Publishing, Cornell University Press, Ithaca.

16. Tumbleson, M.E. (ed.), 1986, *Swine in Biomedical Research,* Volumes 1, 2 and 3, Plenum Press, New York.

17. Miller, E.R., and Ullrey, D.E., 1987, The pig as a model for human nutrition, *Ann. Rev. Nutr.* 7:361-382.

18. Moughan, P.J., and Rowan, A.M., 1989, The pig as a model animal for human nutrition research, *Proc. Nutr. Soc. New Zealand* 14:116-123.

19. Moughan, P.J., Birtles, M.J., Cranwell, P.D., Smith, W.C., and Pedraza, M., 1992, The piglet as a model animal for studing aspects of digestion and absorption in milk-fed human infants, *World Rev. Nutr. Diet* 67:40-113.

20. Romsos, D.R., Allee, G.L., and Leveille, G.A., 1971, In vivo cholesterol and fatty acid synthesis in the pig intestine, *Proc. Soc. Exp. Biol. Med.* 137:570-573.

21. Spady, D.K., and Dietschy, J.M., 1983, Sterol synthesis in vivo in 18 tissues of the squirrel monkey, guinea pig, rabbit, hamster, and rat, *J. Lipid Res.* 24:303-315.

22. Dietschy, J.M., Turley, S.D., and Spady, D.K., 1993, Role of liver in the maintenance of cholesterol and low density lipoprotein homeostasis in different animal species, including humans, *J. Lipid Res.* 34:1637-1659.

23. Janado, M., Martin, W.G., and Cook, W.H., 1966, Separation and properties of pig-serum lipoproteins, *Canad. J. Biochem.* 44:1201-1209.

24. Vitic, J., and Stevanovic, J., 1993, Comparative studies of the serum lipoproteins and lipids in some domestic, laboratory and wild animals, *Comp. Biochem. Physiol.* 106B:223-229.

25. Mahley, R.W., 1978, Alterations in plasma lipoproteins induced by cholesterol feeding in animals including man, in: *Disturbances in Lipid and Lipoprotein Metabolism* (J.M. Dietschy, A.M.J. Gotto, and J.A. Ontko, eds.), American Physiological Society, Bethesda, pp. 181-197.

26. Jackson, R.L., Baker, H.N., Taunton, O.D., Smith, L.C., Garner, W., and Gotto, A.M., 1973, A comparison of the major apolipoprotein from pig and human high density lipoproteins, *J. Biol. Chem.* 248:2639-2644.

27. Chapman, M.J., and Goldstein, S., 1976, Comparison of the serum low density lipoprotein and of its aproprotein in the pig, rhesus monkey and baboon with that in man, *Atherosclerosis* 25:267-291.

28. Stucchi, A.F., Ordovas, J.M., Shwaery, G.T., and Smith, S.C., 1990, Comparative molecular properties of swine and human very low density lipoproteins - apoproteins E and C, *Comp. Biochem. Physiol.* 96B:209-214.

29. Birchbauer, A., Wolf, G., and Knipping, G., 1992, Metabolism of very low density lipoproteins in the pig: An in vivo study, *Int. J. Biochem.* 24:1591-1597.

30. Terpstra, A.H.M., Stucchi, A.F., Foxall, T.L., Shwaery, G.T., Vespa, D.B., and Nicolosi, R.J., 1993, Unidirectional transfer in vivo of high-density lipoprotein cholesteryl esters to lower-density lipoproteins in the pig, an animal species without plasma cholesteryl ester transfer activity, *Metabolism* 42:1524-1530.

31. Greeve, J., Altkemper, I., Dieterich, J.H., Greten, H., and Windler, E., 1993, Apolipoprotein B mRNA editing in 12 different mammalian species: hepatic expression is reflected in low concentrations of apoB-containing plasma lipoproteins, *J. Lipid Res.* 34:1367-1383.

32. Hayes, K.C., Pronczuk, A., and Khosla, P., 1995, A rational for plasma cholesterol modulation by dietary fatty acids: Modeling the human response in animals, *J. Nutr. Biochem.* 6:188-194.

33. Mersmann, H.J., Arakelian, M.C., and Brown, L.J., 1979, Plasma lipids in neonatal and growing swine, *J. Anim. Sci.* 48:554-558.

34. Johansson, M.B.N., and Karisson, B.W., 1982, Lipoprotein and lipid profiles in the blood serum of the fetal neonatal and adult pig, *Biol. Neonate* 42:127-137.

35. Johansson, M.B.N., 1984, Heterogeneity of serum lipoproteins during the fetal and neonatal development of the pig, *Int. J. Biochem.* 16:1359-1366.

36. Hentges, W.L.S., and Martin, R.J., 1987, Serum and lipoprotein lipids of fetal pigs and their dams during gestation as compared with man, *Biol. Neonate* 52:127-134.

37. Jones, P.J.H., Hrboticky, N., Hahn, P., and Innis, S.M., 1990, Comparison of breast-feeding and formula feeding on intestinal and hepatic cholesterol metabolism in neonatal pigs, *Am. J. Clin Nutr.* 51:979-984.

38. Arbuckle, L.D., Rioux, F.M., Mackinnon, M.J., Hrboticky, and Innis, S.M., 1991, Response of (n-3) and (n-6) fatty acids in piglet brain, liver and plasma to increasing, but low, fish oil supplementation of formula, *J. Nutr.* 121:1536-1547.

39. Huff, M.W., Telford, D.E., Edmonds, B.W., McDonald, C.G., and Evans, A.J., 1993, Lipoprotein lipases, lipoprotein density gradient profile and LDL receptor activity in miniature pigs fed fish oil and corn oil, *Biochem. Biophys. Acta* 1210:113-122.

40. Jauhiainen, M., Aro, A., Blomqvist, S.M., Kemppinen, A., Alaviuhkola, T., and Antila, P., 1993, Effect of milk fat, unhydrogenated and partially hydrogenated vegetable oils on serum lipoproteins in growing pigs, *Comp. Biochem. Physiol.* 106A:565-570.

41. Kingman, S.M., Walker, A.F., Low, A.G., and Sambrook, I.E., 1993, Comparative effects of four legume species on plasma lipids and faecal steroid excretion in hypercholesterolaemic pigs, *Brit. J. Nutr.* 69:409-421.

42. Seiquer, I., Manas, M., Martinez-Victoria, E., Huertas, J., Ballesta, M.C., and Mataix, F.J., 1994, Effects of adaptation to diets enriched with saturated, monounsaturated and polyunsaturated fats on lipid and serum fatty acid levels in miniature swine, *Comp. Biochem. Physiol.* 108:377-386.

43. Baldner-Shank, G.L., Richard, M.J., Beitz, D.C., and Jacobson, N.L., 1987, Effect of animal and vegetable fats and proteins on distribution of cholesterol in plasma and organs of young growing pigs, *J. Nutr.* 117:1727-1733.

44. Pfeuffer, M., Ahrens, F., Hagemeister, H., and Barth, C.A., 1988, Influence of casein versus soy protein isolate on lipid metabolism of minipigs, *Ann. Nutr. Met.* 32:83-89.

45. Beynen, A.C., West, C.E., Spaaij, C.J.K., Huisman, J., Van Leeuwen, P., Schutte, J.B., and Hackeng, W.H.L., 1990, Cholesterol metabolism, digestion rates and postprandial changes in serum of swine fed purified diets containing either casein or soybean protein, *J. Nutr.* 120:442-430.

46. Faidley, T.D., Luhman, C.M., Galloway, S.T., and Foley, M.K., 1990, Effect of dietary fat source on lipoprotein composition and plasma lipid concentrations in pigs, *J. Nutr.* 120:1126-1133.

47. Scholz-Ahrens, K.E., Hagemeister, H., Unshelm, J., Agergaard, N., and Barth, C.A., 1990, Response of hormones modulating plasma cholesterol to dietary casein or soy protein in minipigs, *J. Nutr.* 120:1387-1392.

48. Luhman, C.M., Faidley, T.D., and Beitz, D.C., 1992, Postprandial lipoprotein composition in pigs fed diets differing in type and amount of dietary fat, *J. Nutr.* 122:120-127.

49. Kummerow, F.A., Wasowicz, E., Smith, T., Yoss, N.L., and Thiel, J., 1993, Plasma lipid physical properties in swine fed margarine or butter in relation to dietary magnesium intake, *J. Am. Coll. Nutr.* 12:125-132.

50. Fremont, L., Gozzelino, M.-T., and Bosseau, A.F., 1993, Effects of sugar beet fiber feeding on serum lipids and binding of low-density lipoproteins to liver membranes in growing pigs, *Am. J. Clin. Nutr.* 57:524-532.

51. Huff, M.W., Telford, D.E., Woodcroft, K., and Strong, W.L.P., 1985, Mevinolin and cholestyramine inhibit the direct synthesis of low density lipoprotein apolipoprotein B in miniature pigs, *J. Lipid Res.* 26:1175-1186.

52. Marsh, A., Kim, D.N., Lee, K.T., Reiner, J.M., and Thomas, W.A., 1972, Cholesterol turnover, synthesis, and retention in hypercholesterolemic growing swine, *J. Lipid Res.* 13:600-615.

53. Kim, D.N., Lee, K.T., Reiner, R.M., and Thomas, W.A., 1975, Effect of combined clofibrate-cholestyramine treatment on serum and tissue cholesterol pools and on cholesterol synthesis in hyper-cholesterolemic swine, *Exper. Molec. Path.* 23:83-95.

54. Rogers, D.H., Kim, D.N., Lee, K.T., Reiner, J.M., and Thomas, W.A., 1981, Circadian variation of 3-hydroxy-3-methylglutaryl coenzyme A reductase activity in swine liver and ileum, *J. Lipid Res.* 22:811-819.

55. Kim, D.N., Rogers, D.H., Li, J.R., Reiner, J.M., Lee, K.T., and Thomas, W.A., 1977, Effects of cholestyramine on cholesterol balance parameters and hepatic HMG-CoA reductase and cholesterol-7-alpha-hydroxylase activities in swine, *Exp. Molec. Path.* 26:434-447.

56. Kim, D.N., Schmee, J., Baker, J.E., Lunden, G.M., Sheehan, C.E., Lee, C.S., Eastman, A., Solia, O., Ross, J.S., and Thomas, W.A., 1993, Dietary fish oil reduces microthrombi over atherosclerotic lesions in hyperlipidemic swine even in the absence of plasma cholesterol reduction, *Exp. Molec. Path.* 59:122-135.

57. Schmee, J., Kim, D.N., Ross, J.S., and Thomas, W.A., 1993, Exponential relationship between plasma cholesterol levels and atherosclerotic lesion size in hyperlipidemic swine, *Exp. Molec. Path.* 59:177-185.

58. Fuster, V., Badimon, L., Badimon, J.J., Turitto, V., Lie, J.T., and Bowie, E.J.W., 1986, Experimental approach to vascular disease in swine with von Willebrand's disease, in: *Swine in Biomedical Research*, Volume 3 (M.E. Tumbleson, ed.), Plenum Press, New York. pp. 1527-1542.

59. Reddick, R.L., Read, M.S., Brinkhous, K.M., Bellinger, D., Nichols, T., and Griggs, T.R., 1990, Coronary atherosclerosis in the pig: Induced plaque injury and platelet response, *Arteriosclerosis* 10:541-550.

60. Nichols, T.C., Bellinger, D.A., Davis, K.E., Koch, G.G., Reddick, R.L., Read, M.S., Rapacz, J., Hasler-Rapacz, J., Brinkhous, K.M., and Griggs, T.R., 1992, Porcine von Willebrand Disease and atherosclerosis, *Am. J. Path.* 140:403-415.

61. Nichols, T.C., Bellinger, D.A., Tate, D.A., Reddick, R.L., Read, M.S., Koch, G.G., Beinkhous, K.M., and Griggs, T.R., 1990, von Willebrand factor and occlusive arterial thrombosis. A study in normal and von Willebrand's disease pigs with diet-induced hypercholesterolemia and atherosclerosis, *Arteriosclerosis* 10:449-461.

62. Hetzer, H.O., and Harvey, W.R., 1967, Selection for high and low fatness in swine, *J. Anim. Sci.* 26:1244-1251.

63. Mersmann, H.J., 1991, Characteristics of obese and lean swine, in: *Swine Nutrition* (E.R. Miller, D.E. Ullrey and A.J. Lewis, eds.), Butterworth-Heinemann, Boston, pp. 75-89.

64. Patterson, B.W., Wong, W.W., Sheng, H-P., Mersmann, H.J., Insull, W., Klein, P.D., Fiorotto, M.L., and Pond, W.G., 1992, Neonatal genetically lean and obese pigs respond differently to dietary cholesterol, *J. Nutr.* 122:1830-1839.

65. McWhinney, V., Pond, W.G., and Mersmann, H.J., 1994, Dietary suppression of 3-hydroxy-methylglutaryl CoA reductase in neonatal lean and obese pigs, *FASEB J.* 8:A543.

66. Hackman, A.M., Pond, W.G., Mersmann, H.J., Wong, W.W., and Krook, L.P., 1994, Hypercholesterolemia and atherosclerosis despite suppression of HMG-CoA reductase in lean and obese pigs fed high cholesterol from birth to two months, *J. Am. Coll. Cardiol.* 22:99A.

67. Wong, W.W., Hachey, D.L., Feste, A., Leggitt, J., Clarke, L.L. Pond, W.G., and Klein, P.D., 1991, Measurement of in vivo cholesterol synthesis from ^2H_2O: a rapid procedure for the isolation, conbustion, and isotopic assay of erythrocyte cholesterol, *J. Lipid Res.* 32:1049-1056.

68. Rapacz, J., Hasler-Rapacz, J., Taylor, K.M., Checovich, W.J., and Attie, A.D., 1986, Lipoprotein mutations in pigs are associated with elevated plasma cholesterol and atherosclerosis, *Science* 234:1573-1577.

69. Lowe, S.W., Checovich, W.J., Rapacz, J., and Attie, A.D., 1988, Defective receptor binding of low density lipoprotein from pigs possessing mutant apolipoprotein B alleles, *J. Biol. Chem.* 263:15467-15473.

70. Hasler-Rapacz, J.O., Nichols, T.C., Griggs, T.R., Bellinger, D.A., and Rapacz, J., 1994, Familial and diet-induced hypercholesterolemia in swine: Lipid, ApoB, and ApoA-I concentrations and distributions in plasma and lipoprotein subfractions, *Arterioscler. Thromb.* 14:923-930.

71. Prescott, M.F., Hasler-Rapacz, J., Linden-Reed, V.J., and Rapacz, J., 1995, Familial hypercholesterolemia associated with coronary atherosclerosis in swine bearing different alleles for apolipoprotein B, *Ann. New York Acad. Sci.* 748:283-293.

72. Pond, W.G., Mersmann, H.J., and Young, L.D., 1986, Heritability of plasma cholesterol and triglyceride concentrations in swine, *Proc. Soc. Exp. Med.* 182:221-224.

73. Pond, W.G., Mersmann, H.J., Klein, P.D., Ferlic, L.L., Wong, W.W., Hachey, D.L., Schoknecht, P.A., and Zhang, S., 1993, Body weight is correlated with serum cholesterol at 8 weeks of age in pigs selected for four generations for low or high serum cholesterol, *J. Anim. Sci.* 71:2406-2411.

74. Young, L.D., Pond, W.G., and Mersmann, H.J., 1993, Direct and correlated responses to divergent selection for serum cholesterol concentration on day 56 in swine, *J. Anim. Sci.* 71:1742-1753.

75. Harris, K.B., Cross, H.R., Pond, W.G., and Mersmann, H.J., 1993, Effect of dietary fat and cholesterol level on tissue cholesterol concentrations of growing pigs selected for high or low serum cholesterol, *J. Anim. Sci.* 71:807-810.

76. Abbott, T.L., Savell, J.W., Pond, W.G., and Morgan, W.W., 1995, Neonatal dietary cholesterol: Effects on tissue cholesterol and carcass characteristics of growing pigs selected for high or low serum cholesterol, *J. Anim. Sci.* 73(Suppl 1):A106.

77. Schoknecht, P.A., Ebner, S., Pond, W.G., Zhang, S., McWhinney, V., Wong, W.W., Klein, P.D., Dudley, M., Goddard-Finegold, J., and Mersmann, H.J., 1994, Dietary cholesterol supplementation improves growth and behavioral response to pigs selected for genetically high and low serum cholesterol. *J. Nutr.* 124:305-314.

78. Lucas, A., Morley, R., Cole, T.J., Lister, G., and Leeson-Payne, C., 1992, Breast milk and subsequent intelligence quotient in children born preterm, *Lancet* 339:261-264.

79. Dobbing, J. (ed.), 1993, Lipids, learning, and the brain: Fats in infant formulas, Report of the 103rd Ross Conference on Pediatric Research, Columbus, Ohio Ross Laboratories, pp 1-249.

80. Dobbing, J. 1972, Vulnerable periods of brain development, in: *Lipids, Malnutrition, and the Developing Brain.* CIBA Foundation Symposium jointly with Nestle Foundation, Elsevier Medica-North Holland, New York.

81. Zhang, S. Wong, W.W., Hachey D.L., Pond, W.G., and Klein, P.D., 1994, Dietary cholesterol inhibits whole-body but not cerebrum cholesterol synthesis in young pigs, *J. Nutr.* 124:717-725.

82. Davis, A.M., White, B.A., and Wheeler, M.B., 1994, Rapid communications: A *Taq* I restriction fragment length polymorphism at the porcine cholesterol 7 ∝-hydroxylase (CYP7) locus, *J. Anim. Sci.* 72:797.
83. Davis, A.M., White, B.A., and Wheeler, M.B., 1993, A *Taq*I RFLP at the porcine low density lipoprotein receptor (LDLR) locus, *Anim. Genetics* 24:330-333.
84. Davis, A.M., Pond, W.G., and Wheeler, M.B., 1995, Association between genetic markers in seventh-generation pigs selected for high and low plasma cholesterol, *FASEB J.* 9:A4436.

73

GLUCOSE AND FATTY ACID METABOLISM IN THE NEWBORN PIG

Pierre-Henri Duee,[1] Jean-Paul Pegorier,[2] Béatrice Darcy-Vrillon,[1] and Jean Girard[2*]

[1] Institut National de la Recherche Agronomique
Unité d'Ecologie et de Physiologie du Système Digestif
78350 Jouy-en-Josas, France
[2] Centre National de la Recherche Scientifique
Centre de Recherche sur l'Endocrinologie Moléculaire et le Développement
9, rue J. Hetzel
92190 Meudon, France

1. INTRODUCTION

In mammalian species, important modifications in several physiological functions and profound changes of nutrition occur during the perinatal period.[1] *In utero*, the fetus is maintained at a controlled temperature and receives from its mother a continuous supply of nutrients and oxygen used for growth and oxidative metabolism. Fetal nutrition can be classified as a high carbohydrate, low fat diet. After birth, the maternal supply of nutrients ceases and newborns are fed at intervals with milk, which is a high fat, low carbohydrate diet. At birth, the gut becomes the natural route of nutrient delivery to the organism; its postnatal maturation and growth are rapid, especially in the pig,[2] and associated with profound modifications in hemodynamics and a high local oxygenation[3]. As in most species, the newborn pig is exposed to a sudden 15 to 20C decrease in its thermal environment and must maintain its body temperature by shivering thermogenesis, as the piglet is devoid of brown adipose tissue.

Postnatal growth of gut and skeletal muscles is characterized by a high rate of protein synthesis.[4] Because amino acid intake from milk proteins is just sufficient for the accretion of body protein, amino acid catabolism rates and urea excretion are low during this period; amino acids do not play an important role in energy metabolism during the postnatal period.[1] In turn, the successful adaptation of neonates to changes of nutrition and environment requires important modifications of glucose and fatty acid metabolism which are controlled mainly by alterations in hormone secretion. In the pig, regulation of glucose homeostasis is

* Reprint requests to: Dr J. Girard, Centre de Recherche sur l'Endocrinologie Moléculaire et le Développement, 9 rue J. Hetzel, 92190 Meudon, France. (33) 1 45 07 58 50.

Advances in Swine in Biomedical Research, edited by Tumbleson and Schook
Plenum Press, New York, 1996

critical at birth since the unfed newborn pig frequently develops hypoglycemia, similar to the hypoglycemia described in human neonates at risk.[5]

2. FUEL AND HORMONE CHANGES DURING THE POSTNATAL PERIOD

Immediately after birth, the newborn pig has to withstand a brief period of starvation before its first meal; during this period, the neonate is dependent upon its capacity to mobilize energy stores. Under normal conditions, this period does not last more than 20 to 30 min. Then the newborn pig receives nutrients from colostrum and milk. The postnatal period is characterized by changes in circulating fuels reflecting both available nutrients and metabolism.

2.1. Energy Stores during the Postnatal Period

At birth, piglet energy stores consist essentially of glycogen (25 to 30 g/kg bwt) which is present in large amount in liver and skeletal muscle. At term, liver and muscle glycogen concentrations are several times higher than the usual adult concentrations.[6-8] In contrast, piglet body fat content is low (<1% body wt) and is present mostly as structural fat[9,10] not available for mobilization during starvation.[11] The absence of fat stores at birth is due to the limited transfer of free fatty acids from the mother to the fetus[12,13] and to the low capacity for lipogenesis from precursors (glucose, lactate) transported across the placenta.[13-15] Moreover, nutritional or endocrine manipulations of the mother during late pregnancy do not increase glycogen or fat deposition in the fetus at term.[16-19] During the first 24 hr after delivery, liver glycogen concentration decreases in response to the nutritional state of the newborn pig (100 mg/g at birth; 10 mg/g at 24 hr);[7,20] whereas, glycogen concentration in skeletal muscle decreases more slowly. In contrast, body fat content increases rapidly in the suckling pig. Body fat primarily derives from the uptake of colostrum triacylglycerols rather than from lipogenesis.[21,22] Later on, lipogenesis increases in adipose tissue and contributes to the enhanced postnatal fat deposition.[21,22]

2.2. Nutrient Intake during the Postnatal Period

From an energy point of view, milk is a high fat, low carbohydrate diet. Lactose is the predominant carbohydrate in milk; whereas, 95% of milk fats are in the form of triacylglycerols.[23] Since lactose and fat content in sow milk (125 Kcal/100 g) amounts to 5.0 and 8.0 g/100 g, respectively, fat represents a large part of milk energy (60%); whereas, energy from lactose is low (15%). Milk nutrient content also depends upon the stage of lactation. Colostrum is characterized by a lower lactose content (4g/100 g) than mature milk. In contrast, colostrum protein content (10 to 13 g/100 g) is 2 to 3 fold higher than that of mature milk. The proportion of medium and short chain fatty acids in milk triacylglycerols is low in the pig; whereas, the proportion of unsaturated long chain fatty acids in sow milk is 50 to 55%. Lactose is hydrolyzed by lactase, a specific intestinal disaccharidase localized in the brush border of absorptive intestinal mucosal cells. In pigs, intestinal lactase is active at birth and during the suckling period.[24] Lactose hydrolysis produces equal amounts of glucose and galactose which are taken up by the enterocyte via an active transport system which is functional at birth.[25,26] The complete digestion of milk triacylglycerols requires the combined action of gastric lipase, pancreatic colipase dependent lipase and the presence of bile salts. Although the activity of pancreatic lipase is low at birth, colostrum fat digestion and fatty acid absorption are efficient in the newborn pig.[24] Long chain fatty acids (LCFA) generated from the digestion of milk triacylglycerols are absorbed from the small intestine as chylomicrons which are synthesized within epithelial cells. Furthermore, the total carnitine concentration in sow milk is high just after delivery.[27,28] Thus, tissue carnitine levels are not limited by carnitine intake when the newborn pig is fed naturally. Quantification of nutrient intake requires the determination of milk intake, which depends on stage of lactation and litter size.[23] Average milk intake is 200 to 300 g/da within the first day of life and increases gradually during the suckling period.[23]

2.3. Changes in Circulating Fuels

Between birth and the first meal, blood glucose concentration increases slightly and remains at normal values (5 to 6 mmol/l) during the suckling period (Fig. 1). In contrast, blood glucose concentration decreases to reach hypoglycemia values (1 to 2 mmol/l) when the newborn pig remains unfed (Fig. 1).[20,29,30] A significant concentration of blood fructose (3 mmol/l) is present at birth but rapidly decreases within the first day after birth.[31] Blood lactate concentration is high at birth (3 to 3.5 mmol/l) and remains at a high level during the postnatal period, whatever the nutritional state of the newborn.[20] Blood concentrations of alanine (600 µmol/l) and glutamine (300 µmol/l) are high at birth and do not change during the first days of life.[20] Plasma free fatty acids (FFA), blood glycerol and blood ketone bodies are low at birth and remain at a low level during the first days after birth when the pig is fasted,[20,30] (Fig. 1) since the newborn pig is devoid of fat stores. In contrast, plasma FFA and blood glycerol increase progressively in fed piglets. However, blood ketone bodies remain at a low level in suckling animals despite elevated plasma FFA concentrations.[20,30] In the pig, the low ketonemia, despite high levels of plasma FFA, suggests a limitation of liver ketogenesis rather than a high rate of ketone body utilization.

2.4. Changes in Circulating Hormones

The immediate postnatal period is characterized by rapid changes in pancreatic hormones; within 30 min after delivery, plasma insulin decreases and plasma glucagon increases, whatever the nutritional state (fed or fasted, Fig. 2) of the newborn pig.[20,29] Neonatal changes in pancreatic hormones could be related to the stress of birth through an activation of the sympathetic nervous system. High plasma epinephrine and norepinephrine levels prevail, due to an increase of sympathetic nervous system activity and adrenal catecholamine release in response to cold exposure or transient hypoxia, despite an incomplete innervation of the adrenal medulla.[32] Plasma corticosteroid or cortisol concentrations already are high at delivery and increase further within the first 24 hr of extrauterine life. Then, plasma cortisol concentration decreases progressively with age.[33-35] Growth hormone plasma concentration is high at birth and remains elevated during the first postnatal day.[34,36] The early postnatal period is characterized by an abrupt rise in plasma thyroid hormone levels. There is an increase in total and free plasma triiodothyronine (T_3) concentration immediately after birth; whereas, plasma thyroxine (T_4) concentration gradually increases during the first day after birth.[34,37,38] In addition, cold farrowed piglets exhibit higher T_3 and lower T_4 plasma concentrations than controls.[37]

3. GLUCOSE METABOLISM IN THE NEWBORN PIG

Glucose turnover rates have been measured in 2 da old postabsorptive and fasting pigs (Fig. 3).[39] Under steady state conditions, glucose production rate equals glucose utilization rate. When expressed per kg body weight, glucose turnover rates found in 2 da

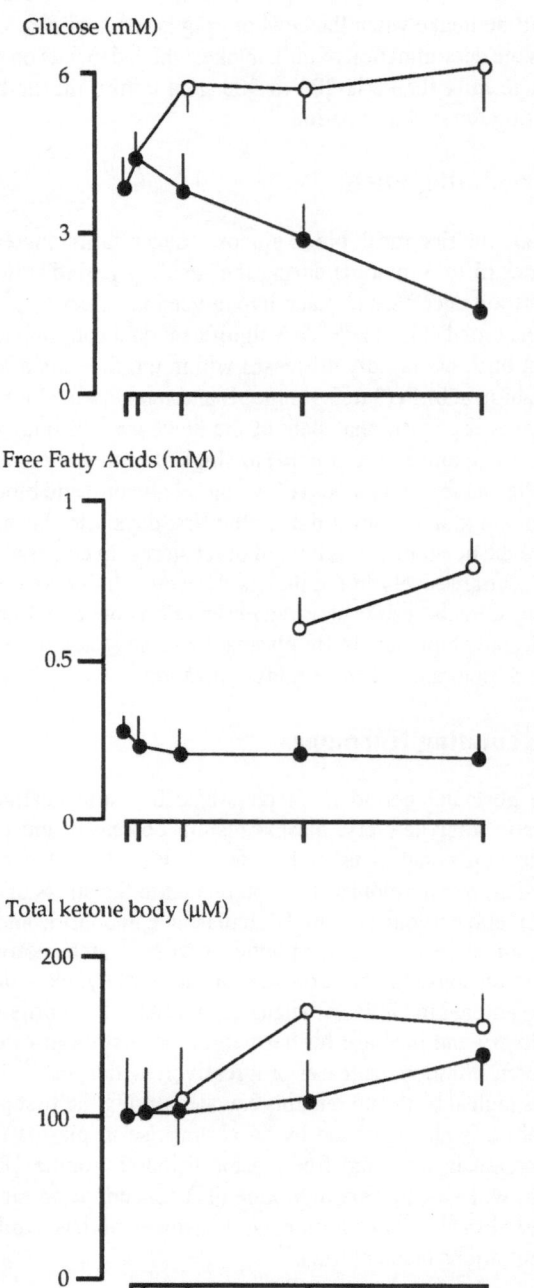

Figure 1. Deveopmental changes in blood glucase, free fatty acids and ketone body concentration in suckling (O) and fasting newborn pigs (●). Pegorier et al. [20]

Insulin (µU/ml)

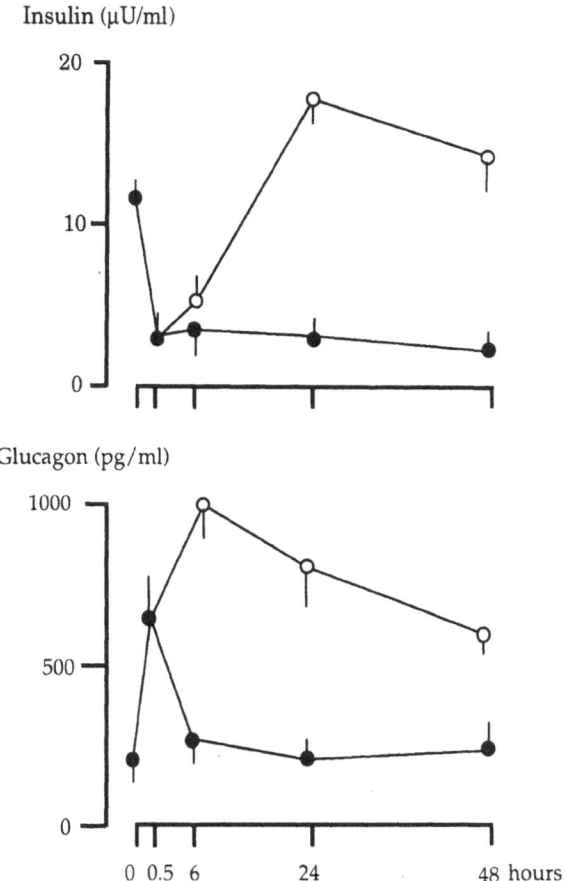

Figure 2. Developmental changes in plasma insulin and glucagon concentrations in suckling (O) or fasting (●) newborn pigs. Pegorier et al. [20]

old postabsorptive pigs (65 µmol/ min/kg bwt with [6³H] glucose; 54 µmol/min/kg bwt with [U-¹⁴C] glucose, according to Pégorier[39] are 2 to 3 fold higher than in postweaning pigs.[40,41] Part of the difference could be due to a proportionately larger brain in the newborn pig. This is supported by the correlation that exists between glucose turnover rate and estimated brain weight during development.[40] Moreover, the rate of glucose turnover is related linearly to blood glucose concentration.[39] The higher rate of glucose turnover in post absorptive newborn pigs, compared with fasting ones, suggests that suckling newborn pigs are able to maintain a normal blood glucose concentration through an increase in glucose production rather than through a decrease in glucose utilization. As liver glycogen stores are exhausted 24 hr after birth in suckling newborn pigs, this suggests that glucose production in post absorptive newborn pigs is supported by hepatic gluconeogenesis. Moreover, the comparison of glucose supplied via colostrum to the rate of glucose utilization (Table 1) allows an estimate of the contribution of hepatic gluconeogenesis to neonatal glucose needs. Glucose supplied by the milk accounts for no more than 30% of glucose requirements of the newborn. Thus, hepatic gluconeogenesis provides 70% of the glucose turnover rate of the newborn pig. Active gluconeogenesis in postabsorptive newborn pigs is sustained partly by the supply

Blood glucose (mM)

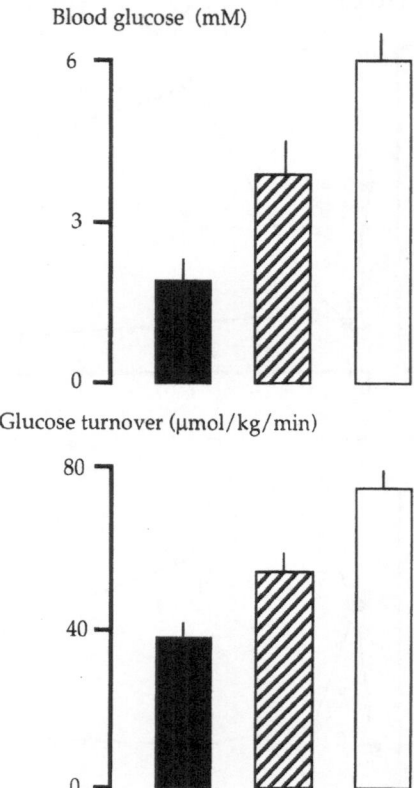

Glucose turnover (μmol/kg/min)

Figure 3. Relationship between blood glucose concentration and glucose turnover rates in 2 da old fasting (■), LCT fed (⊠) and suckling (□) piglets. Prior to LCT feeding, piglets were fasted for 44 hr and had received 3 gastric administrations of LCT emulsion (at 1 hr intervals). Results presented here were collected 1 hr after the last LCT feeding. From Pegorier et al. [39,88]

of potential gluconeogenic precursors from milk constituents, i.e., galactose, glycerol and amino acids.

3.1. Postnatal Development of Hepatic Gluconeogenesis

The pathway of gluconeogenesis involves 4 specific enzymes located in different cell compartments. In the pig, phosphoenolpyruvate carboxykinase (PEPCK) is located inside

Table 1. Glucose balance in 2 da old suckling pigs[39]

	mmol/kg/day
Net glucose need	80
Rate of glucose utilization	95
Rate of glucose recycling	15
Glucose from lactose intake	25
Hepatic glucose production	5
Gluconeogenesis from galactose	25
Gluconeogenesis from other substrates	30

both mitochondria and cytosol. However, cytosolic PEPCK is an adaptive enzyme whose activity is correlated directly with the hepatic capacity for gluconeogenesis.[1] Fructose-1,6-bisphosphatase also is located in cytosol; whereas, pyruvate carboxylase is located in the mitochondrial matrix and glucose-6-phosphatase inside microsomes. At birth, activities of glucose-6-phosphatase, fructose-1,6-bisphosphatase, pyruvate carboxylase and PEPCK in pig liver are 70, 45, 117 and 35%, respectively, of adult values.[42] Furthermore, in suckling pigs, all these activities increase slightly during the first day of life.[42,43] The postnatal increase of PEPCK is much more marked for the cytosolic form than for the particulate form. However, the presence of high activities of key enzymes does not indicate whether gluconeogenesis already is present in pig liver at birth and if its rate is affected by suckling. Rates of glucose production from various substrates have been quantified in hepatocytes isolated from newborn pigs. The capacity to synthesize glucose from lactate, pyruvate, glycerol and alanine is low in hepatocytes isolated from the newborn pig immediately at birth[44] and develops markedly during the first postnatal day.[44-46] In contrast, a net glucose production from galactose already is present at birth.[44] Moreover, postnatal development of hepatic gluconeogenesis from C3 precursors seems to be related to the induction of cytosolic PEPCK since the addition of quinolinate (an inhibitor of cytosolic PEPCK) inhibits by 70 to 75% glucose production rates from 10 mM lactate or 10 mM pyruvate in isolated hepatocytes from 1 or 2 da old suckling piglets (Pégorier, Duée and Girard, unpublished). Therefore, the suckling newborn pig rapidly develops an active gluconeogenesis which is in agreement with the maintenance of a normal blood glucose concentration. In contrast, gluconeogenesis rates from various gluconeogenic precursors are 50 to 70% lower in 2 da old fasting than in suckling newborn pigs.[44] Thus, the low rate of hepatic gluconeogenesis in fasting pigs is responsible for their profound hypoglycemia.

3.2. Regulation of Gluconeogenesis

Although a direct demonstration has not been produced until now in the newborn pig, it seems probable the postnatal induction of cytosolic PEPCK is triggered by an increase in liver cAMP secondarily to the increase in plasma glucagon and the decrease in plasma insulin that occur immediately after birth, as suggested in the rat.[1] As activities of liver gluconeogenic enzymes[42] are appropriate for an active gluconeogenesis in fasting newborn pigs, this suggests that the principal factors controlling the rate of gluconeogenesis at that time could be the 1) supply of gluconeogenic precursors, 2) hormonal milieu or 3) availability of fatty acids to the liver.

The amounts of lactose (glucose + galactose) supplied via colostrum are sufficient to cover a significant part of the glucose need in the newborn pig (Table 1), especially as pig liver is capable to synthesize glucose from galactose at birth. Galactose administration in the postabsorptive newborn pig increases plasma glucose levels.[47] However, 2 lines of evidence suggest the reduced supply of gluconeogenic precursors cannot by itself explain the impairment of gluconeogenesis in the fasting newborn pig. First, concentrations of blood lactate or alanine, which are gluconeogenic precursors, are high during the postnatal period. Second, the rate of gluconeogenesis from a saturating concentration (10 mmol/l) of lactate or pyruvate is lower in isolated hepatocytes from fasting newborns than in suckling newborns.[44]

In isolated hepatocytes of 2 da old fasting pigs, the rate of glucose production from dihydroxyacetone (DHA), which enters gluconeogenesis at the level of triose phosphate, is lower than that measured in suckling pigs.[44,48] Conversely, the rate of lactate and pyruvate production from DHA is higher in fasting newborn pigs suggesting a preferential metabolism of DHA through glycolysis.[48] The addition of glucagon completely reverses the partition of DHA between glycolysis and gluconeogenesis.[48] This suggests that glucagon regulates gluconeogenesis[49] through modulations of L-pyruvate kinase and phosphofructokinase activities. Glucagon stimulates phosphorylation of the bifunctional enzyme, 6-phosphofructo-2-kinase/fructose-2,6-biphosphatase, leading to an inactivation of 6-phosphofructo-1-kinase and an activation of fructose-1,6-bisphosphatase. Moreover, L-pyruvate kinase also is inactivated by phosphorylation and allosteric inactivation secondarily to the decrease in fructose-1,6-bisphosphate concentration. Although plasma glucagon concentration is lower in fasting than in suckling newborn pigs,[20] glucagon infusion during 4 hr in the fasting newborn pig,[50] or addition of cAMP to isolated hepatocytes,[51] does not increase glycemia or gluconeogenesis, suggesting that glucagon concentration alone is not sufficient to control the rate of gluconeogenesis. When the newborn pig is fed a low fat colostrum, it is unable to maintain normoglycemia after birth.[52] Therefore, colostral fat is essential for glucose homeostasis.

3.3. Glucose Metabolism in Extrahepatic Tissues

In contrast to the liver which is the site of glucose production, extrahepatic tissues are the sites of glucose utilization. Recently, *in vivo* rates of glucose utilization in brain, heart, skeletal muscle and intestine were measured using 2-deoxy[^{3}H]glucose, in 2 da old suckling and fasting piglets (Darcy-Vrillon, Pénicaud and Duée, unpublished). The essential role of glucose for the brain is confirmed, since the brain of fasting piglets maintains its glucose utilization, despite a 50% decrease in total glucose turnover rate.[39,53] The rate of glucose utilization was decreased markedly in heart, skeletal muscle and intestine of 2 da old fasted piglets (Darcy-Vrillon, Pénicaud and Duée, unpublished). At birth, the gastrointestinal tract becomes an active organ for absorption and metabolism of nutrients and undergoes marked structural and functional changes. In the pig, most of these changes occur soon after birth and are accompanied by a rapid growth of intestinal mucosa during the first days of life.[54] The metabolism of intestinal mucosal cells has been measured recently in suckling and fasting pigs.[55] Enterocytes are able to metabolize glucose at a high rate, similar to that measured in enterocytes isolated from postweaned pigs.[55] Glycolysis represents the predominant metabolic pathway of glucose in enterocytes, lactate, pyruvate and alanine accounting for at least 70% of glucose uptake. Moreover, the capacity for glucose consumption and glycolysis increases in enterocytes from 2 to 5 da old suckling piglets. Although the contribution of glycolysis to energy metabolism increases between birth and 2 da of age, glucose oxidation remains the major source of ATP production in cells incubated with glucose alone.[56] This increase in the glycolytic capacity is not observed in piglets fasted from birth.[55] Therefore, the increase in glycolytic activity in suckling pig enterocytes depends on colostrum ingestion. A high glycolytic capacity also is observed in cells from 2 da old suckling animals, even in the presence of glutamine, the principal substrate of newborn pig enterocytes.[56] Glycolytic activity, which does not serve for energy purposes, could be responsible for generating specific metabolites in the suckling neonate. This could be a mechanism for conserving glucose as C3 units, which subsequently would be available for hepatic gluconeogenesis. Also, it could provide pyruvate needed for glutamate transamination or precursors for phospholipid synthesis.

Possible regulatory steps in the development of glucose metabolism in newborn pig enterocytes have been investigated recently.[57] Whereas the maximal activity of 6-phosphofructo-1-kinase does not change, there is an increase in hexokinase activity and rate of glucose phosphorylation. Stimulation of glucose phosphorylation plays a key role in the development of a high glycolytic capacity in enterocytes from 2 da old piglets. Future investigations are needed to identify the factors involved in the increase of hexokinase activity. Various growth factors and hormones found in colostrum and in milk are known to affect the maturation of the small intestine.[58]

Besides a high glycolytic capacity, enterocytes from suckling pigs also have the capacity to produce glucose from galactose or DHA[55] suggesting that fructose 1,6-bisphosphatase and glucose 6-phosphatase activities are active in newborn pig enterocytes during the suckling period. Again, the appearance of gluconeogenic capacity in suckling pig enterocytes depends on colostrum ingestion, since glucose formation from galactose is low in 2 da old fasted pigs. It has not been determined whether or not glycolysis and gluconeogenesis coexist in the same cells.

3.4. Glucose Homeostasis during Acute Postnatal Cold Exposure

Maintenance of homeothermy in the newborn pig is a crucial problem since the difference between the external temperature and the temperature of thermal neutrality may be quite large (10 to 15C).[11,59] Maintenance of an efficient thermogenesis in the piglet is dependent upon the availability of energy reserves and substrates in colostrum. As the newborn pig is devoid of fat store, it is dependent entirely upon carbohydrates to cover its energy needs until the first meal.[60] Muscle glycogenolysis is stimulated by cold exposure[61,62] but not by fasting[61] during the first day of life. Fasted newborn pigs exposed to a cold external temperature (20C) are unable to maintain homeothermy.[11,61] Mildly hypothermic newborn pigs have a normal blood glucose concentration 6 hr after birth and an enhanced glucose turnover rate (+25%), i.e., a concomitant increase in hepatic glucose production and peripheral glucose utilization.[61] Increased glucose production probably reflects increased liver glycogenolysis. The increase in glucose utilization may be due to the increase in heat production during acute cold response. This transient increase in peripheral glucose uptake appears to be a mechanism which does not require an increase in plasma insulin concentration and probably involves an hemodynamic effect due to an increased sympathetic nervous system and adrenal catecholamine release.[61,63]

When fasted newborn pigs are exposed to a low external temperature (12C), profound hypothermia is reached rapidly (within 10 to 12 hr after birth) associated with a marked hyperglycemia.[61] Hyperglycemia does not result from a higher glucose production but is the consequence of a decrease in glucose metabolic clearance.[61,64] This reduced glucose utilization may reflect profound changes in muscle tonicity as suggested by the depressed muscular activity related to a modified behavior of the newborn pig during sustained cold exposure. Profound hypothermia also is associated with an impaired insulin secretion in response to glucose in the newborn pig.[64] The infusion of phentolamine, a nonselective alpha-adrenergic antagonist, restores insulin secretion in response to glucose in hypothermic newborn pigs and enhances the rate of glucose disappearance (+40%) during an IV glucose tolerance test (Le Dividich, unpublished). Alterations of peripheral glucose utilization during profound hypothermia may be mediated by sympathetic nervous system activity. In suckling newborn pigs, there is evidence for the thermogenic importance of colostrum, since during cold exposure both temperature and heat production are correlated positively to the amount of colostrum ingested.[65] Moreover, colostral fat content seems essential for the regulation of glucose homeostasis during cold exposure.[62]

4. FATTY ACID METABOLISM IN THE NEWBORN PIG

Blood ketone body concentrations remain low in suckling newborn pigs[20,30,66,67] as well as in 5 da starved adult pigs,[68] despite high levels of plasma FFA. Possible factors responsible for the limitation in hepatic ketogenesis in newborn pig liver will be discussed later. Before considering the regulation of fatty acid metabolism, it will be useful to review briefly the different steps of these pathways (Scheme 1).

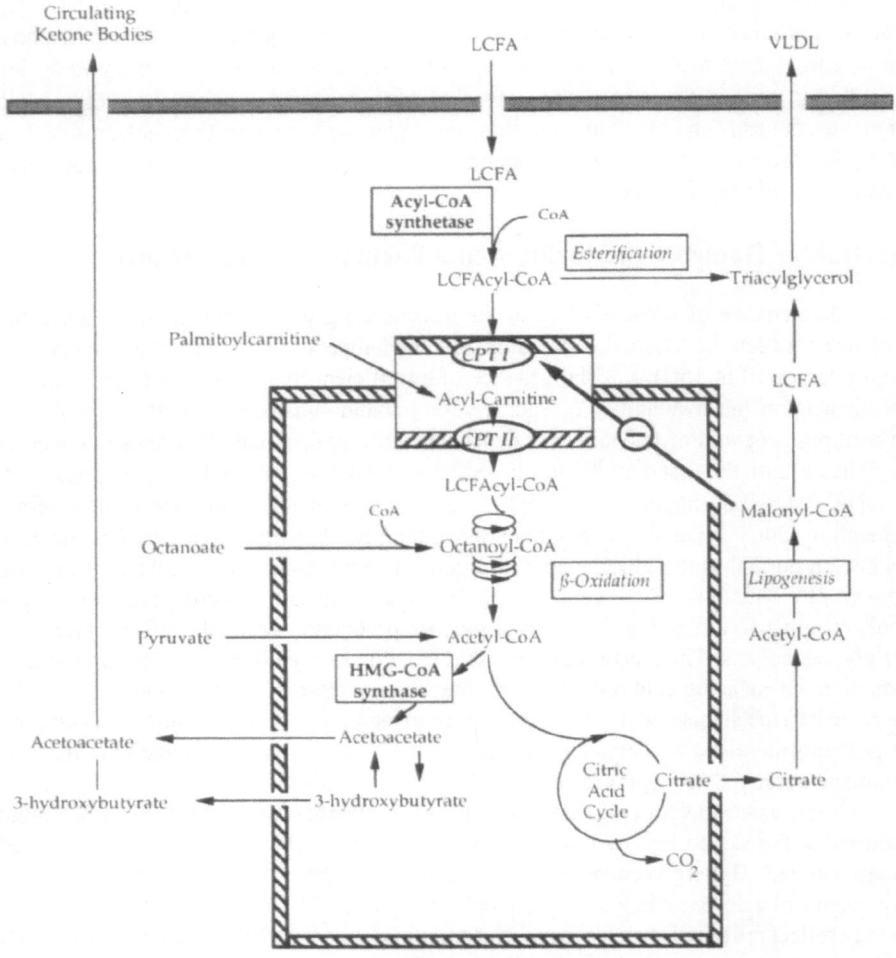

Scheme 1. Metabolic fate of long chain fatty acid (LCFA) and ketone body synthesis in the liver.

4.1. Pathways for Fatty Acid Oxidation and Ketogenesis

FFA taken up by the liver are activated to their corresponding acyl-CoA by acyl-CoA synthetases located either on microsomes and outer mitochondrial membrane for activation of LCFA, (more than 12 carbons) or in the mitochondrial matrix for the activation of medium chain fatty acids (MCFA) or short chain fatty acids (SCFA).[69] LCFAcyl-CoA can be oxidized or reesterified depending on the physiological state. The entry of LCFAcyl-CoA into the mitochondrial matrix is ensured by the carnitine palmitoyltransferase (CPT) system which consists of 3 distinct activities: overt CPT I, a carnitine acylcarnitine translocase and latent CPT II. In contrast to LCFA, MCFA or SCFA readily cross mitochondrial membranes and are activated to their corresponding acyl-CoA by a short or a medium chain acyl-CoA synthetase located in the mitochondrial matrix.[69] Within mitochondria, acyl-CoA (whatever their chain length) undergo the successive steps of ß-oxidation to yield molecules of acetyl-CoA and reduced equivalents (NADH and FADH$_2$). In mitochondria of extrahepatic tissues (mainly heart, skeletal muscles, kidney cortex, lung, small intestine), acetyl-CoA undergoes oxidation into the citric acid cycle to yield CO_2, H_2O, GTP and further reduced equivalents that may enter the electron transport chain to yield ATP. In liver mitochondria, acetyl-CoA produced by ß-oxidation can undergo either the citric acid cycle or the hydoxymethylglutaryl(HMG)-CoA pathway for ketone body synthesis. These ketone bodies are released into the circulation and transported to peripheral tissues where they are used as respiratory fuels or as substrates for lipid biosynthesis.

4.2. Regulation of Fatty Acid Oxidation and Ketogenesis in Liver

In 2 da old suckling newborn pigs, low blood ketone body concentrations result from a limitation in their hepatic production[27,70] that is 90% lower than in other mammalian species.[1] This low capacity of the liver to synthesize ketone bodies seems to be a characteristic of the pig, rather than being due to a precise stage of development, since it persist in the adult pig liver.[71] This is due to a limited capacity for hepatic fatty acid oxidation[51,70,72] since most of the LCFA taken up by isolated pig hepatocytes are converted into esterified fats[70,72] (Scheme 1) whatever the age of the animal or its nutritional state.[70] It seems unlikely this reduced capacity for LCFA oxidation results from a low mitochondrial mass since it increases within the first 24 hr after birth in newborn pig liver.[73,74] In most mammalian species, the main regulatory site for LCFA oxidation is the CPT transfer system[75] (Scheme 1). As the mitochondrial concentration of immunoreactive CPT II protein is similar in newborn pig or adult rat livers,[71] this suggests the major site of control is located at the level of CPT I. Three distinct mechanisms are involved in the regulation of CPT I: 1) changes in its activity, 2) changes in the concentration of malonyl-CoA, its physiological inhibitor, and 3) changes in its sensitivity to malonyl-CoA inhibition.[75] None of these mechanisms seem to be responsible for the limitation of LCFA oxidation in newborn pig liver. Firstly, CPT I protein amount and activity are similar in newborn pig or adult rat liver mitochondria; whereas, LCFA oxidation is 95% lower in newborn pig hepatocytes than in adult rat hepatocytes.[71] Similar conclusions were drawn previously on the basis of total CPT activity.[76] CPT I activity is not limited by carnitine availability since the addition of a saturating concentration of carnitine to isolated hepatocytes from fasting pigs failed to enhance LCFA oxidation.[51,70] Secondly, the concentration of malonyl-CoA is low in isolated newborn pig hepatocytes (5% of that found in rat liver cells[71]) reflecting the low hepatic lipogenic activity[70] due to reduced lipogenic enzymes activities.[21] Thirdly, although CPT I is sensitive to malonyl-CoA inhibition in newborn pig liver mitochondria,[71] the oxidation of LCFA is low in isolated liver pig mitochondria, an *in vitro* system in which fatty acid oxidation is not controlled by malonyl-CoA.[71] Moreover, oxidation of octanoate, whose entry into mitochondria is independent of CPT I, [77] also is low in isolated newborn pig mitochondria.[71] The rate limiting step has been identified by studying the metabolic fate of LCFA inside the mitochondria. When acetyl-CoA, arising from oxidation of oleyl-CoA or palmitoylcarnitine, is channeled into the citric acid cycle (Scheme 1), oxygen consumption rates, due to the oxidation of these LCFA esters, are similar in mitochondria isolated either from newborn pig or from adult rat livers[71] (Fig. 4). In contrast, when acetyl-CoA is directed towards the synthesis of ketone bodies, i.e., into the HMG-CoA pathway (Scheme 1), oxidation of LCFA esters is 75% lower in newborn pig mitochondria than in adult rat ones[71] (Fig. 4). Formation of ketone body from acetyl-CoA is limited as compared with fatty acid oxidation. Synthesis of ketone bodies is depressed in pig liver mitochondria because activities and amounts of immunoreactive HMG-CoA synthase are 95% lower than in rat liver mitochondria.[71] Thus, the metabolic consequence is a limited capacity for fatty acid oxidation that could be due to the accumulation of acetyl-CoA and/or ß-oxidation intermediates each of them being strong inhibitors of acyl-CoA dehydrogenase activities.[69]

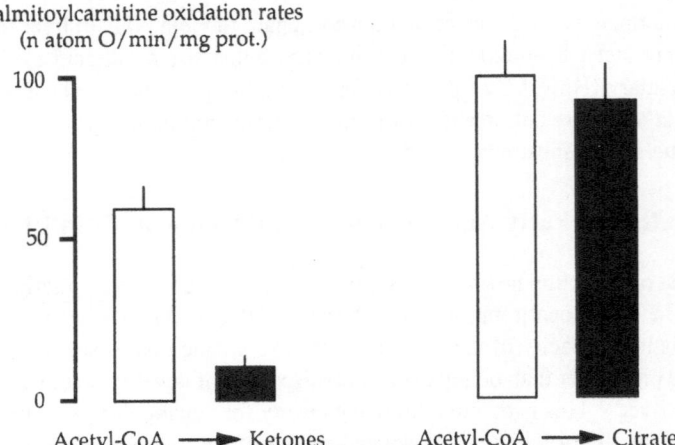

Figure 4. Rates of palmitoylcarnitine oxidation in liver mitochondria isolated from 2 da old fasting pigs (■) or fed adult rats (□). Acetyl CoA arising from β-oxidation was channelled either towards ketone body synthesis (by adding malonate, an inhibitor of succinate dehydrogenase) or into the critic acid cycle (by adding malate, a donor of oxaloacetate). Duee et al. [71]

4.3. Regulation of Fatty Acid Oxidation in Extrahepatic Tissues

In the pig, as in other species, the ability to oxidize FFA develops rapidly after birth in kidney cortex, heart and skeletal muscles. In *in vitro* studies, fetal pig heart and muscle have been shown to have a low capacity for LCFA oxidation, even in the presence of carnitine.[78-80] As mitochondrial oxidation of octanoate, octanoylcarnitine or palmitoyl-carnitine is similar in the heart of fetal, newborn and adult pigs,[79] the reduced capacity of fetal tissue to oxidize LCFA could result from the low activity of CPT system. Moreover, the reduced number of mitochondria in fetal tissues could be responsible for their low capacity to oxidize FFA. Shortly after birth the capacity of the heart and muscle to oxidize FFA increases.[78-81] The increased ability to oxidize FFA in neonatal tissues is not due to the supply of a high fat diet during the postnatal period. Piglets fed a low fat diet after birth develop a normal capacity for fatty acid oxidation in heart, skeletal muscle and kidney.[78] In general, the increase in extrahepatic tissue capacity to oxidize FFA varies in parallel with the activity of CPT.[1] However, CPT activity does not change with age in the heart and skeletal muscles of pigs, although palmitate oxidation increases 2 to 3 fold.[78] The increase in amount of mitochondrial protein during the postnatal period probably is a factor involved in the increased capacity for FFA oxidation.[78] Moreover, a large increase in carnitine concentration has been reported in the pig tissues.[76] After weaning, this mitochondrial oxidative capacity remains elevated in the heart[79] and to a lesser extent in skeletal muscles.

5. INTERRELATIONSHIPS AMONG FATTY ACID AND GLUCOSE METABOLISMS IN NEWBORNS

As mentioned above, development of hepatic gluconeogenesis is essential for maintenance of a normal blood glucose concentration. Although newborn pigs are able to oxidize efficiently exogenous ketone bodies,[82] their oxidation, under physiological conditions, accounts for only 4.5% of the piglet energy requirement.[82] This is due to the low levels of ketone bodies in newborn pig plasma. Thus, a preferential oxidation of fatty acids in some peripheral tissues would appear relevant, since it could spare glucose for tissues or cells, such as the brain, erythrocytes and kidney medulla, that are dependent entirely upon glucose to maintain their functional activity. In this section, we review the role of fatty acid oxidation in the regulation of glucose production and utilization.

5.1. Role of Fatty Acid Oxidation in Hepatic Gluconeogenesis Regulation

As previously described, suckling newborn pigs fed a high fat, low carbohydrate diet remain normoglycemic owing to an active gluconeogenesis. However, the role of FFA in regulation of glucose homeostasis in newborn pig remains controversial. Fat administration to fasting piglets failed to increase blood glucose levels or the conversion of gluconeogenic substrates to glucose by liver slices.[46,83-85] However, it was difficult to draw a clear conclusion from these experiments since plasma FFA concentration did not increase after fat administration. Similarly, the addition oleate or octanoate to isolated hepatocytes from fasting newborn pigs failed to stimulate hepatic glucose production from lactate or pyruvate.[51] In contrast, addition of medium chain triglycerides (MCT) to colostrum[86,87] or feeding newborn pigs with colostrum enriched in its fat component[52] induced an increased in blood glucose concentration. Exogenous fat contributed to the regulation of glucose homeostasis in fasting newborn pig despite a low capacity for fatty acid oxidation. Three lines of evidence supported this hypothesis. Firstly, when endogenous fatty acid oxidation was blocked by an inhibitor of CPT I, it resulted in a 40% decrease in the rate of glucose production from lactate in isolated hepatocytes from 2 da old suckling pigs.[48] Secondly, when hepatocytes from 2 da old fasting pig were incubated in the presence of oleate and carnitine, gluconeogenesis from lactate was 30% higher than in their absence.[48] Thirdly, gastric administration of a long chain triglyceride (LCT) emulsion to 2 da old fasting pigs, resulted in a 2 fold increase in blood glucose concentration after 3 hr.[88] This increase was due to a stimulation of hepatic glucose production as the result of an increased availability of acetyl-CoA and reducing equivalents arising from LCT oxidation[88] (Scheme 2). Hepatic fatty acid oxidation is essential to support an active gluconeogenesis in newborn pig liver, although its contribution seems quantitatively less important than in newborn (rat, rabbit, guinea pig) exhibiting high rates of hepatic fatty acid oxidation.[1]

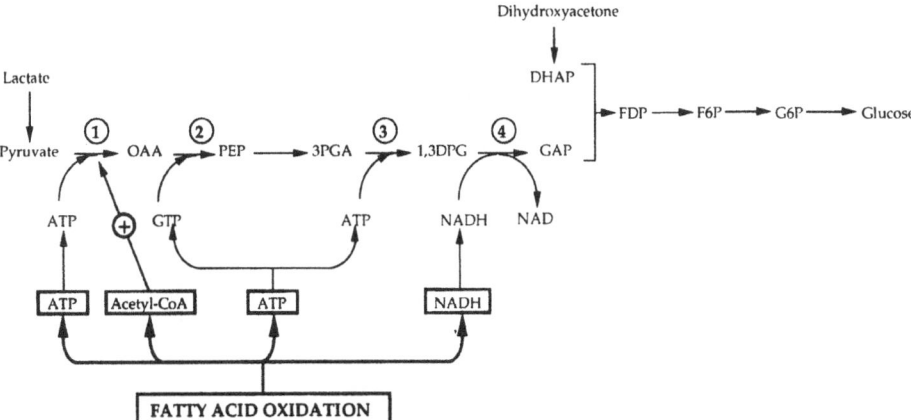

Scheme 2. Interrelationship between fatty acid oxidation and gluconeogenesis in the liver. 1 = Pyruvate carboxylase; 2 = Phosphoenolpyruvate carboxykinase; 3 = Phosphoglycerate kinase; 4 = Glyceraldehyde-3-phosphate dehydrogenase.

5.2. Interrelationships among Fatty Acid Oxidations and Glucose Utilization in Peripheral Tissues

As mentioned before, the capacity of extrahepatic tissues to utilize fatty acids as energetic fuels develop soon after birth. For instance, LCFA are major sources of metabolic energy in neonatal pig hearts.[80] In this tissue, fatty acid utilization spares glucose consumption by 2 mechanisms: 1) a decrease in glucose uptake[80] (Fig. 5) and 2) an inhibition of glucose oxidation[79] secondary to increased concentrations in glucose-6-phosphate (an inhibitor of hexokinase) and of citrate (an inhibitor of phosphofructokinase). Such a decrease in glucose oxidation also was observed in the whole animal as suggested by the increases in blood lactate, pyruvate and alanine concentrations in LCT fed newborn pigs, despite their

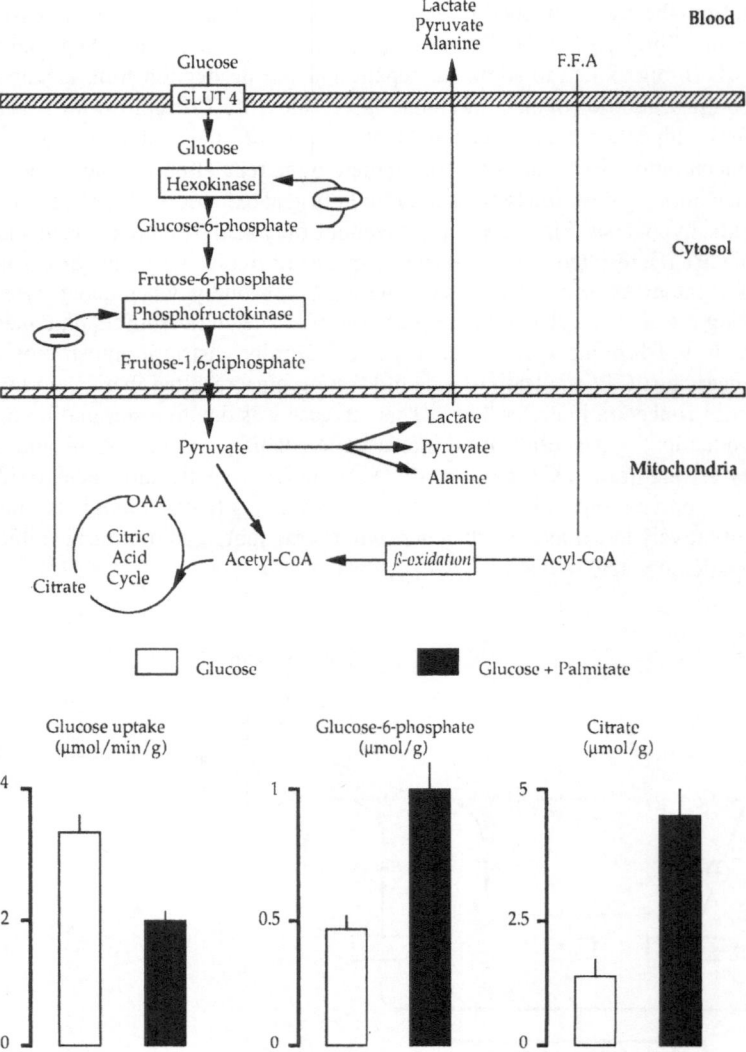

Figure 5. Effects of fatty acid oxidation on glucose metabolism in isolated working newborn pig heart and localisation of the regulatory steps. From Werner et al. [79,80]

enhanced utilization in hepatic gluconeogenesis.[88] In peripheral tissues, LCFA induce a redirection of pyruvate formed into glycolysis, towards lactate and alanine synthesis rather than to acetyl-CoA secondary to an inhibition of pyruvate dehydrogenase[89] (Fig. 5). Glucose sparing mechanisms are of particular importance for the brain of newborn pigs which has no alternative energy substrates, since blood ketone body levels are low.[53]

6. METABOLIC ADAPTATION IN THE HUMAN NEWBORN

Three important conclusions can be drawn from studies performed in the pig. First, pancreatic hormones play a crucial role in metabolic adaptation of the neonate. Second, an active gluconeogenesis is essential to maintain a normal blood glucose level in the newborn. Third, hepatic fatty acid oxidation plays an important role in supporting gluconeogenesis. To what extent can these conclusions be generalized to human babies?

6.1. Role of Pancreatic Hormones

Newborns from poorly controlled diabetic mothers are hyperinsulinemic and do not increase their plasma glucagon concentration in the postnatal period.[90] Neonatal hypoglycemia is of frequent occurrence in infants of diabetic mothers.[5] The reduced hepatic glucose production reported in infants of diabetic mothers[91] could result from both hypoglucagonemia and hyperinsulinemia. In addition, rare newborn infants suffering of congenital glucagon deficiency develop severe neonatal hypoglycemia that is corrected by glucagon administration.[92,93] Pancreatic hormones may play a crucial role in neonatal metabolic adaptation in the human neonate.

6.2. Importance of Hepatic Gluconeogenesis

Using stable isotopes, it has been shown that gluconeogenesis from lactate, pyruvate, alanine and glycerol[94-96] and ketogenesis[94,97,98] were active in the human neonate 12 to 24 hr after birth, thus confirming the crucial importance of these metabolic pathway for glucose homeostasis in the human newborn.

6.3. Importance of Liver Fatty Acid Oxidation in Gluconeogenesis Regulation

Unlike the pig, the normal baby at term has a high body fat content (16% of body weight) and can survive prolonged starvation without hypoglycemia.[99] In contrast, the small for gestational age (SGA) newborn has a low body fat content at birth and develops hypoglycemia after a short fast.[5] A defect in gluconeogenesis has been proposed to explain hypoglycemia in SGA babies;[100,101] infusion of triglycerides into hypoglycemic SGA neonates raises their blood glucose levels.[102-104] The hyperglycemic effect of triglyceride in SGA neonates resulted from a stimulation of hepatic glucose production.[103] We have the opportunity to study an 8 mo old girl who developed severe hypoglycemia after a fast of 24 hr and had low blood ketone body levels despite high plasma FFA concentrations.[105] This suggested she had an inability to oxidize LCFA. Cultured fibroblasts from this girl have a reduced capacity to oxidize LCFA but a normal capacity for oxidation of MCFA or SCFA.[106] The defect in LCFA oxidation results from a CPT I deficiency.[107,108] Administration of MCT to this hypoglycemic girl produced a rapid rise of blood ketone body levels, since MCFA enter mitochondria in bypassing the CPT system. Restoration of normal beta-oxidation was followed by a complete correction of hypoglycemia.[105] Thus, we confirm the essential relationship between fatty acid oxidation and gluconeogenesis in the liver of human neonates.

7. CONCLUSIONS

From these few examples drawn from the physiopathology of the human neonate, we suggest the studies performed in animal species contribute to the understanding of fundamental mechanisms involved in the regulation of glucose homeostasis of the normal infant and can be used as a physiological basis for the treatment of neonatal hypoglycemia.

8. REFERENCES

1. Girard, J., Ferré, P., Pégorier, J.P., and Duée, P.H., 1992, Adaptations of glucose and fatty acid metabolism during the perinatal period and the suckling-weaning transition, *Physiol. Rev.* 72:507-562.
2. Klein, R.M., and McKenzie, J.C., 1983, The role of cell renewal in the ontogeny of intestine. 1-Cell proliferation patterns in adult, fetal and neonatal intestine, *J. Pediatr. Gastroenterol. Nutr.* 2:10-43.
3. Nowicki, P.T., Stonestreet, B.S., Hansen, N.B., Yao, A., and Oh, W., 1983, Gastrointestinal blood flow and oxygen consumption in awake newborn piglets: effect of feeding, *Am. J. Physiol.* 245:G697-G702.
4. Reeds, P.J., Burrin, D.G., Davis, T.A., and Fiorotto, M.L., 1993, Postnatal growth of gut and muscle : competitors or collaborators, *Proc. Nutr. Soc.* 52:57-67.
5. Cornblath, M., and Schwartz, R., 1993, Hypoglycemia in the neonate, *J. Pediatr. Endocrinol.* 6:113-129.
6. Randall, G.C.B., and L'Ecuyer, C., 1976, Tissue glycogen and blood glucose and fructose levels in the pig fetus during the second half of gestation, *Biol. Neonate* 28:74-82.
7. Elliot, J.I., and Lodge, G.A., 1977, Body composition and glycogen reserves in the neonatal pig during the first 96 hours postpartum. *Can. J. Anim. Sci.* 57:141-150.
8. Okai, D.B., Wyllie, D., Aherne, F.X., and Ewans, R.C., 1978, Glycogen reserves in the fetal and newborn pig, *J. Anim. Sci.* 46:391-401.
9. Widdowson, E.M., 1950, Chemical composition of newly born mammals, *Nature* 166:626-628.
10. Manners, M.J., and McCrea, M.R., 1963, Changes in the chemical composition of sow-reared piglets during the first month of life, *Br. J. Nutr.* 17:495-517.
11. Curtis, S.E., Christison, G.I., and Robertson, W.D., 1970, Effects of acute cold exposure and age on respiratoty quotients in piglets, *Proc. Soc. Exp. Biol. Med.* 134:188-191.
12. Elphick, M.C., Flecknell, P., Hull, D., and McFayden, I.R., 1980, Plasma free fatty acid umbilical venous-arterial concentration differences and placental transfert of [^{14}C] palmitic acid in pigs, *J. Develop. Physiol.* 2:347-356.
13. Duée, P.H., Simoes-Nunes, C., Pégorier, J.P., Gilbert, M., and Girard, J., 1987, Uterine metabolism of the conscious gilt during late pregnancy, *Pediatr. Res.* 22:587-590.
14. Reynolds, L.P., Ford, S.P., and Ferrell, C.L., 1985, Blood flow and steroid and nutrient uptake of the gravid uterus and fetus of sows, *J. Anim. Sci.* 61:968-974.
15. Hausman, D.B., Hausman, G.J., and Martin, R.J., 1993, Endocrine regulation of fetal adipose tissue metabolism in the pig. Role of thyroxine, *Biol. Neonate* 64:116-126.
16. Ezekwe, M.O., and Martin, R.J., 1980, The effects of maternal alloxan-diabetes on body composition, liver glycogen and metabolism and serum metabolites and hormones of fetal pigs, *Horm. Metab. Res.* 12:136-139.
17. Pettigrew, J.E., 1981, Supplemental dietary fat for peripartal sows: a review, *J. Anim. Sci.* 53:101-107.
18. Yen, J.T., Eichner, R.D., Arnold, R.J., and Pond, W.G., 1982, Tissue glycogen levels in dams and fetuses as affected by fasting and refeeding pregnant sows, *J. Anim. Sci.* 54:796-799.
19. Ezekwe, M.O., Ezekwe, E.I., Sen, D.K., and Ogalla, F., 1984, Effects of maternal streptozotocin-diabetes on fetal growth, energy reserves and body composition of newborn pigs, *J. Anim. Sci.* 59:974-980.
20. Pégorier, J.P., Duée, P.H., Assan, R., Peret, J., and Girard, J., 1981, Changes in circulating fuels, pancreatic hormones and liver glycogen concentration in fasting or suckling newborn pigs, *J. Develop. Physiol.* 3:203-217.
21. Mersmann, H.J., Phinney, G., Sanguinetti, C., and Houk, J.M., 1973, Lipogenic capacity of liver from perinatal swine (sus domesticus), *Comp. Biochem. Physiol.* 46B:493-497.

22. Mersmann, H.J., 1974, Metabolic patterns in the neonatal swine, *J. Anim. Sci.* 38:1022-1030.

23. Salmon-Legagneur, E., 1965, Quelques aspects des relations nutritionnelles entre la gestation et la lactation chez la truie, *Ann. Zootech.* 14:1-137.

24. Moughan, P.J., Birtles, M.J., Cranwell, P.D., Smith, W.C., and Pedraza, M., 1992, The piglet as a model animal for studying aspects of digestion and absorption in milk-fed human infants, in: *Nutritional Triggers for Health and Disease,* 67 (D.A.P. Simopoulos, ed.), Karger, Basel, pp 40-113 .

25. Smith, M.W., 1988, Postnatal development of transport function in the pig intestine, *Comp. Biochem. Physiol.* 90A:577-582.

26. Puchal, A.A., and Buddington, R.K., 1992, Postnatal development of monosaccharide transport in pig intestine, *Am. J. Physiol.* 262:G895-G902.

27. Coffey, M.T., Shireman, R.B., Herman, D.L., and Jones, E.E., 1991, Carnitine status and lipid utilization in neonatal piglets fed diets low in carnitine, *J. Nutr.* 121:1047-1053.

28. Li, B.U.K., Murray, R.D., Heitlinger, L.A., Hughes, A.M., McLung, H.J., and O'Dorisio, T.M., 1992, Enterohepatic distribution of carnitine in developing piglets: relation to glucagon and insulin, *Pediatr. Res.* 32:312-316.

29. Swiatek, K.R., Kipnis, D.M., Mason, G., Chao, K.L., and Cornblath, M., 1968, Starvation hypoglycemia in newborn pigs, *Am. J. Physiol.* 214:400-405.

30. Gentz, J., Bengtsson, J.K., Hakkarainen, J., Hellstrom, R., and Persson, B., 1970, Metabolic effects of starvation during neonatal period in the piglet, *Am. J. Physiol.* 218:662-668.

31. Aherne, F.X., Hays, V.W., Ewan, R.C., and Speer, V.C., 1969, Glucose and fructose in the fetal and newborn pig, *J. Anim. Sci.* 29:906-911.

32. Stanton, H.C., and Woo, S.K., 1978, Development of adrenal medullary function in swine, *Am. J. Physiol.* 234:E137-E145.

33. Dvorak, M., 1972, Adrenocortical function in foetal, neonatal and young pigs, *J. Endocr.* 54:473-481.

34. Herbein, J.H., Martin, R.J., Griel, L.C., and Kavanaugh, J.F., 1977, Serum hormones in the perinatal pig and the effect of exogenous insulin on blood sugars, *Growth* 41:277-283.

35. Kattesh, H.G., Charles, S.F., Baumbach, G.A., and Gillepsie, B.E., 1990, Plasma cortisol distribution in the pig from birth to six weeks of age, *Biol. Neonate* 58:220-226.

36. Campion, D.R., McCusher, R.H., Buonomo, F.C., and Jones, W.K., 1986, Effect of fasting neonatal piglets on blood hormone and metabolite profile and on skeletal muscle metabolism., *J. Anim. Sci.* 63:1418-1427.

37. Parker, R.O., Williams, P.E.V., Aherne, F.X., and Young, B.A., 1980, Serum concentration changes in protein, glucose, urea, thyroxine and triiodothyronine and thermostability of neonatal pigs farrowed at 25 and 10°C, *Can. J. Anim. Sci.* 60:503-509.

38. Berthon, D., Herpin, P., Duchamp, C., Dauncey, M.J., and LeDividich, J., 1993, Modification of thermogenic capacity in neonatal pigs by changes in thyroid status during late gestation, *J. Develop. Physiol.* 19:253-261.

39. Pégorier, J.P., Duée, P.H., Simoes-Nunes, C., Girard, J., and Peret, J., 1984, Glucose turnover and recycling in unrestrained and unanesthetized 48 hours old fasting or post-absorptive newborn pigs, *Br. J. Nutr.* 52:277-287.

40. Flecknell, P.A., Wooton, R., and John, M., 1980, Total body glucose metabolism in the conscious, unrestrained piglet and its relation to body-organ weight, *Br. J. Nutr.* 44:193-203.

41. Cote, P.J., Wangsness, P.J., Varela-Avarez, H., Griel, L.C.J., and Kavanaugh, J.F., 1982, Glucose turnover in fast-growing, lean and in slow-growing, obese swine, *J. Anim. Sci.* 54:89-94.

42. Robinson, J.L., Duée, P.H., Schreiber, O., Bois-Joyeux, B., Chanez, M., Pégorier, J.-P., and Peret, J., 1981, Development of gluconeogenic enzymes in the liver of fasting or suckling newborn pigs, *J. Develop. Physiol.* 3:191-201.

43. Mersmann, H.J., 1971, Glycolytic and gluconeogenic enzyme levels in pre- and postnatal pigs, *Am. J. Physiol.* 220:1279-1302.

44. Pégorier, J.P., Duée, P.H., Girard, J., and Peret, J., 1982, Development of gluconeogenesis in isolated hepatocytes from fasting or suckling newborn pigs, *J. Nutr.* 112:1038-1046.

45. Swiatek, K.R., 1971, Development of gluconeogenesis in pig liver slices, *Biochim. Biophys. Acta* 252:274-279.

46. Helmrath, T.A., and Bieber, L.L., 1974, Development of gluconeogenesis in neonatal pig liver, *Am. J. Physiol.* 227:1306-1313.

47. Bird, P.H., and Hartmann, P.E., 1994, The response in the blood of piglets to oral doses of galactose and glucose and intravenous administration of galactose, *Br. J. Nutr.* 71:553-561.

48. Duée, P.H., Pégorier, J.P., Peret, P., and Girard, J., 1985, Separate effects of fatty acid oxidation and glucagon on gluconeogenesis in isolated hepatocytes from newborn pigs, *Biol. Neonate* 47:77-83.

49. Helmrath, T.A., and Bieber, L.L., 1975, Glucagon stimulation of hepatic gluconeogenesis in neonatal pigs, *Proc. Soc. Exp. Biol. Med.* 150:561-563.

50. Boyd, R.D., Whitehead, D.M., and Butler, W.R., 1985, Effect of exogenous glucagon and free fatty acids on gluconeogenesis in fasting neonatal pigs, *J. Anim. Sci.* 60:659-665.

51. Lepine, A.J., Watword, M., Boyd, R.D., Ross, D.A., and Whitehead, D., 1993, Relationship between hepatic fatty acid oxidation and gluconeogenesis in the fasting neonatal pig, *Brit. J. Nutr.* 70:81-91.

52. LeDividich, J., Esnault, T., Lynch, B., Hoo-Paris, R., Castex, C., and Peiniau, J., 1991, Effect of colostral fat level on fat deposition and plasma metabolites in the newborn pigs, *J. Anim. Sci.* 69:2480-2488.

53. Flecknell, P.A., Wootton, R., and John, M., 1982, Acute measurement of cerebral metabolism in the conscious, unrestrained neonatal piglet. II glucose and oxygen utilization, *Biol. Neonate* 41:221-226.

54. Widdowson, E.M., Colombo, V.E., and Artavanis, C.A., 1976, Changes in organs of pigs in response to feeding for the first 24 h after birth, *Biol. Neonate* 28:272-281.

55. Darcy-Vrillon, B., Posho, L., Morel, M.T., Bernard, F., Blachier, F., Meslin, J.C., and Duée, P.H., 1994, Glucose, galactose and glutamine metabolism in pig isolated enterocytes during development, *Pediatr. Res.* 36:175-181.

56. Posho, L., Darcy-Vrillon, B., Blachier, F., and Duée, P.H., 1994, The contribution of glucose and glutamine to energy metabolism in newborn pig enterocytes, *J. Nutr. Biochem.* 5:284-290.

57. Posho, L., Darcy-Vrillon, B., Morel, M.T., Cherbuy, C., Blachier, F., and Duée, P.H., 1994, Control of glucose metabolism in newborn pig enterocytes: evidence for the role of hexokinase, *Biochim. Biophys. Acta* 1224:213-220.

58. Sheard, N.F., and Walker, W.A., 1988, The role of breast milk in the development of the gastrointestinal tract, *Nutr. Rev.* 46:1-18.

59. Mount, L.E., 1959, The metabolic rate of the newborn pig in relation to environmental temperature and to age, *J. Physiol. (London)* 147:333-345.

60. Mellor, D.J., and Cockburn, F., 1986, A comparison of energy metabolism in the newborn infant, piglet and lamb, *Quart. J. Exp. Physiol.* 71:361-379.

61. Duée, P.H., Pégorier, J.P., Le Dividich, J., and Girard, J., 1988, Metabolic and hormonal response to acute cold exposure in newborn pigs, *J. Develop. Physiol.* 10:371-381.

62. Herpin, P., LeDividich, J., and Van Os, M., 1992, Contribution of colostral fat to thermogenesis and glucose homeostasis in the newborn pig, *J. Develop. Physiol.* 17:133-141.

63. Mayfield, S.R., Stonestreet, B.S., Brubakk, A.M., Shaul, P.W., and Oh, W., 1986, Regional blood flow in newborn piglets during environmental cold stress, *Am. J. Physiol.* 251:G308-G313.

64. Close, W.H., LeDividich, J., and Duée, P.H., 1985, Influence of environmental temperature on glucose tolerance and insulin response in the new-born piglet, *Biol. Neonate* 47:84-91.

65. Herpin, P., and LeDividich, J., 1995, Thermoregulation and environment, in: *The Neonatal Pig: Development and Survival* (M.A. Varley, ed.), pp. 57-95.

66. Bengtsson, G., Gentz, J., Hakkarainen, J., Hellstrom, R., and Persson, B., 1969, Plasma levels FFA, glycerol, 3-hydroxybutyrate and blood glucose during the postnatal development of pig, *J. Nutr.* 97:311-315.

67. Adams, S.H., and Odle, J., 1993, Plasma beta-hydroxybutyrate after octanoate challenge - Attenuated ketogenic capacity in neonatal swine, *Am. J. Physiol.* 265:R761-R765.

68. Müller, M.J., Paschen, U., and Seitz, H.J., 1982, Starvation-induced ketone body production in the conscious unrestrained miniature pig, *J. Nutr.* 112:1379-1386.

69. Bremer, J., and Osmundsen, H.,1984,Fatty acid oxidation and its regulation, in: *Fatty Acid Metabolism and its Regulation* (S. Numa, ed.), Elsevier Science Publishers B.V., Amsterdam, pp 113-154.

70. Pégorier, J.P., Duée, P.H., Girard, J., and Peret, J., 1983, Metabolic fate of non esterified fatty acids in isolated hepatocytes from newborn and young pigs: evidence for a limited capacity for oxidation and an increased capacity for esterification, *Biochem. J.* 212:93-97.

71. Duée, P.H., Pégorier, J.P., Quant, P.A., Herbin, C., Kohl, C., and Girard, J., 1994, Hepatic ketogenesis in newborn pig is limited by low mitochondrial 3-hydroxy-3-methylglutaryl-CoA synthase activity, *Biochem. J.* 298:207-212.

72. Odle, J., Benevenga, N.J., and Crenshaw, T.D., 1991, Postnatal age and the metabolism of medium- and long-chain fatty acids by isolated hepatocytes from small-for-gestational-age and appropriate-for-gestational-age piglets, *J. Nutr.* 121:615-621.

73. Bischoff, M.B., Richter, W.R., and Stein, R.J., 1969, Ultrastructural changes in pig hepatocytes during the transitional period from late fetal to early neonatal life, *J. Cell Sci.* 4:381-395.

74. Mersmann, H.J., Goodman, J., Houk, J.M., and Anderson, S., 1972, Studies on the biochemistry of mitochondria and cell morphology in the neonatal swine hepatocyte, *J. Cell. Biol.* 53:335-347.

75. McGarry, J.D., Woeltje, K.F., Kuwajima, M., and Foster, D.W., 1989, Regulation of ketogenesis and the renaissance of carnitine palmitoyltransferase, *Diabetes/ Metab. Rev.* 5:271-284.

76. Bieber, L.L., Markwell, M.A.K., Blair, M., and Helmrath, T.A., 1973, Studies on the development of carnitine palmitoyltransferase and fatty acid oxidation in the liver mitochondria of neonatal pigs, *Biochim. Biophys. Acta* 326:145-154.

77. McGarry, J.D., and Foster, D.W., 1980, Regulation of hepatic fatty acid oxidation and ketone body production, *Annu. Rev. Biochem.* 49:395-420.

78. Wolfe, R.G., Maxwell, C.V., and Nelson, E.C., 1978, Effect of age and dietary fat level on fatty acid oxidation in the neonatal pig, *J. Nutr.* 108:1621-1634.

79. Werner, J.C., Whitman, V., Vary, T.C., Fripp, R.R., Musselman, J., and Schuler, G., 1983, Fatty acid and glucose utilization in isolated working newborn pig heart, *Am. J. Physiol.* 244:E19-E23.

80. Werner, J.C., Sicard, R.E., and Schuler, H.G., 1989, Palmitate oxidation by isolated working fetal and newborn pig hearts, *Am. J. Physiol.* 256:E315-E321.

81. Ascuitto, R.J., Ross-Ascuitto, N.T., Chen, V., and Downing, S.E., 1989, Ventricular function and fatty acid metabolism in neonatal piglet heart, *Am. J. Physiol.* 256:H9-H15.

82. Tetrick, M.A., Adams, S.H., Odle, J., and Benevenga, N.J., 1995, Contribution of D-(-)-3-hydroxy-butyrate to the energy expenditure of neonatal pigs, *J. Nutr.* 125:264-272.

83. Cornblath, M.,1968, Neonatal hypoglycemia, in: *Fetal Homeostasis,* 4 (R.M. Wynn, ed.), Century Crofts, New York, pp. 122-131.

84. Bruegger, S.J., and Conrad, J.H., 1972, Effects of orally adminitred albumin and corn oil, on blood constituents, survival and weight gain in neonatal pig, *J. Anim. Sci.* 34:411-415.

85. Bieber, L.L., Helmrath, T., Dolanski, E.A., Olgaard, M.K., and Belanger, L.L., 1979, Gluconeogenesis in neonatal piglet liver, *J. Anim. Sci.* 49:250-257.

86. Lepine, A.J., Boyd, R.D., Welch, J.A., and Roneker, K.R., 1989, Effect of colostrum or medium-chain triglyceride supplementation in the pattern of plasma glucose, non-esterified fatty acids and survival of neonatal pigs, *J. Anim. Sci.* 67:983-990.

87. Odle, J., Benevenga, N.J., and Crenshaw, T.D., 1989, Utilisation of medium-chain triglycerides by neonatal piglets. II Effects of even- and odd-chain triglyceride consumption over the first two days of life on blood metabolites and urinary nitrogen excretion, *J. Anim. Sci.* 67:3340-3351.

88. Pégorier, J.P., Simoes-Nunes, C., Duée, P.H., Peret, J., and Girard, J., 1985, Effects of intragastric triglyceride administration on glucose homeostasis in newborn pigs, *Am. J. Physiol.* 249:E268-E275.

89. Wieland, O.H., 1983, The mammalian pyruvate dehydrogenase complex: structure and regulation, *Rev. Physiol. Biochem. Pharmacol.* 96:123-170.

90. Bloom, S.R., and Johnston, D.I., 1972, Failure of glucagon release in infants of diabetic mothers, *Br. Med. J.* 4:453-454.

91. Kalhan, S.C., Savin, S.M., and Adam, P.A.J., 1977, Attenuated glucose production rate in newborn infants of infants of insulin-dependent diabetic mothers, *N. Engl. J. Med.* 296:375-376.

92. Kollee, L.A., Monnens, L.A., Cezka, V., and Wilms, R.H., 1978, Persitent neonatal hypoglycemia due to glucagon deficiency, *Arch. Dis. Child.* 53:422-424.

93. Denne, S.C., and Kalhan, S.C., 1986, Glucose carbon recycling and oxidation in human newborns, *Am. J. Physiol.* 251:E71-E77.

94. Bougnères, P.F., 1987, Stable isotopes tracers and the determination of fuel fluxes in newborn infants, *Biol Neonate* 52(Suppl 1):87-96.

95. Patel, D., and Kalhan, S.C., 1992, Glycerol metabolism and triglyceride-fatty acid cycling in the human newborn. Effect of maternal diabetes and intrauterine growth retardation, *Pediat. Res.* 31:52-58.

96. Bougnères, P.F., Rocchiccioli, F., Nurjhan, N., and Zeller, J., 1995, Stable isotope determination of plasma lactate conversion into glucose in fasting infants, *Am. J. Physiol.* 268:E652-E659.

97. Bougnères, P.F., Lemmel, C., Ferré, P., and Bier, D.M., 1986, Ketone body transport in the human neonate and infant, *J. Clin. Invest.* 77:42-48.

98. De Boissieu, D., Rocchiccioli, F., Kalach, N., and Bougneres, P.F., 1995, Ketone body turnover at term and in premature newborns in the first two weeks after birth, *Biol. Neonate* 67:84-93.

99. Wilkinson, A.W., 1969, The starving newborn baby, *Proc. Nutr. Soc.* 28:61-66.

100. Haymond, M.W., Karl, I., and Pagliara, A.S., 1974, Increased gluconeogenesis substrates in the small-for-gestational age infant, *N. Engl. J. Med.* 291:322-328.

101. Mestyan, J., Soltesz, G., Schultz, K., and M., H., 1975, Hyperaminoacidemia due to accumulation of gluconeogenic amino acid precursors in hypoglycemic small-for gestational age infants, *J. Pediatr* 87:409-414.

102. Sabel, K.G., Olegard, R., Mellander, M., and Hildingsson, K., 1982, Interrelation between fatty acid oxidation and control of gluconeogenic substrates in small for gestational age (SGA) infants with hypoglycemia and with normoglycemia, *Acta Paed. Scand.* 71:53-61.

103. Bougnères, P.F., Castano, L., Rocchiccioli, F., Pham Gia, H., Leluyer, B., and Ferré, P., 1989, Medium chain fatty acids increase glucose production in normal and low birth weight newborns, *Am. J. Physiol.* 256:E692-E697.

104. Sann, L., 1990, Neonatal hypoglycemia, *Biol. Neonate* 58(Suppl 1):16-21.

105. Bougnères, P.F., Saudubray, J.M., Marsac, C., Bernard, O., Odievre, M., and Girard, J., 1980, Decreased ketogenesis due to deficiency of hepatic carnitine acyltransferase, *N. Engl. J. Med.* 302:123-124.

106. Saudubray, J.M., Coudé, F.X., Demaugre, F., Johnson, C., Gibson, K.M., and Nyhan, W.L., 1982, Oxidation of fatty acids in cultured fibroblasts : a model system for the detection and study of defects in oxidation, *Pediatr. Res.* 16:877-881.

107. Bougnères, P.F., Saudubray, J.M., Marsac, C., Bernard, O., Odievre, M., and Girard, J., 1981, Fasting hypoglycemia resulting from carnitine palmitoyltransferase deficiency, *J. Pediatrics* 98:742-746.

108. Demaugre, F., Bonnefont, J.P., Mitchell, G., Nguyen-Hoang, N., Pelet, A., Rimoldi, M., Di Donato, S., and Saudubray, J.M., 1988, Hepatic and muscular presentations of carnitine palmitopyltransferase deficiency: two distinct entities, *Pediatr. Res.* 24:308-312.

AUTHOR INDEX

SUBJECT INDEX